STO

ACPL ITEM
DISCARDED

621.3893 Sy64p
Symko, Orest G.
Physics of hi-fi

PHYSICS OF HI-FI

Analog to Digital

Orest G. Symko
University of Utah

KENDALL/HUNT PUBLISHING COMPANY
4050 Westmark Drive Dubuque, Iowa 52002

Allen County Public Library
900 Webster Street
PO Box 2270
Fort Wayne, IN 46801-2270

Copyright © 1995 by Kendall/Hunt Publishing Company

ISBN 0-8403-8823-3

All rights reserved. No part of this publication may be reproduced, stored in a retrieval system, or transmitted, in any form or by any means, electronic, mechanical, photocopying, recording, or otherwise, without the prior written permission of the copyright owner.

Printed in the United States of America
10 9 8 7 6 5 4 3 2 1

Contents

PREFACE

This textbook is based on a course which has been offered for the past 15 years in the Liberal Education Program at the University of Utah. The course, the Physics of Hi–Fi, was designed as an Introduction to Physics, within the framework of the hi–fi so as to take advantage of this popular subject which interests a broad spectrum of students, even the ones with essentially no background in science. For that reason, the emphasis is toward a qualitative treatment of the subject rather than a quantitative one. In teaching, it is important to motivate the students as this makes the learning experience more efficient and certainly more pleasant. The hi–fi presents a platform to do this by exploring fundamentals of physics, from waves to electricity and magnetism to mechanics, in a manner which is relevant to the student. The physics is presented within a subject which is familiar and interesting to the student, and this makes the topic of physics more acceptable and enjoyable.

The course is presented as a series of lectures with many demonstrations to illustrate the principles and to maintain the student's interest. The demonstration aspect of the course is very important to the students as it provides a better involvement in the lecture; after the course is over the demonstrations leave pleasant impressions that remain for a long time. With the lectures this course has labs, where each student does four experiments, and they typically are on: Fourier analysis of musical sounds, speaker dispersion, frequency response of various hi–fi components, and principles of radio broadcasting. The labs provide the students with hands–on experience in dealing with equipment and taking measurements; for many this is a unique experience.

This textbook begins with a definition of hi–fi followed by a presentation of its components and their function in the system. Although hi–fi units vary with manufacturer and with the date when they were built, a general characterization in terms of basic physics principles and an appreciation for this field can be achieved. The book deals with a perspective of the physics principles involved in the hi–fi. The approach taken is to first present the physics principles and then to show how they are applied in the hi–fi. Fundamentals of waves are presented, first in general terms, and then they are related to the various aspects of sound dealt with in the hi–fi. Continuing with the presentation of concepts as encountered in the hi–fi, a chapter is devoted to decibels; this is important for knowing what the specifications mean and for getting a quantitative feeling for the scope of the acoustic phenomena involved in the hi–fi. At this point the reader has been presented with enough background to apply it to one of the most fascinating topics in the hi–fi, loudspeakers. The chapter on loudspeakers covers the principles of speakers and the details of how they perform. The main types of enclosures are presented with specific characteristics of each. An over–view of speaker goals, performance, and limitations is presented to the reader.

The chapter on electricity starts off with aspects of this subject which are found in the hi–fi. It then leads to important applications such as how to fuse a speaker and how to connect speakers in parallel or in series. Speakers are used again as an example and this time to illustrate the concept of electrical impedance as applied to cross–over networks. This chapter ends with an introduction to the topic of amplifiers. Their role in the hi–fi and behavior are discussed, analyzed, and related to their specifications.

The next important section of the textbook deals with electromagnetism. Following its introduction, examples from the hi–fi such as the drive action of an electromagnetic speaker, transformers, and planar speakers are used to discuss the principles. This chapter leads to electromagnetic waves. After a description of their nature and how they are produced, the emphasis is on their application toward communications and radio waves. Modulation, AM and FM, and radio broadcasting are discussed together with the methods of radio wave detection and antennas. The last part of this chapter deals with tuner specifications and world communications.

The chapter on analog recording and playback introduces the subject with phono records as an example, and then it deals with some of the details of magnetic recording using tape decks. Central to this chapter is the concept of hysteresis of magnetic materials, such as a magnetic tape and head core, in a magnetic field. The recording/playback process is presented, as well as some of its limitations. Next comes the chapter on digital recording with systems based on optical read–out. The how and why of the digital process is covered, with a presentation of the Sampling Theorem as used in existing systems. The advantages of digital over analog recording are discussed. The compact disc (CD) system is then explored from how information is stored to the application of physics principles for the read–out using interference of an infra–red laser between the pits and lands of the disc. The chapter on magnetic digital recording on tapes and small discs deals with principles used in the digital audio tapes (DAT), and in the recent developments the digital compact cassette (DCC) and the Mini–Disc. The quest for high information density in recording makes use of new fascinating physics applications.

The textbook ends with two topics which play an important role in the hi–fi, heat and mechanics. After an introduction, the chapter on heat deals with examples where heat production causes problems and how it can be removed effectively; but under control, it is used for recording, as in the recordable Mini–Disc. The chapter on mechanics deals with pressure, speed, motion, force, and rotation; here as well examples from the hi–fi are used to illustrate the principles.

The approach taken in this book consists of exploring the physics involved in the hi–fi and thus to capture the interest of the student. Since the student is naturally attracted to the hi–fi and is interested in it, learning physics becomes enjoyable. Many concepts presented in the book have been simplified since they were meant to be used as an introduction to the subject. The instructor teaching the material in this book should encourage students to go beyond what is presented here and ask questions about the phenomena explored. One of the goals of the book is to promote further discovery of our everyday use of physics principles.

In order to help students master the material in this textbook, there are questions and exercises at the end of each chapter. Because of the multiple–choice nature of the questions, the precise meaning at times may be open for discussions, and this should also be encouraged. This type of question has been used quite effectively in tests and exams given with this course; with the large number of students taking the class and their wide range of backgrounds, the multiple–choice type of exams have worked very well.

This textbook is meant to give the reader a general idea of how physical principles are used in the hi–fi and an appreciation of the scientific method used in developing a subject so useful to all of us, the storage and playback of sound, for communication or for our pleasure. Its importance is evident from the rapidly changing developments and progress made. This book has two uses: to teach physics principles using the hi–fi and to show how the hi–fi works. Both goals bring out the fascinating aspects of physics, and encourage further exploration of the world of science around us.

ACKNOWLEDGEMENTS

Many people have contributed to the completion of this book and to the development of the Physics of Hi–Fi course. I take this opportunity to thank them for their help.

To the thousands of students who have taken this course, I thank them for all the questions and discussions that they have put forward; they were all excellent.

Dr. Jackson Newell, Dean of Liberal Education at the time of the development of this course, has provided constant encouragement in establishing this course. I thank him for his confidence and support.

Dr. Reba Keele, Dean of Undergraduate Studies, has continuously been supportive of this course and its developments. I thank her for all her help.

The figures were computer-reproduced by Jeff F. Gold. Thank you for the interesting artwork.

I am grateful to Christine McDonough for typing the manuscript and her contributions in the editing. It was a pleasure to work with her.

I am particularly indebted to my wife Ivanna for her understanding and enthusiastic support during the writing of this book, and my daughters Sophie and Martha for their encouragement.

CHAPTER 1

Introduction to Hi-Fi

1.1 Definitions

This text will present a study of acoustics, the physics of sound by examining the acoustic phenomena that exist in the recording and playback of audio information. This presentation will integrate terms defining the basic principles of physics and jargon as it is used in the field of hi–fidelity, known simply as hi–fi, which will give the reader an understanding of the terms. Many terms can be overwhelming to the uninitiated, and the concepts can then become incomprehensible. The approach that will be taken will be the presentation of basic physics principles and then they will be related to the hi–fi. Because this field has grown tremendously in the last few years, the readers are not expected to become experts of one particular system; rather they will be able to understand from the general principles covered in this book the audio systems presently used. Essentially they will be able to assess any system by application of the concepts learned in this book.

This text is also presented in a way that students will be able to learn some basic physics pleasantly. A learning experience becomes more interesting to students when they can relate to the ideas and the concepts. Here, the hi–fi system presents a unique opportunity to learn some basic physics by exploring the hi–fi, which is easily accessible, because most people own a hi–fi system. In fact, students can go to their hi–fi systems and find out which parts or components were discussed in the lecture and what is their function. The lecture then becomes a topic for the student to explore, and perhaps it can even stimulate them for further investigation of certain topics. Students learn best when they are interested in a subject, and this book tries to take advantage of this premise. For students who wish to go deeper into the subject, it is hoped that they will have achieved enough proficiency in the subject and that they will have gained enough self–confidence to go and explore the subject further on their own. The presentation of physics around a popular subject, the hi–fi, has pedagogical value which will help the student in learning some basic science or, at the very least, appreciating the scientific approach. Exercises are presented at the end of each chapter to familiarize the students with the concepts and elicit understanding of the subject. Student learning can be ineffective without trying out concepts on their own.

The approach taken in this text is to first present the concepts, then to apply them directly to the hi–fi, and finally to learn about the specifications of hi–fi components. At first it is important to give some definitions of words that will be recurring in this book. One such word is hi–fi. A simple definition is:

***Hi–Fi* = faithful reproduction of original sound.**

The seemingly simple definition of hi–fi, or high–fidelity, is the goal toward which this book is aimed. It turns out to be a very difficult goal to achieve. There is a less stringent definition of hi–fi and it can be given as:

***Hi–Fi* = ability to reproduce sound with excellent tonal qualities (so that we can enjoy it!).**

Of course we have to define what excellent tonal qualities are and what they mean. This relates to the expectations of our auditory senses and to the characteristics of a live musical performance.

It is useful to introduce another definition of a word which quite often is used incorrectly. It is the word "stereo" and sometimes it gets confused with hi–fi. The word "stereo" is a method of recording and playback, both done in two channels; it can be defined as:

***Stereo* = sound recording and reproduction through two channels.**

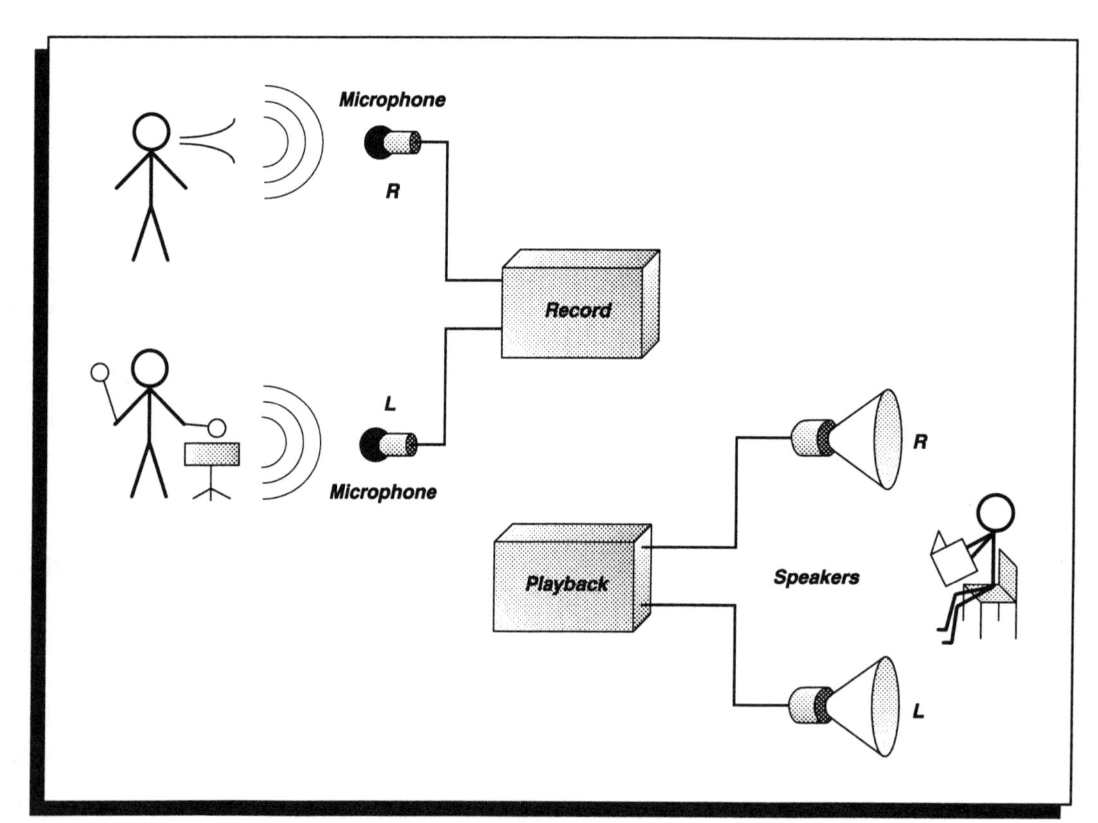

Figure 1.1 Stereo process in recording and playback.

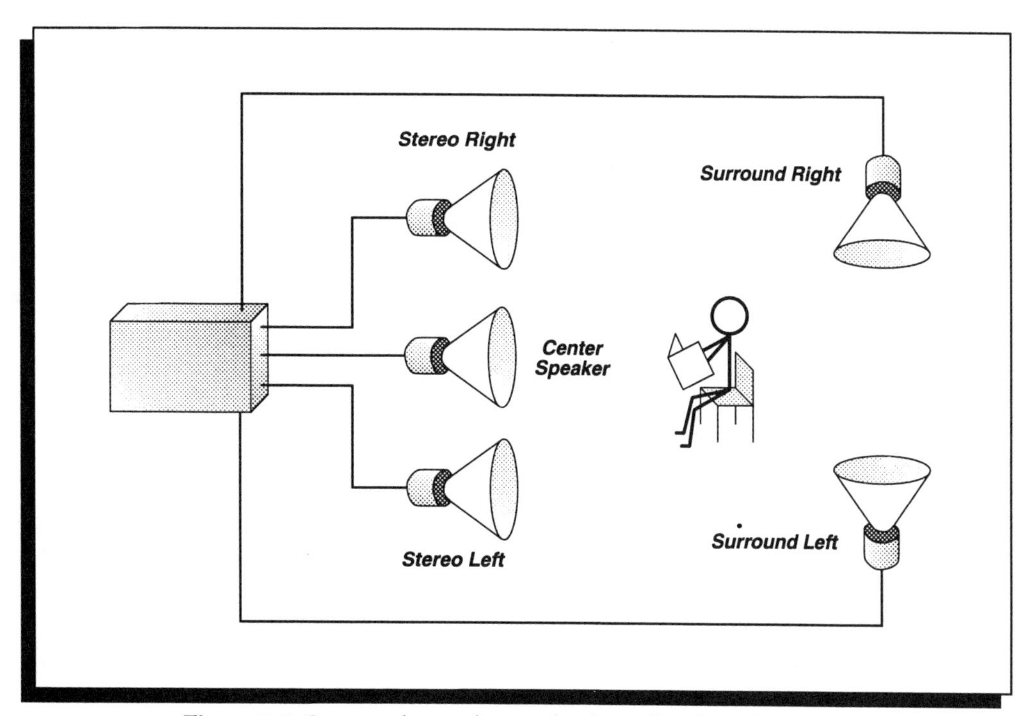

Figure 1.2 Surround sound reproduction of audio information.

The channels are usually referred to as the LEFT and RIGHT channels. Figure 1.1 explains what is meant by stereo. Not all stereo is hi–fi, but it would be nice if it were.

It is interesting to note that in the 1970s, a system known as quadraphonic sound was introduced in the recording and reproduction of sound. It was based on using four channels in the recording and playback; this required sophisticated techniques in the storage part and playback. The ideas had much merit, but due to incompatibility between the manufacturers, it was not successful.

Although stereo hi–fi has brought some excellent music reproduction, a new approach, surround sound, is gaining in popularity and provides music enhancement by adding depth and realism to the stereo sound. It provides an extra spatial dimension for music and for video reproduction. It makes you feel like you are actually at a concert or movie theater. This is achieved by placing a center speaker between the left and right speakers, and adding two more speakers at the rear of the room with surround sound information. Figure 1.2 illustrates this by using five speakers. The rear speakers provide the ambience, which is the surround sound component necessary to complete the acoustic image. Hence:

Surround Sound **= mode of reproducing sound by using channels in addition to the stereo channels to provide a feeling of ambience, of the "you–are– there" illusion. This is performed usually with five speakers, three at the front and two at the rear of the room for the surround–sound effect.**

For ultimate performance of this effect, a special processor is needed to decode and channel the information, and to provide the delay necessary in creating the ambience feeling. While stereo sound tends to be acoustically two–dimensional, surround sound is more three–dimensional.

At present the word hi–fi does not seem to have as much importance as when it was first introduced in the 1950s and 1960s. All systems are now expected to perform with high–fidelity, thanks to the tremendous achievements in electronics. The advancements in electronics have brought out a variety of units, which record and reproduce sound with excellent tonal qualities. This book tries to place a general perspective on "what goes where and why" in the hi–fi. In this chapter the basic hi–fi units and components will be introduced and labeled where they belong in the acoustic chain, from the source of sound to its reproduction.

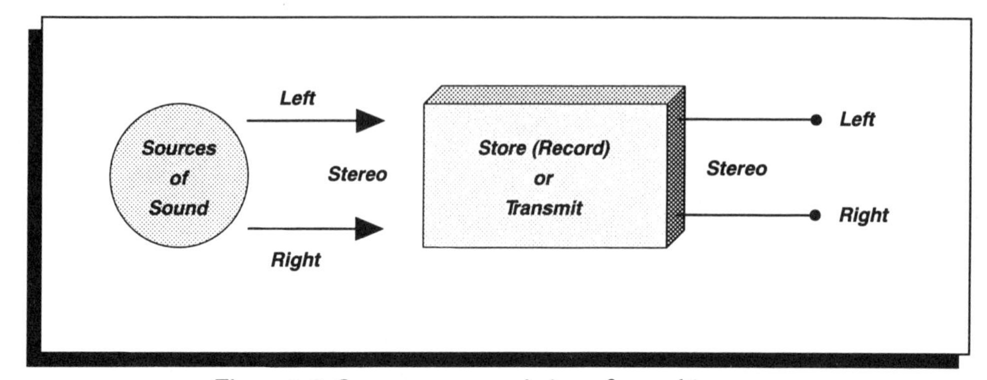

Figure 1.3 Storage or transmission of sound in stereo.

1.2 GOALS

In order to understand and to appreciate a complete hi–fi system, it is useful to start with a simple approach to what this is all about. First, sound or music is recorded for storage or it is transmitted over various media. Second, the various sources of stored sound are played back so that we can hear them at home, in a studio, in a car, or while jogging. Figure 1.3 shows the basic approach taken in the hi–fi for storage or broadcasting.

It is not as simple as it appears because all along this transmission it is necessary to maintain high–fidelity. Basically, the storage or transmission process is in stereo, with a LEFT channel and a RIGHT channel. Recently, this has evolved with the addition of surround–sound, to give sound extra dimensionality and make it appear as if it came from a live source during playback.

The reason for storing or broadcasting is to repeat the performance at a later time, for example in our home, maybe even to play the recording many times. This is an important part that deals with convenience and efficiency of storage. Tare different approaches that can be taken for storage and/or broadcasting.

The playback part exists for our convenience and pleasure. This is shown in Figure 1.4 for the stereo system.

With all the different varieties of systems that exist and their respective sources of sound, the reproduction part of the cycle ultimately ends at two or more speakers (or it may even end at headphones) that produce the sound (of hi–fidelity quality!).

We already see what a formidable task lies ahead. The speakers have to reproduce exactly the original sound after going through the complicated process of storing, broadcasting, and playback. Is it possible that the speakers can reproduce the original sound? One can be skeptical as to whether or not two channels of information that are going to two speakers can bring a whole orchestra to a living room or to headphones. How closely a system

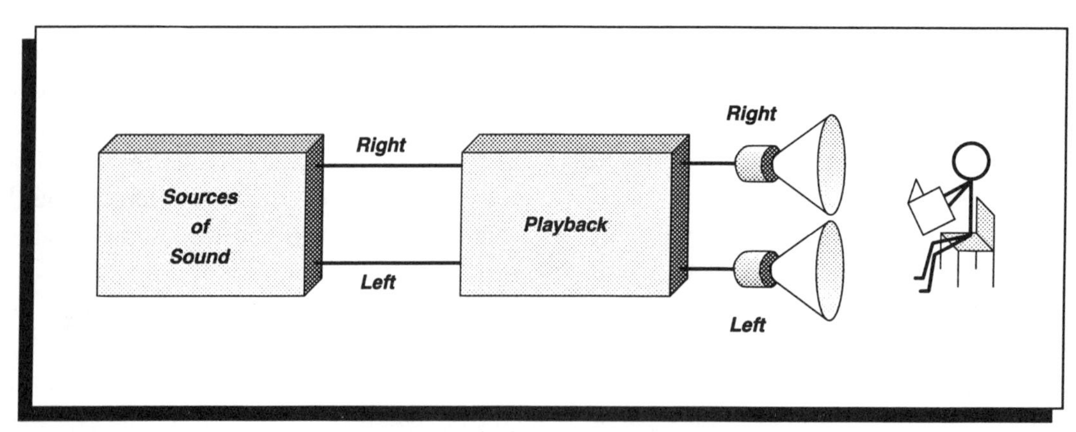

Figure 1.4 Playback process in stereo.

approaches this goal is part of the presentation of this book. Listen to your system, compare it to a live performance, and then go on to the next chapters of this book to form your own opinion.

1.3 COMPONENTS

Shopping in a hi–fi store for a sound system can be an intimidating or even confusing experience. It is not clear what is necessary for an acceptable system, and never mind which brand is the most appropriate. A question may arise as to why there are so many components (audio electronic boxes) and why they are not all in one unit packaged neatly together. The answer to this, in part, is that there are many different approaches that can be taken in setting up a sound unit. For example, it is possible to produce sound (music) only from a variety of sources, or/and it is possible to copy music from one recording medium to another and then play it. This situation can be compared to that of buying a car: there are so many models and makes, and each comes with a variety of different options. It is just a matter of what you want to do and what you wish to have in the car.

Some of the reasons for having so many diverse components available in the selection of hi–fi are:

- — it is a popular approach
- — it provides a great deal of flexibility
- — it provides possibilities for exploring sound effects and upgrading the system or just adding new particular components
- — it is a traditional approach. Going back to the early days when tubes were used in audio systems these components had to be kept separate in order to cope with heat dissipation problems caused by the tubes

Common components in a hi–fi are:

■ SOURCES OF SOUND

These are the units that send audio signals to the system so that we can hear them. The list keeps on increasing as new possibilities are developed, but here are some of them:

turntable or phono player: A unit that is quickly disappearing from the market, but for many decades it was one of the most important sources of sound. It spins a record that has mechanically imprinted audio information on it in the form of waves. These cause a stylus to vibrate, producing electrical signals that resemble the original information in the record grooves.

CD player: A popular unit that reads imprinted information with a laser beam on a compact disc, also known as the CD. It is the unit, started in 1983, that began the digital approach to recording sound.

tuner: Probably the oldest unit in the hi–fi. It picks up electromagnetic waves, or radio waves, that carry the audio information and delivers an audio signal. It is like a radio except that it does not have enough power on its own to drive speakers, nor does it have speakers.

tape deck: This unit delivers the audio signal from information stored on a magnetic tape. It is a convenient and popular unit for combining, storing, editing, and playing back a variety of selections.

microphone: A basic unit for converting sound to an electrical signal that resembles the original acoustic signal.

mini–Disc player: A new unit that resembles a CD but which can be used for reading audio information stored digitally on a magnetic film, or as bits in a disc. The information can be erased and new signals can then be recorded. In both cases a laser beam is used for the read–out.

DCC player: This is also a new unit where the audio information is stored digitally on a magnetic tape. It can be played and it can be erased.

The digital compact cassette (DCC) player is a competitor of the mini–Disc system as both have similar goals, but they use different approaches.

DAT player: A unit that was supposed to complement the CD player. In this the audio information is recorded digitally on a magnetic tape using a variety of options. The Digital Audio Tape (DAT) had long delays in introduction to the public, which resulted in the loss of interest by consumers and subsequently caused the development of the DCC unit.

audio from VCR: VCR audio can now be processed by hi–fi systems so that high–quality audio can be heard with speakers of the system, rather than with the speakers of the TV display (which tend to be of poor quality).

audio part of TV programs received from airwaves or by cable: TV programs can now be processed and delivered by hi–fi systems to provide high quality sound along with the video display.

laser disks: Audio information that is stored along with the video on a laser disk can be reproduced by a hi–fi system, rather than by the TV speakers to add audio realism to the video presentation.

PRE–AMPLIFIER

A basic unit that deals with the electrical signals produced by the various sources of audio signals just discussed. It has three functions:

1. to raise the level of electrical signal from some of the sources.
2. to control the audio signals: the volume, treble, bass, loudness, and filters for audio disturbances.
3. to route the signal from a source to various parts of the hi–fi system. For example, it can take an audio signal from a radio program and send it to the tape deck for recording.

The pre–amplifier is only a signal distribution center, it does not have enough power to drive speakers. Electrical signals arrive at the input to the pre–amplifier and they are sent out from the output(s).

POWER AMPLIFIER

This receives the relatively weak electrical signals from the pre–amplifier and raises their level so that they have enough power to drive speakers. Speakers need power to produce sound. The amount of power that can be delivered to the speakers depends on the amplifier and the speakers.

LOUDSPEAKERS

They serve the very important function of converting the electrical signals to sound. Such a task is very difficult to accomplish well, and for that reason manufacturers have tried many different approaches. When dealing with stereo there is a loudspeaker for the left channel, and there is an identical speaker for the right one. The need for better imaging of the sound in the room or around the television has led to extra speakers which are added for a "surround sound" effect. This provides an extra dimension and leads to a greater reality in the projected sound. Usually two speakers are used as surround speakers, a left one and a right one in addition to the regular LEFT and RIGHT speakers (and even a middle speaker). Extra speakers can also be used for sound production in other rooms.

ACCESSORIES

A variety of units can be incorporated into a hi–fi system. For example, a more refined control on the amount of sound produced at different frequencies can be achieved by a unit called the equalizer. Also, delays in the signal or parts of the signal can be produced by reverberation units. These are examples of popular units used to enhance the performance of hi–fi systems. New accessories are also being introduced for special effects in recording or playback. Surround sound is created by signal processing units such as Dolby Pro Logic or other THX processors which come combined with receivers, integrated amplifiers, or as separate units.

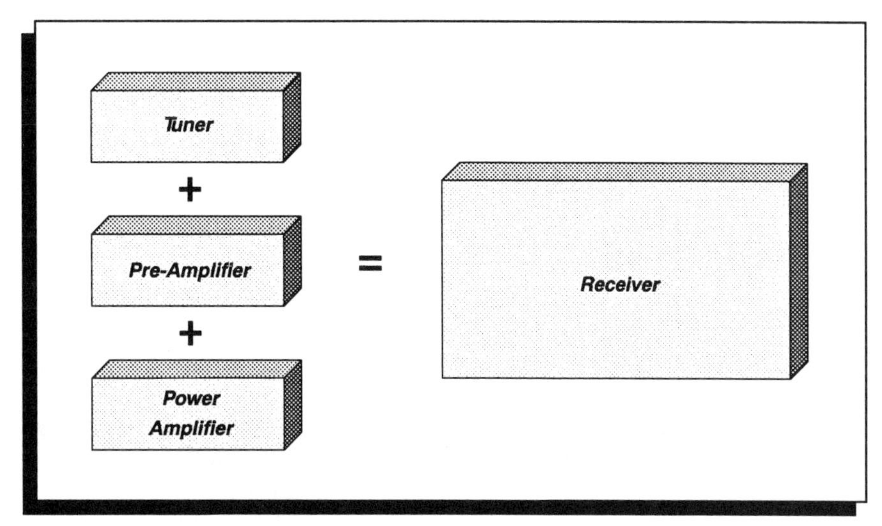

Figure 1.5 Elements of a receiver.

1.4 COMBINATIONS

Because the units above come in modules it is interesting to look at the various ways of combining them. The module approach is popular because it provides the chance to combine different brands, to change the units when upgrading is necessary, or to explore the effect produced by different combinations of units. There are three basic approaches:

■ RECEIVER APPROACH

Here the pre–amp and the tuner are combined with a power amplifier into a single unit called the receiver. It is probably the most popular approach. Figure 1.5 illustrates the elements of this unit.

The advantages of such an approach are:

— Compactness. Everything is contained in one chassis box.

— Reasonable price. This presents a considerable saving in the parts needed, such as the single box, power cables, power supplies, and switches, which otherwise would have been separate elements.

In Figure 1.6 an example is given of how this unit would be incorporated into a hi–fi system.

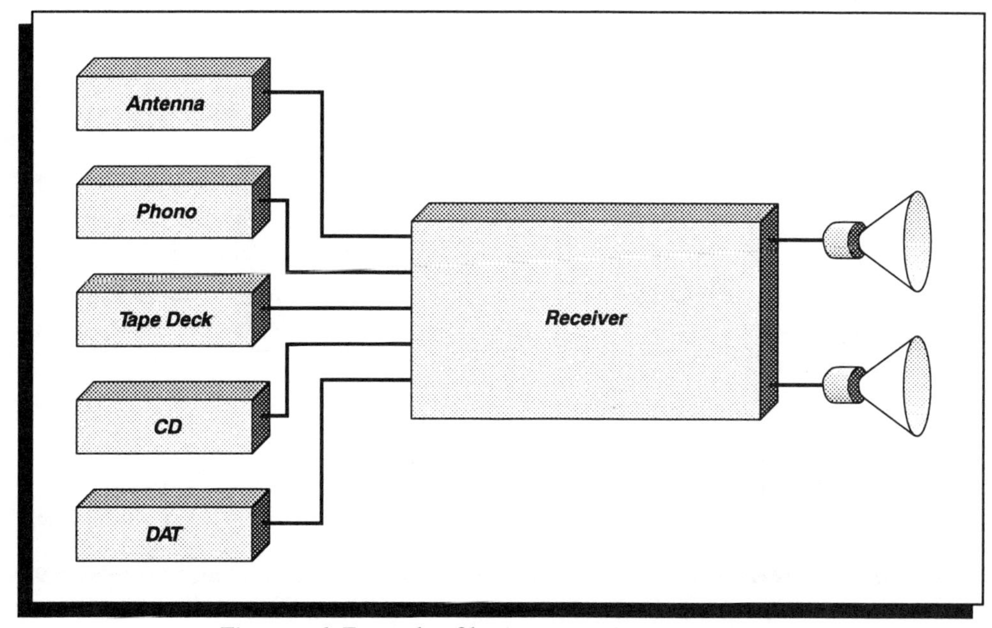

Figure 1.6 Example of basic connections to a receiver.

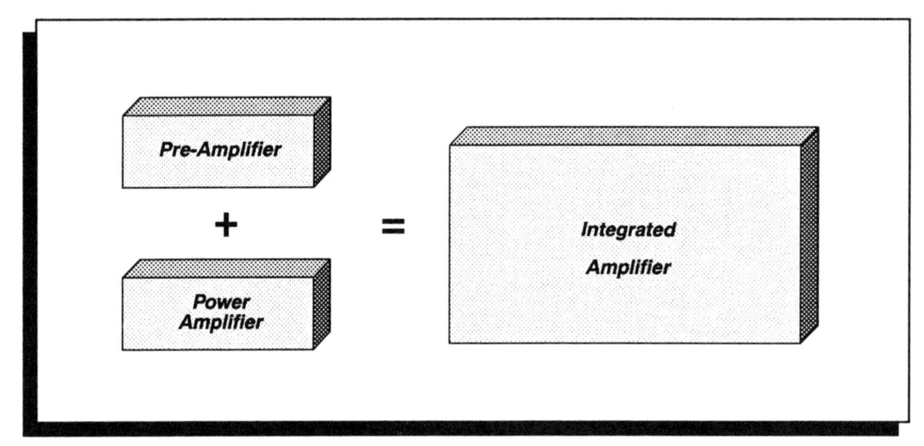

Figure 1.7 Elements of an integrated amplifier.

INTEGRATED AMPLIFIER APPROACH

This is another popular approach that differs from the previous one because the tuner is not included in the unit. It consists only of a pre–amp, plus a power amplifier. The basic unit is shown in Figure 1.7.

This is a popular approach especially for those who have their own tuner, home–built or bought, or for those who are not interested in listening to the radio. A simple connection to some of the components in a hi–fi can be as in Figure 1.8, with sound sources at the input.

ALL SEPARATE APPROACH

In this approach every unit comes separate in the tradition of early hi–fi. In those days it was necessary to keep each component separate so that there would be sufficient heat dissipation from each unit (vacuum tubes were used and they produced a large amount of heat!). In this approach

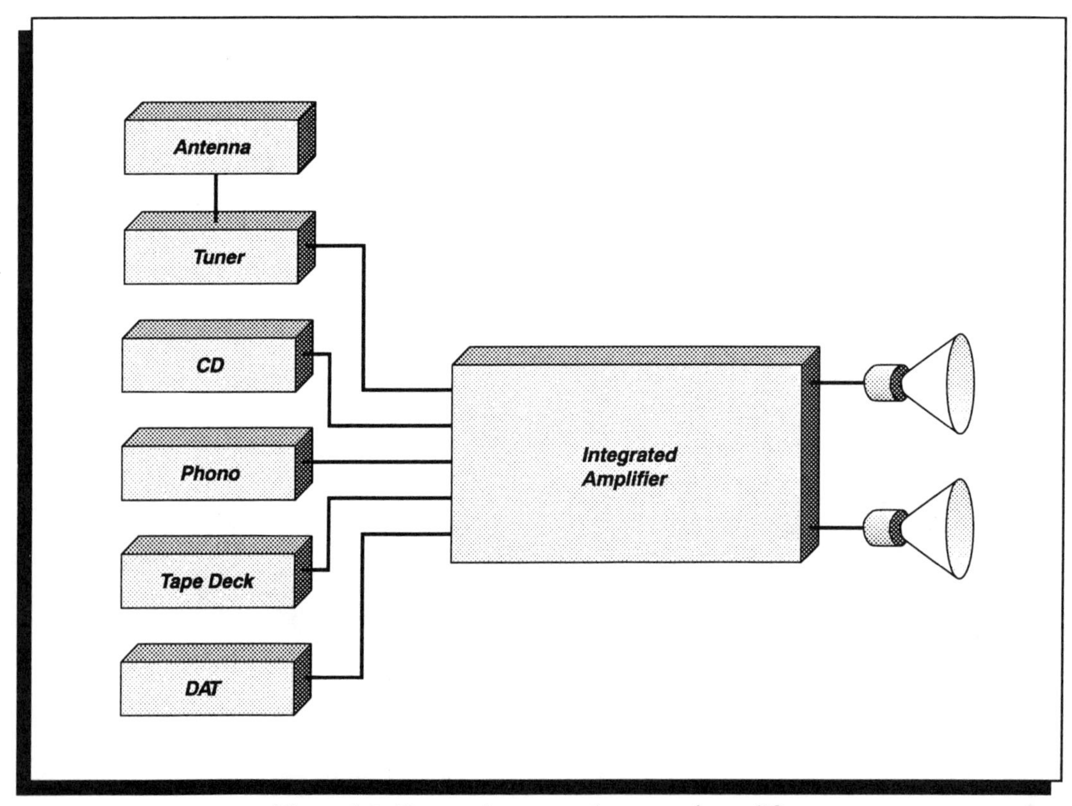

Figure 1.8 Connections to an integrated amplifier.

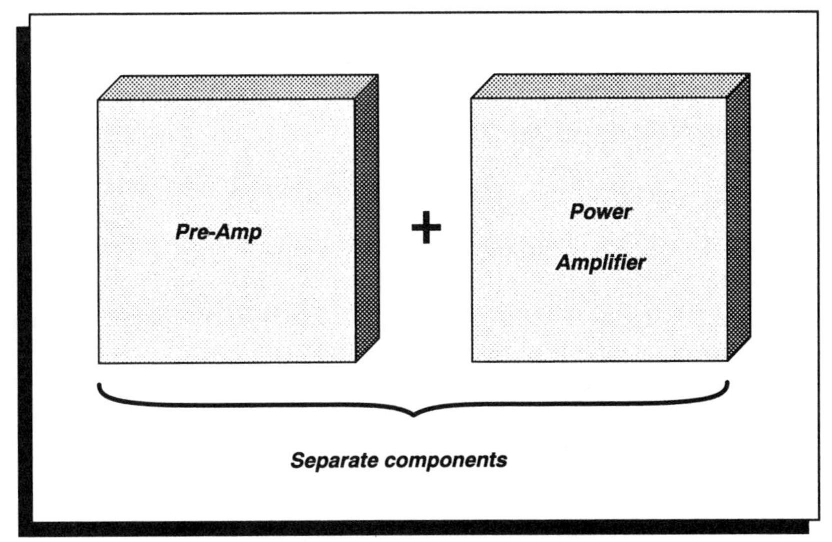

Figure 1.9 All-separate approach.

there is a separate pre–amp and a separate power amplifier. Figure 1.9 shows all the elements separately. Incorporation of these units into the hi–fi could follow an approach as shown in Figure 1.10 with some sound sources at the input. The main advantage of this approach is the ability to use a power amplifier with a substantial amount of power, as necessary for a large hall, garden, or for multi–room connections.

AUDIO-VIDEO (A/V) SYSTEMS APPROACH

The impressive technical achievements in recording and playback of audio just for listening can be extended to the audio that goes with video information. Why not have hi–fi quality audio with the video? This approach has led to the development of the A/V (audio/video) receiver system. It is not economical to have a special audio system

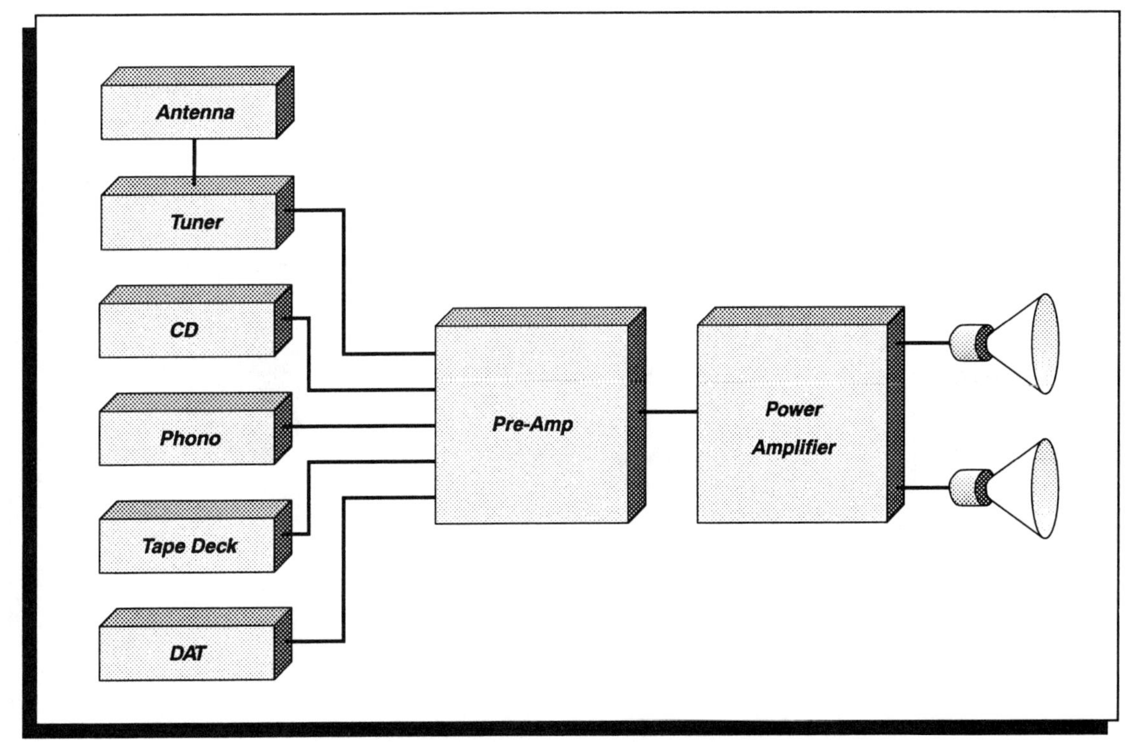

Figure 1.10 Connections in all-separate approach.

for the video and then use a separate one for just the audio. The two are combined into one system. Figure 1.11 shows the most basic A/V system. The A/V receiver drives a set of speakers, and it receives the audio signal that was recorded on the video tape at the auxiliary input from a VCR, preferably a hi–fi VCR. The video signal from the VCR then goes directly to the video monitor, the TV.

Because many videos deal with scenes of action it would be interesting to enhance this effect by creating an audio surround sound effect where the listener is located, acoustically, in the middle of the action. This can be achieved by using auxiliary channels and speakers. In this case there would be two or more speakers, in addition to the regular two–channel stereo speakers. Most A/V receivers can provide surround–sound to the surround speakers, as well as to the regular stereo speakers, and perhaps even to a subwoofer. In this case, a scheme for connecting the system is shown in Figure 1.12. There are two speakers for the stereo effect, a middle one for locating the source of action, and two rear or side speakers for the surround sound effect.

1.5 PRE-AMPLIFIER

The pre–amplifier has many sections and controls. It is interesting to explore them in some detail. Important parts are listed below with a description of their function.

■ SELECTOR SWITCH

This switch sets the selection of the sound source, which comes from one of the inputs. There are no difficulties here as it is self–evident. The position of the switch determines which source will come in. A fact to remember is that it is impossible to listen to more than one source at the same time. Some receivers have multi–zone capabilities where listeners in a room can select a different program from the one playing in the main room.

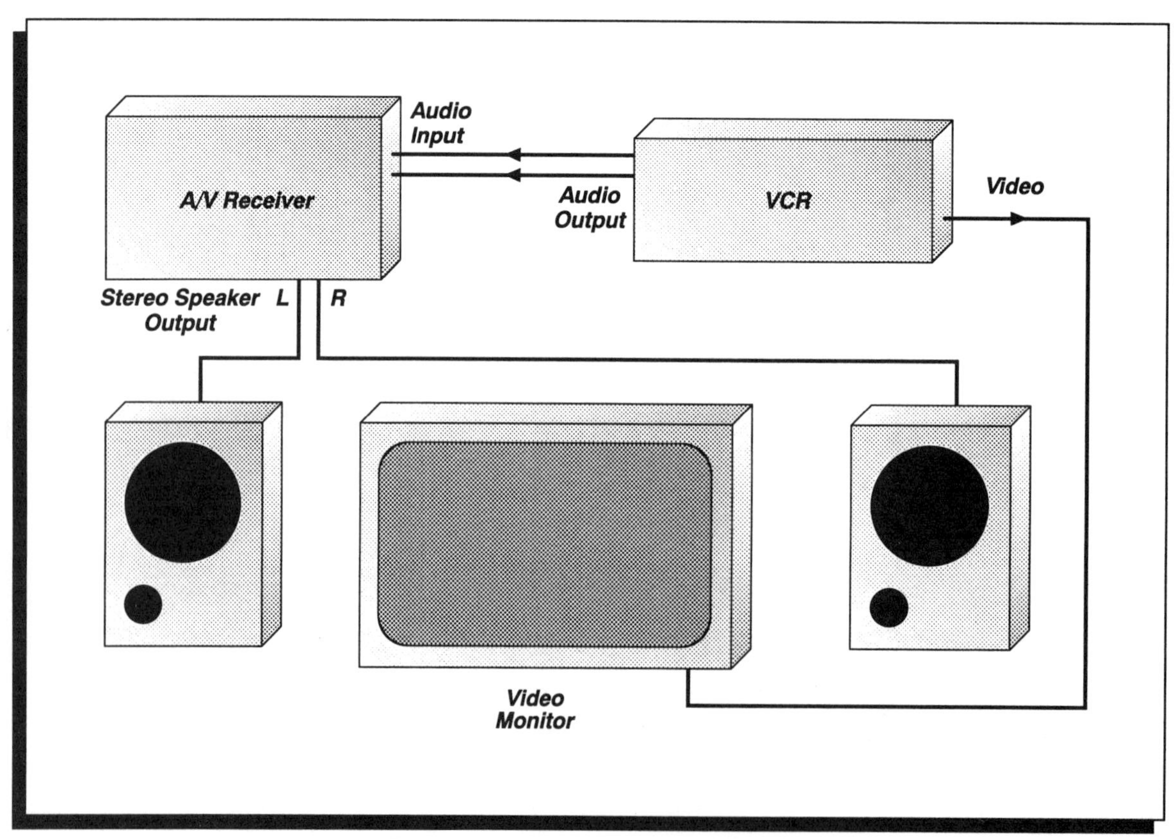

Figure 1.11 Simple basic A/V system.

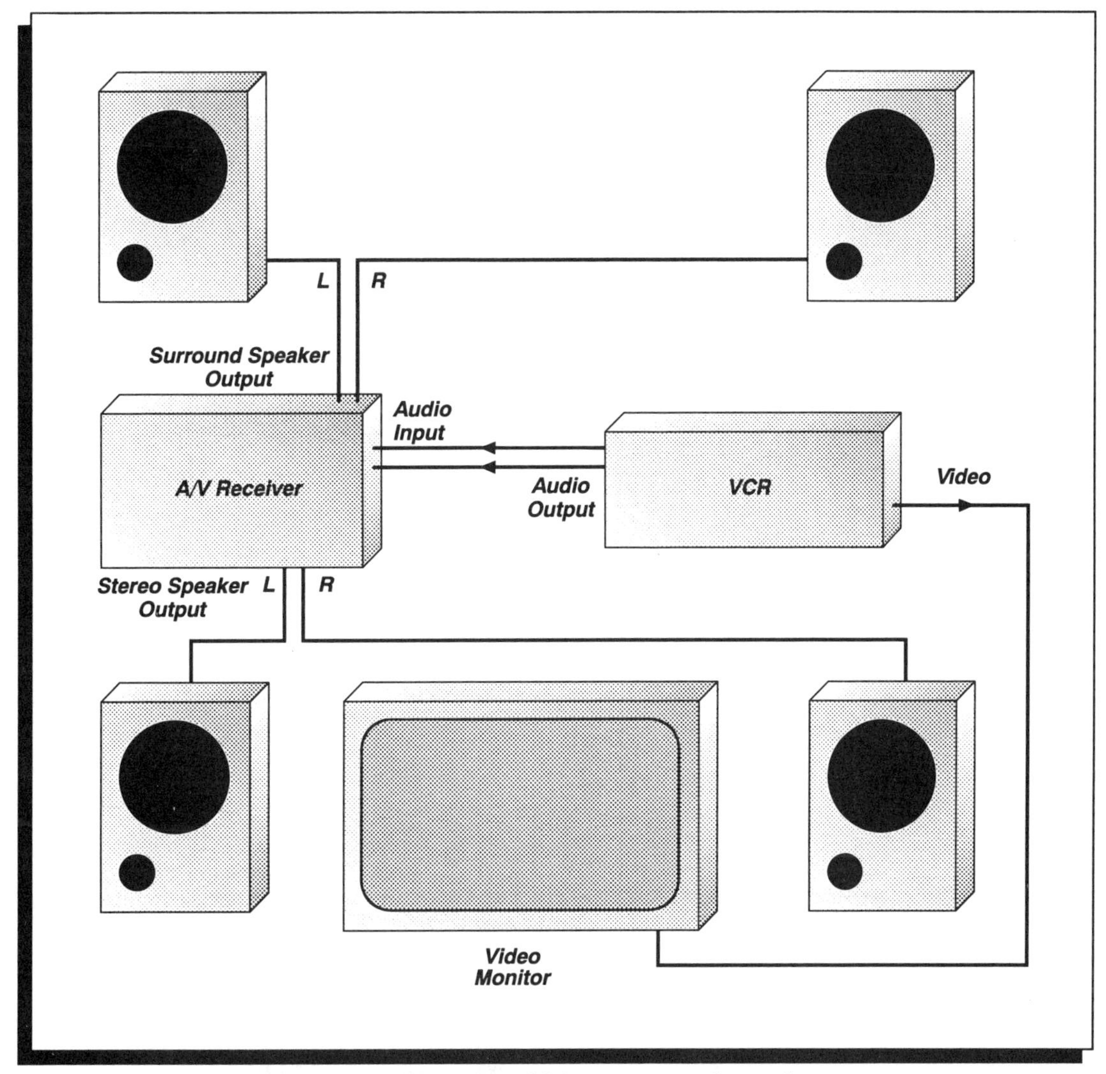

Figure 1.12 A/V receiver driving a surround-sound system.

The inputs are:

— phono: the signal from a record player
— auxiliary: a useful input that can be the signal from a CD, tape deck, tuner, or just testing equipment
— tuner: the signal from an AM or FM tuner
— CD: the signal from a CD player

■ TAPE MONITOR SWITCH

This switch has to be well understood in order for the system to be functional. It has interesting applications. For example, it might be useful to record on a magnetic tape the source of sound that we are listening to, and maybe even to listen to it just after it was recorded. The tape monitor switch provides this capability. There are three possibilities when using the tape monitor switch:

— the sound source can be listened to directly
— the sound source can be recorded on a tape at the same time as one is listening to it, or even without listening to it
— the sound source can be recorded on a magnetic tape (digital or analog) and it is possible to listen to it as soon as it is recorded to see if the recording is proceeding well.

However, it is impossible to listen to the sound source, and the recorded signal at the same time.

When the tape monitor switch is in the SOURCE position, one hears the sound source.

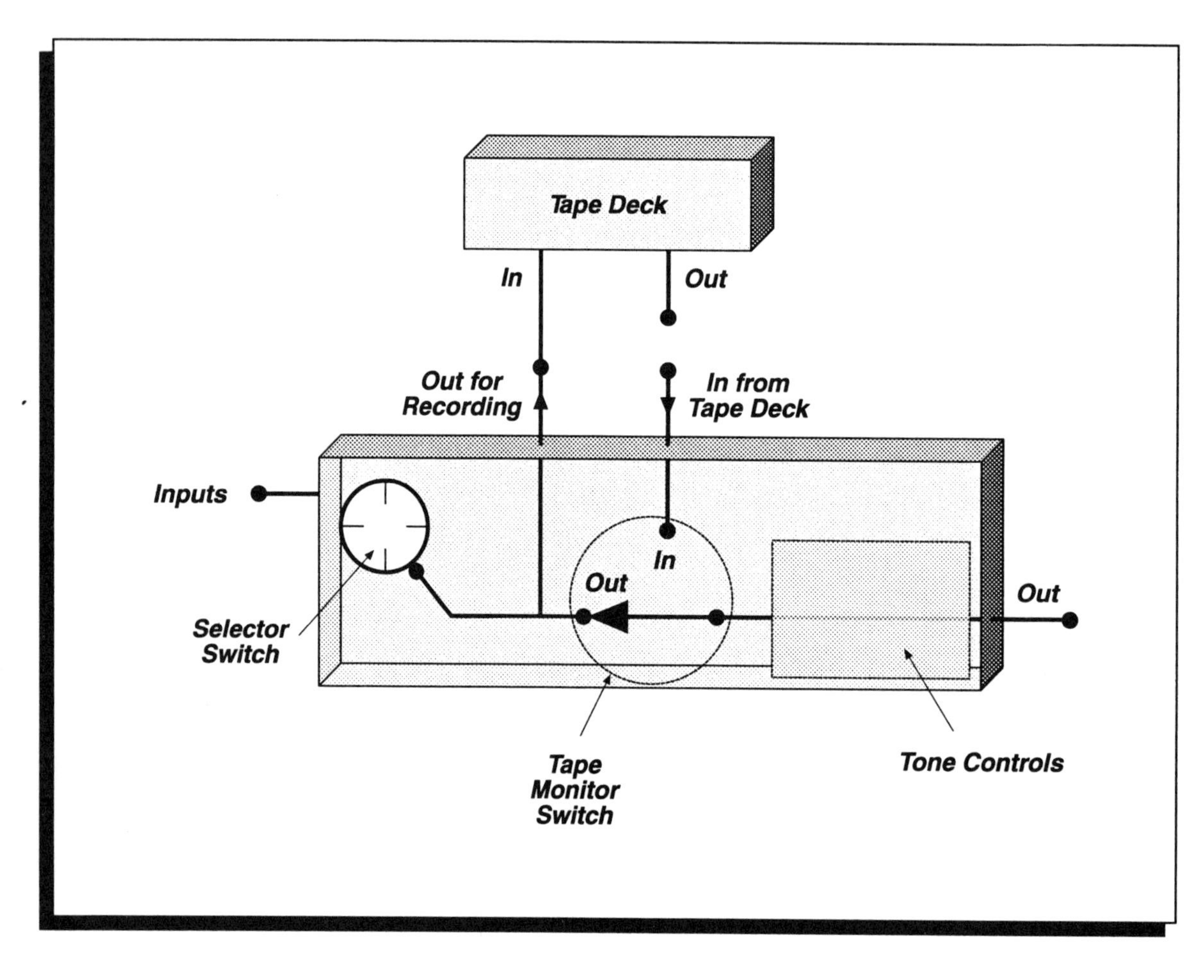

Figure 1.13 Details of the tape monitor switch when listening to a sound source with available tape recording

Of course, it can always be taped at the same time. When the tape monitor switch is in the IN position, one listens to the recorded signal on the tape, a fraction of a second later after the recording or at any other time. Figure 1.13 details this unit. Figure 1.14 explains how the tape monitor switch can be used to listen to the information just recorded on a tape or coming from a tape.

The tape monitor switch can be labeled in a slightly different way depending on the unit or the make of the pre–amp. Usually it is labeled in one of the ways presented below:

	music from sound source	**music from tape**
switch position:	source	play
or	monitor source	play
or	source	tape
or	out	in

■ Tone Controls

The pre–amplifier has controls such as: VOLUME, BASS, TREBLE, BALANCE, FILTERS, LOUDNESS, and others. They are placed after the tape monitor switch and hence have no effect on what is being recorded at the tape OUT terminal by a tape deck.

1.6 A/V System

Receivers, integrated amplifiers, and special systems are now available to produce hi–fi quality sound and ambience with regular audio sources or with movies and videos. This special effect is called surround sound. Surround sound creates ambience by placement of two surround speakers (each producing the same sound) at the side, or rear, of the room. The goal of these speak-

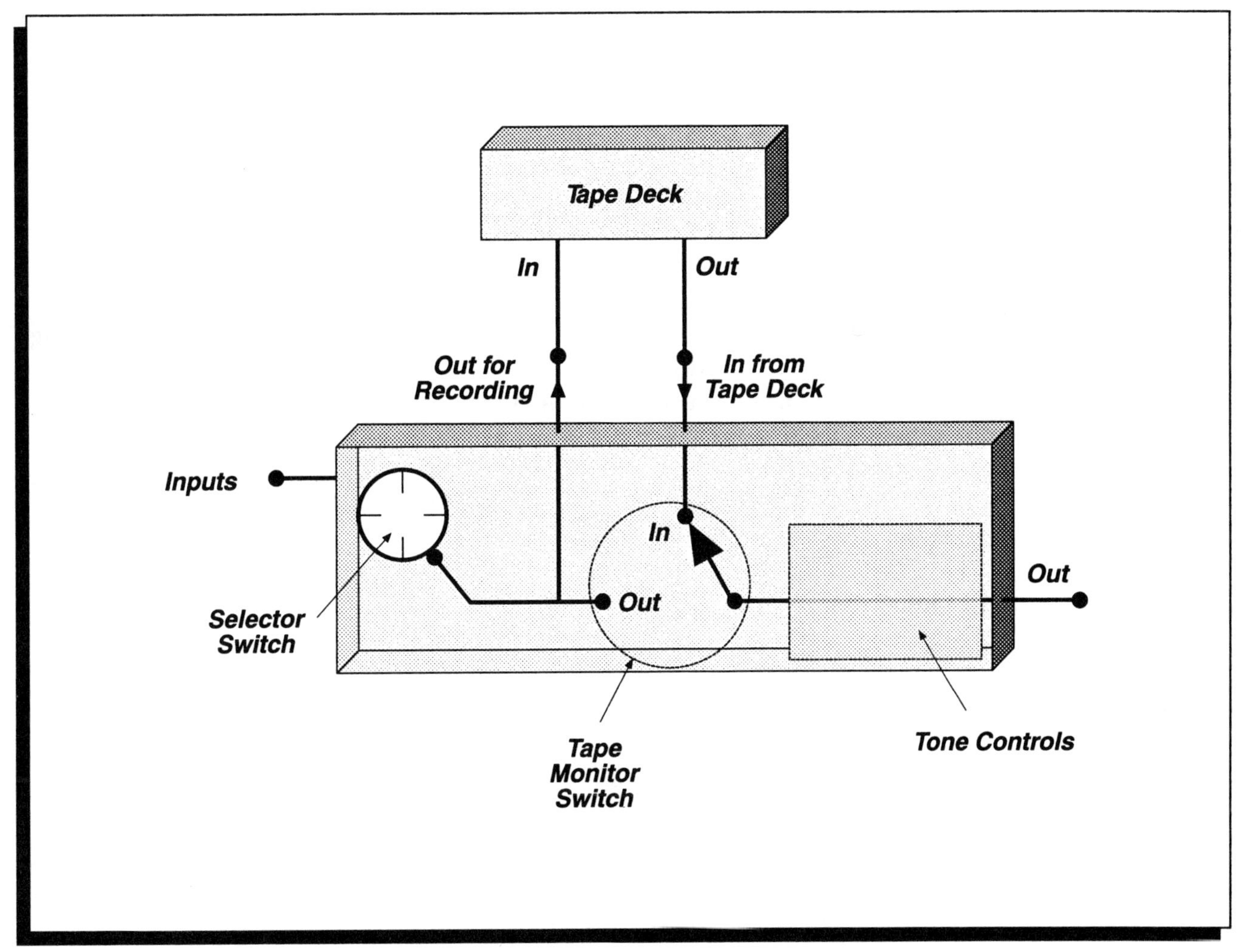

Figure 1.4 Listening to a tape; tape switch in.

ers is to produce an "image" of sound, as if it is coming from everywhere and nowhere simultaneously.

This is achieved by using an audio/video system that has a processor to decode the information recorded on tapes, laserdiscs, TV programs, and even CDs. All recording for home use are in two–channel form, stereo. The stereo signals consist of a LEFT (L) channel and a RIGHT (R) channel. This information can be analyzed as consisting of mono signal information (L+R) which is the information from both channels added together, and of a difference signal consisting of stereo information (L–R) which is the information of one channel subtracted from the other. In fact (L–R) information tells how the left differs from the right. The (L–R) information contains the ambience part of the signal important for imaging in a surround system and it can also contain the surround information encoded in a Dolby Surround recording. The surround portion (L–R) of a stereo signal is handled by the two rear speakers each reproducing the same (L–R) signal, and reflecting it off walls or ceiling to produce an ambience that is fairly uniform in all directions.

Dolby Stereo soundtracks are encoded with surround sound information on tapes, laserdiscs, and even TV broadcasts. An A/V system can decode this information and send it to the appropriate channels.

The most popular system is the Dolby Pro Logic which decodes the signal into five speakers: the LEFT stereo speaker, the RIGHT stereo speaker, the center speaker with (L+R) information, and the surround sound speakers, left and right, with (L–R) information. This is shown in Figure 1.15.

The center speaker, situated near the video monitor, lends realism to the acoustic scene by ensuring that on–screen sound appears to come from the screen for viewers, even those seated off–center from the LEFT and RIGHT speakers. In general, the front speakers should get equal power and the rear ones (with surround–sound information) should get one–half to one–quarter of the total value. The surround–sound signals are delayed by 0.02 to 0.03 seconds.

For tapes and CDs that are not encoded with Dolby Surround, many of them contain ambience (L–R) signals that can also be extracted by a processor such as the Dolby Pro Logic. In fact, a surround–sound processor provides a certain amount of music enhancement in addition to adding reality to video soundtracks. The sounds that you hear become life–like with greater realism and three–dimensionality. The five–channel approach is an improvement over the straight stereo approach.

Summary of Terms

Ambience: collection of sonic cues that tell your ear–brain system that you are in a particular place.
Antenna: input to tuner that captures electromagnetic waves, such as radio waves.
Audio/Video System (A/V): a system that can be used to produce ambiance and surround sound with video tapes or sound. Adds an extra dimension to sound from and extra channel of surround sound.
Hi–Fi: faithful reproduction of sound or reproduction of sound with excellent tonal qualities.

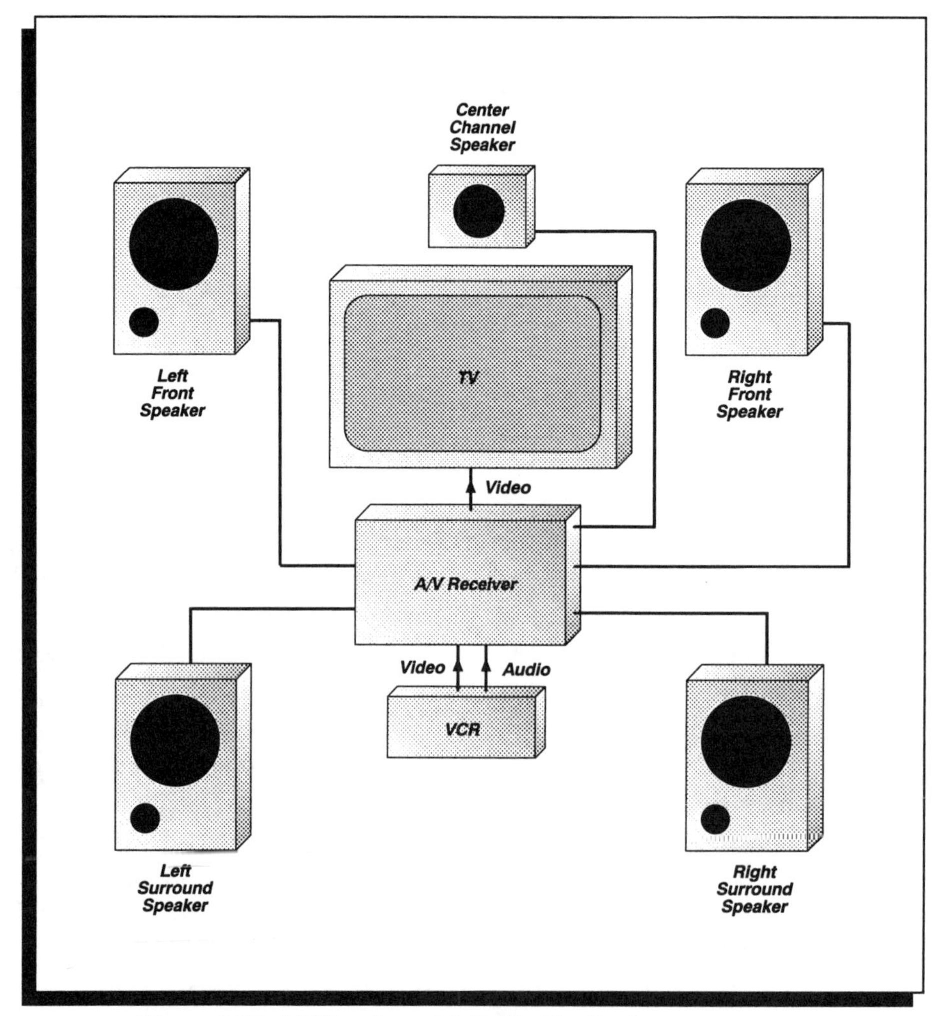

Figure 1.15 A/V Receiver with Dolby Pro Logic processor.

Integrated amplifier: electronic unit that consists of a pre–amplifier and a power amplifier.
Loudspeaker: a unit that converts electrical signals to sound.
Microphone: electronic device that converts sound into electrical signals.
Pre–amplifier: electronic unit that raises the level of certain signals, has all the controls, and routes the signal between different components.
Power Amplifier: electronic unit that boosts the level of the signals from a pre–amp to drive the speakers.
Receiver: electronic unit containing a tuner, a pre–amplifier and a power amplifier.
Selector Switch: determines which specific source of sound is used as a signal input to the hi–fi.
Stereo: processing of sound in two channels.
Tape Deck: unit used for recording on magnetic tapes and playing the information back.
Tape In Jacks: provides signals into the pre–amp from recorded sources such as a tape deck or a signal processing unit.
Tape Monitor Switch: a switch in the pre–amp that accepts signals from sources of sound or from a tape deck.
Tape Out Jacks: provides signals from sound sources as selected by the selector switch for recording or for further processing.
Tuner: electronic unit that picks up radio stations and extracts the audio information from radio waves.

NAME ____________________ DATE ____________

QUESTIONS FOR REVIEW

1 What is the difference between the signals picked up by a microphone and the signals going into a speaker?

2 Explain what would happen if a source of sound were connected to a pre–amplifier and then to a speaker.

3 What are the functions of a pre–amplifier?

4 In a pre–amplifier what are the effects of the tone controls on the signals recorded by a tape deck?

5 Discuss the goals of a hi–fi system and how well it achieves these goals.

6 When will a stereo system not be considered hi–fi?

7 What is the difference between a stereo system and a surround–sound system?

8 What is the purpose of a tape monitor switch?

9 Where does the surround–sound information come from?

10 What is the significance of the information obtained when, in stereo, the signal from the LEFT channel is subtracted from that of the RIGHT channel?

NAME____ DATE ____________

EXERCISES

Please select the best answer for each question.

1. The combination of a pre-amplifier, power amplifier, and a tuner contained in one unit is called:
   ___A. a radio
   ___B. a receiver
   ___C. a hi-fi
   ___D. an integrated amplifier
   ___E. a sound source

2. While recording a CD on a magnetic tape from the TAPE OUT jack of a pre–amplifier, more bass can be achieved by means of what control on the pre–amp?
   ___A. the BASS control
   ___B. the VOLUME control
   ___C. the TREBLE control
   ___D. the BALANCE control
   ___E. no control on the pre–amp can boost the bass on the tape

3. If you wanted to listen to a CD, the SELECTOR switch must be set to _________ and the TAPE MONITOR switch to _________ .
   ___A. AUX, TAPE
   ___B. PHONO, SOURCE
   ___C. AUX, SOURCE
   ___D. FM, TAPE
   ___E. AUX, FM

4. The combination of a pre–amp and a power amplifier is called:
   ___A. an integrated amplifier
   ___B. a receiver
   ___C. a tuner
   ___D. an audio/video receiver
   ___E. an amplifier

5. An old receiver differs from regular receivers in that it has no tape monitor switch. If you are interested in recording a tape with this system, what will you encounter?
   ___A. tape cannot be recorded
   ___B. tape cannot be played back
   ___C. tape recording cannot be monitored almost instantly after the recording
   ___D. tape cannot be recorded in stereo
   ___E. information going to the tape deck cannot be heard

6. An electronic unit that produces large currents that can drive speakers is called:
   ___A. a pre–amplifier
   ___B. a tuner
   ___C. a microphone
   ___D. a receiver
   ___E. a tape deck

7. The tone controls are usually found in the:
   ___A. tuner
   ___B. power amplifier
   ___C. pre–amplifier
   ___D. sound source
   ___E. CD player

8. If you only bought a CD player and a pair of speakers, would you be able to listen to them?
   ___A. yes
   ___B. no, because the signal from a CD player is too large for the speakers
   ___C. no, because a CD player puts out an electrical signal and speakers need sound to drive them
   ___D. no, because a pre–amp is needed
   ___E. no, because an integrated amplifier is needed

9. The term high-fidelity means:
 ___A. stereo recording
 ___B. a total faithfulness to the original sound
 ___C. surround sound music
 ___D. ability of a system to accept a CD player, a DAT system, and a micro–phone
 ___E. being able to record loud sounds

10. Ambience means:
 ___A. sound in every part of the room is the same
 ___B. sound in every part of the room is quiet
 ___C. sound at one end of the room can be different from that at the other end
 ___D. recording in stereo
 ___E. recording with high fidelity

11. While recording a CD on a magnetic tape, the volume on the pre–amp is turned down. What effect will it have on your recording?
 ___A. nothing will happen, because the volume control is after the TAPE OUT jacks
 ___B. the recording will be very quiet
 ___C. only very loud sounds will be recorded
 ___D. the left minus the right (L–R) information on the tape will be small
 ___E. the tape cannot be recorded with high fidelity

12. When the volume on a receiver is turned down, what effect will it have on the sound coming out of the system?
 ___A. only one channel, will be turned down, the other one will keep on playing at the same level.
 ___B. both channels will sound quieter.
 ___C. one channel will become quieter, while the other one will be louder.
 ___D. there will be no effect on the sound since the tone controls come after the tape monitor switch.
 ___E. only the mono information (L+R) will be affected.

13. If you wanted to listen to a prerecorded tape through an input other than the TAPE IN, where would you connect it in a pre–amp?
 ___A. phono input
 ___B. auxiliary input
 ___C. antenna input
 ___D. directly to the speakers
 ___E. output of pre–amplifier

14. In a surround–sound system, roughly how much power should be sent to the rear left surround–sound speaker, when the left stereo speaker gets 100 watts?
 ___A. 100 watts
 ___B. 200 watts
 ___C. 300 watts
 ___D. 30 watts
 ___E. 120 watts

15. Consider the various waveforms.

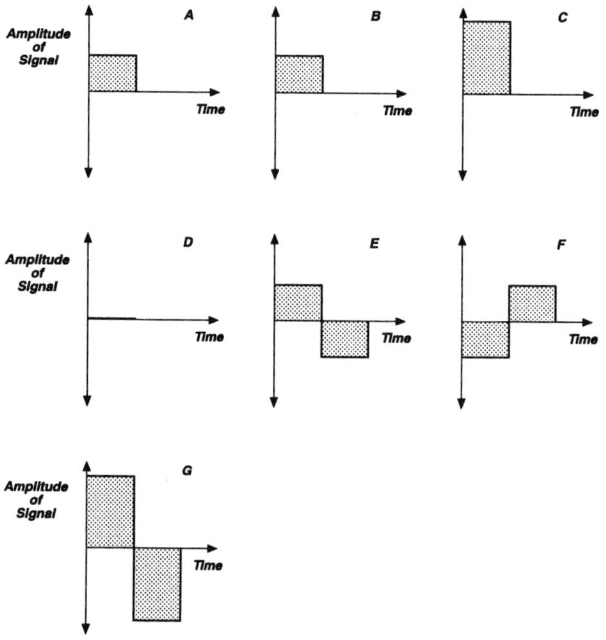

If A and B are the left and right channels respectively, then which one is (L+R)?

____

NAME ______________________________ DATE ____________

16. In the previous example which one is (L–R)?

____

17. If E and F are left and right channels, which one is (L–R)?

____

18. Which one of the following represents a correctly assembled hi–fi system?
 ___A. tape deck→tuner→power amp→speakers
 ___B. tape deck → pre–amp → power amp → speakers
 ___C. pre–amp → tape deck → speakers
 ___D. tape deck → pre–amp → speakers
 ___E. tape deck → power amp → pre–amp → speakers

19. If you wanted to listen to a CD connected to the AUX input of a receiver with speakers and no sound was being produced, then:
 ___A. tape monitor switch was in IN position
 ___B. selector switch was set to AUX
 ___C. tape monitor switch was in OUT position
 ___D. VOLUME control switch was set too high
 ___E. (L–R) information from CD was zero

20. In a hi–fi system, the Dolby Pro Logic processor ensures that all the five speakers put out:
 ___A. the same frequencies
 ___B. the same volume
 ___C. same sound at the same time
 ___D. sound which consecutively goes from one speaker to the next
 ___E. none of the above

21. Any sound source applied to the SELECTOR switch of a pre–amp can be monitored without involving the tone control section from the:
 ___A. TAPE IN jack
 ___B. PHONO jack
 ___C. AUX jack
 ___D. TAPE OUT jack
 ___E. OUTPUT jack

22. If you wanted to test the performance of your hi–fi system, where would you connect the test signals?
 ___A. PHONO input
 ___B. AUX input
 ___C. antenna input
 ___D. pre–amp input
 ___E. video input

23. A tape deck cannot drive speakers directly because:
 ___A. it does not have enough power
 ___B. signals are magnetic but speakers need electrical signals
 ___C. signals are magnetic but speakers need large sound power
 ___D. it needs a pre–amp
 ___E. it needs a tuner

24. The advantages of playing audio on a hi–fi when watching a video is:
 ___A. hi–fi speakers are usually of higher quality than the TV speakers
 ___B. surround sound can be produced by a processor
 ___C. the audio can have extra dimensionality
 ___D. ambience can be produced
 ___E. all of the above are correct

25. The principles of surround sound are based on:
 ___A. four to five speakers each producing the same sound equally shared
 ___B. four to five speakers producing sounds of different intensities and all at the same time
 ___C. four to five speakers producing different sound intensities at different times
 ___D. four to five speakers with stereo sound in front of the room and with stereo sound in the back of the room
 ___E. four to five speakers producing stereo sound with two speakers and the others existing as spare speakers

CHAPTER 2 WAVES

The hi–fi system deals with all sorts of waves; sound waves, radio waves, and even light waves, and hence waves are fundamental in the hi–fi. In this chapter general properties of waves will first be presented and then they will be extended to sound. Properties such as reflection, refraction, and interference will be studied and then they will be related through examples to the hi–fi. Complex waves will be built up from the various standing waves produced by musical instruments. The response of the hi–fi system to the wide range of complex waves produced by instruments will provide an evaluation of the hi–fi performance.

2.1 Waves in General

Since this book will deal with sound waves and other types of waves, it is important to start with an introduction to the subject in general. The simplest definition of a wave is:

> ***wave* = a disturbance which propagates.**

The specific name given to the wave depends on the nature of the disturbance. For example, it is possible to have a water wave, a sound wave, an audio wave on a moving magnetic tape.

It is important to note that in the definition above of a wave it is the *disturbance* which propagates, and not the medium. In a speaker producing sound the vibrating air molecules do not go from the speaker to our ear drums, it is the disturbance which propagates. There is a transport of energy that can activate our ear drums, drive a microphone or be simply absorbed in a hall.

The concept of a wave becomes clear and easy to understand when examples are given and a picture of a wave is presented. In the hi–fi sys-

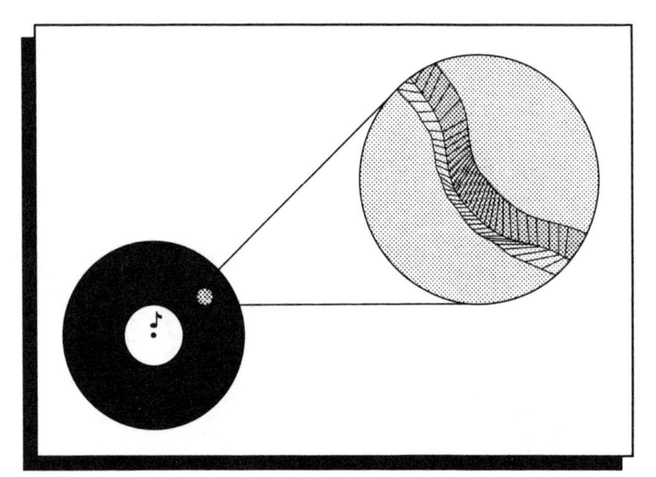

Figure 2.1 Phono record and an enlarged groove showing engraved wave representing sound.

tem, there are many examples of waves and some of them even show the wave directly. The old–fashioned record is one such example because there is a mechanical engraving of a wave in the record plastic. Figure 2.1 is an enlargement of a section of a phono record groove.

Another example of a wave that is perhaps more familiar to most of us is a water wave. Imagine going to the beach and taking a picture of a water wave. What would it look like? Well, something is waving; the water. The important features of this photo could be represented by Figure 2.2. Water is displaced, up and down, from its equilibrium position, and the displacements vary with position

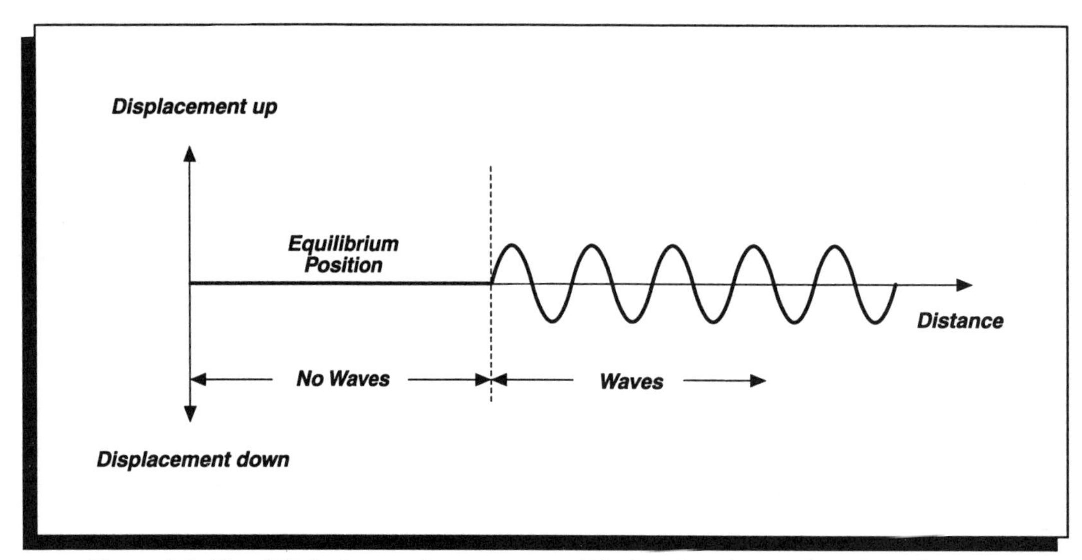

Figure 2.2 Simplified picture of a water wave. Displaced water as a function of position

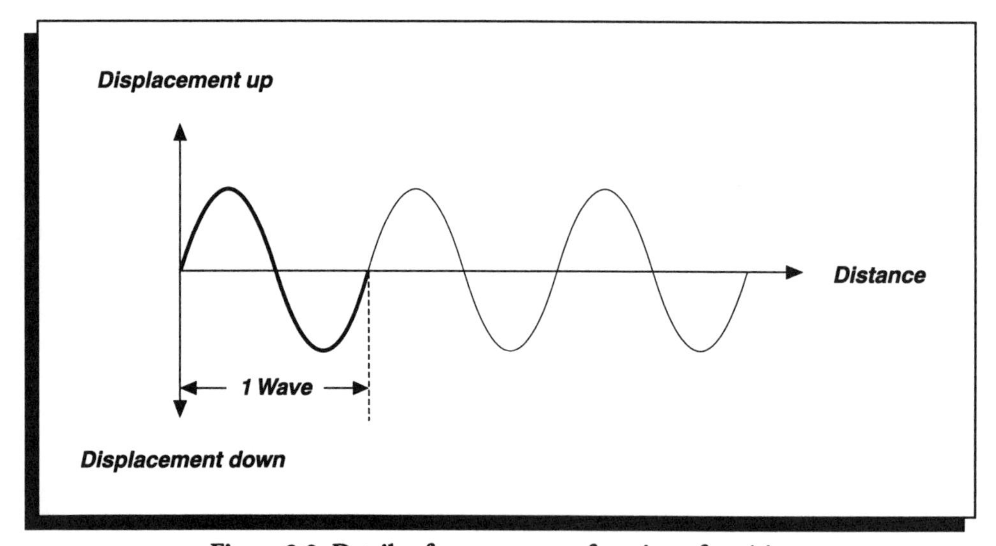

Figure 2.3 Details of one wave as a function of position.

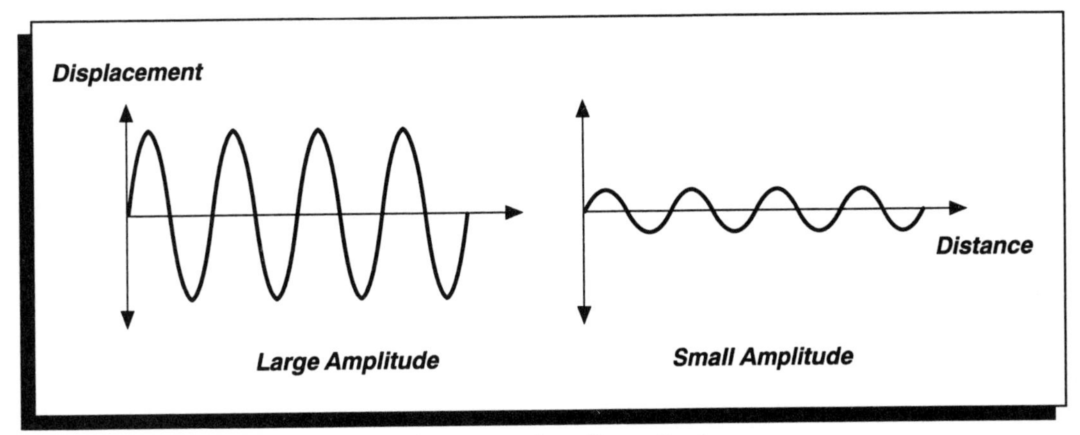

Figure 2.4 Large and small amplitude waves.

along the water wave. The picture of displaced water from equilibrium as a function of position looks like Figure 2.2.

Let us look in detail at one wave, as shown in Figure 2.3. The water is displaced, more and more, as we go along the wave until it reaches a maximum and then it is displaced, less and less, approaching the equilibrium position. Then the water is displaced below the equilibrium position reaching a maximum displacement, and then it comes back toward its equilibrium position.

This figure provides the following information:

— amplitude: maximum displacement from equilibrium, up or down
— wavelength: how long one wave is

Note that a wave consists of up and down displacements, and a subsequent return to its original starting point. The wave drawn in Figure 2.3 is arbitrary. It is possible to have all sorts of situations, such as a large amplitude wave, a small amplitude wave, as in Figure 2.4, along with a variety of diverse amplitudes.

It is clear from this figure that the large amplitude wave has more power than the small one (a large amplitude water wave can knock you down, while a small amplitude wave will not have enough power). Likewise, a large amplitude sound wave will have more power than a small amplitude sound wave.

The simple picture of a wave that we described was just one event in a series that occurs over time. Had we looked at the displacements in one position, we would see the variation with time. In fact, if we had taken a movie of the wave, we would be able to see the variations of the displacement with time and at any given point, as seen in Figure 2.5.

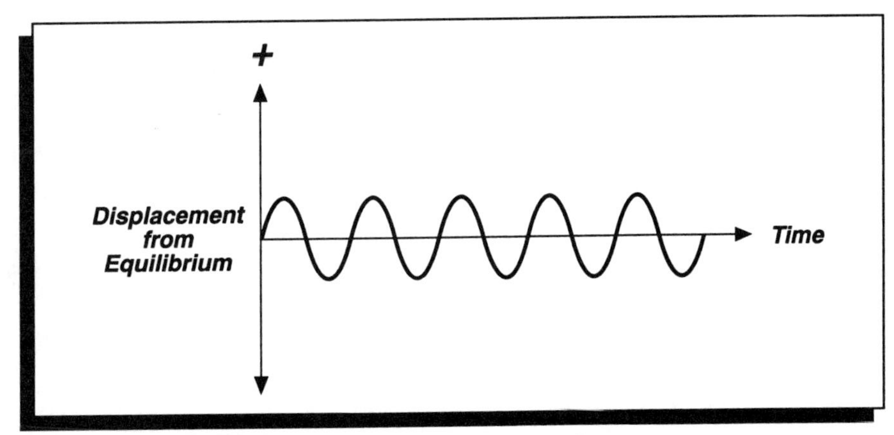

Figure 2.5 Time dependence of displacement of a point on a water wave.

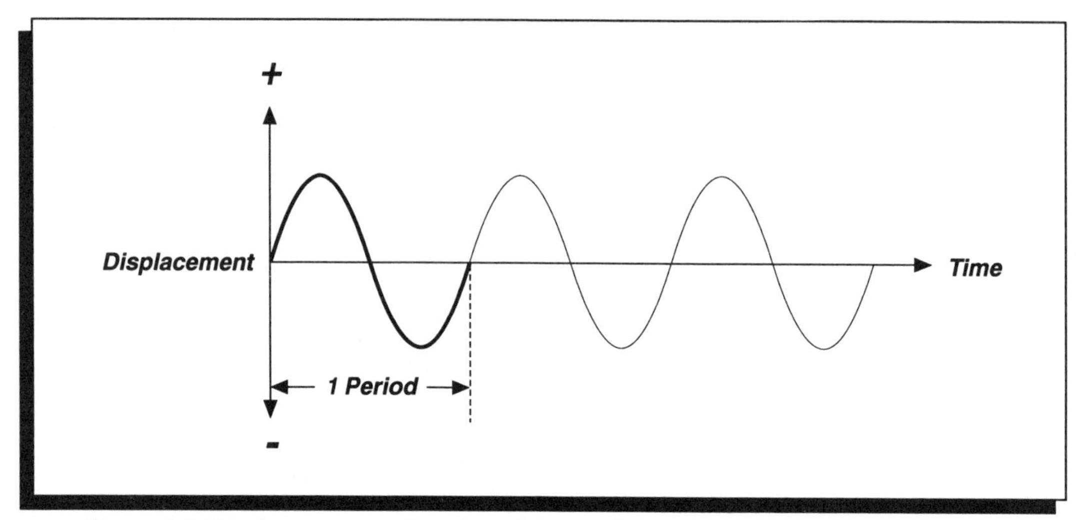

Figure 2.6 Displacement as a function of time and time required to complete one wave.

Concentrating on one wave, the important details are presented in Figure 2.6, which shows the variation of the displacement at a given point and as a function of time. It takes a certain amount of time to complete one wave, and this is called the period. Hence,

Period **= how long it takes to complete one full wave.**

There is no need to consider other waves, because they all repeat themselves. Once the time for one wave is known it applies to any wave of that wave form, assuming the wave pattern does not change. A rapidly varying wave will have a short period, while a slowly changing wave will have a long period.

Because a wave repeats itself (this is obvious especially when taking a cruise, there are just waves and waves on the water all around) we can ask how frequently the waves repeat themselves. The answer to this question gives a very important characteristic of a wave, known as frequency. Its definition is:

Frequency **= how many waves (or wavelengths) go by a point in one second.**

The unit of frequency is the Hertz, which is actually the number of waves/second. As an example, consider a frequency of 1,000 Hz. This means that there are 1,000 waves that go by in one second. When frequencies are high, it is useful to use the following definitions:

1,000 Hz ≡ 1 kHz (kiloHertz)
1,000,000 Hz ≡ 1 MHz (MegaHertz)

Now, from the definitions of period and frequency it appears that they should be related. If you know how many waves there are per second, we could figure out how long it takes for one wave to go by? For example, consider a wave with a frequency of 1,000 Hz, how long does it take to complete one wave? It must be 1/1,000 of a second. Hence, from this example we can say that:

Period **= 1/frequency**

Consider another example to illustrate this. If it takes a wave (1/2) sec. to go by a point, then in one second there must be two waves (i.e. its frequency is 2 Hz). This is generally presented as:

Frequency **= 1/period**

Continuing with the characteristics of a wave, what is its speed? From every day experience, especially driving a car, speed means distance covered in a certain amount of time (i.e. distance/time). In describing the characteristics of a wave there must then be a relationship between speed, frequency, and wavelength. Frequency is related to the inverse of time, while wavelength relates to distance. In order to explore these concepts let us try an experiment, where we multiply frequency by wavelength to determine the speed. Hence, we conclude a very important relation:

***Speed* = frequency × wavelength**

It is true for any wave. This is a very useful formula that will be used throughout this book.

For example, knowing two of the quantities in this formula can help us in calculating the third. To illustrate this, consider a few examples of waves that will be important to us such as: sound waves, light waves, radio waves, mechanical waves in grooves of a record, or magnetic information on a moving magnetic tape.

Example 1. Sound wave in air. The speed has been measured to be 344 meters/sec. What is the wavelength of sound in air at a frequency of 1,000 Hz? Using the important formula which relates speed, frequency and wavelength, we can find the wavelength. Rearranging the equation leads to:

wavelength = speed/frequency
= (344 m/sec.) / (1,000 Hz)
= 0.344 meter

Example 2. The speed of a radio wave is almost 3×10^8 m/sec. What is the wavelength of a radio wave at 1,000 kHz? Again, using the same formula:

wavelength = speed/frequency
= (3×10^8 m/sec.)/ 1,000,000 Hz
= 3×10^2 meters

Example 3. A wave of 1,000 Hz on a guitar string that has a wavelength of 0.10 meter, will have a speed of:

Speed = frequency × wavelength
= 1,000 Hz × 0.10 meter
= 100 m/sec

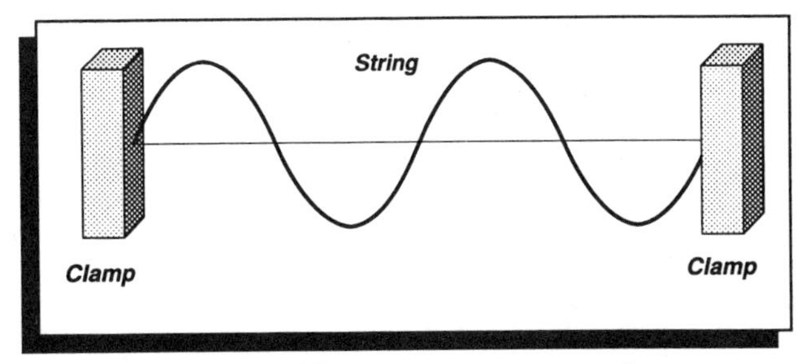

Figure 2.7 Transverse wave on a string.

In the general properties of waves, two types of waves exist depending on the direction of the displacement of the wave relative to the direction of propagation. When the disturbance is perpendicular to the direction of wave propagation, it is known as a transverse wave. An example of this wave would be the transverse vibrations on a string, as illustrated in Figure 2.7.

In certain systems, vibrations can be set up so that the displacements are parallel to the direction travelled of the wave. Such waves are known as longitudinal. An example would be a metallic bar struck on one end, so that vibrations are excited parallel to the bar axis. Figure 2.8 illustrates this concept:

Some examples of the two types of waves are:

Sound waves in air or in a gas. Longitudinal because a gas is not rigid and it cannot sustain transverse waves.

Sound waves in a solid. Can be longitudinal or transverse, depending on how they are generated.

Light waves. Always transverse.

Magnetic recording on a moving analog tape. Longitudinal since it is easier to achieve technically.

Undulations in grooves of a rotating phono record recorded in mono. Transverse since the displacements are perpendicular to direction travelled of the grooves.

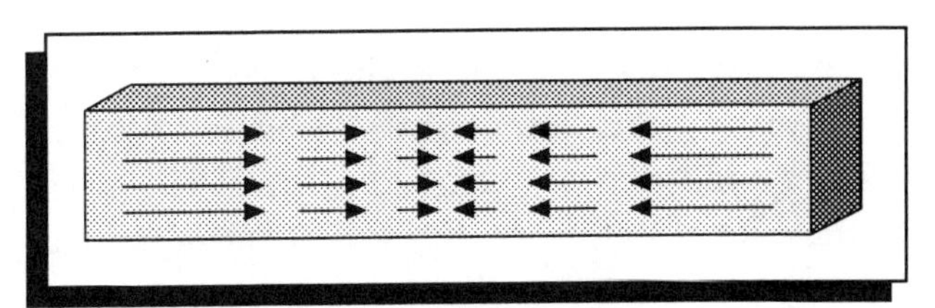

Figure 2.8 Longitudinal waves along a solid bar.

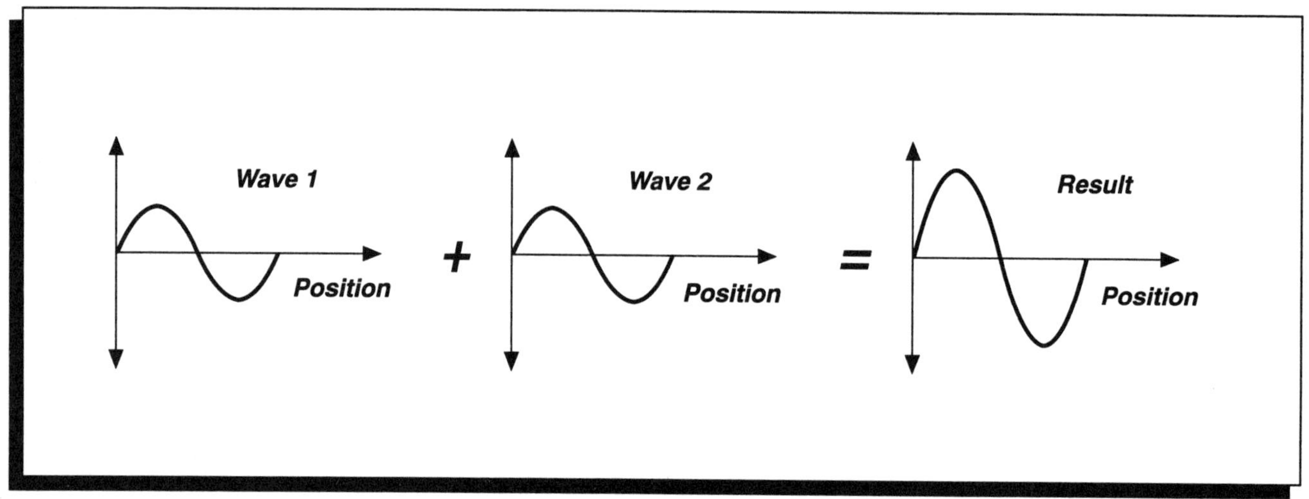

Figure 2.9 Addition of two waves.

Water waves are complicated due to rotary motion. Although the list of particular waves have just been introduced here to illustrate the ideas, these waves have not yet been defined. Further development of the subjects will provide a definition and in–depth treatment.

So far simple concepts of waves have been introduced. However, the hi–fi system deals with a diverse mixture of waves from each instrument and all the instruments collectively in an orchestra or band. What really happens when there is more than one wave present, as it is very likely to occur? There is a very important principle which tells us what to do in such cases. It is known as the **Superposition Principle**, and it states that if there are many waves, they should all be added together.

Superposition Principle **= if two or more waves are present they should be added together.**

For example, in Figure 2.9 there are two waves played together. The Superposition Principle tells us that the net effect of the two waves will be the result of adding them.

Of course, this is a simple example. Later, more complicated situations will be introduced.

There are other important properties of waves which should also be introduced here. However, for the sake of simplicity they will be introduced as the subject develops, with each specific case being related to the hi–fi.

2.2 SOUND

This wave is the central theme of this text. We have already used sound in the examples, but without its definition, nor an explanation of what it is. Before presenting its definition, a few facts need to be introduced. To produce sound, two conditions are necessary:

— a vibrating object
— a medium in which it can propagate.
The medium can be a liquid, gas, or a solid. Sound does not propagate in a vacuum, as there is no medium.

A classic example illustrating this concept is in Figure 2.10, with a source of sound in a bell jar. The speaker within the bell jar produces sound when the speaker is excited and there is air inside the bell jar. When the air is pumped out, there is hardly any sound produced since the medium has been removed. The small amount of sound transmitted through the glass walls of the jar comes from the weak coupling to the speaker supports. This experiment clearly shows that there is no propagation of sound in vacuum.

When a loudspeaker is driven at various frequencies, the disturbances thus created are known as sound. There is a wide range of frequencies to which our ears can respond. The majority of people would agree that, on the average, sound can be

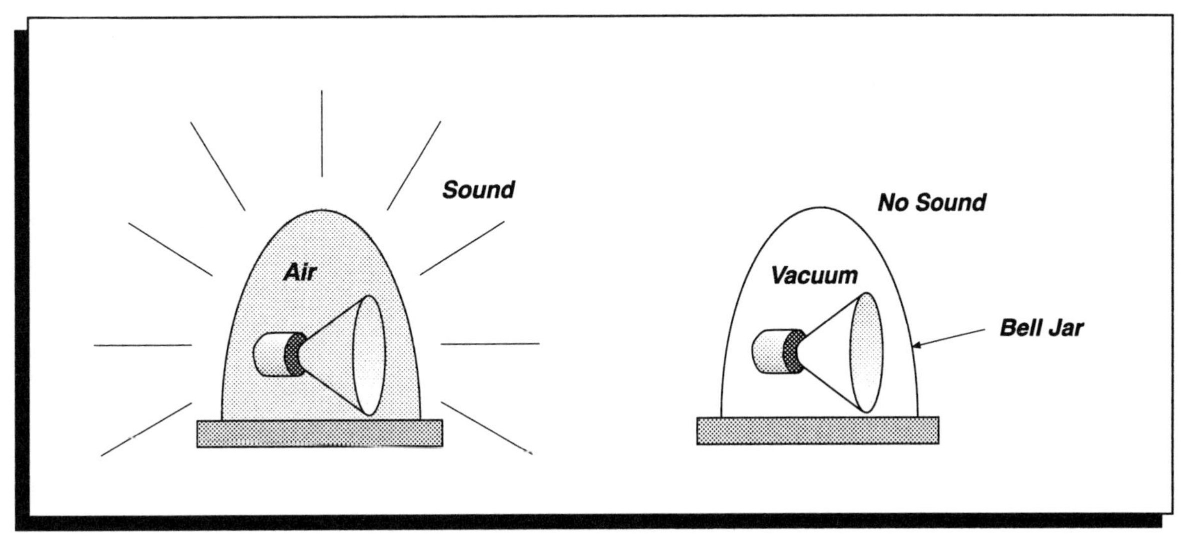

**Figure 2.10 Sound requires a medium to propagate.
In a vacuum, there is no sound propagation.**

heard for frequencies ranging from 20 Hz to 20,000 Hz. Therefore, one can say that:

> ***Sound*** **= vibrations of air or other media at frequencies ranging from 20 Hz – 20,000 Hz.**

To gain some insight into the nature of the vibration of air and sound, consider a loudspeaker, shown in Figure 2.11. It acts like a drumhead in that it can move a volume of air, back and forth, when set into vibration.

When it is electrically activated at some frequency, the diaphragm of the speaker will move forward and backward. It will keep on repeating this motion at the frequency of the electrical signal input. Let us analyze this motion as it will provide an understanding of the nature of sound and its production. When the diaphragm moves to the

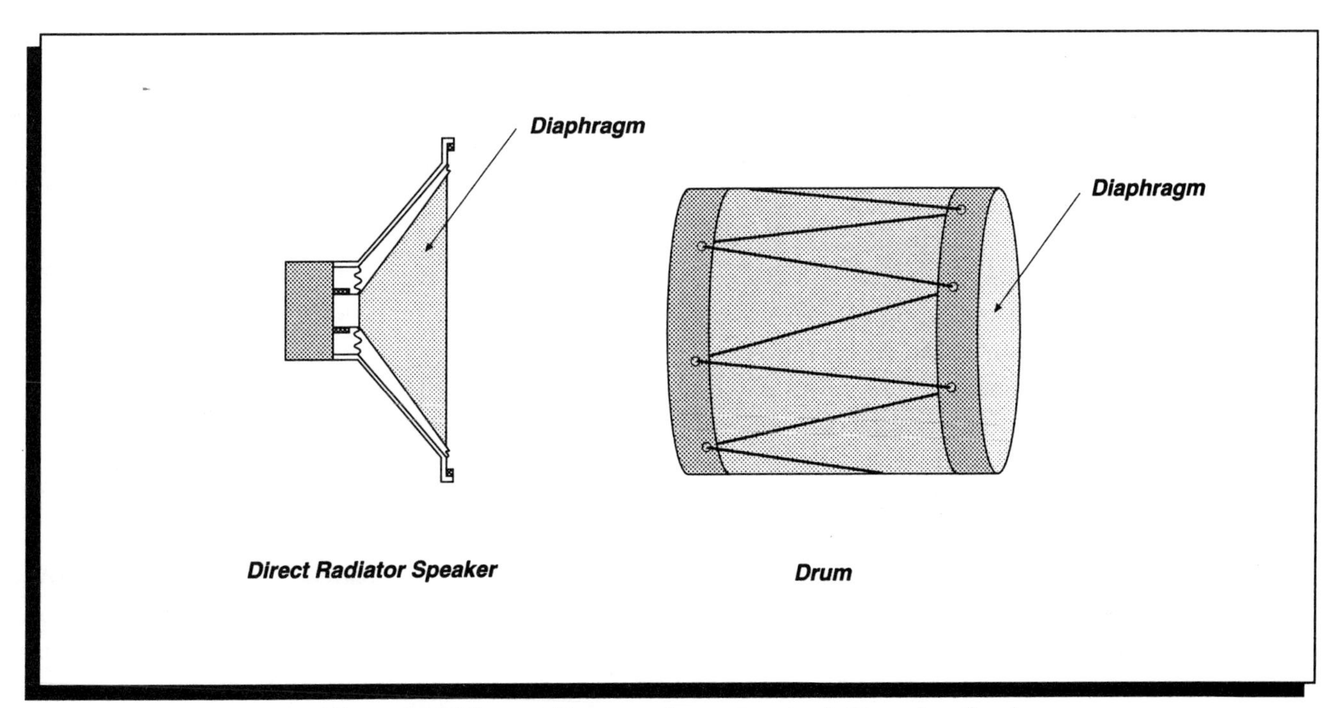

Figure 2.11 Direct radiator speaker can move air like a drumhead.

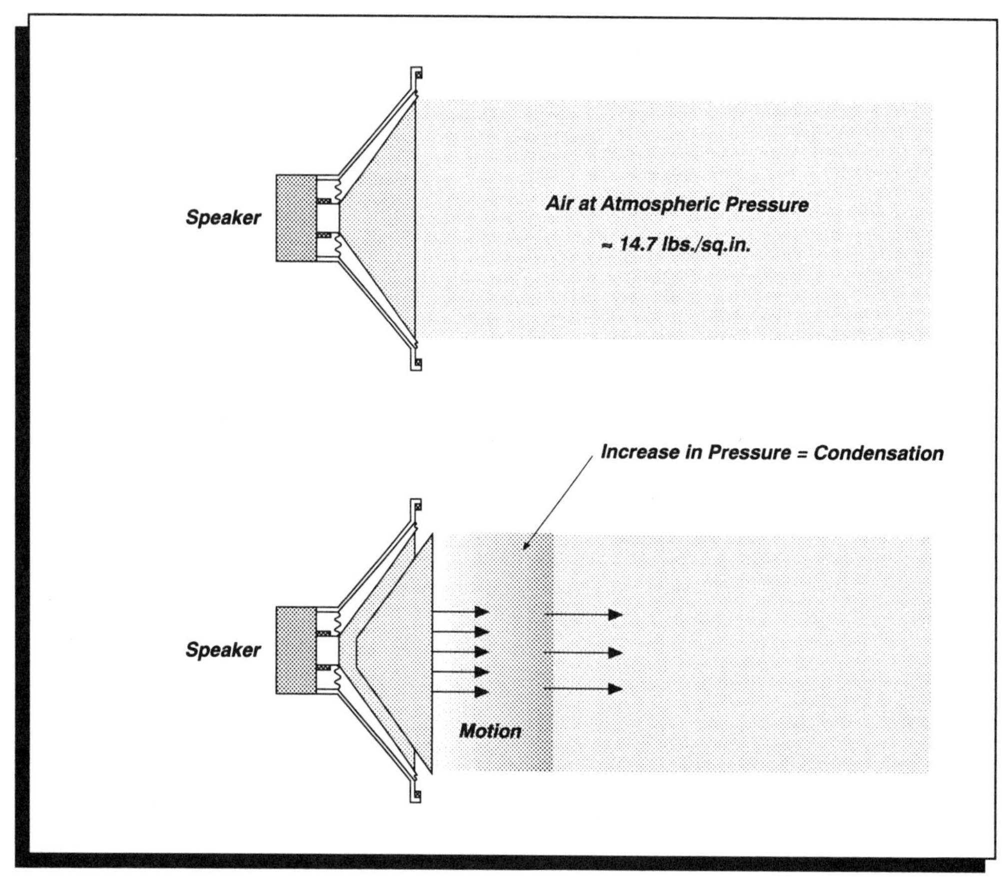

Figure 2.12 Generation of sound by loudspeaker.

right, as in Figure 2.12, the air in front is compressed. Such an increase in pressure is called a condensation.

When the loudspeaker moves back, the air in front will be decompressed (i.e. the pressure will decrease slightly). This is called a rarefaction. Such back and forth motion will continue as long as there is an input signal to the speaker, causing a series of condensations and rarefactions in the air around the speaker. These disturbances will propagate at a certain speed, depending on the properties of the air. The propagation is caused by the pressure change affecting the medium around it which gets passed on. If the pressure in the air around the speaker were to be measured, it would consist of condensations and rarefactions, shown in Figure 2.13. A graph representing these pressure changes as one moves away from the speaker is shown just below the disturbances.

The equilibrium pressure is the atmospheric pressure of air, which is about 14.7 pounds per square inch. Vibrations of the loudspeaker cause the pressure in air to increase and decrease. Such pressure changes are extremely small for the usual levels of speaker vibrations we hear. We call such disturbances sound.

Of course, the speaker could be driven at all sorts of frequencies, but for the range of 20 Hz to 20,000 Hz our ears will respond to these pressure changes. Hence:

> ***Sound*** **= pressure variations at frequencies in the range of 20–20,000 Hz.**

This definition tells us what is "waving" in sound: the pressure changes. Also, it is clear now why a medium is necessary for sound propagation. Pressure changes have to move on, and a medium is necessary to have pressure changes.

The disturbances created in the air around the speaker are longitudinal, as they are parallel to the direction of wave travel. The complete picture in 2–dimensions of waves, created around the vibrating speaker are also shown in Figure 2.14.

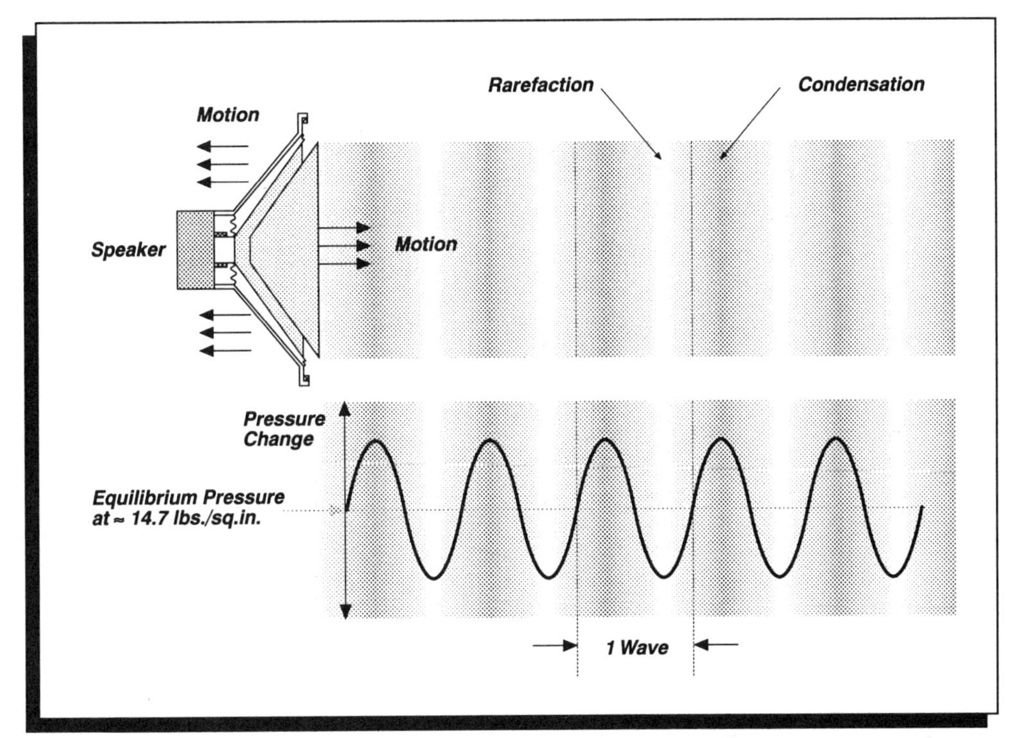

Figure 2.13 Disturbances created by loudspeaker. Pressure changes cause sound.

Figure 2.14 explains how the pressure varies in front of a vibrating speaker. Louder sounds correspond to larger changes in pressure. The sound disturbances created by the speaker propagate at a certain speed. In air, the speed of sound is 344 meters/second at a room temperature of 20° C. This speed is dependent on the temperature because as air warms up its density decreases, causing the speed of sound to increase. Since sound waves propagate as a result of collisions between molecules, one can expect an increase in speed when the speed of molecules increases due to a temperature increase. This increase in the speed of sound will amount to 0.6 meters per second for each degree centigrade temperature rise.

The speed of sound will depends on the medium that it is traveling through. By going to a medium of smaller density, like helium gas whose density is much less than that of air, the speed of sound will be higher. Indeed, the speed of sound in helium gas is 2.9 times faster than in air. The speed also depends on the rigidity of a material.

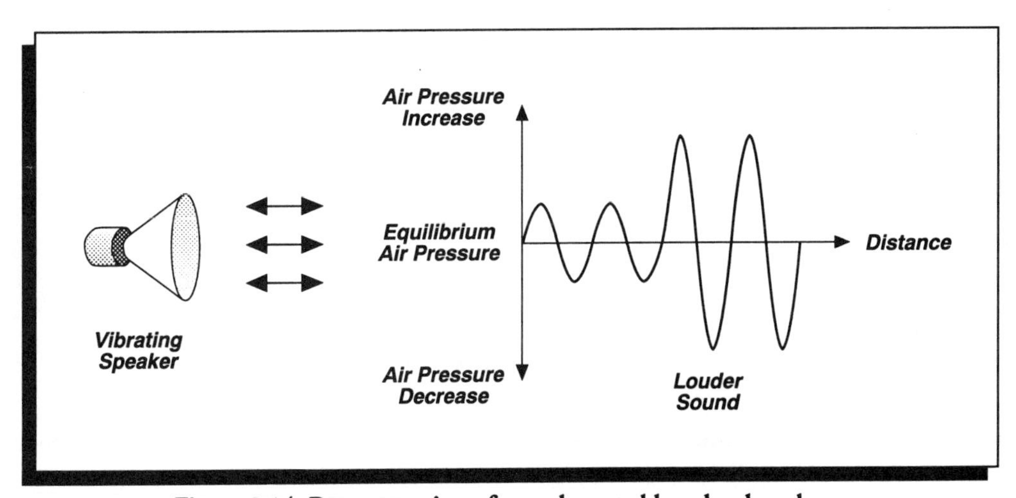

Figure 2.14 Representation of sound created by a loudspeaker.

Consequently, the speed of sound in solids; rigid objects like steel or aluminum, is much higher than in air. Table 2.1 provides a selected list of the speed of sound in various media.

Table 2.1 Speed of sound in various media

Medium	Speed (in m/sec)
vulcanized rubber	54
air(20° C)	344
helium (20° C)	1001
water	1432
glass	5500
aluminum	5100
steel	5060

It is interesting to note that solids, unlike gases, can sustain longitudinal and transverse sound waves.

Our ears use the speed of sound to help us locate the origin of a sound. When you listen to a sound coming from one side, it must travel further to reach the ear on the opposite side of the head. This causes a small delay between the time of arrival to one ear and the other one. This information is used to pinpoint the source. It is not the only method used by our brain, since there is also a difference in amplitude, which is also used for locating the source. The brain uses all of these information processors to place the source of sound in three dimensions and tell us where the sound originates.

POWER

In order for sound to be produced, work has to be performed in a certain period of time by the sound source. This means that power must be applied to the sound generating system which will then produce a certain amount of sound. The sound wave can do work at a certain rate (i.e. it has a certain amount of power), since it causes the membranes in our eardrums to vibrate or else it can cause a microphone to have electrical output. A common example that waves can do work can be demonstrated with the evident power of a water wave suddenly crashing on a beach, perhaps knocking down everything in its way. From your experience with water waves, you know that waves with a large amplitude will have a large amount of power. However, since a wave goes up and down about the equilibrium position (i.e. positive and negative displacement about equilibrium), it does work in the up as well as in the down part of the wave. To demonstrate that a wave does work in the up as well as the down part of the wave, its power depends on the wave's amplitude. This can be seen as:

$$\textbf{Power depends on (Amplitude)}^2$$

This means that by increasing the amplitude of a wave by a factor of four, its power will be $(4)^2$ times

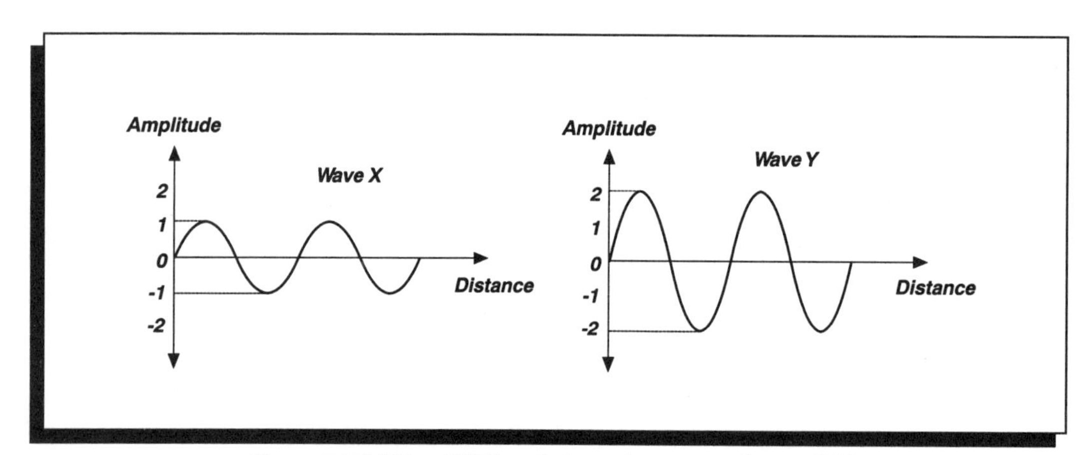

Figure 2.15 Wave "Y" has 4 times the power of wave "X", but their amplitudes differ only by a factor of 2.

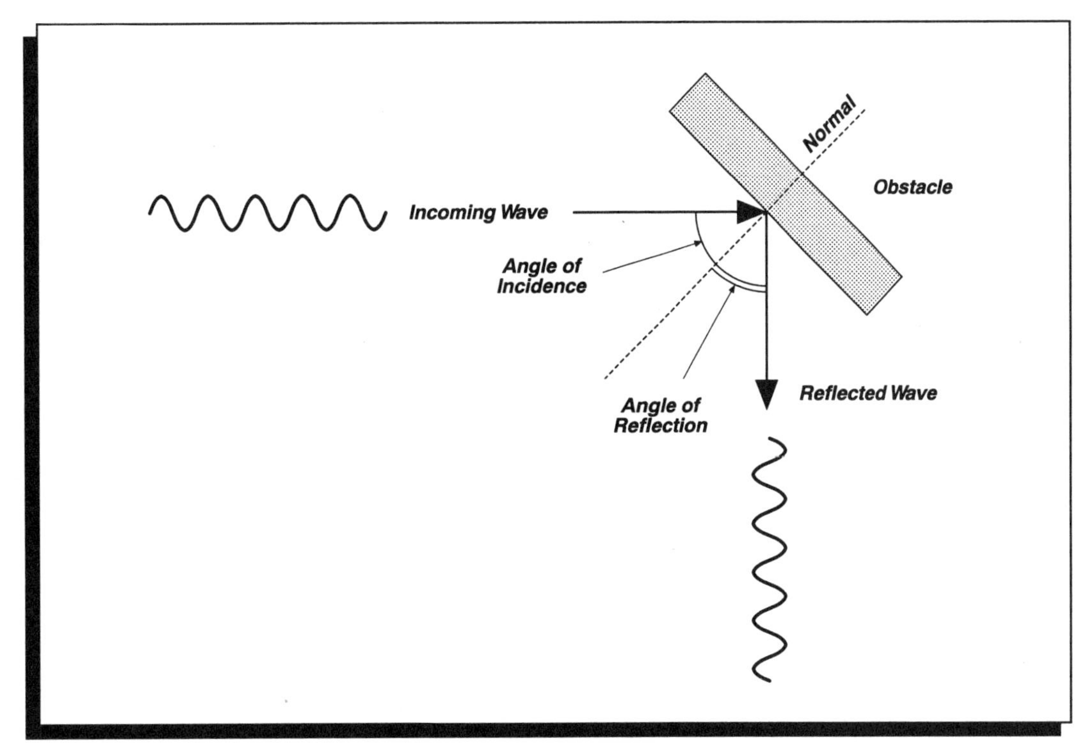

Figure 2.16 Reflection of a wave by an obstacle or a different medium.

larger (i.e. 16 times larger). Or by decreasing the amplitude of a wave by a factor of 3, its power will be down by a factor of $(3)^2$, which is nine times smaller. The amplitude-squared dependence ensures that a wave does work on the up and down part of its displacement, as evidenced that the same amount of power is needed to create a wave going up (positive) as one of same amplitude going down (negative). By increasing the amplitude of a wave by a factor of 10, its power will have increased by a factor of $(10)^2$, which is 100 times. This is illustrated in Figure 2.15, where wave **Y** has double the amplitude of **X**, but its power is 4 times that of **X**.

REFLECTIONS AND REVERBERATIONS OF SOUND

When a wave encounters an obstacle or a medium of different properties, part of the wave will be reflected. Figure 2.16 shows a wave being reflected by an obstacle. The angle at which the wave is reflected is given by the Law of Reflection. This law states that:

Law of Reflection **= angle of reflected wave with respect to normal line is equal to angle of incidence with respect to normal line.**

For example, when the angle of incidence is 50°, the angle of the reflected wave is 50° with respect to the normal. The normal line is an imaginary line drawn perpendicular to obstacle at the point where the wave hits the obstacle to help us determine the direction of a reflected wave. Not all of the wave is reflected. Part of it is absorbed by the obstacle upon incidence.

When a sound wave hits a wall, part of it will also be reflected according to the Law of Reflection. The direction of the reflected sound wave can then be calculated. It is possible that the reflected wave can hit another wall and be reflected again, always in a direction given by the Law of Reflection. Such multiple reflections can complicate the situation. In fact the acoustic properties of a room or a concert hall are determined by the reflections of sound waves and their absorption by the obstacles. This topic is of great importance in the production and reproduction of sound, and its resultant quality.

It is interesting to apply the Law of Reflection to sound in a concert hall or a room to see how this will affect the quality of sound in such an environment. There will always be reflection by walls, chairs, furniture, and people. Since at each reflec-

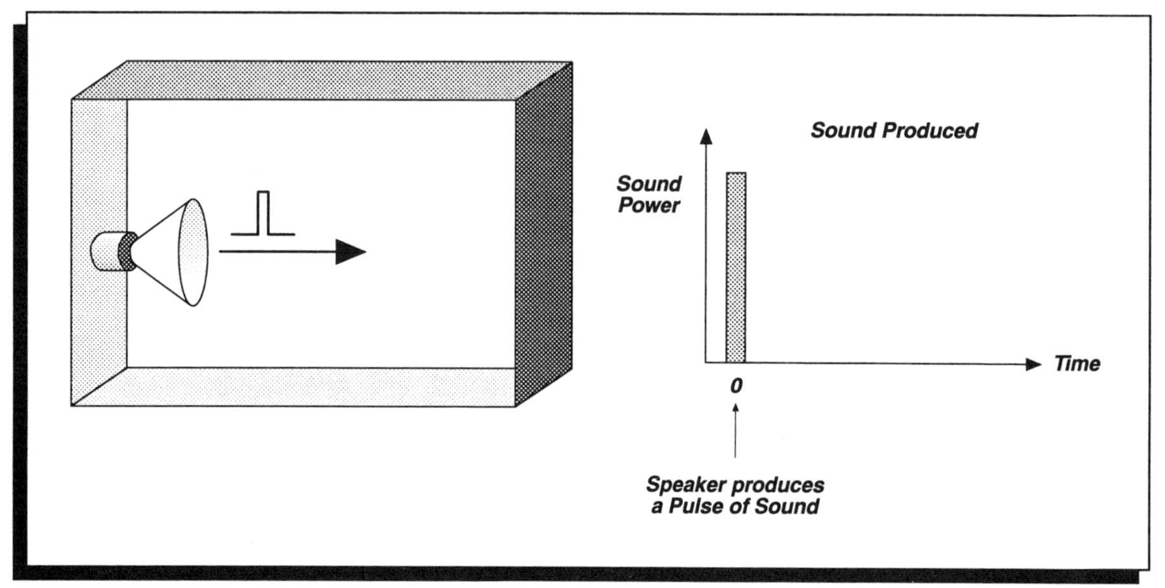

Figure 2.17 Speaker producing a pulse of sound in a hall.

tion part of the sound wave is absorbed, eventually the wave will become inaudible. In fact the multiple reflections in the hall or the room where the sound is produced will have a major effect on the quality of the sound that is heard there. To visualize this, consider a hall with a speaker emitting a pulse of sound. Figure 2.17 illustrates this situation.

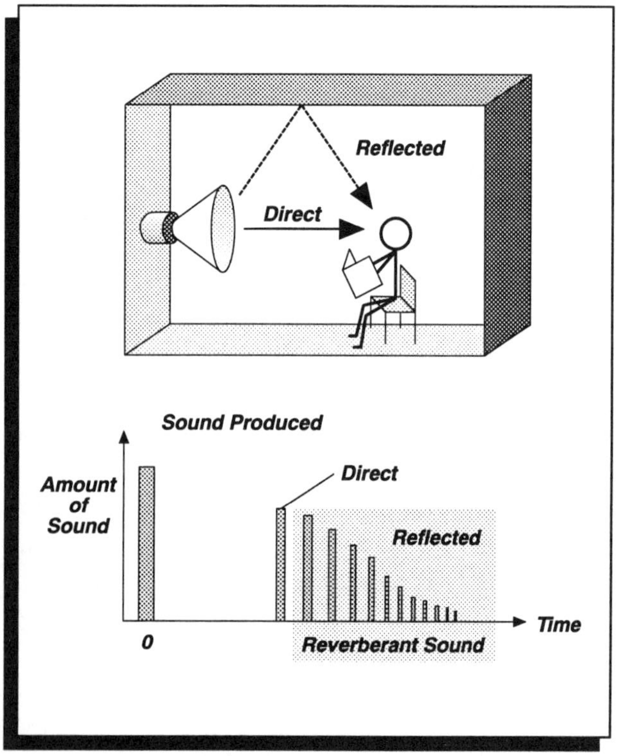

Figure 2.18 Paths of direct and reflected sound in a hall.

A listener is located in the hall. Let us analyze what he hears when the pulse of sound is produced. Sound travels in all directions and Figure 2.18 shows the various paths that this pulse of sound will take as the sound waves eventually reach the listener.

The listener will hear two kinds of sound:

direct sound: sound that travels directly to listener from speaker. From the speed of sound in this hall and the distance between the listener and speaker, it is possible to calculate how long it will take for this pulse to arrive at the listener.

reflected sound: sound reflected by the walls, ceiling, floor, and by the furniture, that will arrive at the listener, though slightly behind the direct sound. This will consist of a series of sound pulses that will become more diffuse with time as they overlap and will eventually die away by absorption inside the hall.

The direct sound is useful as it helps the listener to locate the source of sound. Its main function is to acoustically orient the listener. The reflected sound is caused by the acoustic properties of the hall and what is inside it. It persists for a certain amount of time and this is how it can provide quality to the sound that the listener hears. In fact, the hall adds richness to the sound. Since each hall is different there has to be a criterion for determining how much the hall should affect the sound. If the reflected waves die away rapidly, the hall is not lively. However, if the reflected waves take a long time to die away, there may be too much reflection, and the sound becomes lively, but blurry.

To estimate how much reflection is useful, one first defines reverberant sound. Reverberant really means a diffuse mixture of reflected sound, as in a concert hall or room. The duration of the reflected sound can be measured before it decays; the reverberation time is a very important characteristic of the hall. It is defined as:

> ***Reverberation time*** **= time for reverberant sound which is a diffuse mixture of reflected sounds to decay from its original energy level to a level where its energy is 1/ 1,000,000 of the original value.**

In practice this duration corresponds to an almost audible extinction of a loud sound. In a lecture hall seating about 200 students, the reverberation time could be about 0.8 second. In a large hall, like the Salt Lake Tabernacle, the reverberation time is:

–1.5 seconds with 8,000 people
–1.8 seconds with 6,000 people

This example shows that people also absorb the sound and hence, they contribute to the characteristics of a hall. A simplified general rule can be given:

Reverberation time depends on:

a. volume of hall. The larger the hall, the longer it takes to get to a wall, and hence it will take longer to decay.
b. on 1/absorption. A large absorption will cause the reverberation time to be short. Contributing factors to the absorption are usually people, seats, carpets, drapes, and walls. (This explains the different values of the reverberation time in the example just given.)

The absorption will also depend on the frequency, which is usually larger at high frequencies. Figure 2.19 shows how sound from a source is divided into direct and reflected sound in a hall. The optimal value for the reverberation time depends on the type of music and the instruments one is listening to. Over the centuries music was written for different room acoustics and occasions.

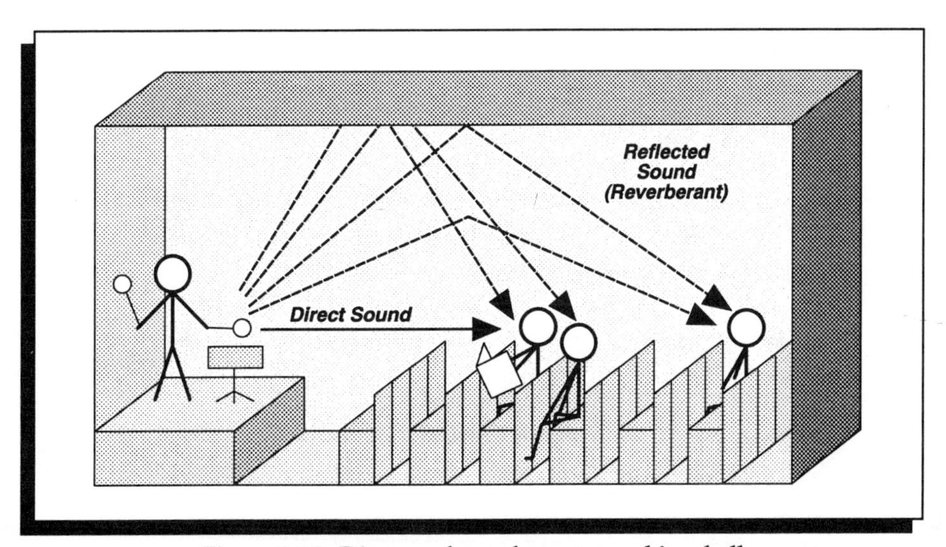

Figure 2.19 Direct and reverberant sound in a hall.

Music and Room Acoustics

Music was written over the centuries for all sorts of acoustic environments with specific goals. It is important that the acoustic goals be met and that the hall reverberation times are appropriate for the type of music played. Let us look at a few types of music and the acoustic environment they would need.

Chamber music: Written to create an intimate atmosphere where the audience, usually a small group of people, is in close proximity to the musicians. Such an atmosphere will be created by a short reverberation time, about 1.2 seconds.

Opera: Tells a story where the audience wants to understand the words and follow the plot. The room acoustics require warmth and clarity. This can be achieved with a reverberation time of about 1.4 seconds.

Romantic symphony: The composer does not intend the audience to hear the individual notes of each instrument, but rather the tones and colors of the orchestra blending together. This requires a longer reverberation time, about 2.0 seconds.

Gregorian chant: As performed in Gothic cathedrals. It requires long reverberation time, 2.2 to 2.4 seconds. Because it is a prayer the sound is meant to last for some time.

For speech, clarity is important, and hence it requires a rather short reverberation time. Delays due to a long reverberation time would cause the sound to be blurry and difficult to understand. This sometimes occurs in convention halls.

The acoustic quality of a hall is based on a blend of direct sound and reverberant sound. Measurements in one of the best halls, Boston Hall, shown in Figure 2.20, demonstrates how the sound is distributed as a listener moves away from the stage. At a distance of 6 meters from the stage, the level of direct sound is the same as reverberant sound, and it decreases as the listener moves away from the stage. Most of the sound in a hall is reverberant, and it remains at a constant level no matter where the position in the hall.

Figure 2.20 summarizes the importance of direct and reverberant sound in a hall, similar to Boston Hall.

Since most of the sound that one hears in a hall is reverberant sound, hall and room acoustics is extremely important for the appreciation of music. With the wide range of music that one listens to on

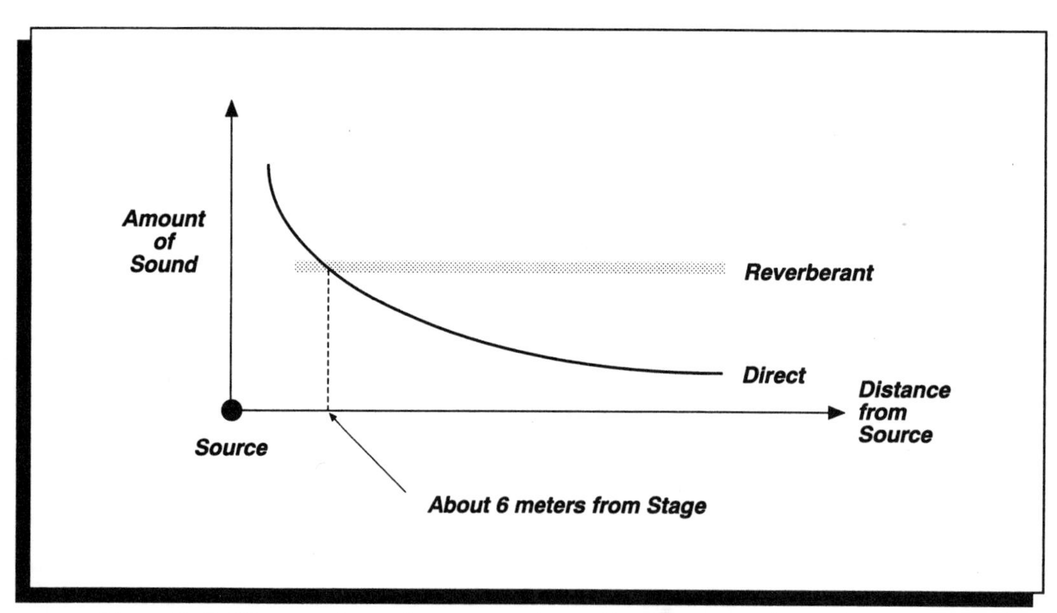

Figure 2.20 Direct and reverberant sound contributions to total sound in a hall.

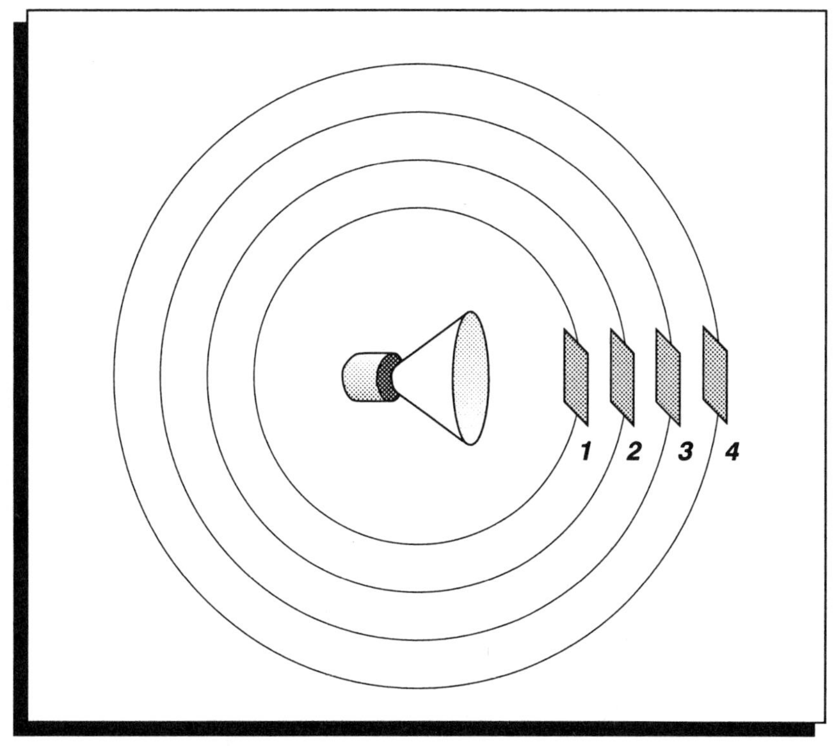

**Figure 2.21 Sound radiated by a speaker.
The intensity decreases as one moves away.**

a hi–fi, what can be done with the reverberation time of a room? It is quite different from a hall. Fortunately, there is an electronic unit that addresses this question. It is a reverberation unit. It tries to mimic the acoustic characteristics of a hall by delaying parts of the sound and then injecting them into the room. There is a control on the delay times of the reverberation unit, which provides the extension of room acoustics for all sorts of music.

In designing speakers, it is important to perform tests in an environment that does not introduce reflected sound. This is difficult to achieve, as there is usually some reflection in the test room. Special acoustic rooms have been built for this purpose, so there is essentially no reflection. The walls and ceiling are padded with acoustically absorbing material. Such rooms are called "anechoic" as all echoes (i.e. reflections) are suppressed. The closest example in nature, to such a room would be a large field covered with fresh powder snow; most of the sound would not be reflected, it would be absorbed by the large porous surface of snow.

The subject of room acoustics is a very important field that deserves special consideration when designing and building concert halls.

■ Intensity of Sound

Consider a source of sound, like a speaker. As sound is emitted a certain amount of acoustical power is radiated by the speaker. The question is, how much acoustic power is received by a surface at a certain distance from the source? To visualize the situation consider Figure 2.21.

For the sake of simplicity let us assume that the sound is radiated equally in all directions. Inspection of Figure 2.21 shows that the small surface 4 will not get the same amount of acoustic power as the surface 1, even though the two surfaces have the same area. It is common experience that as one moves away from a source of sound, the sound becomes fainter (here there are no reflections). The reason for this is that the sound power from the speaker is spread out over the entire surface of the sphere passing through surface 1. That same power has to pass also through the surface of the sphere passing through surface 2. Since the spheres become larger moving away from the

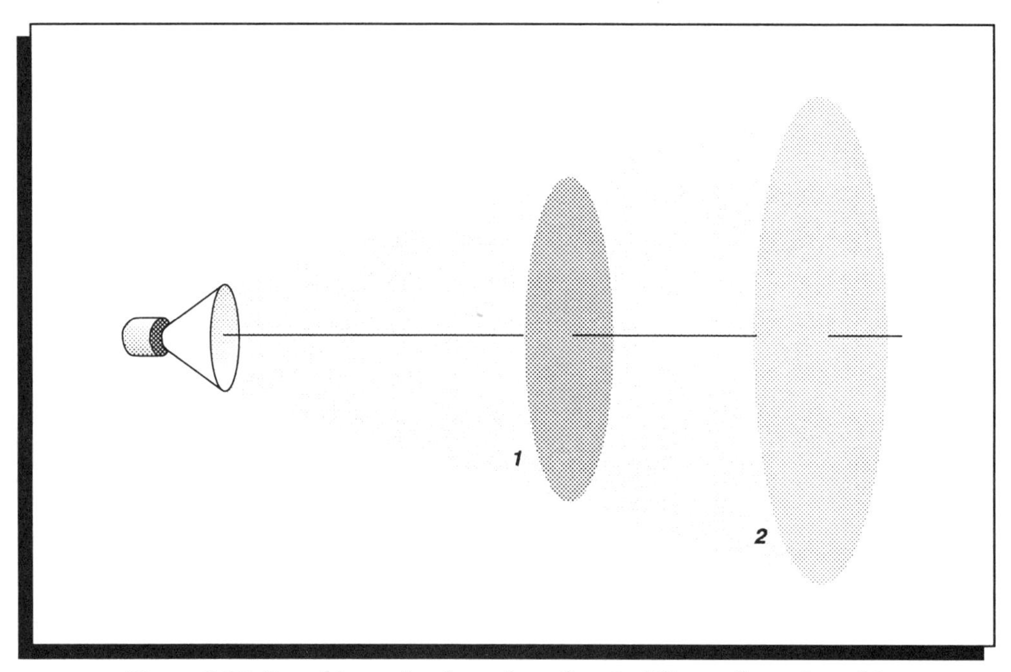

Figure 2.22 Sound intensity through surface 2, different from that of 1.

source, their surface areas also increase and there will be progressively less sound power passing through surfaces 2 than 1. The total surface of the sphere passing through surface 3, for example, will have more power passing through it than the small surface 1. It is a question of power spread out over a surface area. Hence, we define a very useful concept, intensity, which will tell us how much sound power is passing through a surface. It is measured in watts/m^2.

***Intensity* = power/ unit area**

Looking at a section of Figure 2.21, presented in Figure 2.22, we see the usefulness of intensity.

The power passing through surface 2 is the same as that passing through surface 1. However, the intensity at surface 2 is less than at surface 1. This means that the sound at 2 will be quieter than at 1. To illustrate this concept, consider an example:

Example: Consider sound passing through a surface at position A and through one at B.

When 12.5×10^{-3} watt of power crosses an area of $2.5m^2$ at position A, the intensity is:

Intensity $= 12.5 \times 10^{-3}$ watt / $2.5m^2$
$= 5 \times 10^{-3}$ watt /m^2

At location B when 15×10^{-3} watt cross an area of $6m^2$, the intensity is:

Intensity $= 15 \times 10^{-3}$ watt /$6m^2$
$= 2.5 \times 10^{-3}$ watt /m^2

The intensity at A is larger than at B. Hence, the sound will be quieter at B than A.

DOPPLER EFFECT

We know that the frequency of a sound wave is the same as that of the source. When a speaker vibrates at 500 Hz, it produces a sound of 500 Hz. Likewise, a singer who is producing a tone at 500 Hz, will produce a sound at 500 Hz. The frequency of the sound will be the same in all directions around the source of sound. An interesting effect occurs when there is relative motion between the source and the observer. An experiment can be performed to study this. In Figure 2.23, a source of sound and an observer are at rest. The observer detects the sound at the same frequency as the source emits it. When there is relative motion

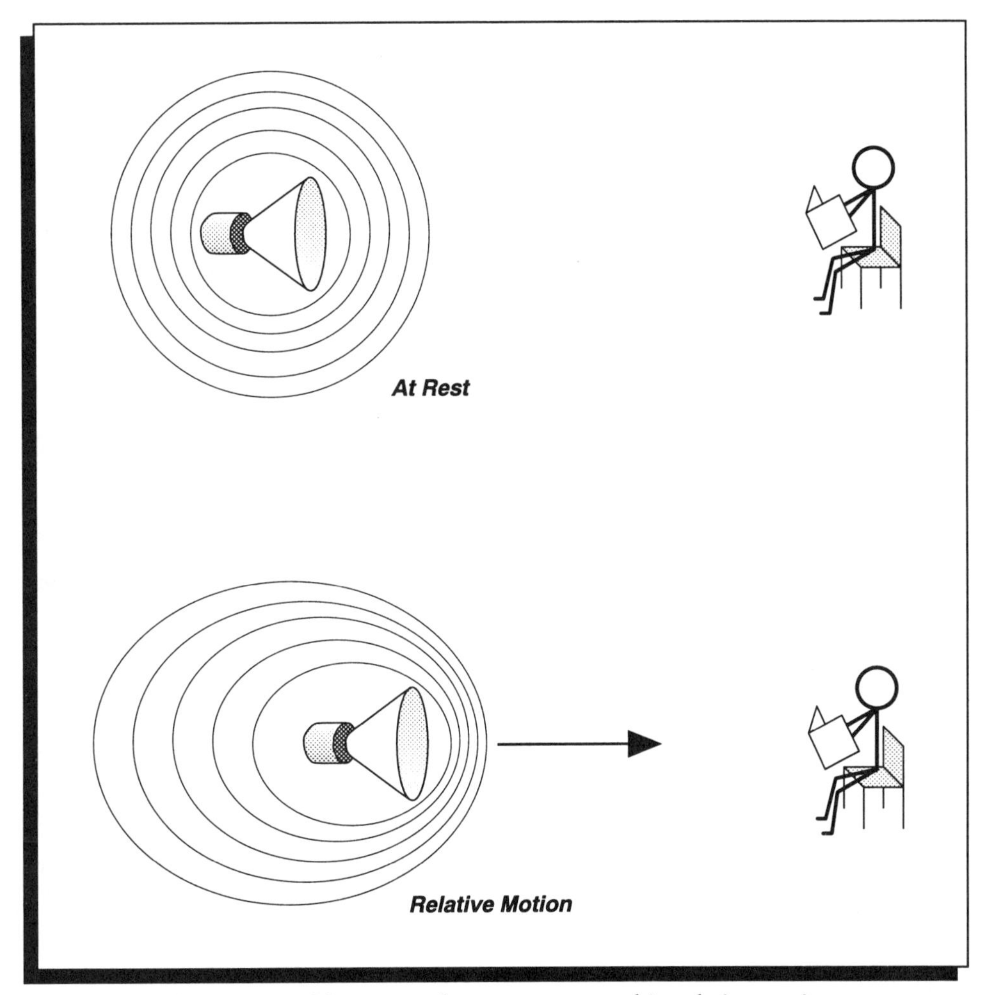

Figure 2.23 Observer and source at rest and in relative motion.

between the source of sound and the observer, there will be a change in frequency. To understand this we consider the fact that the source emits the sound, and that it also moves toward the observer (for example). Hence, the observer will be getting more waves per second (i.e. he will be getting waves at a higher frequency). Had the observer been in a position where the relative motion between source and observer was away from each other, the number of waves per second reaching the observer would be less, and hence the frequency of the received sound would be lower. In both cases, the change in frequency depends on the relative speed of source and/or observer. This effect was discovered by Doppler and demonstrated by a band of musicians playing on a train which was approaching observers. As they approached the observers, the frequency of the emitted sound increased, and as they receded from the observers, the frequency went down. Hence one observes that:

— when the relative motion between observer and source is toward each other, the received frequency is higher.
— when the relative motion between observer and source is away from each other, the received frequency is lower.

This effect is important in many areas of physics, but it is particularly important in the application and study of hi–fi when we consider the motion of a loudspeaker. Assume that the speaker produces two tones, one at 100 Hz and one at 1,000 Hz. Will the observer not detect the Doppler effect in this case? After all, the speaker moves at 100 Hz, while at the same time it also moves at 1,000 Hz.

This concept is depicted in Figure 2.24. During the first half of the wave at 100 Hz, there are 3 ½ waves at 1,000 Hz emitted, and these will be heard as increased in frequency when there is forward motion of the speaker toward observer. During the other half of the wave, the speaker is moving back while producing waves at 1,000 Hz. The observed frequency of the 1,000 Hz will then be decreased. Of course, this depends on how much the speaker moves during the 100 Hz tone. Large amplitude means a large speed, and hence a larger change of frequency. The resulting sensation caused by this increase and decrease in frequency of the 1,000 Hz tone is a blurring of the tone. Such an effect will occur for all the frequencies of interest.

The Doppler effect can be reduced in speakers with some simple modifications:

- — low relative speed by keeping the speaker from moving with large amplitude.
- — restrict the range of frequencies driving speaker, i.e. break up the whole frequency range into sections so that one speaker handles only the low frequencies, another speaker handles the midfrequencies, and another one handles the high frequencies, for example. A convenient approach is:
  - — low frequencies produced by woofer speaker.
  - — midfrequencies produced by midrange speaker.
  - — high frequencies produced by tweeter speaker.

All of these speakers could be housed in one enclosure per channel.

REFRACTION

When a wave enters a medium whose properties are different from the present medium, there will be a change of speed. This can cause a change in direction of the wave when it hits the medium at an angle. Such an effect occurs for all wave phenomena, but here we will discuss it for sound only.

Consider a layer of hot air over a layer of cold air, shown in Figure 2.25. When a sound wave hits the warm air two things will happen:

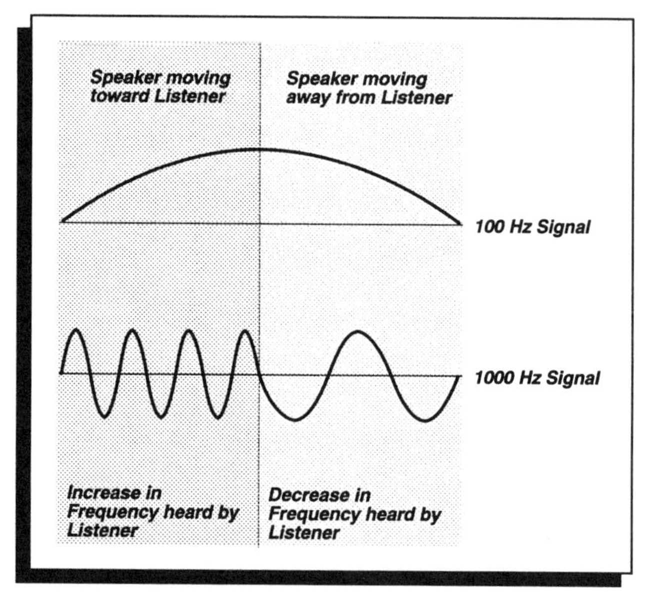

Figure 2.24 Doppler Effect produced by speaker emitting simultaneously 100 Hz and 1,000 Hz sound waves.

1. part of the wave will be reflected because it enters a medium with different acoustic properties.
2. part of the wave will be transmitted, but it will change direction.

Let us see in which direction the wave will go. The situation can be analyzed in Figure 2.26.

Because the wave front hits the cold air_hot air interface at an angle, part of the wave will enter the hot air. This part will start going faster, because sound travels faster in hot air. The part of the wave still in the cold air will keep on going at its original speed. As part of the wave front goes faster in the new medium, the other part is going at its regular speed, and the wave will change direction. The cold air part of the wave drags the higher speed part of the wave. In this case, the new direction will be away from the normal. The angle of the direction change will depend on the speed in the new medium, relative to that in the old one and the angle at which it came in.

This phenomenon of change of direction of a wave entering another medium is known as refraction, defined as:

***Refraction* = the bending of a wave when it enters a medium of different properties**

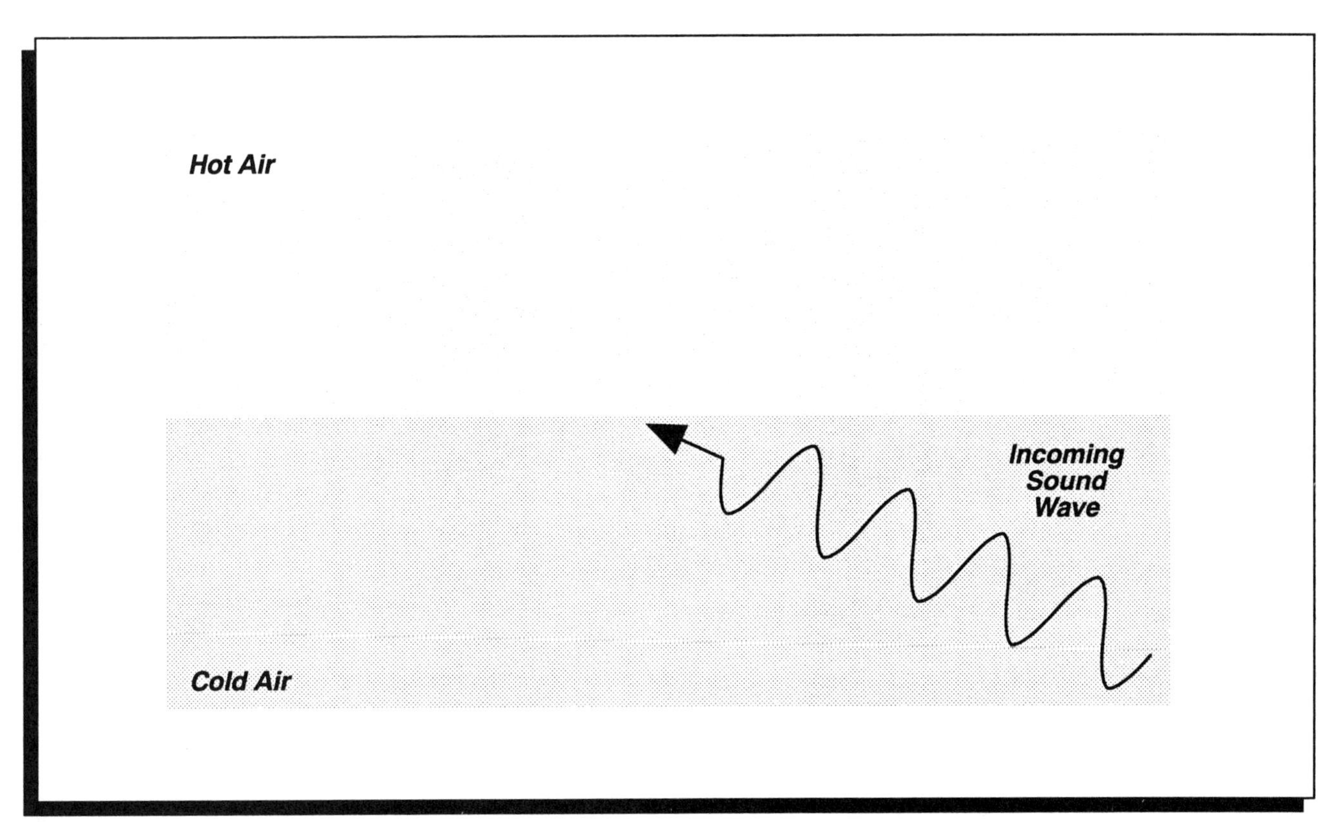

Figure 2.25 Sound wave in cold air as it enters hot air.

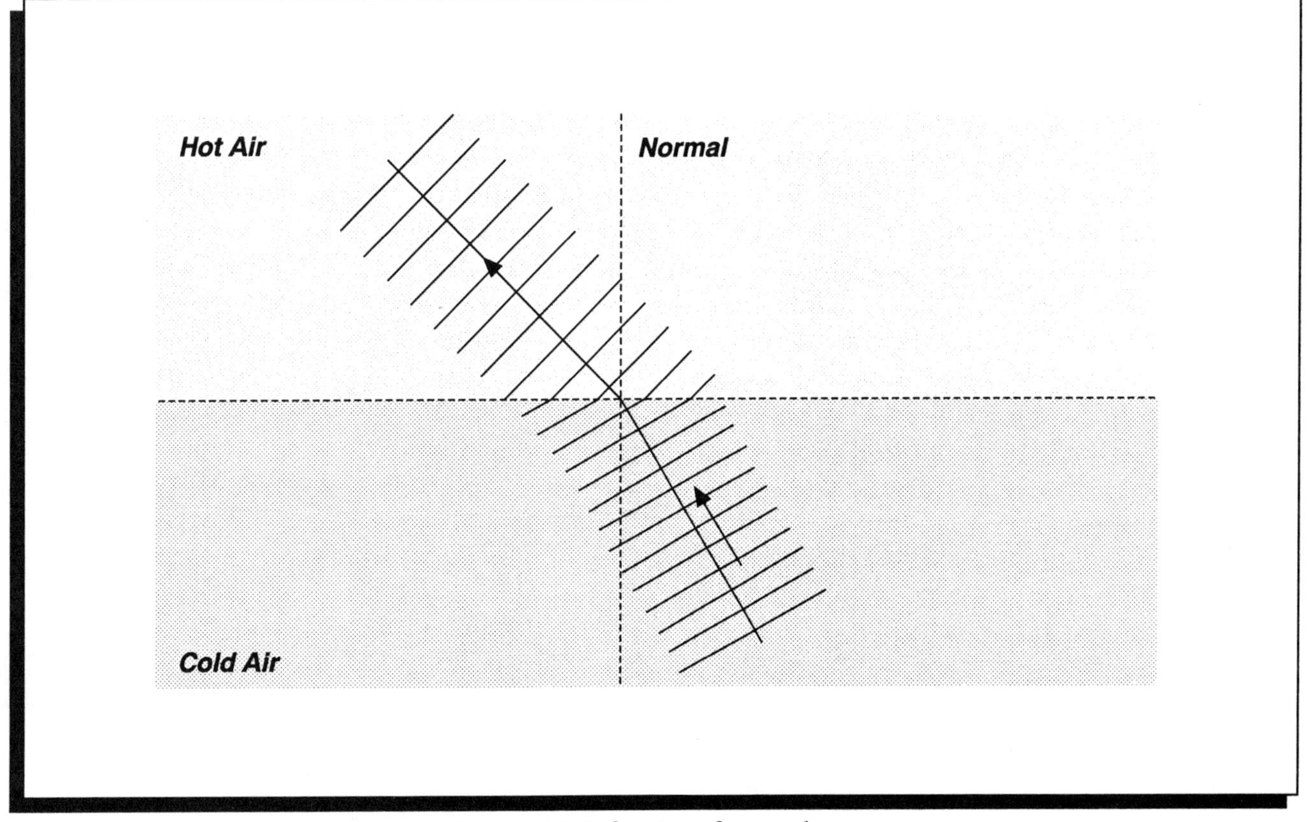

Figure 2.26 Refraction of a sound wave.

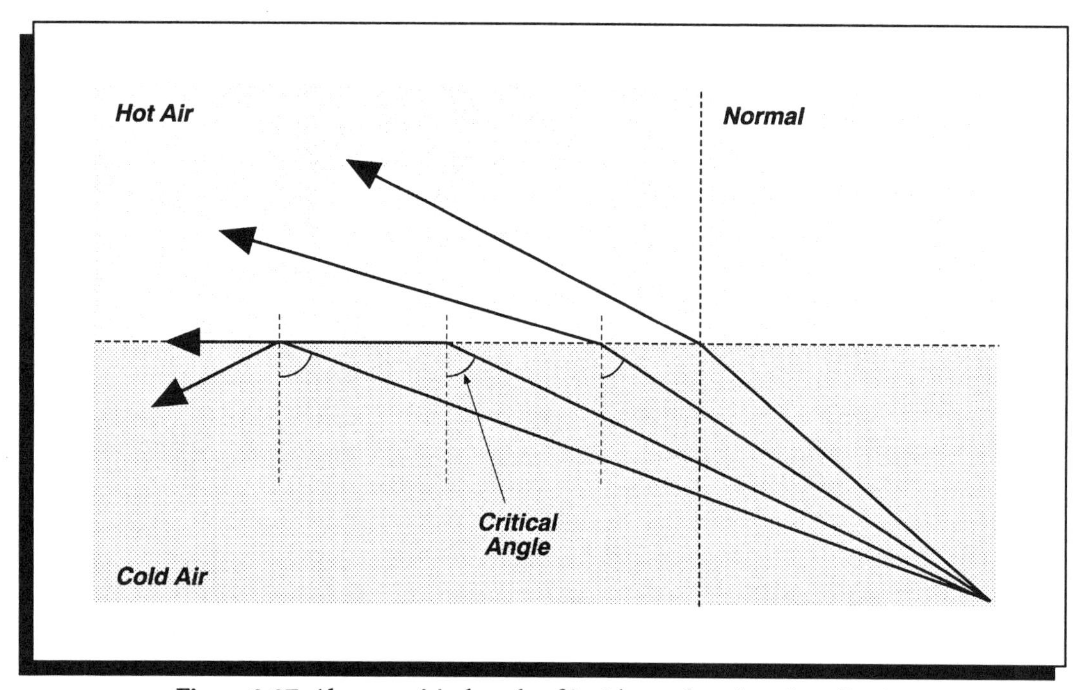

Figure 2.27 Above a critical angle of incidence there is only reflection.

An interesting situation occurs when the angle of incidence is increased. This is shown in Figure 2.27 for varying angles of incidence. At some angle, known as the critical angle, the wave moves parallel to the interface between the two media. Above this angle, there is only reflection, known as internal reflection. This works only when the speed in the second medium is larger than in the first one. There is an interesting application of this effect and it is known as a wave pipe. It is well known for light and light pipes have many practical applications. For sound, an interesting application is the sound pipe which is used in funneling sound in a commercial airplane to the passenger's headphones from the seat outlet. Figure 2.28 shows this. The application to audio entertainment on a passenger airplane is well known. Multiple reflections of sound in the plastic tube will direct the sound to the passenger's ears.

2.3 Complex Waves

The purpose of a hi–fi system is its ability to allow a listener to experience sounds produced by all sorts of instruments, including the human voice. At any instant of time you are hearing a mixture of sound waves which have different frequencies. It is certainly complicated, but in order to understand this phenomena, we will take a simple approach, at least as a start. Figure 2.29 shows a possible sound wave produced by a musical group. It has all sorts of frequencies and the amplitude varies in a complicated manner.

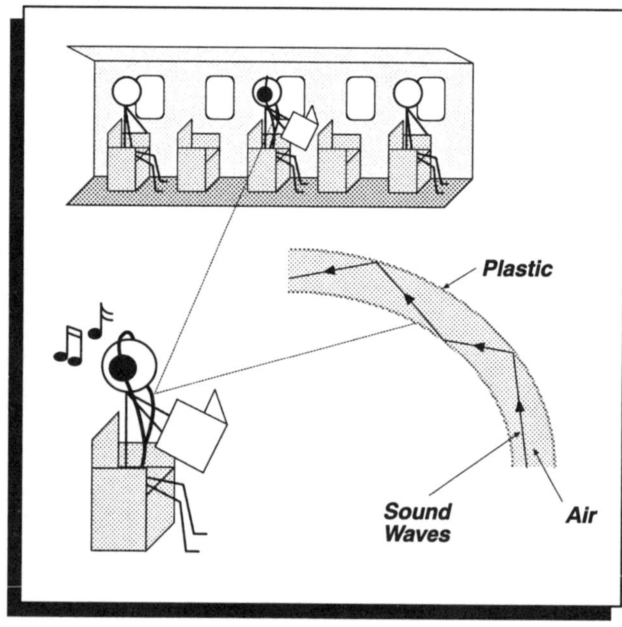

Figure 2.28 Sound travels in a curved hollow plastic tube by multiple reflections.

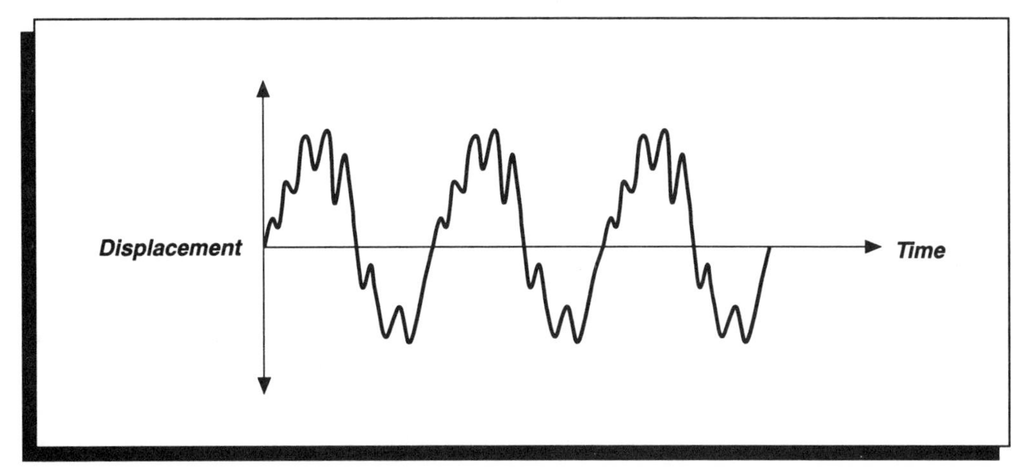

Figure 2.29 Sound wave produced by a musical group. A complex wave.

It is a complicated wave and we call it a complex waveform. Closer examination of this waveform will reveal many different frequencies, but there is a repetition of the complex waveform. Analysis of this wave would have been much easier if the waveform were a simple sine wave of one frequency, as shown in Figure 2.30. We could easily measure the frequency and amplitude of a sine wave. Also, it would be easy to add or subtract many waves of the same frequency. Perhaps an understanding of how the complex wave is formed would help in analyzing such a waveform.

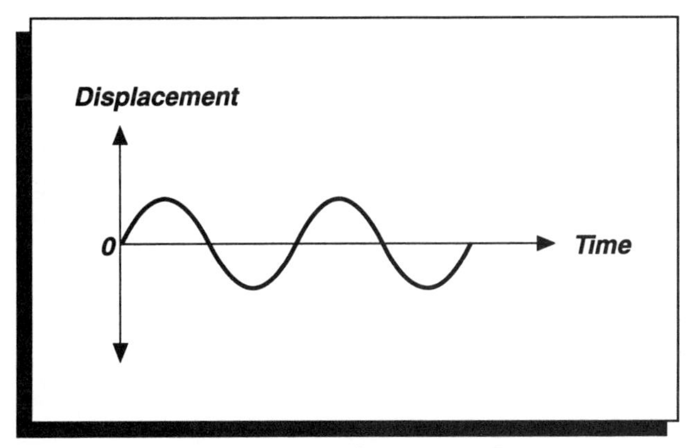

Figure 2.30 Simple sine waveform.

Since a wave keeps on repeating itself, we consider only one wave or one cycle, where the displacement starts at some value and after a complete cycle comes back to the same value. Because this is just like going around a circle, it is useful to make a comparison of the two. Hence, going through one wave is like going around a circle by 360°. Likewise going through a half–wave (i.e. a half–cycle) is like going around by 180°. A quarter of a wave corresponds to 90°. When a wave advances by ½ a wave, one could say that it has advanced by 180°. Figure 2.31 shows the comparison between a cycle of a wave and a full rotation of a circle.

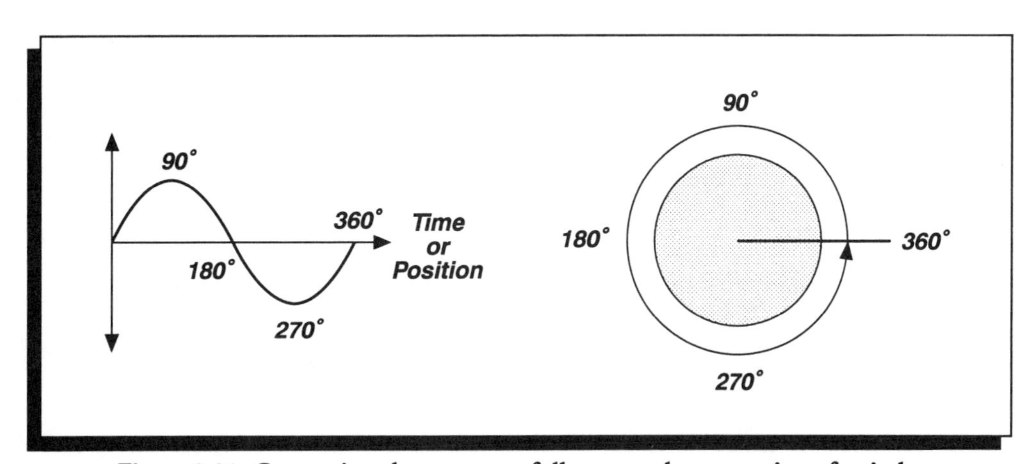

Figure 2.31 Comparison between one full wave and one rotation of a circle.

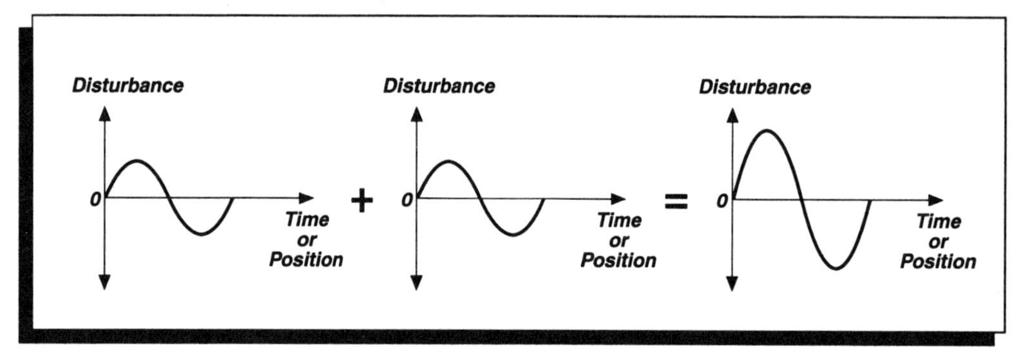

Figure 2.32 Addition of two waves.

To determine where one is on a wave, the concept of phase is very useful. It is defined as:

***Phase* = fraction of a wave that is completed, or how many degrees a wave has completed.**

Now we can consider what will happen when two or more waves are added. To do this we use the Principle of Superposition. Figure 2.32 shows two waves that are combined to produce a larger amplitude wave. This is simple because both waves are in phase. They both start and stop at the same point of their respective cycle. Such a case is known as constructive interference because the waves interfere with each other causing a resultant reinforcement in amplitude. Hence:

***Constructive interference* = addition of two or more waves in phase**

The result will be very different if the phase difference between the waves is not zero. This is shown in Figure 2.33, where two waves are added that are out of phase with each other by 180°. In this example of the Superposition Principle, the addition causes a cancellation of the waves. The result is no wave. This is expected since both waves of same amplitude do the opposite of each other. This is known as destructive interference. Therefore:

***Destructive interference* = addition of two or more waves out of phase by 180°.**

These are two extreme cases that have just been discussed about the addition of waves; however, it is possible to have all sorts of phase differences between the waves being added. As before, the waves are simply added together. To illustrate constructive and destructive interference, one can simply use two speakers, driven at the same frequency and in phase. In Figure 2.34, a listener is in front of the two speakers. For the case illustrated there, constructive interference, the two waves will arrive in phase to the listener. In Figure 2.35, the opposite situation will occur because the listener is closer to one speaker than the other by $\frac{1}{2}$ of a wave. The sound waves at the position of the listener are out of phase. In this case there will be almost no sound.

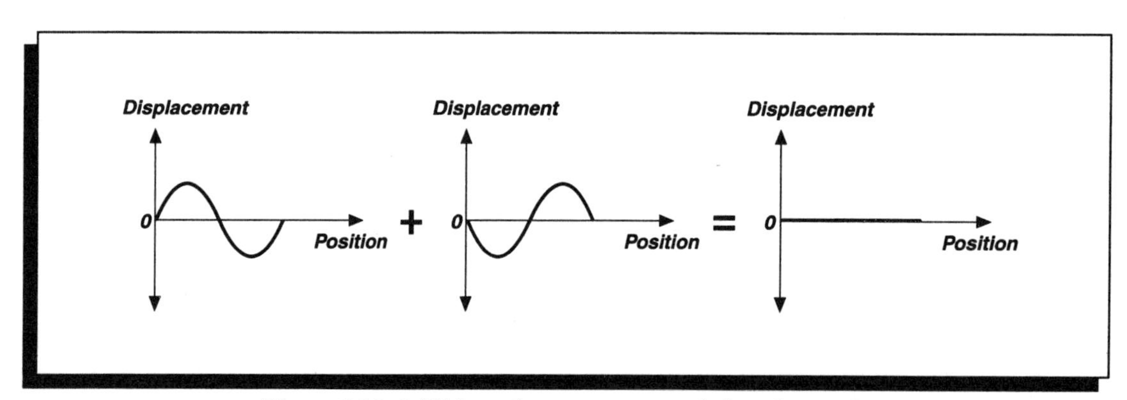

Figure 2.33 Addition of two waves out of phase by 180°.

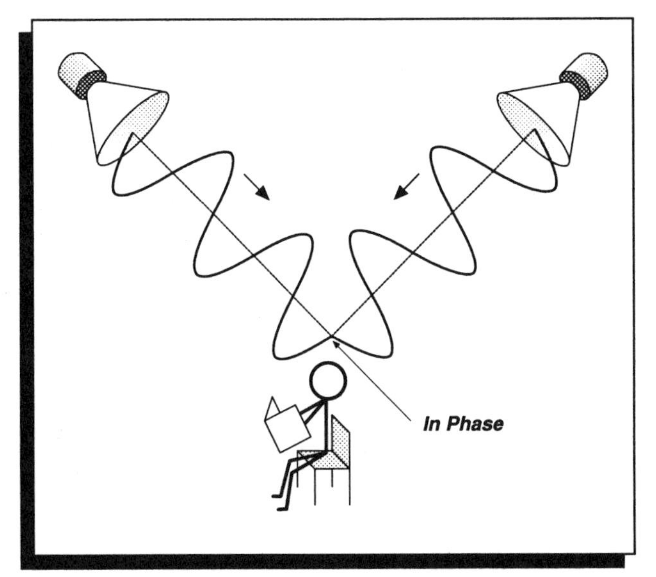

Figure 2.34 Constructive interference.

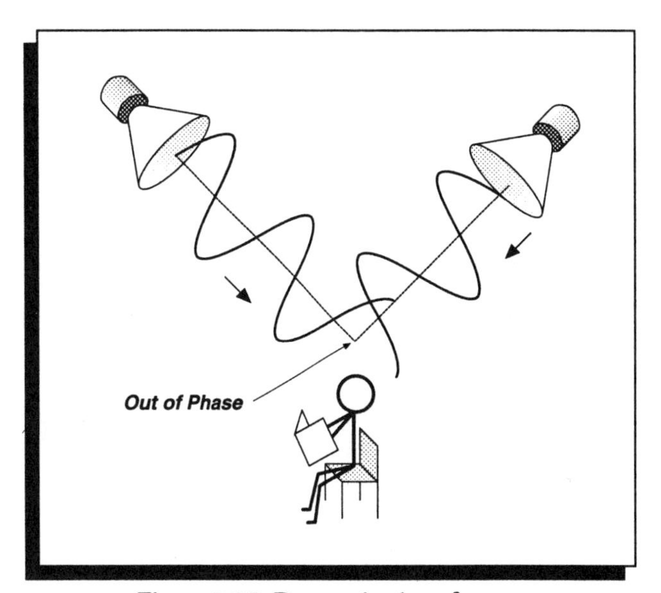

Figure 2.35 Destructive interference.

This is an example of destructive interference. Interference is an important characteristic of waves and it will appear in many topics in this book.

■ DIFFRACTION AND DISPERSION

The wave nature of sound leads to a variety of interesting phenomena which must be considered in the hi–fi system. Consider the case of a sound wave from a speaker as it goes through an aperture. This could be a door, as shown in Figure 2.36. High frequency waves are generated by the speaker. A portion of the wave front passes through the aperture leaving a strong "acoustic" shadow of the obstacle. The sound is mainly in front of the aper-

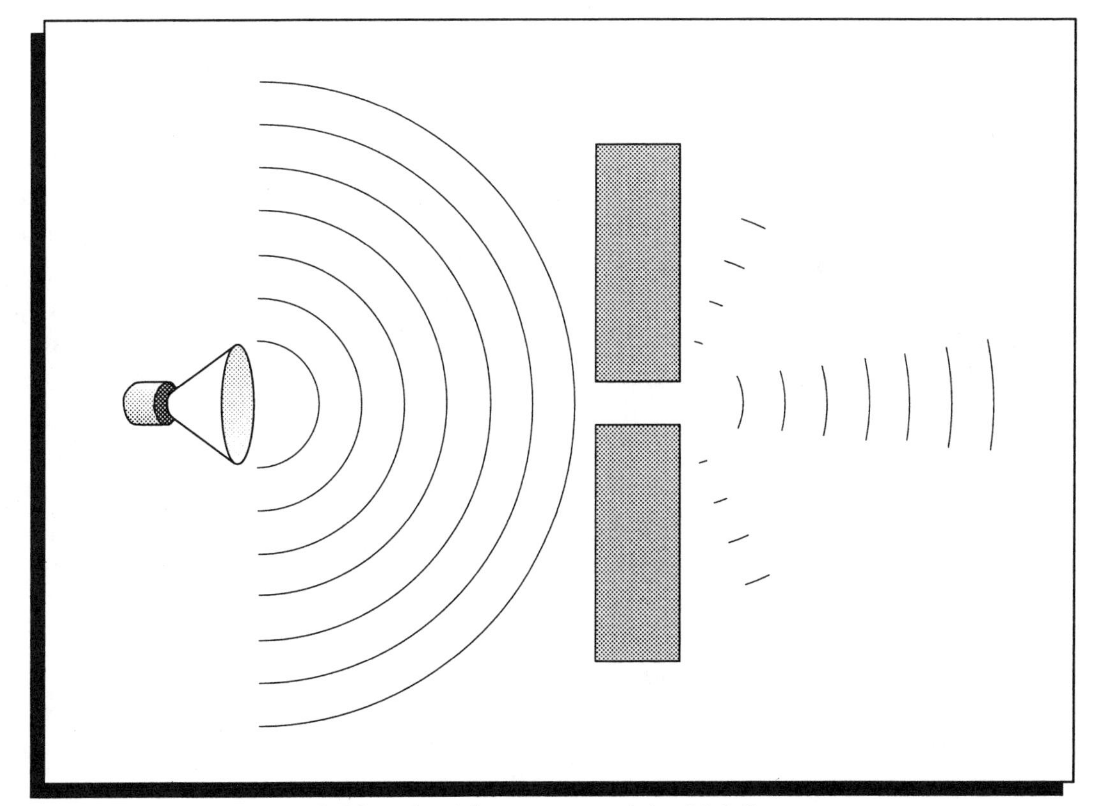

Figure 2.36 Obstacle with aperture receiving high frequency waves.

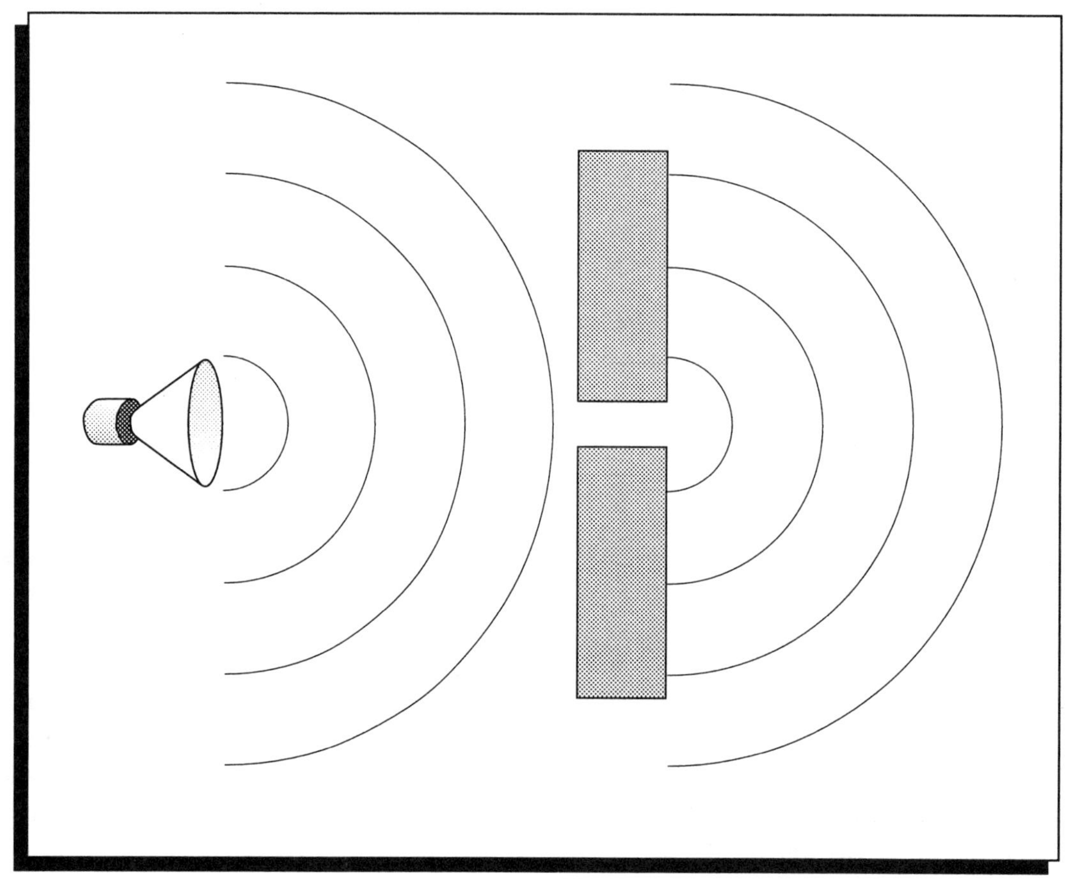

Figure 2.37 Low frequency behavior of obstacle and aperture.

ture. When this experiment is repeated by generating sound at low frequencies, another situation results. Figure 2.37 depicts this situation. The waves go through the aperture, but they also go around the corners of the obstacle. In this, the waves essentially wave their way around the corners. This behavior of a wave is known as diffraction and we define it:

> ***Diffraction*** **= the ability of a wave to bend around an obstacle.**

The degree that a wave will bend depends on the frequency. Low frequencies bend more than high frequencies. It also depends on the size of the obstacle, relative to the wavelength of the wave. The fact that such behavior can be observed is not surprising since the wave fronts act as wave sources which show the interference phenomena between them as they emerge from the various parts of the waveform.

Diffraction is an important manifestation of wave behavior, well known for all sorts of waves. The interest here is in one very important case, a loudspeaker. When the example shown in Figure 2.36 is shrunk in one direction, the limit reached is really a loudspeaker. Figure 2.38 shows this. Hence diffraction effects will dominate the sound distribution of a loudspeaker. In fact, because of this, the sound radiation pattern of a speaker is strongly frequency dependent. This is an important aspect of speaker behavior and is known as dispersion, which refers to how sound is dispersed in all directions by a speaker. Diffraction by the speaker will cause the sound to beam forward at high frequencies, while at low frequencies the sound will radiate in all directions. This is shown in Figure 2.39. We will analyze it when we cover the topic of loudspeakers.

■ STANDING WAVES

In order to elicit a discussion of complex waves, all sorts of wave phenomena were introduced, in particular, interference. This provided a back-

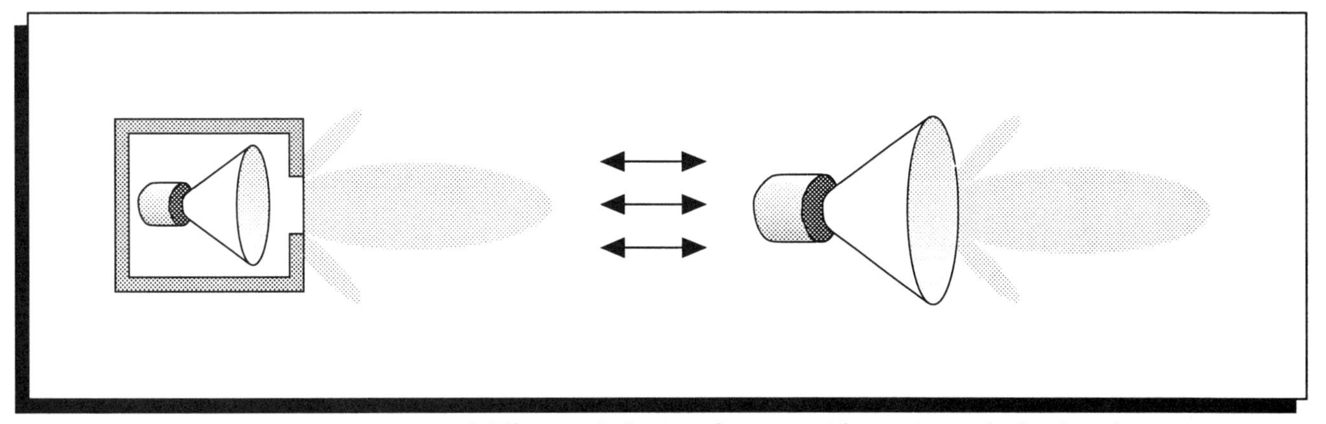

Figure 2.38 Comparison of diffraction behavior of a room with opening and a loudspeaker.

ground for discussion of complex waves, to see how they are generated and what they consist of. Recalling that a wave will be reflected by a medium whose acoustic properties differ from those of the medium in which the wave exists, we will now consider examples of reflected waves and how they interact with the incoming waves. Such effects arise in the excitation of musical instruments and in their sound production. Musical instruments can be divided into three general categories: string instruments, woodwinds, and percussive instruments. They all sustain standing waves.

In order to understand how standing waves are produced, consider an example of a continuous train of waves going to the right; when it hits an obstacle it will be reflected by the obstacle. The net result is a train of waves going to the right and another train of waves going to the left, both at the same frequency and of the same amplitude. Using the Superposition Principle, the two wave trains are added together leading to a resultant wave. This wave is known as a standing wave. It is named a standing wave because it does not appear to go to the right or the left, it seems to be standing, only going up and down. Hence:

> ***Standing wave*** **= interference of two waves of same frequency and amplitude passing through each other in opposite directions.**

The waves are at times in phase, at other times out of phase, producing regions of constructive and

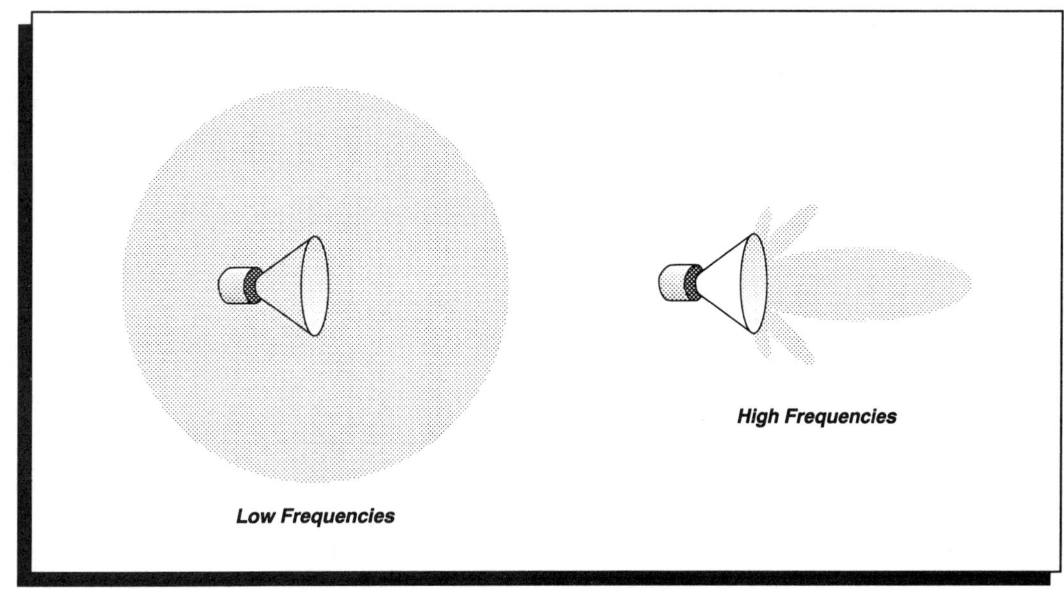

Figure 2.39 Dispersion characteristics of a speaker.

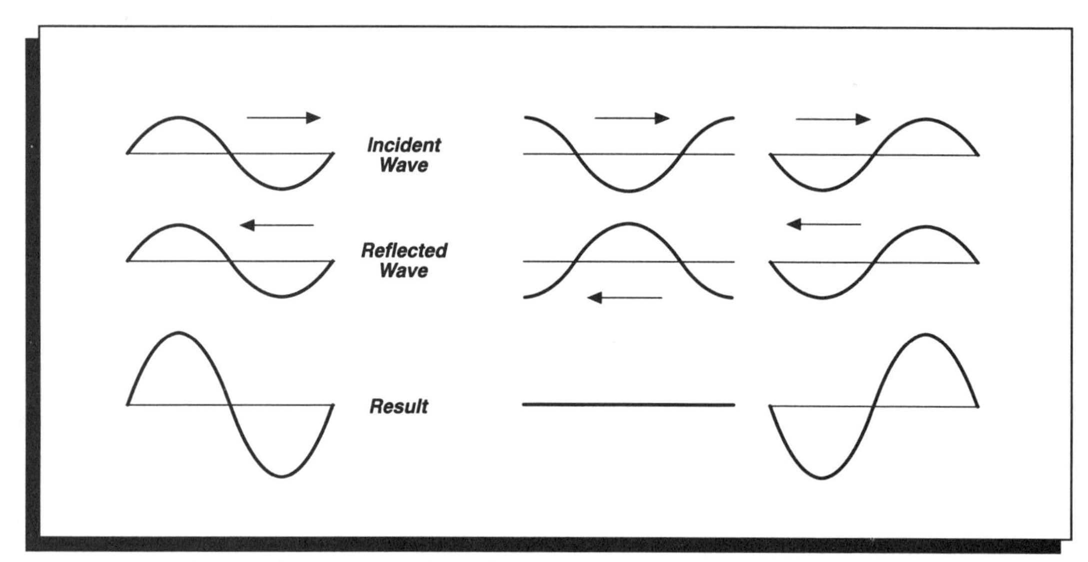

Figure 2.40 Standing wave produced by incident and reflected waves at same frequency.

destructive interference. Figure 2.40 shows an example of three extremes in a standing wave cycle. Although the incident and reflected waves are travelling (and we even know in which direction), the result is a wave which vibrates up and down, without going to the right or the left.

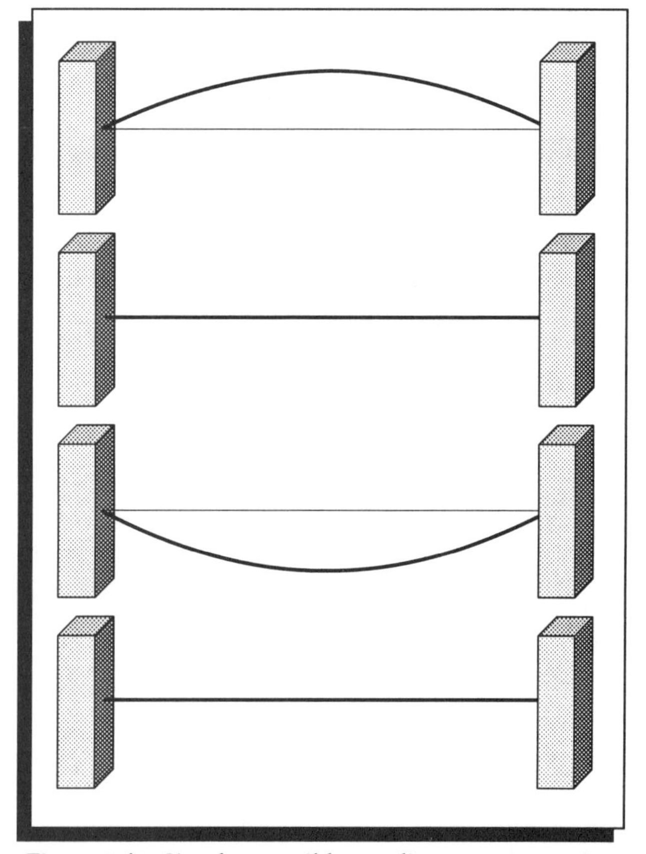

Figure 2.41 Simplest possible standing wave on a string.

Let us do a few important examples of this type of wave.

Transverse Waves on a String

Consider a string clamped at each end. When excited to vibrate by rubbing it with a bow, plucking it, or hitting it with a hammer, disturbances will be set up on the string. The disturbances propagate along the string, and upon hitting the ends are reflected back. The combination of incident and reflected waves on the string leads to standing waves. Figure 2.41 illustrates this for different times of a cycle. The pattern looks like this because we clamped the ends of the string, and they cannot move. The figure is usually simplified as shown in Figure 2.42 and it presents the whole cycle. Because this is the simplest standing wave that can exist on a string, it is called the fundamental. It is also called the first harmonic. The string length corresponds to ½ wave. The next standing

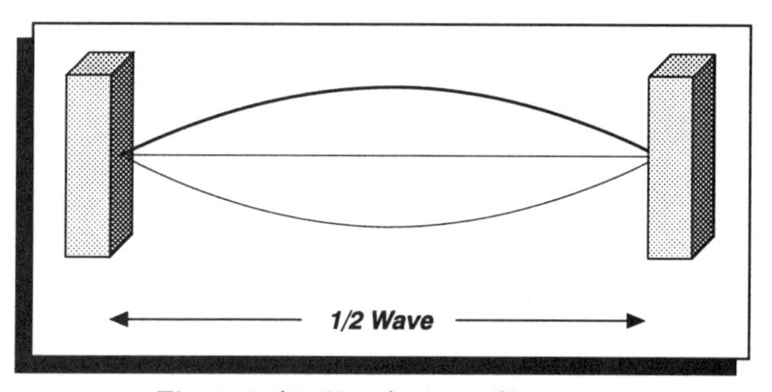

Figure 2.42 Simplest standing wave on a string during one cycle.

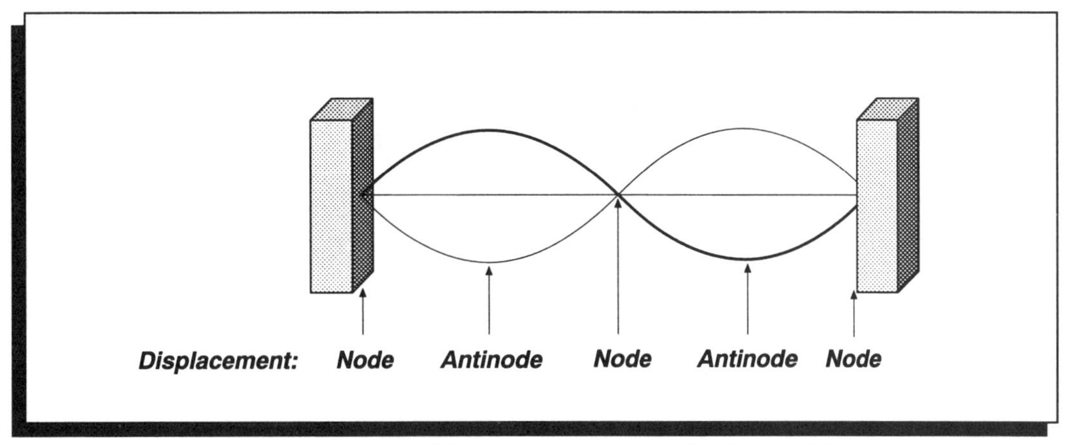

Figure 2.43 Second harmonic on a string showing position of displacement nodes and antinodes.

wave is more complicated, shown in Figure 2.43. The ends are still clamped, and this standing wave is known as the second harmonic. Since the string goes up and down twice during the same time that the fundamental goes up and down once, its frequency must be double that of the fundamental. Now it is clear why this waveform is called the second harmonic.

The third harmonic is a continuation of the same pattern. This happens because the ends of the string are clamped. Figure 2.44 shows the pattern of the third harmonic. Because three half–waves go up and down in the same time as the fundamental goes up and down, its frequency is three times that of the fundamental. It is known as the third harmonic. We can continue this, but it becomes very clear that the string can vibrate only at fixed frequencies which follow this pattern:

fundamental	→	frequency = f
2nd harmonic	→	frequency = 2f
3rd harmonic	→	frequency = 3f
:		:
:		:
nth harmonic	→	frequency = nf

where n = 1, 2, 3, 4, etc.
= integer

From this we can determine that a string clamped at each end can vibrate only at certain frequencies, known as harmonics. A string can be "shaken" at all sorts of frequencies, but it settles into one or more of the standing waves frequencies given above.

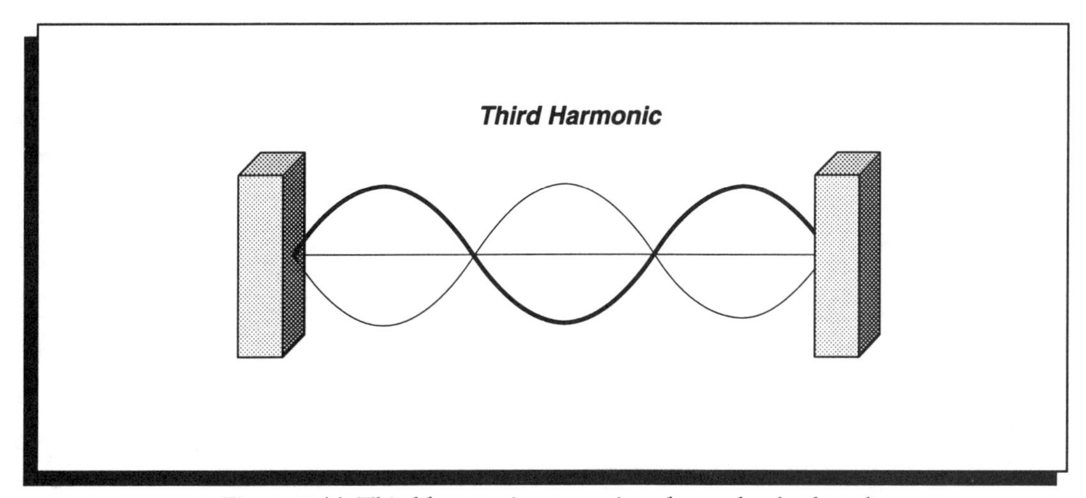

Figure 2.44 Third harmonic on a string clamped at both ends.

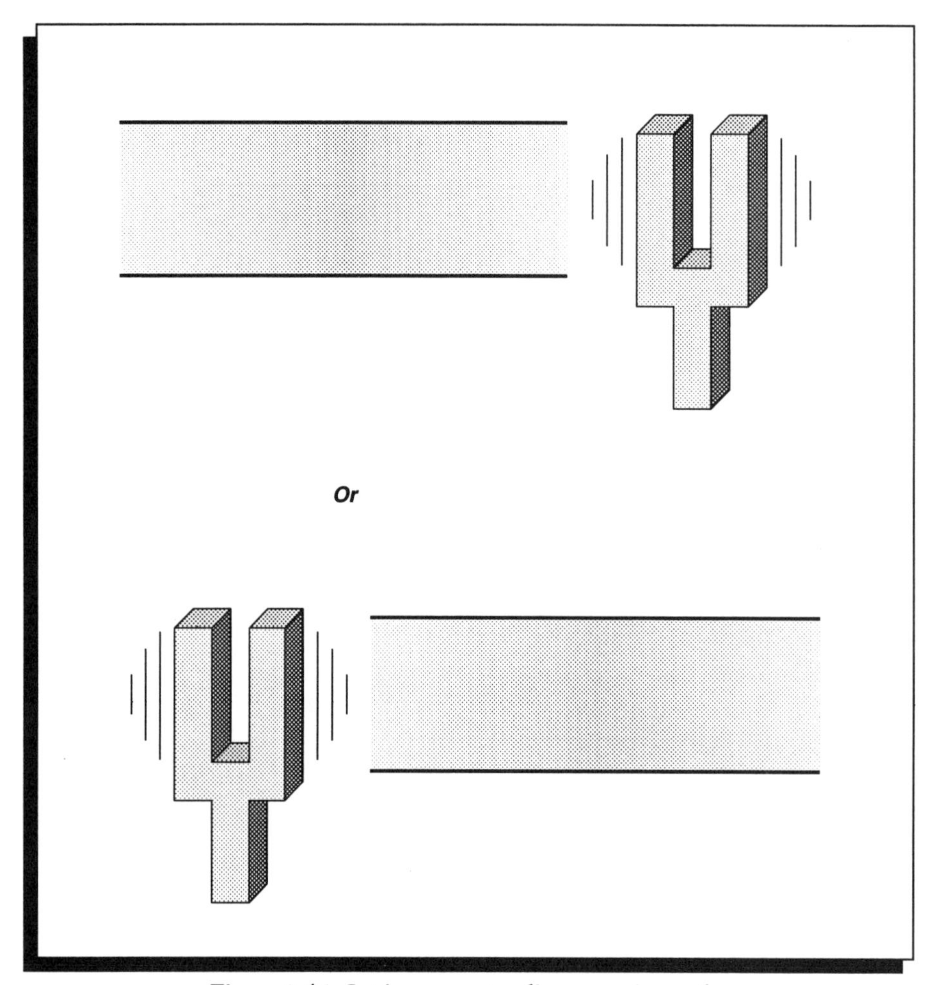

Figure 2.45 Setting up a standing wave in a tube.

LONGITUDINAL WAVES IN A TUBE

Another example of standing waves can be examined in a tube. Oscillations of the air inside the tube can set up standing waves. Consider a tube with air, open at both ends, where a disturbance can be created, as shown in Figure 2.45. A tuning fork can be used to create a disturbance. Since this is a symmetrical situation, with both ends open, the tuning fork can excite it at either end. Because the tuning fork can move air in and out at the tube opening, the air displacement can be large. This disturbance propagates down the tube at the speed of sound and at the end of the tube it is reflected back up. The interference of a sound wave going one way and another one going back produces a standing wave of sound inside the tube. Because the ends are open, the pattern of air displacement inside the tube can be illustrated as in Figure 2.46.

This pattern is known as the fundamental (first harmonic) and it contains a half–wave. It does not look like the pattern of a string because the system is different. Nevertheless, there are some similarities. The center has a displacement node, while the

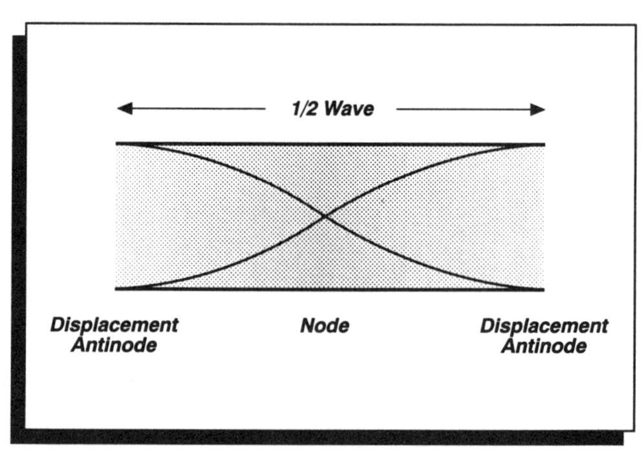

Figure 2.46 Simplest standing wave in a tube open at both ends.

ends are displacement antinodes. The ends are displacement antinodes because they are open to the outside air, and air at the ends can be pushed in and out quite easily (hence a displacement antinode). The next more complicated pattern is described in Figure 2.47. It pattern contains two half–waves, and hence will cause two ups and downs during the same time that the fundamental does one up and down. The name given to this waveform is the second harmonic, and its frequency is double that of the fundamental. The pattern repeats itself with more complicated harmonics. We conclude from this that a tube open at both ends will vibrate only at the following frequencies:

fundamental → frequency = f
2nd harmonic → frequency = 2f
3rd harmonic → frequency = 3f
nth harmonic → frequency = nf
where n = integer

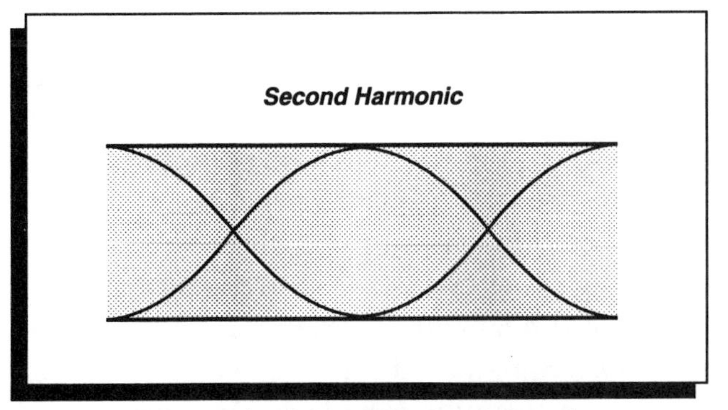

Figure 2.47 Second Harmonic in tube open at both ends.

Here, as on a string, only certain frequencies are allowed. The reason is that since the ends are open to the air, they must always be displacement antinodes.

Example: Calculate the fundamental frequency of a 1 meter tube open at both ends, with air.

Solution: First draw the standing wave as in (Figure 2.48).

Second, we know from the figure that:

½ wave = 1 meter
therefore 1 wave = 2 meters

To find the frequency, use:
frequency × wavelength = speed
Here the speed of sound for air is 344 m/sec.
frequency = speed / wavelength
= 344m/sec / 2 m
therefore,
= 172 Hz
The fundamental *f* is 172 Hz.
The second harmonic is 2*f*
2 × 172 Hz = 344 Hz

TUBE OPEN AT ONE END

Here we consider a tube with air, but that is open only at one end. Now we are not in a symmetrical situation, and differences with the tube open at both ends are expected.

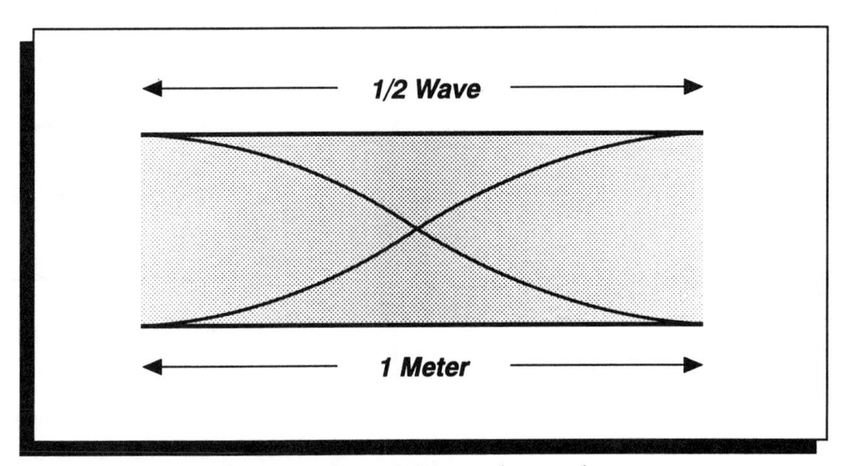

Figure 2.48 Fundamental in a tube.

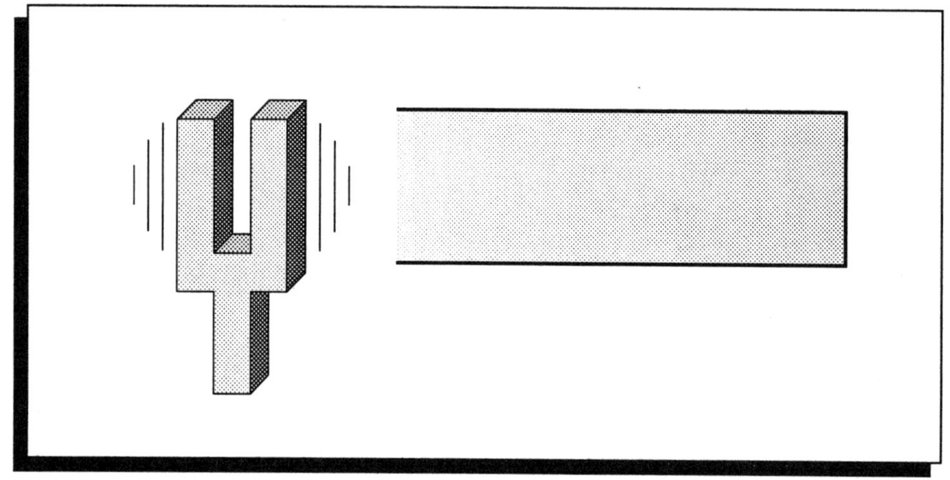

Figure 2.49 Tube open at one end and excited by a tuning fork.

Consider such a tube in Figure 2.49. In this a disturbance is created at the open end by a tuning fork.

The disturbance will travel down the tube and be reflected by the closed end.

We know what to expect for this case:

— a standing wave will be the set up.
— open end of tube will have a displacement antinode because air can move in and out.
— closed end of tube will have no displacement since it is a solid end which does not move (i.e. it will be a displacement node).

Let us draw the simplest standing wave pattern, consistent with one end that cannot move (node) and the other end that can move (antinode) Figure 2.50. By comparing the waveform inside this tube with a regular wave, it is clear that the fundamental pattern is a ¼ wave.

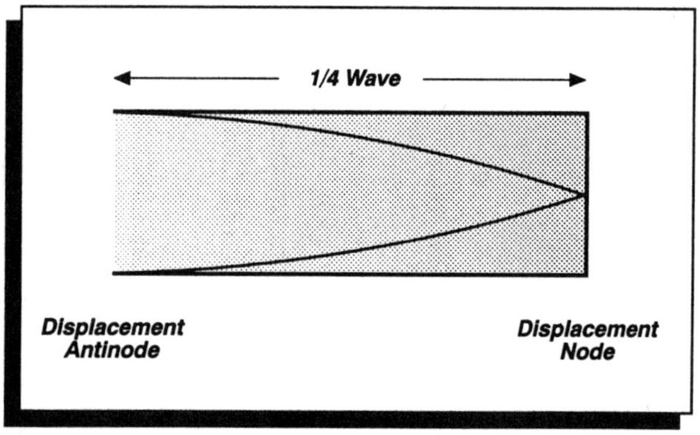

The next more complicated standing wave is a bit more subtle. Remember that:

open end = displacement antinode
closed end = displacement node

Hence, the standing wave now has a pattern depicted in Figure 2.51. Because the tube has now three ¼ waves, the standing wave is a series of three ¼ waves. Hence during the time that it took ¼ wave to go through one cycle, three ¼ waves will go through one cycle (i.e. its frequency is three times the fundamental). Figure 2.52 shows the next more complicated standing waveform which can exist in this tube. The fifth harmonic, named because it is a series of five ¼ waves.

1/4 Wave
Displacement Antinode
Displacement Node

Figure 2.50 Fundamental wave in tube open at one end.

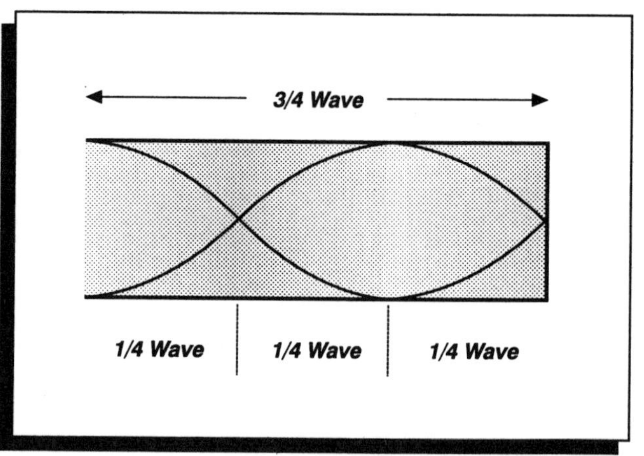

Figure 2.51 Next more complicated standing wave, the third harmonic.

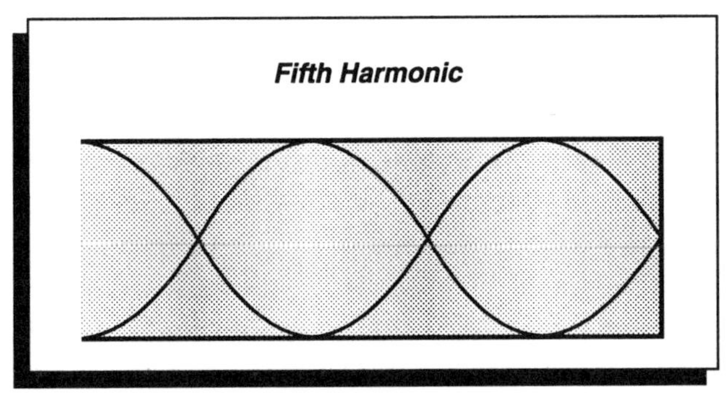

Figure 2.52 Fifth harmonic.

Hence, from the above analysis we conclude that a tube open at one end can have only the following harmonics:

fundamental	→	f
3rd harmonic	→	3f
5th harmonic	→	5f
7th harmonic	→	7f
nth (odd) harmonic	→	nf
		where n= 1, 3, 5, ...

This system can support only odd harmonics.

Exercise: What is the fundamental of a tube 1 meter long, open at one end?

Solution: First draw the standing wave as in Fig. 2.53.

Since ¼ wave = 1 meter
1 wave = 4 meters
but frequency = speed / wavelength
= 344 m/sec / 4 meters
= 86 Hz

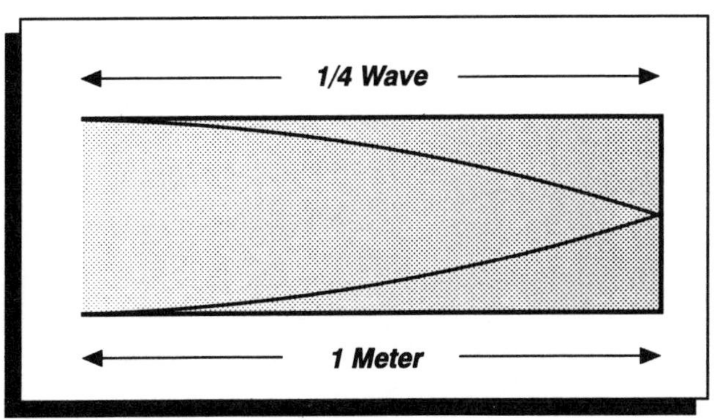

Figure 2.53 Standing wave in tube 1m long. Fundamental.

It is interesting to compare the fundamental wave of the tube open at one end to that of a tube open at both ends, when both tubes are of the same length. The fundamental of the tube open at both ends is twice the fundamental for the tube open at one end. This makes sense, as the tube open at both ends has ½ wave for the fundamental, while the tube closed at one end has ¼ wave for the fundamental. A comparison of a flute (which behaves like a tube open at both ends) and a clarinet (which can be approximated as a tube closed at one end) shows that the flute has a higher fundamental frequency than the clarinet, even though both are approximately of the same length.

TUBE CLOSED AT BOTH ENDS

This is an interesting case. It may not be convenient to create disturbances in it, but it has interesting applications in the hi–fi. Figure 2.54 shows the tube closed at both ends. Construction of standing waves in it is instructive.

Figure 2.54 Tube closed at both ends.

Let us assume that such a tube can be excited from inside (perhaps by a small speaker located inside it); standing waves will then be set up. Because the ends cannot move they will be displacement nodes. The fundamental will correspond to a ½ wave as shown in Figure 2.55. The ends are displacement nodes, and the antinode is at the center.

The next more complicated standing wave will maintain the zero displacement condition at the ends of the tube and it will be the second harmonic; its frequency is twice that of the fundamental. Higher harmonics can be set up as in previous

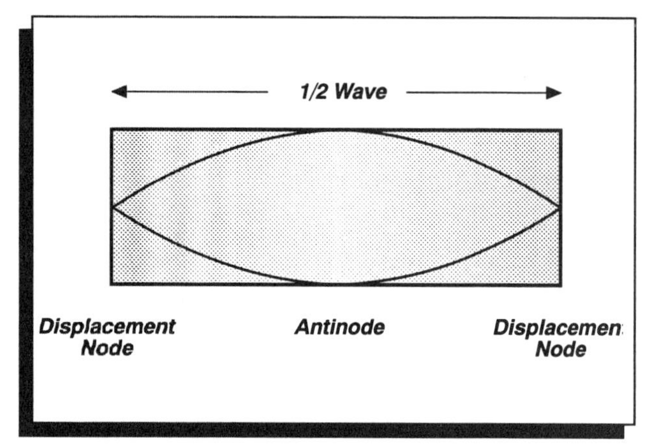

Figure 2.55 Fundamental of a tube closed at both ends.

examples. In summary, the frequencies that can be excited for the above conditions correspond to:

fundamental	→	f
2nd harmonic	→	2f
3rd harmonic	→	3f
:		:
:		:
nth harmonic	→	nf

In fact the results are very similar to those of a string clamped at each end. The reasons for doing this are:

— It is instructive to see how the ends of the tube determine its behavior when excited.

— It can be used as a simple model to understand the behavior of more complicated cases.

A good example is the setting up of standing waves in a room, which is a 3–dimensional extension of a tube closed at both ends. Consider the room shown in Figure 2.56. The walls will cause reflection of sound inside the room and this will lead to standing waves. The walls in the x direction will set up standing waves. The walls in the y direction will set up standing waves, as will the ceiling and floor along the z direction. Consider the y direction. If the walls are separated by a distance, Y, the fundamental of the standing wave is easily calculated:

since Y = ½ wave
2Y = 1 wave
frequency = speed/wavelength
= 344 m/sec / 1 wave

For a room where facing walls are separated in the y direction by 5m, the fundamental frequency for the standing wave in that direction will be:

frequency = 344 m/sec / 10m
= 34.4 Hz

Sound at 34.4 Hz will then be reinforced. In a similar manner the fundamental waves in the x and

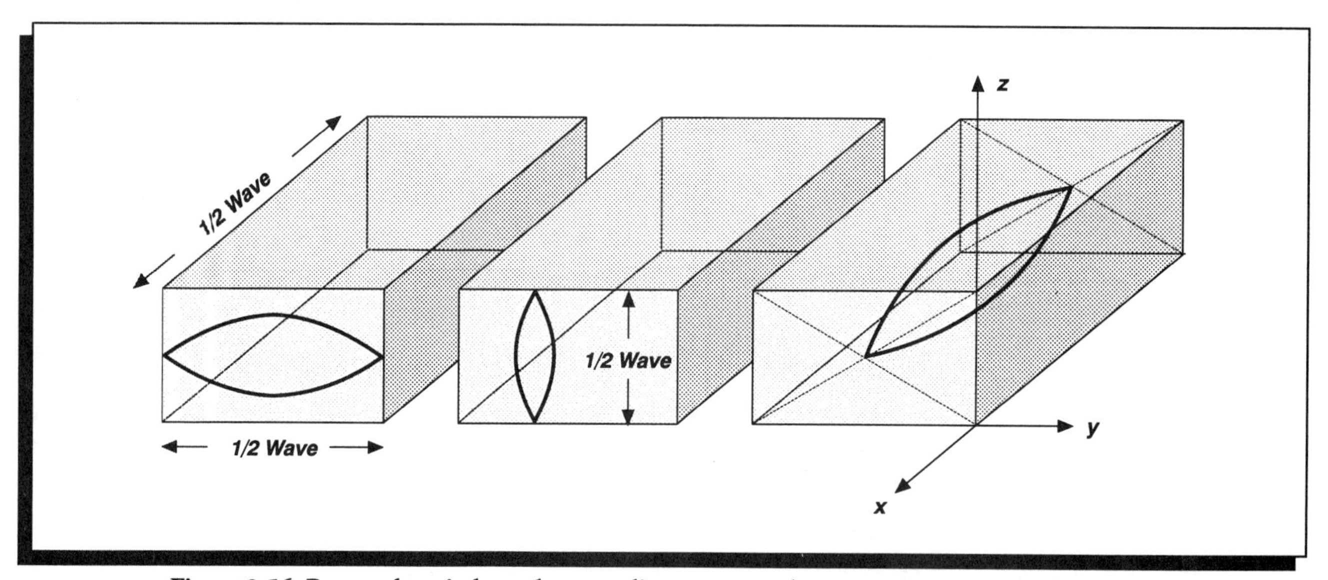

Figure 2.56 Room where independent standing waves can be set up the x, y, and z directions.

the y directions can be calculated. The harmonics can then be derived from the fundamental frequencies.

The examples presented here of longitudinal standing waves in various tubes are the basic elements for a wide range of musical instruments, from woodwinds to brasses. The formation of standing waves in any instrument will determine the sound they will emit.

■ STANDING WAVES ON PLATES

Percussion instruments are based on the vibrations of plates of all shapes and sizes, even deformed plates. When a plate is excited, wave pulses will travel toward the boundaries causing reflections and hence standing waves are set up. Of course, the behavior is expected to be more complicated than that of tubes or strings since the plates are essentially 2–dimensional objects.

Consider a circular plate, a drumhead clamped at its edges as in Figure 2.57. When the plate is struck at some point, waves will propagate to the edges and be reflected, causing a standing wave in 2–dimensions. The fundamental frequency is determined by the material, shape, and dimensions.

More complicated standing wave forms are created as well but they usually do not correspond to

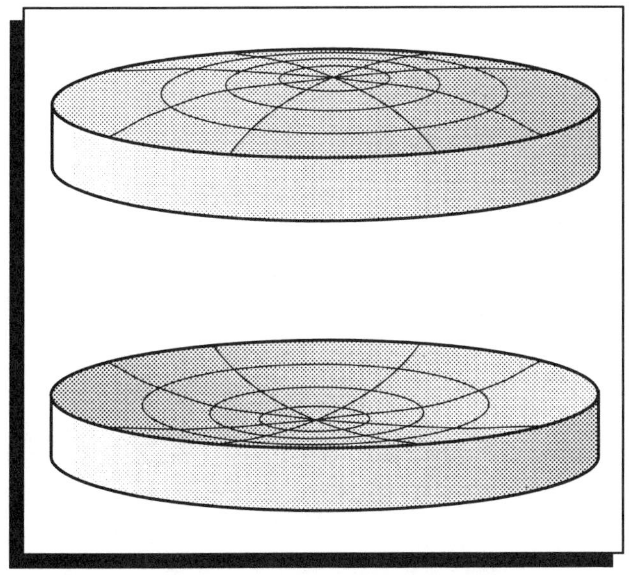

Figure 2.57 A drumhead fixed at its edges and its fundamental mode of vibration.

integer multiples of the fundamental frequency. For this reason they are not called harmonics, rather they are known as overtones. Figure 2.58 displays an overtone of a drumhead. An example for a particular case can be as follows:

fundamental	→	f
first overtone	→	1.59 f
second overtone	→	2.29 f
etc		...

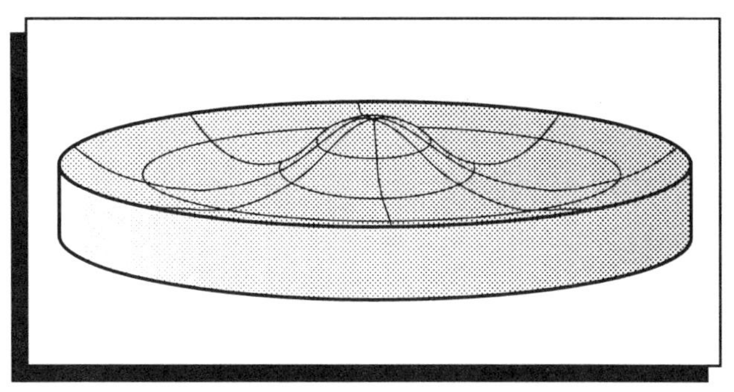

Figure 2.58 Overtone on a drumhead.

There are no harmonics. The value of the overtone will depend on the geometry and the material. Sometimes the word overtone is applied to the description of harmonics and in this case it means:

1st overtone	=	second harmonic
2nd overtone	=	third harmonic
....		...

Some examples of plates used in instruments and in the hi–fi are:

— drumheads
— bells (they are deformed plates)
— our eardrums
— soundboard in a piano
— diaphragm in cone speakers
— diaphragm in microphone

An interesting example of standing waves on plates is the Chladni plate. It is a plate fixed at its center and excited by rubbing a bow against its edge. By sprinkling salt on the plate and exciting

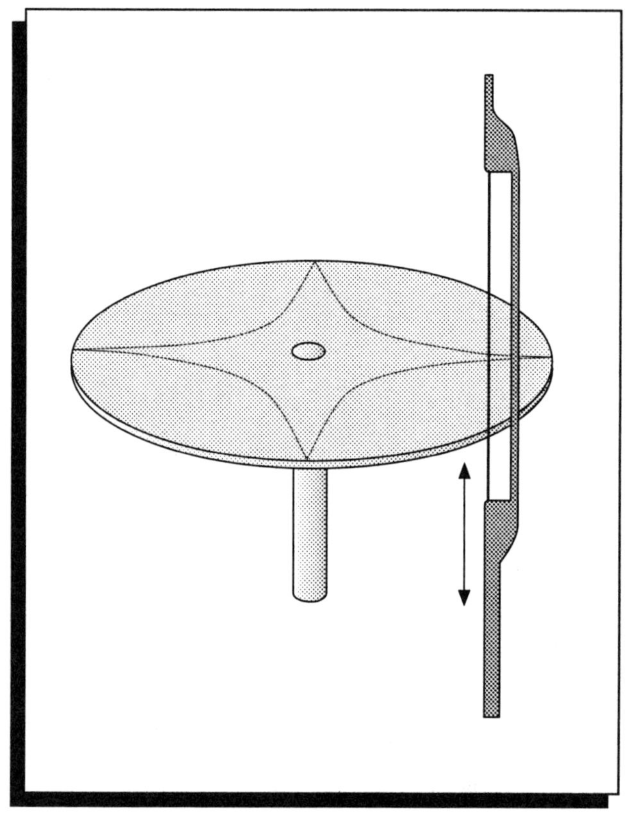

Figure 2.59 Standing wave pattern on a Chladni plate.

the plate with a bow, the standing wave pattern stands out, shown in Figure 2.59. At the center there is a clamp which causes the center to be a displacement node. Where the bow moves up and down there is displacement and hence a displacement antinode.

Complex Waves

Having introduced the concepts of harmonics, interference, and the Superposition Principle, we can apply them to the analysis of the sound waves produced by individual instruments and groups of instruments. Such waves are complex waves. Consider an instrument which at some instant produces a fundamental and its fourth harmonic, as shown in Figure 2.60. The superposition of these two harmonics creates a complex wave. Indeed, every instrument produces complex waves when it is excited. When a bow is pulled across a string as in a violin or a cello, harmonics are excited and these create a complex wave. When that string is plucked by a finger, harmonics are excited and a complex wave is produced. Figure 2.61 shows a string being plucked and some of the harmonics produced. The resultant is shown in Figure 2.62 and this is what is being heard.

Which harmonics are produced will depend on how the string is excited. We would not expect to have the same harmonics for a string that is plucked and one that is bowed. After all, they do sound different.

By adding harmonics consisting of sine waves we have shown that a complex wave can be produced. Therefore, we define:

complex wave **= a wave that contains harmonics**

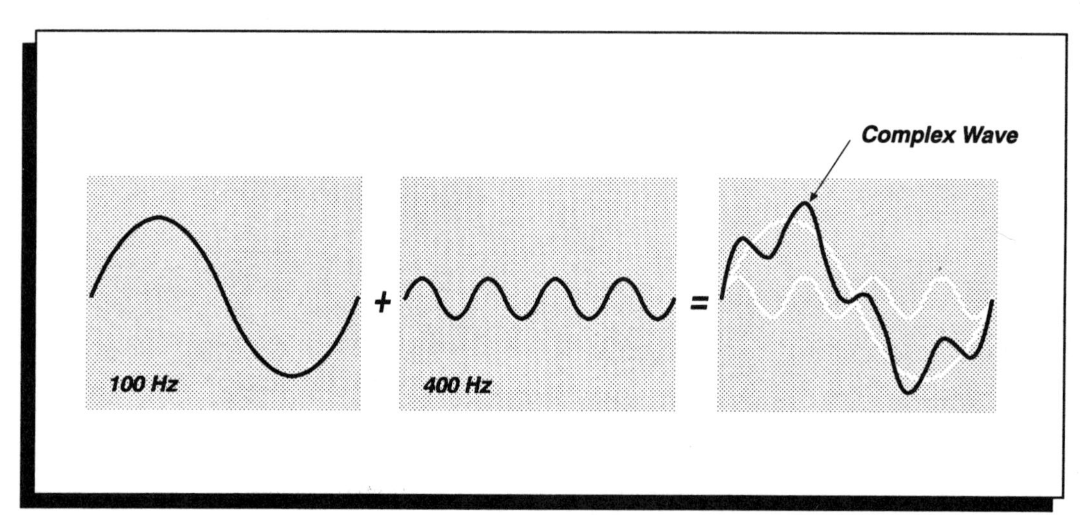

Figure 2.60 Complex wave created by the superposition of a 100 Hz fundamental and its fourth harmonic.

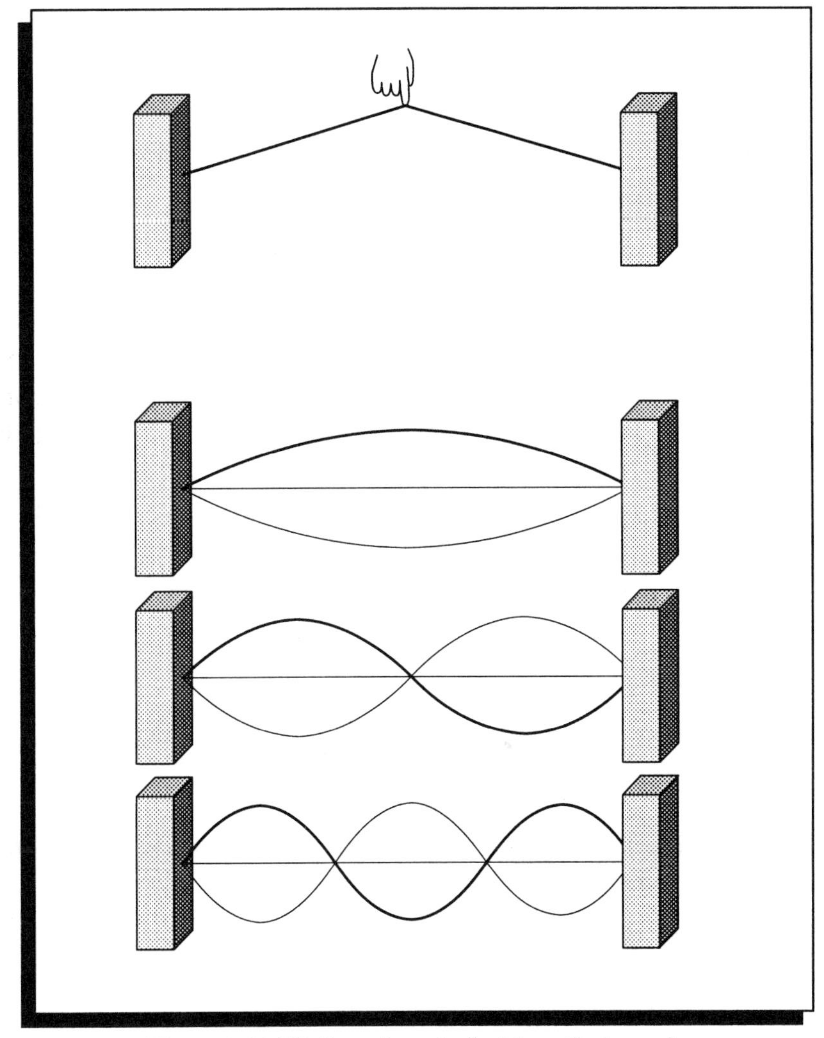

Figure 2.61 Violin string plucked by a finder and producing all sorts of harmonics.

This definition provides us with a tool for dealing with complex waves. This was done a long time ago by J. Fourier, who developed a method known as Fourier Analysis, which is for breaking up complex waves into simple sine waves.

Fourier Analysis **= breaking up of complex waves into simple ones**

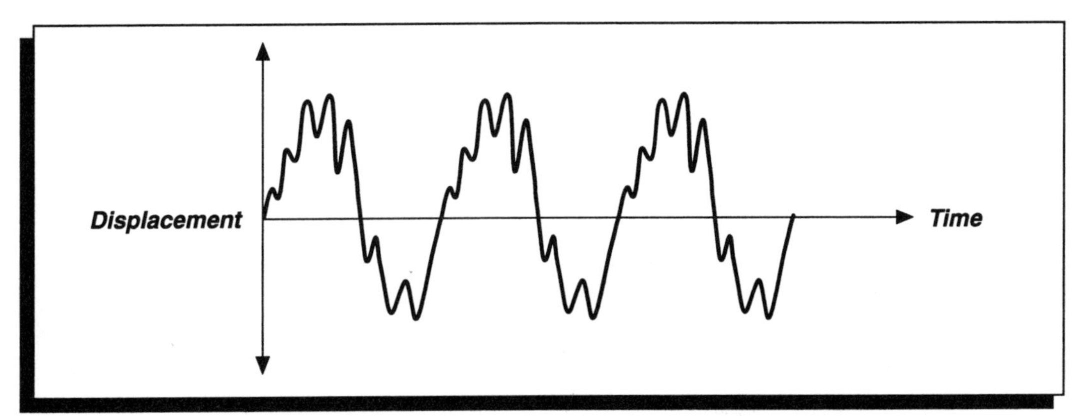

Figure 2.62 Complex wave generated by plucking a string.

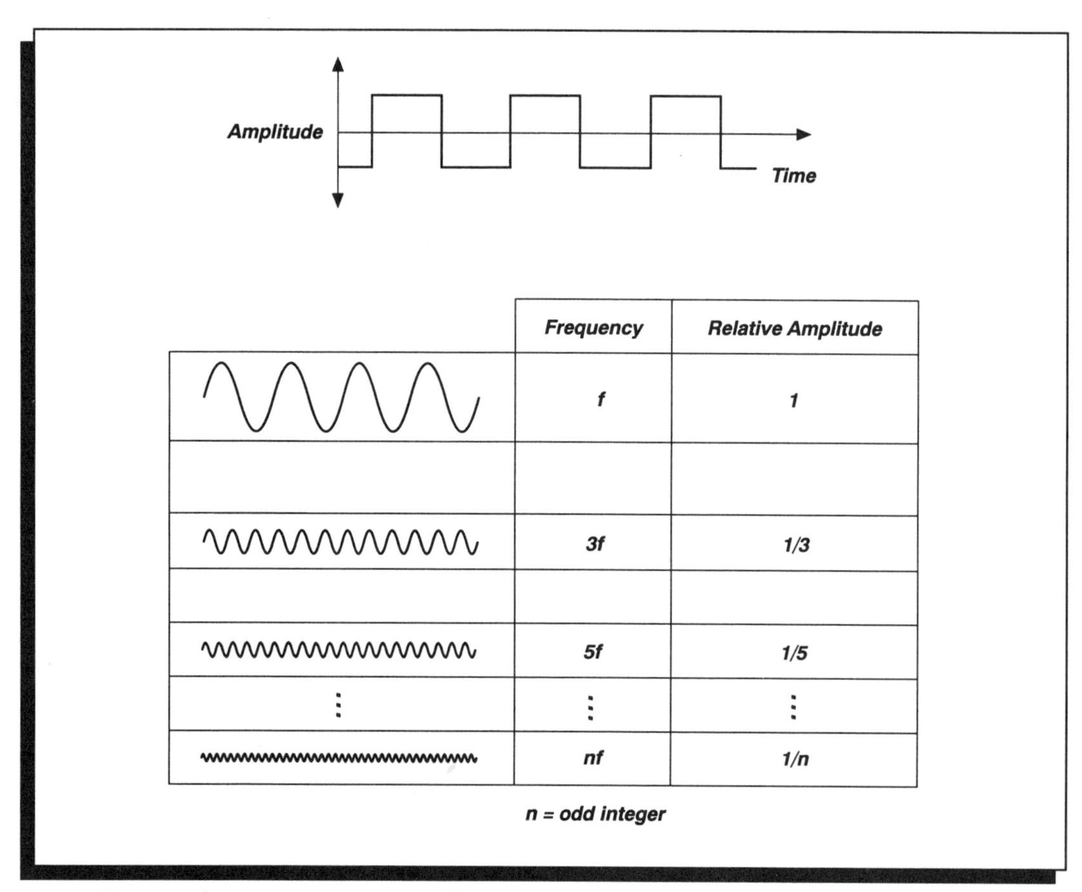

	Frequency	Relative Amplitude
	f	1
	3f	1/3
	5f	1/5
⋮	⋮	⋮
	nf	1/n

Figure 2.63 A square wave is made up of odd harmonics with decreasing amplitudes.

Consider a square wave as in Figure 2.63. A certain combination of harmonics and their specific amplitudes builds a square wave. It is possible to write down a recipe that would specify which harmonics and what their amplitudes should be to produce a square wave. A square wave consists of odd harmonics. Their amplitude decreases as the number of the harmonic increases.

Instead of writing down the recipe for making a square wave, it is possible to put all that information on a graph known as a spectrum of the complex wave. It shows which harmonics are present and what their amplitude is (Figure 2.64) for a square wave.

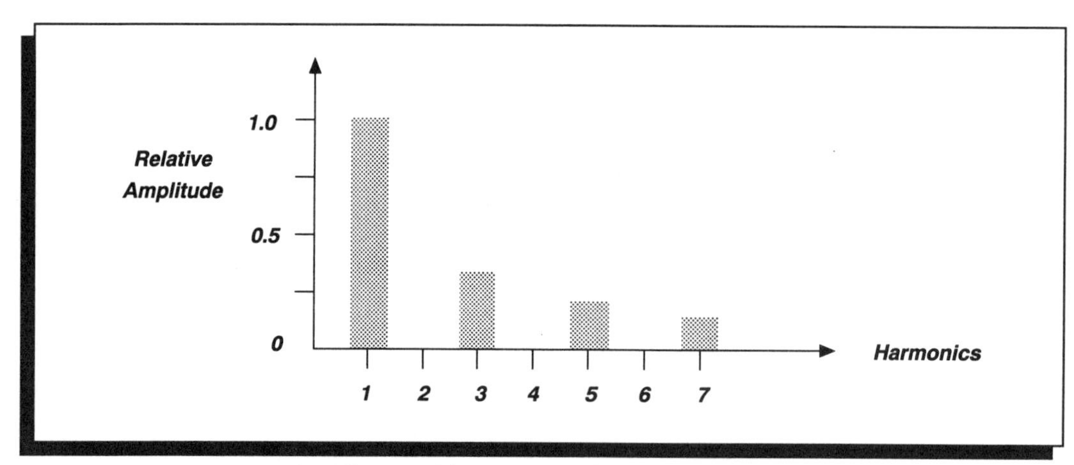

Figure 2.64 Spectrum of a square wave.

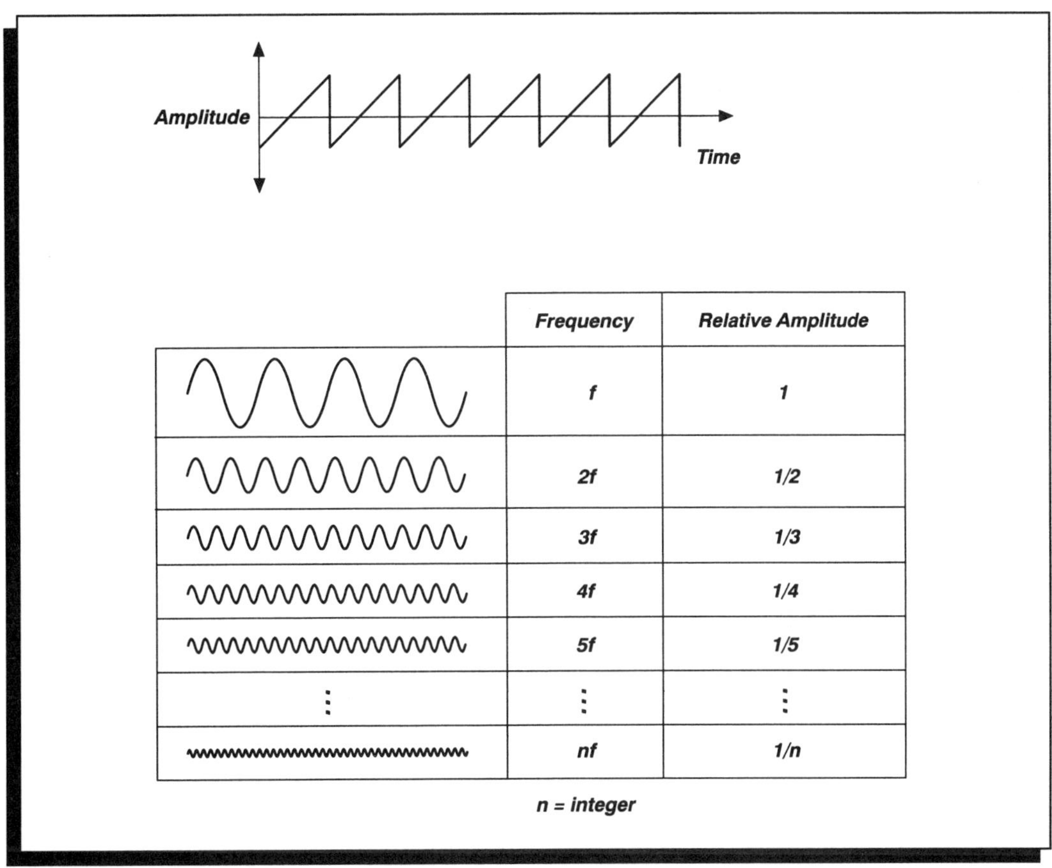

Figure 2.65 Sawtooth wave: its harmonic content and relative amplitudes.

Frequency	Relative Amplitude
f	1
3f	1/3
5f	1/5
7f	1/7
nf	1/n
	where n = an odd integer

Another example of a complex wave is a sawtooth as shown in Figure 2.65. It can be represented by a spectrum as shown in Figure 2.66. The spectrum consists of all the harmonics with the amplitude of each decreasing as the number of the harmonic increases. We have:

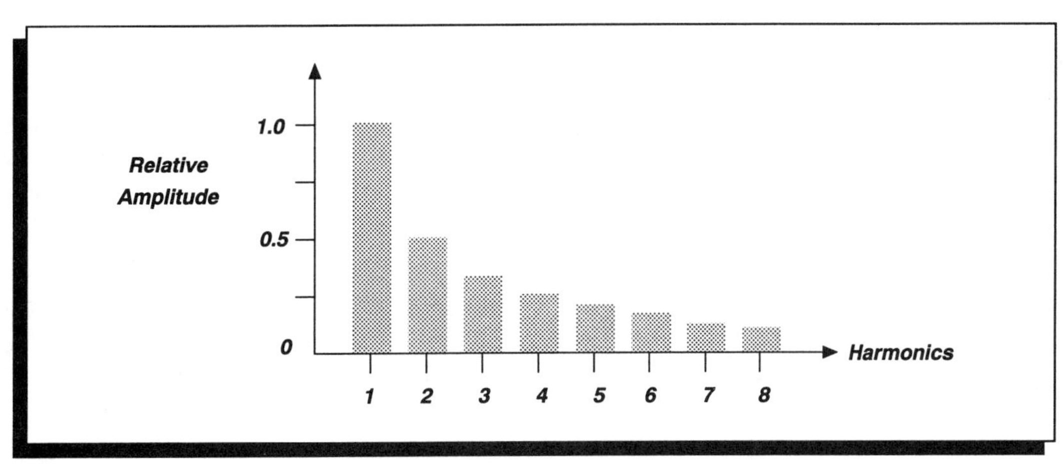

Figure 2.66 Spectrum of a sawtooth wave.

Frequency	Relative Amplitude
f	1
2f	1/2
3f	1/3
4f	1/4
5f	1/5
	etc.

This process can be repeated for any complex wave, be it a note produced by your voice, a flute, or any other instrument.

Such an analysis makes it clear why two different instruments playing the same note sound different. They have the same fundamental, but the harmonic content and their amplitudes are different. Even the same instrument played by two different musicians will sound different because the instruments are excited differently. To some extent our brain does a rapid Fourier Analysis and in this way we recognize the various instruments or the voice of someone we know.

Because each instrument produces complex tones and we often listen to more than one instrument, a hi–fi has to be able to reproduce all the harmonics correctly. A formidable task.

The facts just presented about the characteristics of a tone based on Fourier Analysis are not complete. Otherwise, it would be possible to program the characteristics on a computer for each instrument. When a tone is produced, there is the attack, the build–up, and the decay of that tone and these characteristics also come with the description of a tone.

A simple example can be used to illustrate how the method of exciting a system determines which harmonics will be produced. Figure 2.67 shows what would happen if a string clamped at each end were bowed at the middle.

The point where the string is excited by the bow cannot be a displacement node. The harmonics that can be excited here are shown; they are all the odd harmonics. As the bow moves closer to one end more even harmonics will be excited and the tone will become brighter.

Another example of how the method of exciting a string determines which complex wave will be produced, and which harmonics are excited, is a piano string when struck by a piano hammer. The hammer excites the string by striking it at a specific place. Consider Figure 2.68, where a string is struck by a hammer at a distance of 1/10 of the string length from the end. The action of the hammer will cause a displacement of the string (hence this point will not be a node in the displacement).

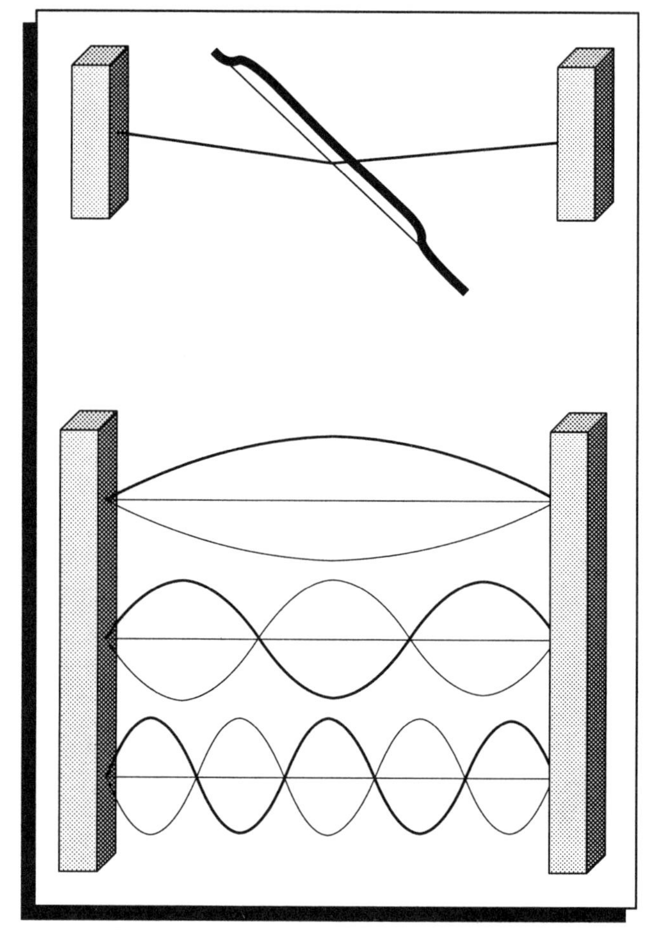

Figure 2.67 A string bowed at its middle and the harmonics which are excited.

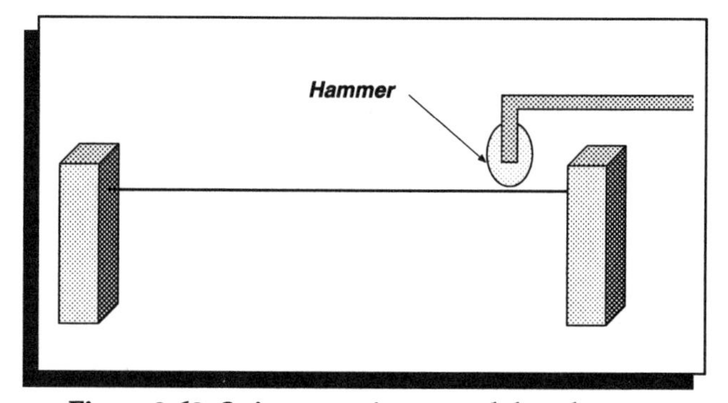

Figure 2.68 String on a piano struck by a hammer at a distance 1/10 from the string end.

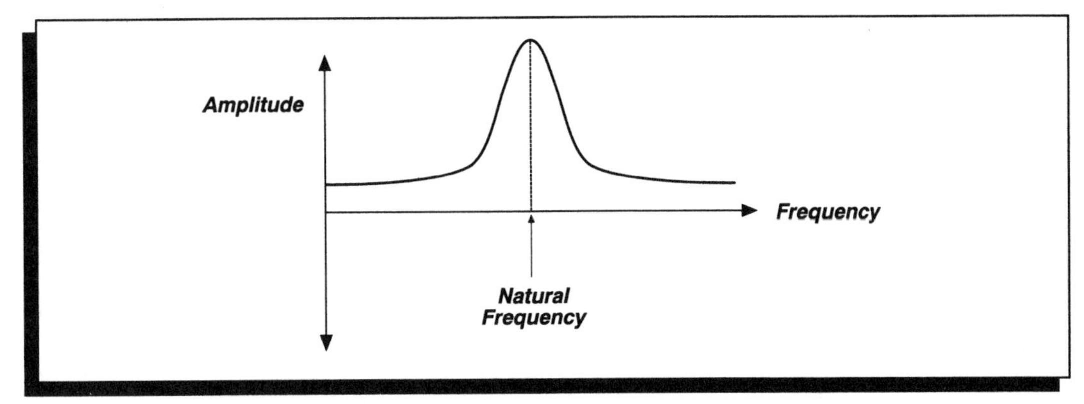

Figure 2.69 Vibrations of an object at different excitation frequencies.

But the 10th harmonic, if it were produced, would have a node at 1/10 the distance from the end. Hence, by striking the string at 1/10 the distance from the end, the 10th harmonic is greatly suppressed.

All instruments use much mechanisms to enhance or suppress certain harmonics.

2.4 OTHER PROPERTIES RELEVANT TO HI–FI

■ RESONANCE

All objects when excited will vibrate at a certain frequency. This frequency is called its fundamental frequency or natural frequency, and it depends on factors such as:

— material
— geometry
— dimensions
— state of tension

For the string clamped at each end, its fundamental frequency (i.e. its natural frequency) will depend on:

— the speed of waves on the string
— the length of the string

The speed of waves on a string will depend on its tension, density, and elastic properties.

In the case of tubes, the natural frequency will depend on the length of the tube and the speed of sound in the gas inside the tube.

A cone in a speaker will also have a natural frequency which will depend on its mass and the stiffness of its suspension.

By exciting an object (striking it with another object, shaking it, plucking it, etc...) it will vibrate at a preferred frequency. This is shown in Figure 2.69, where the amplitude of vibration of the object is shown at various frequencies of excitation.

When the excitation frequency equals the natural frequency of the system, large amplitude oscillations of the object are then observed. The object likes to vibrate at a specific frequency, its natural frequency. This is known as a resonance. Hence:

> ***Resonance* = when excitation frequency equals the natural frequency of an object its oscillations have a maximum amplitude.**

The amplitude at resonance depends on how much friction there is. By increasing the friction (damping), the amplitude on resonance is reduced. This is illustrated for a mass at the end of a spring, shown in Figure 2.70 when it is made to oscillate in air and in oil. The oscillations in oil will have a much smaller amplitude than in air, where the motion is damped.

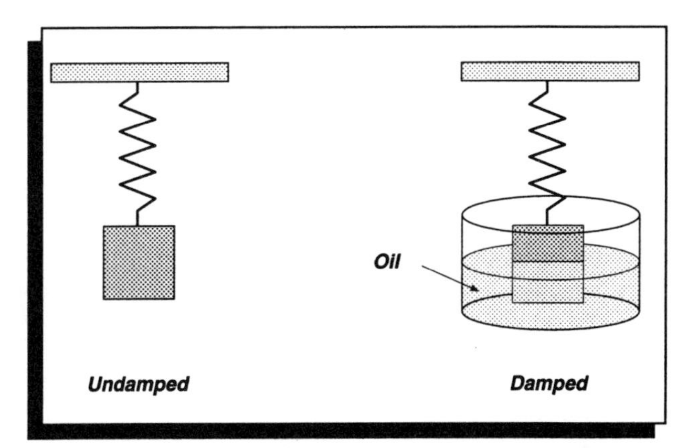

Figure 2.70 Oscillations of a mass on a spring, undamped and damped when submerged in oil.

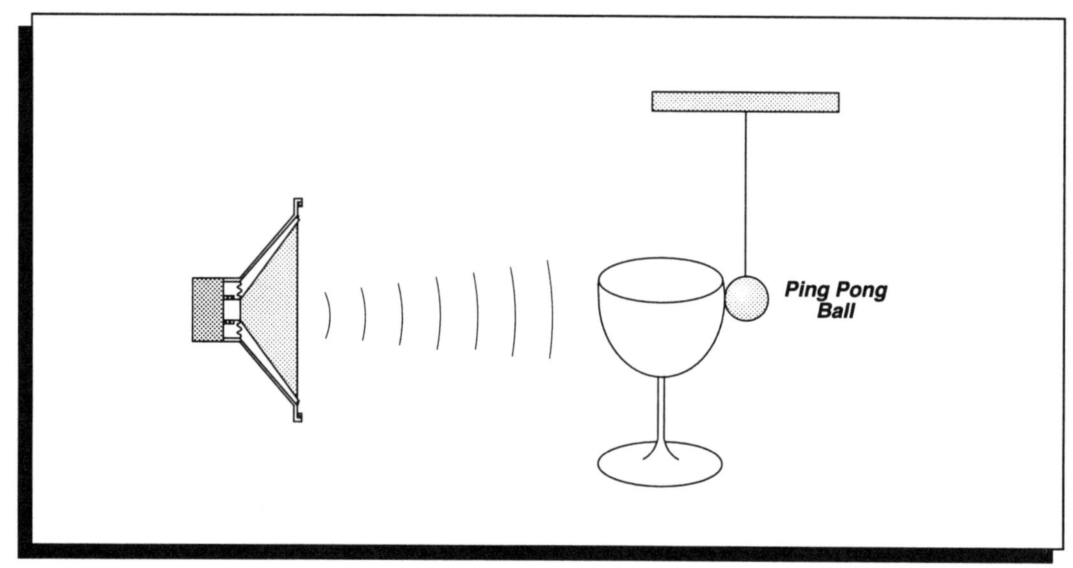

Figure 2.71 Resonance of wine glass excited by sound.

A spectacular demonstration of a resonance is the well–publicized case of a wine glass being shattered by sound. Consider Figure 2.71 where a wine glass is in the presence of sound produced by a speaker, and its frequency can be varied. In the presence of sound the wine glass will vibrate. To show the oscillations of the wine glass, a ping–pong ball lightly touches its rim. The frequency of the speaker is then varied. On resonance, the amplitude of oscillations of the wine glass rim will become large. This is shown by the ping–pong ball as it bounces around violently. The amplitude of the glass builds up so much that eventually the wine glass shatters (i.e. the displacements of the glass rim went beyond the elastic properties of the glass).

In the hi–fi there are many examples of resonance. Some are useful in the hi–fi, while others have to be minimized. A few examples of resonance as related to the hi–fi are:

— speaker
— speaker enclosure
— tuner
— radio antenna
— listening room
— our ears
— laser
— filters
— equalizer

Details of the resonance in each case as related to the hi–fi will be presented later on, as well as resonance in the human ear and the listening room.

■ BEATS

This is another example of interference, but deals with time rather than spatial position. It occurs when two sources of sound (or other waves) at slightly different frequencies are played together. There will be constructive and destructive interferences which will vary with time, and this will cause variations in the resultant sound intensity. Figure 2.72 demonstrates the concept. The variations in intensity will occur at a certain rate, with so many changes per second. This corresponds to a frequency known as beat frequency. It is defined as:

Beat Frequency **= (frequency of source 1) – (frequency of source 2)**

For example, speaker 1 produces 155 Hz, while speaker 2 produces 163 Hz. When the two are played together there will be a variation in intensity at the beat frequency:

Beat Frequency **= 163 Hz – 155 Hz**
= 8 Hz

Or, there will be 8 intensity variations per second. The resultant frequency will be an average frequen-

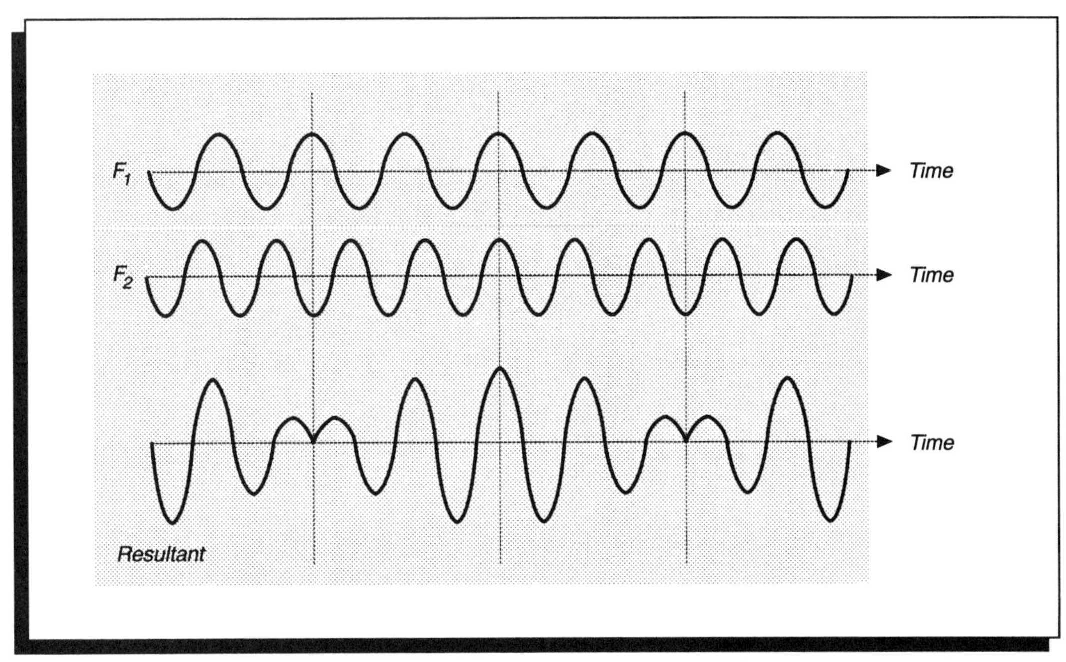

Figure 2.72 Beats caused by the combination of two waves with slightly different frequencies; interference in time.

cy of the two tones. The concept of beats will be very useful later on when we discuss tuners and how they operate.

Before applying all the knowledge learned so far to the hi–fi, there is another important chapter that has to be introduced for dealing with sound intensities and sound levels. It is the concept of decibels, which will be presented in the next chapter.

SUMMARY OF TERMS

Amplitude: maximum deviation or displacement from equilibrium in a wave.
Antinode: maximum deviation in a standing wave.
Beats: intensity variations per unit time when two waves of slightly different frequencies are produced together.
Complex wave: wave made up of harmonics.
Condensation: increase in pressure in a sound wave.
Doppler Effect: change of frequency when there is relative motion between source of waves and observer. When the motion between source and observer is toward each other, there is a frequency increase. When the motion between source and observer is away from each other, there is a frequency decrease.
Frequency: number of waves per second, measured in Hertz (Hz).
Intensity: power from a wave crossing a given surface area; its unit is watts /m^2.
Law of Reflection: angle of incidence with respect to normal is equal to angle of reflection with respect to normal when a wave hits an obstacle.
Longitudinal wave: wave where displacements are parallel to direction of travel. Sound in air is an example of such a wave.
Node: point on a standing wave where the displacement is zero.
Power: rate at which work is done or rate at which energy is given or used up; its unit is watts.
Rarefaction: decrease in pressure in a sound wave.
Reflection: return of a wave from an obstacle or from a medium with different properties than the medium the wave is in.
Refraction: change of direction of a wave when it enters a medium with different properties, causing its speed to change.
Resonance: increase in amplitude of vibrations of a body when it is caused to oscillate at its natural frequency.

Reverberation: diffuse mixture of reflected sounds in a hall or a room.
Reverberation Time: time for sound power in a hall to decay by a factor of 1,000,000 from its original value when sound power is switched off. Important acoustic characteristic of a hall.
Sound: pressure changes at frequencies in the range of 20 Hz to 20,000 Hz.
Speed: displacement per unit time.
Transverse Wave: wave where displacements are perpendicular to direction of travel.
Wavelength: how long one wave is.

NAME ____________________ DATE __________

QUESTIONS FOR REVIEW

1 List and describe five different types of waves which exist in the hi–fi.

2 Discuss the sound produced by (a) a tube open at both ends and (b) a tube open at one end.

3 Name parts or components of a hi–fi where resonance occurs.

4 Explain why musical instruments sound differently even when they play the same note.

5 How do hall acoustics contribute to the quality of sound?

6 What is Fourier analysis and what do you learn from such analysis?

7 What are standing waves? List three examples.

8 Name two methods for changing the fundamental of a tube open at both ends.

9 When are two waves in phase and when are they out of phase?

10 How does a square wave differ from a sawtooth wave when they both have the same fundamental?

NAME______________________ DATE ____________

Please select the best answer for each question.

1. A wave in which the displacements are perpendicular to the direction in which the wave travels is called a:
   ___A. longitudinal wave
   ___B. sound wave
   ___C. radio wave
   ___D. transverse wave
   ___E. light wave

2. The wavelength of a wave is:
   ___A. the number of waves per second
   ___B. the time for one wave to go by
   ___C. a quantity which depends on the amplitude
   ___D. distance a wave travels before it starts to repeat itself
   ___E. how long a ½ wave is

3. A sound wave in air has a wavelength of 0.5 meter. What is its frequency?
   ___A. 344 Hz
   ___B. 688 Hz
   ___C. 172 Hz
   ___D. 1720 Hz
   ___E. 1.5×10^8 Hz

4. The distance between two successive identical parts of a wave is called the:
   ___A. amplitude
   ___B. wavelength
   ___C. frequency
   ___D. intensity
   ___E. period

5. The frequency of a sound wave tells you:
   ___A. how spread apart in space the pressure fluctuations are
   ___B. how fast the pressure fluctuations are moving through space
   ___C. how large the pressure fluctuations are
   ___D. the distance between successive condensations
   ___E. how long it takes to produce a pressure fluctuation

6. If you increase the frequency of a sound wave by a factor of 4, its wavelength will:
   ___A. decrease by a factor of 4
   ___B. increase by a factor of 4
   ___C. remain the same
   ___D. increase by a factor of two
   ___E. you cannot tell because the speed is not given

7. A speaker produces 250 sound waves in ¼ of a second. The frequency of the sound will be:
   ___A. 250 Hz
   ___B. 1,000 Hz
   ___C. 62.5 Hz
   ___D. 100 Hz
   ___E. you cannot tell because we do not know the speed of speaker

8. If a longitudinal wave has a frequency of 500 Hz, and its wavelength is 0.25 meter, calculate the wave's speed.
   ___A. 125 m/sec
   ___B. 2,000 m/sec
   ___C. (500 + 0.25) = 500.25 m/sec
   ___D. 500 m/sec
   ___E. a longitudinal wave has no speed

9. The human ear can hear the sound produced by a vibrating object when the object is vibrating at frequencies in the range of:
___A. 20 Hz to 20 kHz
___B. 200 Hz to 22 kHz
___C. 15 Hz to 15 kHz
___D. 20 kHz to 200 kHz
___E. 10 Hz to 20 kHz

10. A speaker is producing sound at a certain pressure amplitude. How much more power is required in order for the speaker to produce a sound that has a pressure amplitude four times greater?
___A. four times more power is required
___B. eight times more power is required
___C. sixteen times more power is required
___D. two times more power is required
___E. the same amount of power is required

11. A sound wave of intensity is 5×10^{-5} watts/m^2 crosses a surface of 3 m^2. What is the total sound power that crosses the area?
___A. 5×10^{-5} watts
___B. $(5 \times 10^{-5}$ watts/m$^2) \times 3$ m$^2 = 15 \times 10^{-5}$ watts
___C. $(5 \times 10^{-5}$ watts/m$^2)$ / 3 m$^2 = 1.67 \times 10^{-5}$ watts
___D. (3 m^2) / $(5 \times 10^{-5}$ watts/m$^2) = 0.6 \times 10^5$ watts
___E. $(3 + 5) \times 10^{-5}$ watts

12. Reflection is a property of a wave in which:
___A. all the wave energy is absorbed by the surface
___B. the angle of reflection equals the angle of incidence
___C. the wave is reflected along the normal
___D. refraction occurs at a surface
___E. the wave reflects off a surface at an angle to the surface which is different than its incoming angle

13. A 1 kHz sound wave is travelling through air and then it travels through water. The speed of sound in water is 4 times that in air. The sound wave in air will have a wavelength:
___A. which is 4 times shorter than in water
___B. which is 4 times longer than in water
___C. which is the same as in water
___D. which is 16 times longer than in water
___E. which is 16 times shorter than in water

14. Suppose that sound "X" is produced with 25 times more power than sound "Y." The pressure amplitude of "Y" is ___________ than the pressure amplitude of "X."
___A. 5 times greater
___B. 5 times less
___C. 25 times greater
___D. 25 times less
___E. $25 \times 25 = 625$ times less

15. A speaker puts out a 50 Hz sound wave and a 6,000 Hz sound wave. During the time that the speaker produces a ½ wave of the 50 Hz tone, moving towards you, what happens to the frequency of the 6,000 Hz tone that you hear?
___A. its frequency will increase due to the Doppler Effect
___B. its frequency will remain the same
___C. its frequency will decrease due to the Doppler Effect
___D. its frequency will decrease by a factor of 2 because it happens when the speaker produces a ½ wave.
___E. its frequency will increase by a factor of 2 because it happens when speaker produces ½ wave

16. A speaker placed outdoors produces a sound whose intensity is 4×10^{-4} watts/m^2 at a distance of 5 meters from the speaker with almost no reflections. What will happen to the intensity in general when it is moved indoors into a room as measured at a distance of 5 meters from the speaker? The intensity will:
___A. increase
___B. decrease
___C. remain the same
___D. decrease because the surface area receiving the sound will increase

17. How many waves of a 5 kHz sound will be produced during the time it takes 1 wave of a 250 Hz sound to be completed?
___A. 5,000
___B. 20
___C. 1/20
___D. 250,000
___E. 1

NAME ______________________________ DATE ____________

18. The reverberation time of a hall will ________ when the audience is increased from 1,000 people to 2,000 people.
    ___A. increase
    ___B. remain the same
    ___C. decrease
    ___D. be doubled
    ___E. not change since the volume of the hall did not change

19. The Doppler Effect is produced when:
    ___A. sound is refracted
    ___B. sound is reflected from a smooth surface
    ___C. the sound wave changes its speed
    ___D. two waves interfere with each other
    ___E. there is relative motion between the sound source and the observer

20. When there is a lot of reflections in a room:
    ___A. the reverberation time will decrease
    ___B. the individual notes of music become blurred
    ___C. the room does not reflect the sound much
    ___D. the room has large absorption
    ___E. the room is ideal acoustically for chamber music

21. In a concert hall, the sound that you hear is:
    ___A. mainly reverberant with some direct sound
    ___B. all reverberant with no direct sound
    ___C. all direct with a small amount of reverberant sound
    ___D. all direct with no reverberant sound
    ___E. half reverberant and half direct sound

22. Adding together two waves which are always in phase causes:
    ___A. standing waves
    ___B. destructive interference
    ___C. antinodes
    ___D. constructive interference
    ___E. nodes

23. A standing wave results from:
    ___A. two identical waves travelling together in phase
    ___B. two waves which pass through each other and hence, have different frequencies
    ___C. two identical waves travelling in opposite directions and passing through each other
    ___D. two identical waves travelling together out of phase
    ___E. two waves beating against each other

24. The distance between two successive antinodes of a standing wave on a string is equal to:
    ___A. $1/4$ of a wavelength
    ___B. $1/2$ of a wavelength
    ___C. one wavelength
    ___D. 2 wavelengths
    ___E. 3/2 of a wavelength

25. Why does a middle C played on a piano and on a violin sound different?
    ___A. because the periods of the harmonics may be different
    ___B. because the fundamental frequencies may be different
    ___C. because they both have exactly the same number and relative amplitudes of the harmonics
    ___D. because the number and relative amplitudes of the harmonics may be different
    ___E. because they produce different beat frequencies

26. The method of breaking up a complicated wave into a series of sine waves is called:
    ___A. Fourier analysis
    ___B. standing wave
    ___C. resonance
    ___D. Superposition Principle
    ___E. constructive interference

27. For a standing wave, the region of maximum displacement is called the:
    ___A. displacement antinode
    ___B. displacement node
    ___C. condensation
    ___D. amplitude
    ___E. wavelength

28. "Beats" are an example of:
    ___A. reverberation
    ___B. resonance
    ___C. standing wave
    ___D. interference
    ___E. diffraction

29. What is the fundamental frequency of a tube in air open at both ends, when its length is 2 meters?
    ___A. 172 Hz
    ___B. 86 Hz
    ___C. 344 Hz
    ___D. 43 Hz
    ___E. 688 Hz

30. The natural frequency of an object is determined by:
    ___A. its shape
    ___B. its size
    ___C. the material from which it is made
    ___D. the tension on it
    ___E. all of the above answers are correct

31. Two identical sound sources differ in distance from a listener by ½ wavelength. The result will be:
    ___A. no sound at the listener
    ___B. constructive interference at the listener causing louder sound
    ___C. constructive interference at the listener causing quieter sound
    ___D. beats
    ___E. sound which is twice as loud as one source

32. What is the frequency of the third harmonic of a tube open at one end, when its length is 1 meter?
    ___A. 344 Hz
    ___B. 86 Hz
    ___C. 258 Hz
    ___D. 172 Hz
    ___E. 688 Hz

33. What is the distance between a consecutive node and an antinode on a string clamped at both ends at its second harmonic when the length of the string is 0.5 meter?
    ___A. 0.5 meter
    ___B. 0.25 meter
    ___C. 0.125 meter
    ___D. 1.0 meter
    ___E. 0.0625 meter

34. For a tube closed at one end, how many antinodes will the 7th harmonic have?
    ___A. 7
    ___B. 6
    ___C. 5
    ___D. 4
    ___E. 14

35. What is the beat frequency between the 5th harmonic of a tone whose fundamental is 125 Hz and the 6th harmonic of a tube whose fundamental is 98 Hz?
    ___A. 25 Hz
    ___B. 27 Hz
    ___C. 1 Hz
    ___D. 37 Hz
    ___E. 588 Hz

36. Sitting in front of your speakers, you do not hear certain frequencies well, even though your system works well. In this case it could happen because of:
    ___A. destructive interference
    ___B. travelling waves
    ___C. constructive interference
    ___D. a pressure antinode at your position
    ___E. beats

NAME ______________________________ DATE ____________

37. A siren produces sound by blowing air through a series of holes drilled into a rotating disc. What will be the frequency of the sound if the disc contains 60 holes (spaced uniformly) and is rotating at a speed of 50 revolutions per second?
    ___A. 3000 Hz
    ___B. 50 Hz
    ___C. 60 Hz
    ___D. 300 Hz
    ___E. 110 Hz

38. The lowest fundamental frequency for a standing wave that could be set up in a hall of the following dimensions: 8 meters × 3 meters × 4 meters would be ____________ .
    ___A. 43 Hz
    ___B. 21.5 Hz
    ___C. 86 Hz
    ___D. 114.7 Hz
    ___E. 57.3 Hz

39. A room should have some reflection of sound for proper sound reproduction. However, excessive reflections:
    ___A. produce a "quiet" room
    ___B. cause the reverberation time to decrease
    ___C. cause a decrease in the sound intensity in the room
    ___D. cause individual notes to become blurred together, causing an unnatural sound

40. When a sound wave travelling in warm air enters a layer of cold air at an angle, the new direction of the sound wave in the cold air will be:
    ___A. away from the normal
    ___B. toward the normal
    ___C. the same
    ___D. a backward retracing of the original
    ___E. along the normal

CHAPTER 3 DECIBELS

Intensity of sound, which is the sound power over a surface, varies over a very wide range. In music that we listen to, a concert hall for example, the intensity changes by many factors of ten between the quietest sound and the loudest one. Since our ears respond over a very large range of intensity values, it is convenient to cover such a range in decibels (dB). This chapter introduces the subject of decibels and relates it to the response of our ears at various levels of intensity. Decibels set a scale of intensities that are extended to the performance of a hi–fi system, from speakers to amplifiers, which are both used for presenting the specifications of a hi–fi system.

3.1 DEFINITIONS

The fundamental characteristics of a sound wave are its frequency (or frequencies) and intensity level. The range of frequencies our ears respond to, or what we call sound, is from 20 Hz to 20,000 Hz (this has been discussed the last section in some detail). The intensity of a sound wave was defined without setting any limits to the range of this characteristic. There must be an upper boundary to a sound wave intensity, otherwise we will feel pain, as well as a lower boundary, below which our ears do not respond. Experiments show that, on the average, the human ears respond to a wide range of intensities, between the limits set by the following:

Threshold of hearing at 1,000 Hz, where the intensity = 10^{-12} watts / m^2

and

Threshold of pain, where the intensity = 10 watts / m^2

That is a very wide range of intensities. The ratio of these two limits obtained by dividing the threshold of pain intensity by the threshold of hearing intensity is 10^{13}. It is an extremely large number! In other words the threshold of pain has an intensity which is 10^{13} (this is 1 with 13 zeros) times larger than that of the threshold of hearing. Since we will be dealing with such a large range, it is convenient to define a scale that will make it easier to cover such a range of numbers. That scale is known as the **decibel** scale. As an introduction to the definition of decibel, a few facts dealing with sound intensity will be presented and a few simple definitions will be given.

SIMPLE DEFINITION

The term decibel was invented to describe the smallest noticeable change in the intensity level of a sound. Hence, one definition is:

***One decibel (dB)* = smallest audible change in sound level.**

For example a change of 5 decibels (dB) means that the sound level has increased or decreased by 5 audible steps.

Another example would be a change of 8 decibels; this means that the sound level has changed by 8 audible steps. This is a qualitative definition of decibel, and most people would agree that it corresponds to their perception of sound level changes.

Let us look now at the scale of intensity levels which our ears respond to, starting with the threshold of hearing that we define as the zero decibel level. Examples of some intensity levels in decibels are:

Audible Sound		**Decibel Level**
— threshold of hearing	← start of scale →	0 decibels
— rustle of leaves		10 decibels
— whisper		30 decibels
— normal conversation		50 decibels
— siren, rock concert		120 decibels
— machine gun		130 decibels
— threshold of pain		130 decibels
— nearby jet airplane taking off		150 decibels

The last example of the jet taking off means that the sound level is 150 audible steps above the threshold of hearing. This definition is subjective and will vary from person to person, but it provides a general guideline.

SCIENTIFIC DEFINITION

The scientific definition of decibel can be introduced by the following example. Consider an amplifier delivering 0.001 watt to a speaker. The volume control is now turned up so that the amplifier delivers 100 watts to the speaker. How many times more power is the amplifier delivering now? The question deals with a comparison which is:

$$\text{Power now/Power before} = 100 \text{ watts}/0.001 \text{ watt} = 100{,}000 = 10^5$$

The amplifier is delivering 100,000 times more power now than before. That is a very large factor in the sound change. We could simplify the statement in view of the importance of the exponent 5 in this comparison by simply saying that now the amplifier delivers 50 more decibels of power. Here is how this number of decibels was obtained:

1. take power ratio, (power now divided by power before), and express it as a factor of 10.
2. take the exponent of 10 and multiply by 10
3. this gives us the number of decibel change in sound level

This operation can be restated mathematically as:

***number of decibels (dB)* = 10 log (Power now/Power before)**

The number of decibels does ***not*** mean how many watts are being delivered now, it only means how many times more power is being delivered.

Similarly, intensities can also be compared, so the definition of decibels in terms of intensity can be stated as:

***number of decibels (dB)* = 10 log (Intensity now/Intensity before)**

From the decibel change in sound level, it is always possible to calculate what is the power or intensity delivered. An example to illustrate this is:

— If a sound level is increased by 80dB above the threshold of hearing, what is its intensity?
There are two parts to this answer.
- The first part deals with what 80 dB means. A change of 80dB means a factor of 10^8.
- The second part deals with the fact that the sound is 10^8 times that of the threshold of hearing, which is an intensity of:

$$(10^{-12} \text{ watts/m}^2) \times 10^8 = 10^{-4} \text{ watt/m}^2$$

All these computations can be done quite simply on any calculator which has the logarithm function. To illustrate how this works, Table 3.1 presents simple power or intensity ratios and the corresponding decibel values.

Table 3.1
Conversion of power ratios or intensity ratios to decibels

Power or Intensity Ratio	Difference in Decibels
1.00	0 dB
2.00	3 dB
5.00	7 dB
10.00	10 dB
20.00	13 dB
50.00	17 dB
100 (or 10^2)	20 dB
1000 (or 10^3)	30 dB
10,000 (or 10^4)	40 dB
100,000 (or 10^5)	50 dB
1,000,000 (or 10^6)	60 dB
10,000,000 (or 10^7)	70 dB
100,000,000 (or 10^8)	80 dB
1,000,000,000 (or 10^9)	90 dB
etc.	etc.

At this point it is interesting to introduce an important fact, which deals with what sort of auditory sensation an increase of 10 decibel will create. It causes a sensation that is two times louder. Similarly, when the sound level is reduced by 10dB the auditory sensation created is two times quieter. This fact comes from the way our ears respond to sound levels; they respond to changes by powers of 10. Thus,

a 10dB increase corresponds to 2 times louder sound

and

a 10dB decrease corresponds to 2 times quieter sound

Hence a 20dB increase causes sound to be 2×2 louder (i.e. four times louder). A 40dB increase causes sound to be 2×2×2×2 times louder (i.e. 16 times louder). Decibel changes can be measured directly by a decibel meter, as shown in Figure 3.1. It basically is a special microphone that responds to sound intensities and displays them directly in decibels.

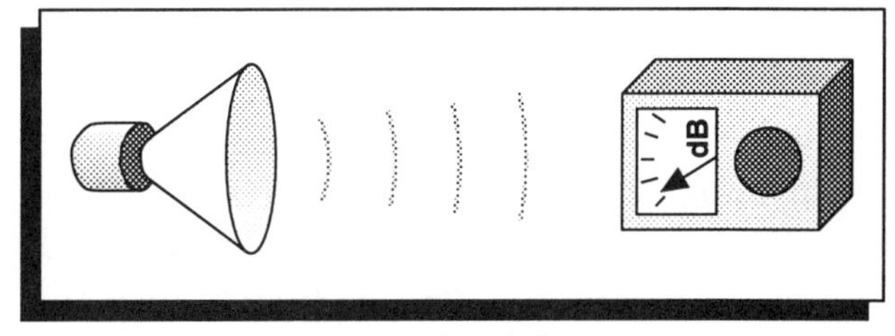

Figure 3.1 Decibel meter.

At first the decibel scale may seem complicated. However, it really simplifies calculations when dealing with large power changes of relative intensities. The importance of such scale will become clearer when dealing with hi–fi specifications, as they extensively use the decibel scale.

3.2 EXAMPLES OF DECIBELS IN HI–FI

From its definition, decibel refers to a comparison of sound powers or intensities; it expresses how many times larger or smaller a given quantity is when compared to another one. In specifying the performance of the components in a hi–fi system comparisons will be made in decibels of how many times more power will be delivered at one frequency compared to another, or how many times more sound power will be delivered in one direction compared to another direction, etc. The comparisons are expressed in decibels (dB). By retracing back from the definition of decibel, a decibel spec can tell us what the power ratio (Power now/Power before) is (remembering that it was multiplied by 10).

As an example, consider a 50 watt amplifier being replaced by a 100 watt amplifier. The power is 2 times larger and hence, the decibel increase will be:

number of decibels change = 10 log (Power now/ Power before)
= 10 log 2
= 3dB

which does not appear to be a large increase.

On many pre–amps the volume control is marked in terms of 0dB to –70dB or even –90dB. This means that the volume goes from a maximum value which is at 0dB, down to the minimum level since there is a negative value in front of the dB. Figure 3.2 shows a volume control marked in dB.

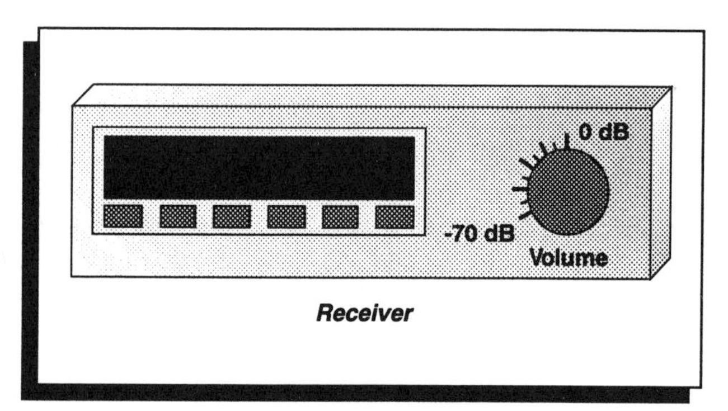

Figure 3.2 Receiver with volume control marked in dB.

3.3 RESPONSE OF THE HUMAN EAR

Before assessing the specifications of hi–fi components, it is instructive to find out how human ears respond to different sound frequencies and also to various sound intensities. Questions like, "Do we hear equally well at all frequencies?" are interesting and relevant here. An experiment can be performed to verify this.

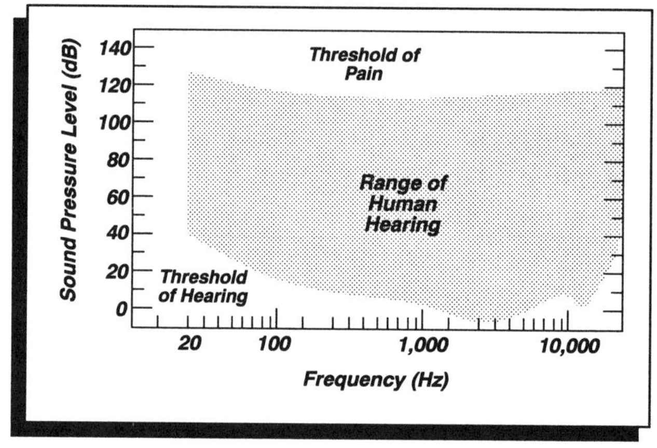

Figure 3.3 Response of human ears at the threshold of hearing.

Let us produce a sound of a frequency of 1,000 Hz at the threshold of hearing, this threshold means that on the average, most people would just barely hear this sound. This experimental data point is plotted on a graph of sound intensity needed to be heard as a function of frequency, Figure 3.3. When the frequency is decreased we observe that the intensity has to be increased by so many dB in order for the lower frequency sound to be heard as well as the 1,000 Hz sound. By continuing the experiment for frequencies below and above 1,000 Hz, we get a curve as in Figure 3.3. This important curve shows how much sound we need to hear frequencies from 20 Hz to 20,000 Hz at the threshold of hearing. The curve is not flat; we need more sound at low and high frequencies to hear as well as frequencies in the range of 1 kHz to 5 kHz.

The experiment is continued for different levels above the threshold of hearing. For example, the sound level at 1,000 Hz is set at 10dB above the threshold of hearing and the experiment is performed relative to this level. Here as well, we need more sound at low and high frequencies to hear just as well as the level at 1,000 Hz. A curve is plotted for this set of measurements. Then the 1,000 Hz is set to 20dB and the experiment is repeated. Measurements are made for levels up to the threshold of pain. We obtain a series of curves, shown in Figure 3.4. They are known as the Fletcher–Munson curves. Each curve corresponds to an equal level of loudness and it shows how much more power is needed at all the frequencies, relative to the 1,000 Hz level, to be heard as well as at 1,000 Hz.

This figure shows that at very low sound levels, we do not hear well at low and at high frequencies. As the sound level is raised the response of our ears becomes flatter. At high intensity levels we hear all the frequencies almost equally. We conclude from Figure 3.4 that:

— the frequency response of human ears is not flat at low levels

— the frequency response of human ears is almost flat at high levels

Figure 3.4 raises two questions: 1) Why do our ears behave like this, and 2) What can we do about this when we listen to music at low levels?

Let us address the first question.

At around 1,000 Hz to 5,000 Hz, we hear very well. The explanation for this is that perhaps there is a resonance of our ears. Let us approximate our outer ear by a tube closed at one end, as in Figure 3.5. It may be a poor approximation, but at least it gives us a start for a simple calculation. A standing

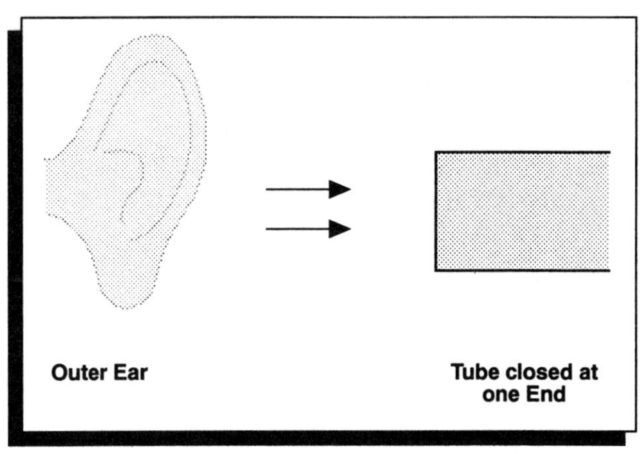

Figure 3.5 Outer ear approximated by a tube closed at one end.

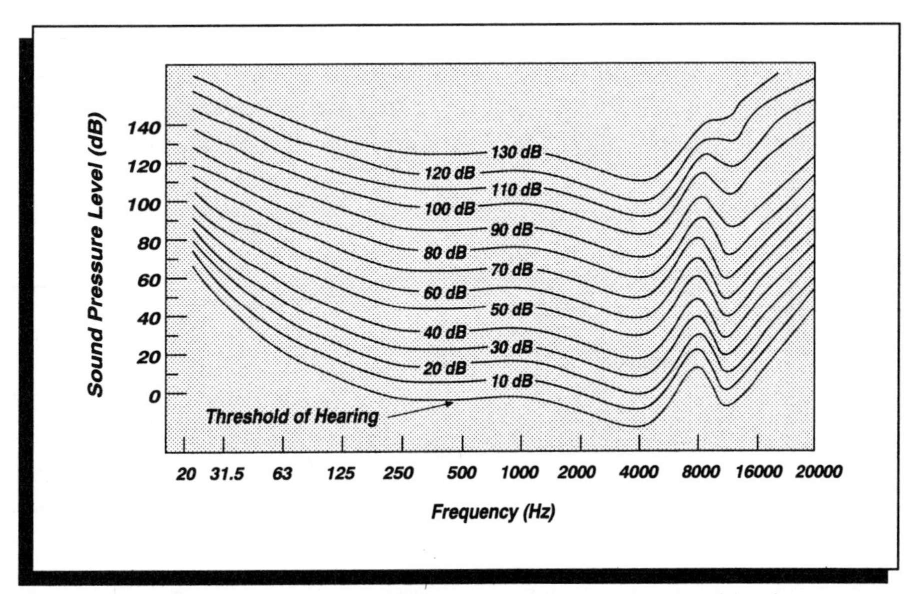

Figure 3.4 Response of human ears for various sound levels:

wave can be set up in the tube, used as a model of the ear. For a tube closed at one end, when the tube is 3 cm long (corresponding to an estimate of the length of our outer ear), we have a standing wave, which corresponds to a ¼ wave. Hence,

$$\begin{aligned} \tfrac{1}{4}\text{ wave} &= 3.0\text{ cm} \\ 1\text{ wave} &= 12.0\text{ cm} \\ \text{frequency} &= \text{speed/wavelength} \\ &= 344\text{ m/sec}/0.12\text{ m} \\ &\approx 3{,}000\text{ Hz} \end{aligned}$$

This simple calculation shows that a standing wave resonance could explain the high sensitivity in this frequency range.

— At low frequencies, we do not hear well at low levels. This may be due, in part, to the fact that many of our body functions (such as a growling stomach after a meal) which are at low frequencies, could be heard and this would bother us. Hence, we have evolved to not hear low frequency sounds.

— At high frequencies, our ears have problems in responding to such rapid variations. It is asking a lot to expect our ear drums to vibrate at 20,000 times/sec (they can barely do it!)

Let us consider the second question, that of listening to music at low level.

There is an interesting control in the pre–amp section which takes care of problems that arise from our poor ear response at low levels. It is the loudness switch, which electronically makes up for such losses. Its function is:

***Loudness switch* = this raises the level of low frequency signals and high frequency signals at low sound levels.**

It is a clever circuit that automatically adjusts to the Fletcher–Munson curve, of course being ineffective at high sound levels, since the response of the ear is essentially flat then. It is useful when we listen to music played on the hi–fi for low levels.

3.4 FREQUENCY RESPONSE

From our definition of hi–fi, we expect that all frequencies in the audio range coming in are treated equally in a hi–fi system. It is possible to make a test to check this for any unit or component in the hi–fi system. The response of interest is by how much the output varies in amplitude with frequency when signals of constant amplitude are fed into the system. This is known as the frequency response of the system. To perform such a test, signals of constant amplitude over the frequency range of 20 – 20,000 Hz are fed into the system and the output is monitored with a decibel meter or more sophisticated equipment. Figure 3.6 shows how this would be done for the frequency response of a speaker.

An audio signal generator puts out signals at selected frequencies; the output amplitude of this unit is kept constant while it is fed to the auxiliary

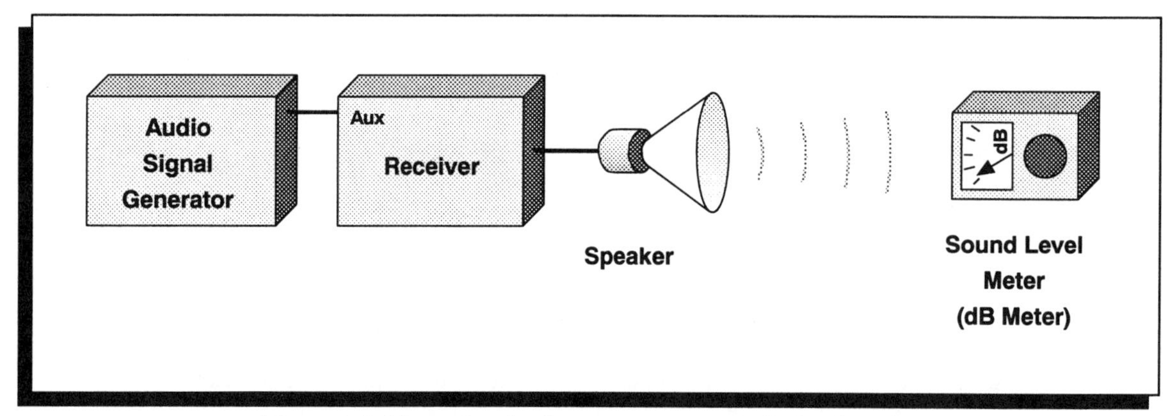

Figure 3.6 Measuring the frequency response of a loudspeaker.

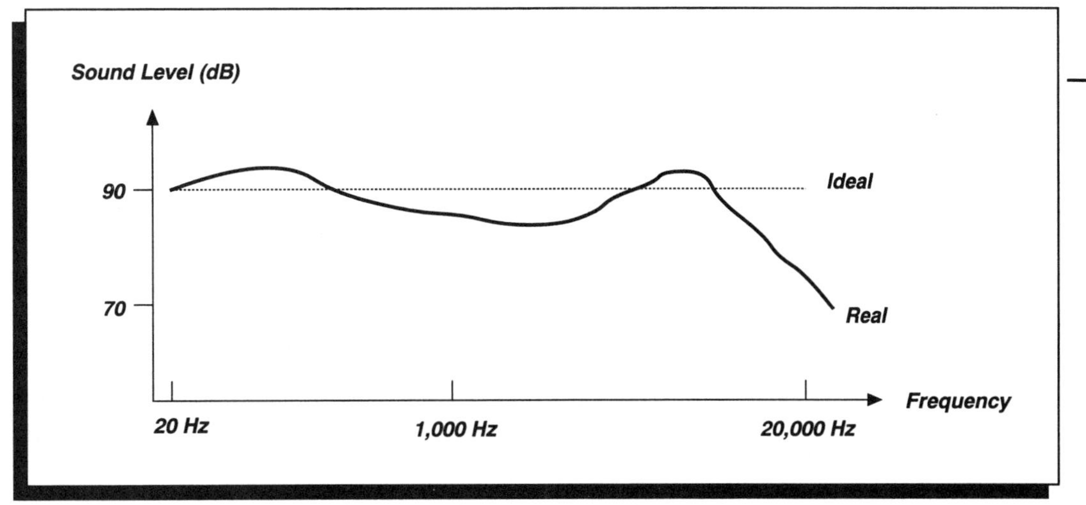

Figure 3.7 Frequency response of a speaker.

input of a receiver. The audio signal generator is a unit that puts out a wide range of frequencies in the audio range, with different waveforms, the simplest being a sine wave. It is used for testing. The speaker is driven by the receiver and a sound level meter is used to monitor the output level. A graph of the speaker output as a function of frequency is shown in Figure 3.7.

No speaker is perfect, and hence the frequency response in Figure 3.7 shows some variation. From this measurement the speaker performance could be given as:

frequency response **= 20 Hz to 20,000 Hz ± 2dB**

The ± 2dB refers to how much the response varies from the average value for this particular speaker. Ideally the ± dB should be zero or as small as possible. Here its maximum variation above the average value is +2dB and below it is –2dB.

The frequency response of the speaker is very important as it provides an indication of how well the loudspeaker can reproduce each frequency relative to all the others. The speaker should have no preference for any frequency and the frequency response ideally should be flat.

Similar tests can be made on the other components in the hi–fi; the receiver, the CD player, the tape deck, etc. Although many units perform over a wide range of frequencies, we are interested in the range of 20 Hz to 20,000 Hz, since that is the range we can hear well.

3.5 DYNAMIC RANGE

Our interest in music does not deal only with sound frequencies, it also deals with the various levels of intensity of the sound. Music played for a long time at the same intensity, whether high or low, can become boring, and hence music varies from very quiet passages to extremely loud passages. The range of sound levels that we hear or that can be recorded and reproduced is known as dynamic range. Its definition is:

dynamic range **=**
power of loudest sound of interest
power of quietest sound of interest

Since this is a comparison and it covers a wide range, it is convenient to express the dynamic range in decibels (dB).

For our ears, the dynamic range is about 130dB, as it starts at the threshold of hearing and goes up to the threshold of pain, which is 130dB higher.

For instruments, the dynamic range would correspond to the comparison of the power of the loudest sound it can produce to the power of the quietest sound. In recording, the range would cover the highest power of a wave recorded to the lowest power recorded. A few examples will now be presented to illustrate this point.

The dynamic range of intensity that a large orchestra can cover is above 60dB. This number is a comparison (i.e. a ratio of the intensity at full power to that of the faintest intensity the orchestra can produce) . This is a large number corresponding to a power ratio of one million.

In phono recording, the dynamic range is typically around 50dB corresponding to the loudest recorded passage compared to that of the quietest passage. We can now make an estimate for this range of how much variation there will be in the amplitudes of the recorded audio information. Remembering that power varies as the amplitude squared, a dynamic range of 40dB would mean a power ratio of 10^4, which corresponds to an amplitude ratio of 10^2. For this range the loudest passages on a phono record have amplitudes which are 100 times that of the quietest passage.

On a CD, since it is recorded digitally, the dynamic range is over 90dB and can approach 100dB. The reason for this will be discussed later, but we can apply simple calculations to find out what this means. Assume the dynamic range is 100dB. What does it mean? A 100dB range corresponds to the power level differences between the quietest sound and the loudest one; it means that the power ratio is 10^{10}. Because power varies with amplitude squared, the ratio of amplitudes (i.e. the amplitude of the loudest sound to the quietest one) would in this case be a factor of 100,000. This is a substantial improvement over the phono record dynamic range, which is 50dB. This is one of the reasons why the digital methods are superior to the analog methods of recording and playback.

SUMMARY OF TERMS

Audio Signal Generator: electronic test unit putting out waveforms in the audio frequency range, the simplest one being a sine wave.

Decibel (dB): quantity used to describe the comparison of two powers or two intensities. It has no unit since it is a comparison. One decibel is approximately the smallest change in loudness level which can be heard.

Dynamic Range: comparison of largest signal to weakest signal that a system can respond to or store.

Fletcher–Munson Curve: frequency response of the human ear at various levels of loudness.

Frequency Response: behavior of a system at different frequencies.

LOUDNESS switch: control in pre–amplifier which makes up for the deficiencies of our ears at low levels by increasing the level of the low and high frequencies.

Threshold of Hearing: level at which the quietest sound can be heard. It is called the 0dB level and corresponds to 10^{-12} watts/m^2 at 1kHz. It varies with frequency.

NAME ______________________________ DATE ____________

QUESTIONS FOR REVIEW

1 Why is it necessary to have a decibel scale?

2 What does frequency response mean? How would you measure it for a tape deck?

3 What is a LOUDNESS switch and when is it used?

4 Is the frequency response of the human ears flat? Why?

5 What is the intensity of a sound that is 100dB above the threshold of hearing at 1kHz?

6 How many times louder sound can a CD produce compared to a phono record player?

7 Is it possible to produce sound waves below the threshold of hearing?

8 Give a model for why the human ears are very sensitive around 3kHz.

9 To produce sound two times louder, how many times more power is necessary? How many more decibels are needed?

10 If a sound level is decreased by 20dB, how many times smaller will the sound amplitude be?

NAME ______________________________ **DATE** ______________

Please select the best answer.

1. The threshold of pain at 1kHz is __________ above the threshold of hearing.
   ___A. 0dB
   ___B. 10dB
   ___C. 80dB
   ___D. 130dB
   ___E. 150dB

2. When a sound level is increased from 60dB to 100dB, what sort of sensation does it create?
   ___A. The sound is 40 times louder
   ___B. The sound is 16 times louder
   ___C. The sound is 100 times louder
   ___D. The sound is 10,000 times louder
   ___E. The sound is 10 times louder

3. Approximately what is the sound level of two people talking?
   ___A. 10dB
   ___B. 30dB
   ___C. 60dB
   ___D. 90dB
   ___E. 120dB

4. A speaker is receiving 0.5 watt. In order to produce sound two times louder, the speaker needs to receive:
   ___A. 1.0 watt
   ___B. 2.0 watts
   ___C. 10 watts
   ___D. 5 watts
   ___E. 50 watts

5. When a sound level is increased from 50dB to 100dB, it is made:
   ___A. 2 times louder
   ___B. 32 times louder
   ___C. 50 times louder
   ___D. 16 times louder
   ___E. 64 times louder

6. When a sound level is changed from 70dB to 50dB, the amplitude of the sound is:
   ___A. 10 times less
   ___B. 20 times less
   ___C. 100 times less
   ___D. 4 times less
   ___E. 2 times less

7. The LOUDNESS switch in a pre–amplifier:
   ___A. makes all sounds louder
   ___B. increases loudness of all frequencies at low volume levels
   ___C. controls the volume in the pre–amplifier
   ___D. increases the loudness of only the low and high frequencies at low volume levels
   ___E. increases the loudness of mid–frequencies only at low volume levels

8. A 100–watt amplifier when compared to a 50–watt amplifier is:
   ___A. only slightly louder
   ___B. 2 times louder
   ___C. 50 times louder
   ___D. 32 times louder
   ___E. 20 times louder

9. Phono record player X has a dynamic range of 50dB while phono record player Y has a dynamic range of 60dB. This means that the loudest sound on Y will be:
___A. two times louder than on X
___B. 10 times louder than on X
___C. slightly louder than on X
___D. 64 times louder than on X
___E. 1,000,000 times louder than on X

10. Please rank the frequency response of the four amplifiers, from best to worst.
X: 55Hz to 15,000Hz, +4dB, –2dB
Y: 20Hz to 20,000Hz, +0.4dB, –0.6dB
Z: 10Hz to 20,000Hz, +5dB, –4dB
W: 20Hz to 21,000Hz, +0.2dB, –0.3dB

___A. X, Y, Z, W
___B. W, Y, Z, X
___C. W, Z, Y, X
___D. Z, W, Y, X
___E. W, Y, X, Z

11. On the threshold of hearing curve, how many more decibels are required for a 100Hz sound to be heard as well as a 1.0 kHz sound?
___A. 40dB
___B. 0dB
___C. 10dB
___D. 16dB
___E. 20dB

12. How does a LOUDNESS switch differ from a VOLUME control?
___A. both are the same.
___B. LOUDNESS is in dB while VOLUME is in watts.
___C. LOUDNESS switch controls certain frequencies by certain amounts while VOLUME switch controls all the frequencies by any amount.
___D. LOUDNESS switch makes the sound 2 times louder, 4 times louder, etc.
___E. they do not differ since they are both on all the time.

13. If speaker X puts out 5×10^{-8} watts/m^2 of sound intensity, and speaker Y also puts out the same, they differ in loudness level by:
___A. 0dB
___B. 1dB
___C. 3dB
___D. 10^{+16} watts/m^2
___E. 5dB

14. The smallest noticeable change in the loudness level is:
___A. 0dB
___B. 1dB
___C. 3dB
___D. 0.5dB
___E. 10dB

15. When a sound is made 128 times louder, this corresponds to a decibel increase by:
___A. 70dB
___B. 128dB
___C. 7dB
___D. 60dB
___E. 80dB

16. An amplifier puts out 1.5 watts. When the power is reduced by 30dB, the new power will be:
___A. 1.5 watts/30 = 0.05 watt
___B. 1.5 watts/8 = 0.1875
___C. 1.5 watts/1,000 = 0.0015 watt
___D. 1.5 watt – 30 = –28.5 watt
___E. 1.5 watt × 30 = 45 watts

17. A system is capable of recording audio information with amplitudes ranging from 1 to 100,000. Its dynamic range will be:
___A. 10^5dB
___B. 50dB
___C. 100dB
___D. 2 × 50dB
___E. 100,000dB

NAME ______________________________ **DATE** __________

18. The LOUDNESS switch is not necessary at high volume levels because:
   ___A. the frequency response of the human ears is almost flat then
   ___B. the volume control can be used to get good frequency response
   ___C. reverberation effects of a room help to boost the sound level
   ___D. the signals are already large
   ___E. many harmonics around 3kHz are produced then

19. In general, why do human ears not respond to frequencies above 20,000 Hz?
   ___A. waves are too short
   ___B. ear drums cannot move that many times per second
   ___C. waves are too long
   ___D. the power level is too low
   ___E. the period is so short that the sound cannot penetrate the ears

20. The maximum dynamic range of a violin is 40dB. This means that violin sound have amplitudes which range from 1 to ________ .
   ___A. 40
   ___B. 100
   ___C. 10,000
   ___D. 16
   ___E. 4

CHAPTER 4 LOUDSPEAKERS

Probably one of the most demanding tasks in the hi–fi is performed by the loudspeaker. Essentially, its goal is to take electrical signals from the amplifier and convert them to sound waves. These waves must sound exactly like the original, and they must create the same acoustical sensations as the original. That is such a difficult goal to achieve that all sorts of approaches have been tried. In fact, there are so many designs in loudspeakers that only the main types will be presented here. This chapter deals with speaker characteristics, cone speakers, enclosures for speakers, horns, loudspeaker placement, and a few representative loudspeaker systems. We will extensively make use of the first three chapters in developing the subject of loudspeakers. Can a loudspeaker really reproduce all the complex waves of each instrument in an orchestra? This chapter shows what are the expectations and what methods are used to achieve that goal.

The principal goal of a loudspeaker is to reproduce all the frequencies that have been recorded (or broadcast) with high–fidelity. Figure 4.1 shows the role of the loudspeaker. An electrical signal representing a sound wave causes the speaker to move back and forth according to the electrical signal. This motion creates pressure changes and hence, sound. The ability of a loudspeaker to accurately reproduce the electrical waveforms into sound waves is a formidable task.

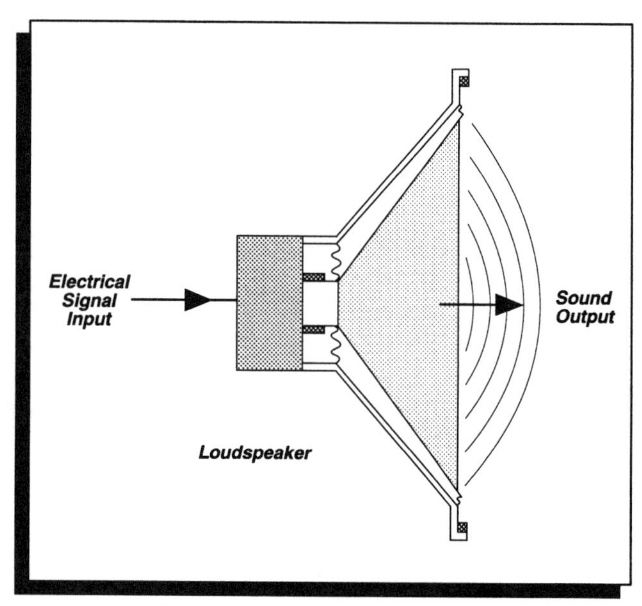

Figure 4.1 Role of loudspeaker.

4.1 CHARACTERISTICS OF SPEAKERS

Let us start with a sound tone. At some instant its spectrum could be as shown in Figure 4.2. If the frequency response of the speaker were as shown in the middle of Figure 4.2, then the sound produced will have reinforcement of the 6th and 7th harmonics; this did not exist in the original signal. It must be avoided in order to preserve high–fidelity of the original sound. Such a change of the signal is known as distortion. Here its meaning is:

Distortion **= modification of original sound**

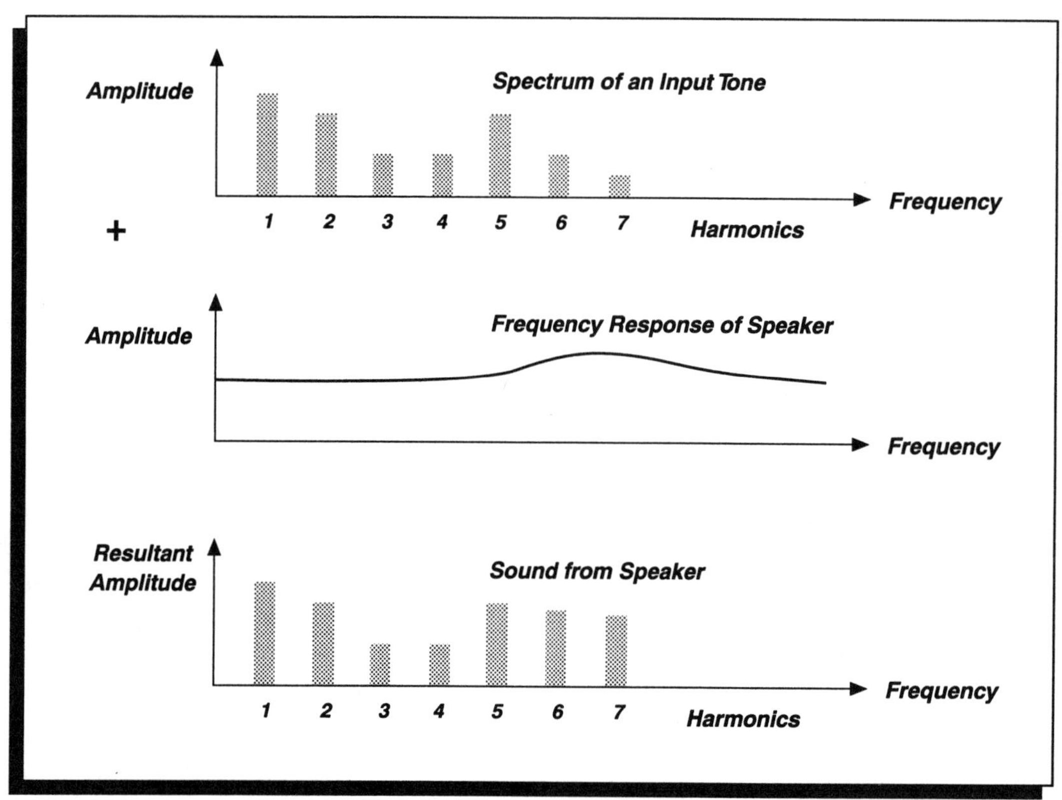

Figure 4.2 Distortion of spectrum of original waveform by non-flat frequency response of loudspeaker.

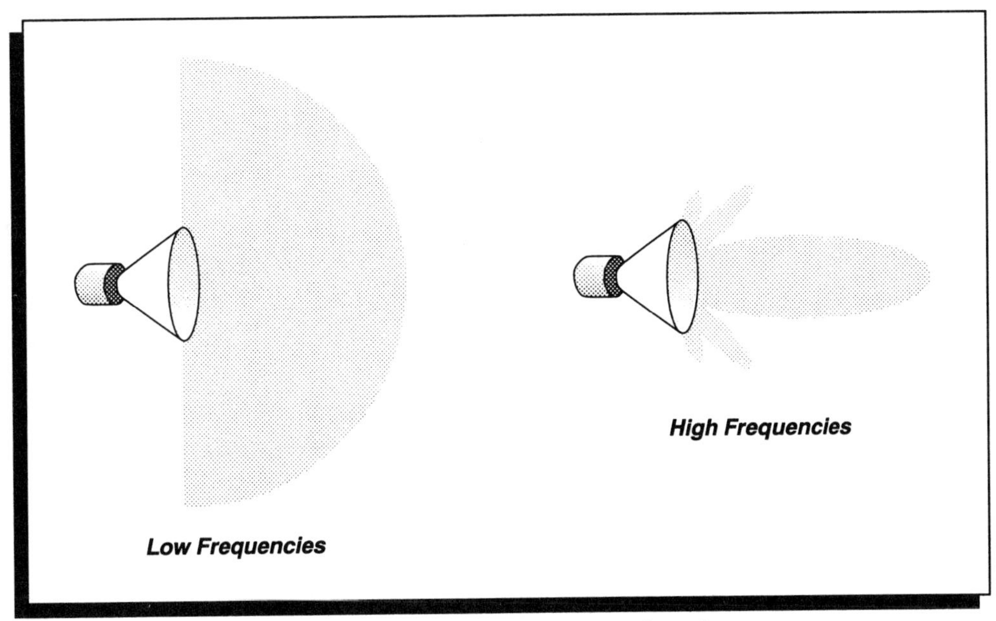

Figure 4.3 Dispersion properties of speakers.

The frequency response characteristics of a loudspeaker must be flat; they must not distort the waveform that is being reproduced.

■ DISPERSION OF SPEAKERS

In the second chapter a brief mention was made of the fact that at low frequencies a speaker disperses sound in all directions, while at high frequencies it beams mainly forward. This is due to the diffraction of the loudspeaker. Figure 4.3 shows these extremes. In designing loudspeakers it is important to know and understand such behavior. What causes such dispersion? Consider the speaker in Figure 4.4 which is driven at low frequencies, say 20 Hz. The speaker size is typically 0.3 m in diameter. Let us look at waves 2 and 1 originating at

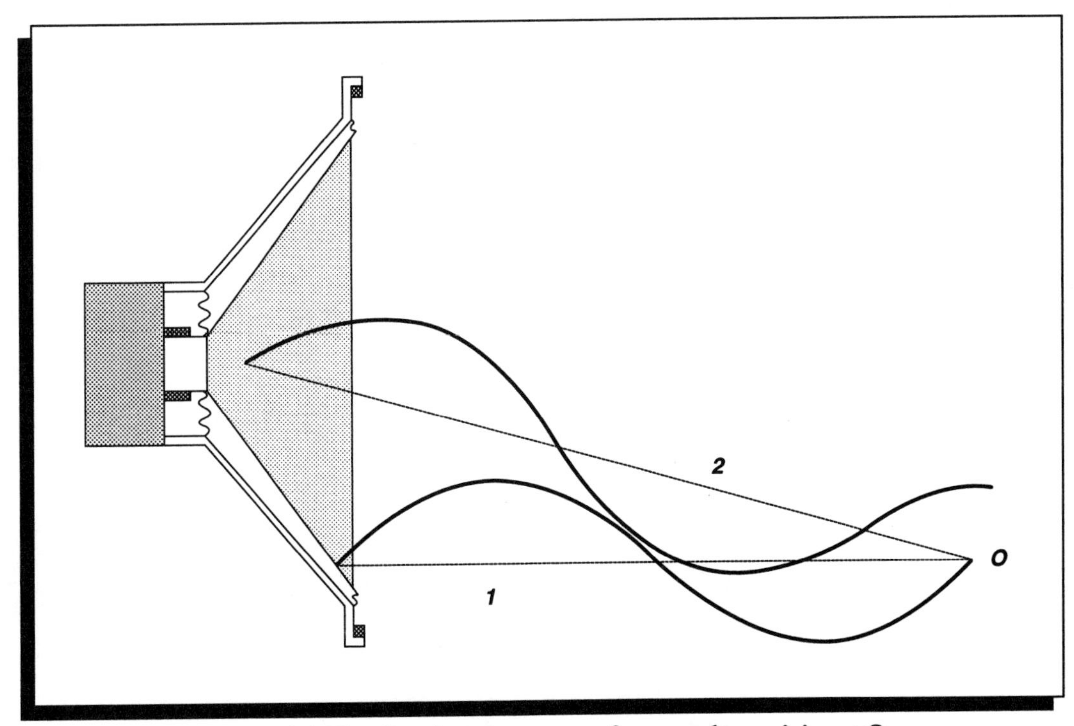

Figure 4.4 Two low frequency waves from speaker arriving at O.

two the extremes, the center of the speaker and one edge. They arrive at position O. At 20 Hz, the wavelength is given by:

$$\begin{aligned} \text{wavelength} &= \text{speed/frequency} \\ &= 344 \text{ m/sec} / 20 \text{ Hz} \\ &= 17.2 \text{ m} \end{aligned}$$

This wavelength is much larger than the diameter of the speaker. Is it possible that the two waves considered arrive out of phase at O? Analyzing this carefully it is clear that since the wavelength of 17.2 meters is much larger than the speaker diameter (here 0.3 meter), the path difference between 2 and 1 is very small, certainly much less than a wave or $\frac{1}{2}$ wave. Hence, for this case the two waves arrive almost in phase at O and there will be constructive interference.

Now let us analyze the behavior of the same speaker when driven at 1,000 Hz, as shown in Figure 4.5. Looking closely in detail at Figure 4.6 it is shown that the two waves can be out of phase. When the path difference between waves 2 and 1 is equal to $\frac{1}{2}$ wave there will be destructive interference between the two waves at point O. Hence:

$\frac{1}{2}$ wavelength ≈ $\frac{1}{2}$ diameter
→ destructive interference

This means that destructive interference will set in when:

1 wavelength ≈ diameter of speaker

At higher frequencies the waves get shorter and things will get worse. At some location off axis, the center wave will arrive out of phase with the wave from the edge. When all the waves from every part of the speaker are summed at some position in front of the speaker, we get the high frequency dispersion characteristics shown in Figure 4.7. The solution to the speaker dispersion problem is:

- use speaker until 1 wavelength ≈ diameter of speaker
- switch over then to smaller speaker
- when the condition of 1 wavelength ≈ diameter of speaker is reached, again go to an even smaller speaker .

Hence, the sound spectrum is divided into frequency ranges and each range is fed into the speaker drive in the loudspeaker system. Each speaker drive is called the driver. One way of doing this is by subdividing the loudspeaker system:

— low frequencies are sent to a large driver, or the woofer

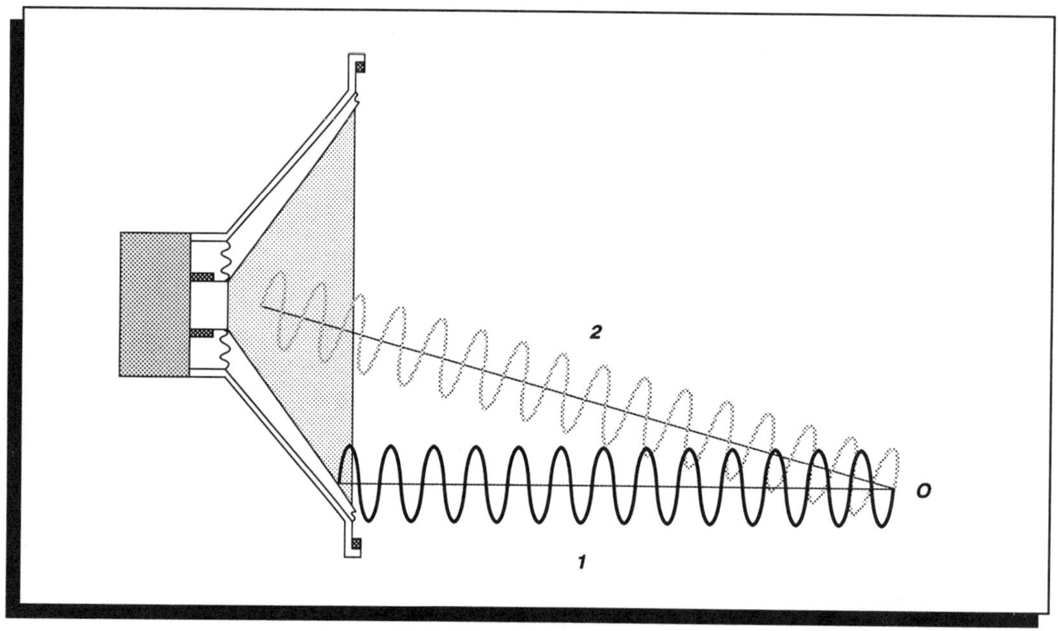

Figure 4.5 Two high frequency waves from speaker arriving at O.

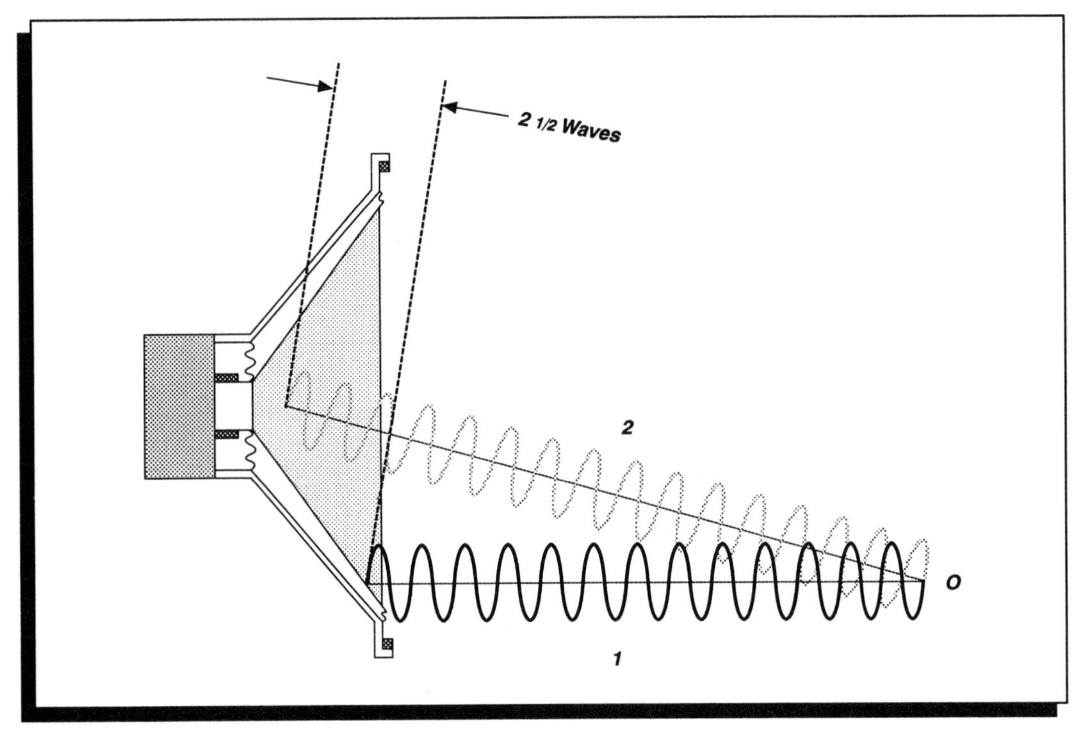

Figure 4.6 Details of waves 2 and 1 at high frequencies.

— mid frequencies are sent to small driver, or the midrange
— high frequencies are sent to very small driver, or the tweeter

This is a reasonable approach, because whenever the wavelength becomes comparable to the speaker diameter, dispersion effects start becoming important. One has to go to a smaller driver. When that driver shows dispersions, at high enough frequency, an even smaller driver has to be used, and so on. Figure 4.7 shows the sound patterns from a driver as the frequency is increased. There is no accepted rule for how many drivers there should be

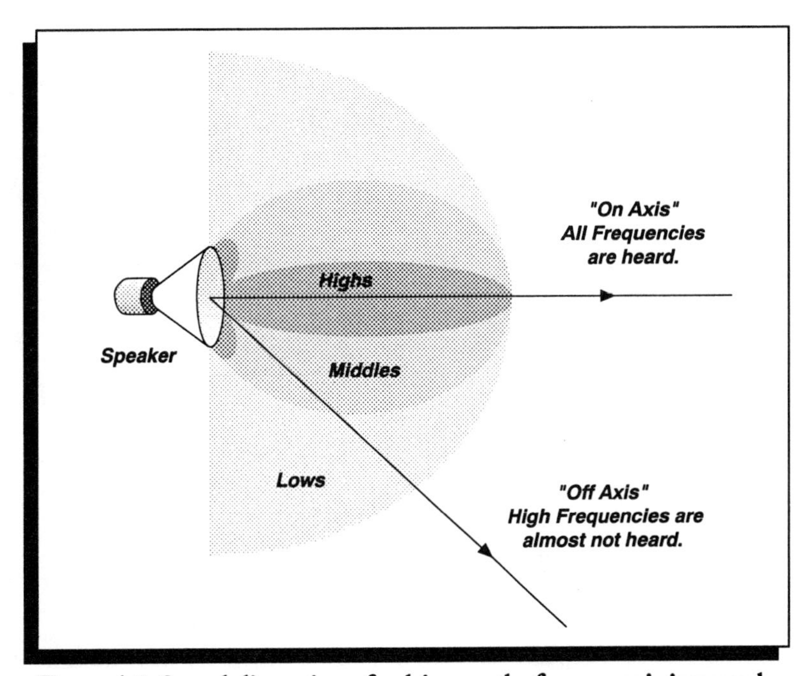

Figure 4.7 Sound dispersion of a driver as the frequency is increased.

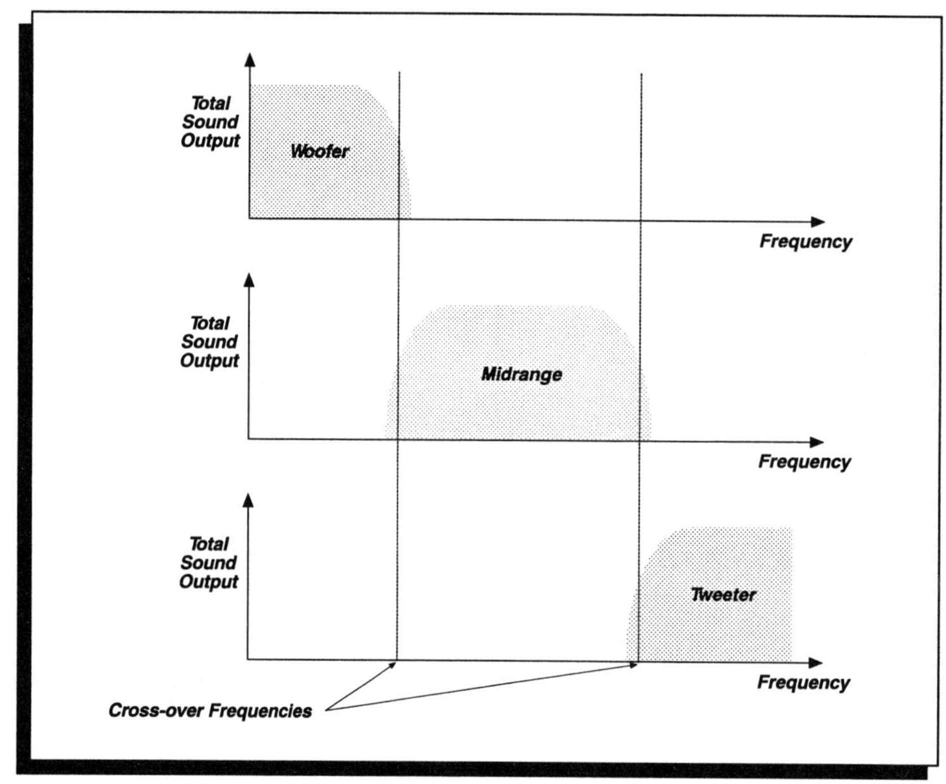

Figure 4.8 Division of audio spectrum for a three-way loudspeaker.

per loudspeaker. Many have three, as in the example above. Some have two, while others use four or even five drivers.

MULTIPLE DRIVERS

The dispersion problem of a driver is one of the reasons for using multiple drivers in a loudspeaker system. Other reasons are:

— to reduce Doppler effect (this was discussed in the previous chapter)

— it is easier to design loudspeakers when the frequency range of 20 – 20,000 Hz is divided into small sections instead of the wide range

The division of the whole audio frequency range has to be done very carefully depending on the drivers, their characteristics, and the design. For example, for division of the sound range into three drivers, we have a three–way system consisting of frequency ranges, shown in Figure 4.8. The result of such divisions is shown in Figure 4.9.

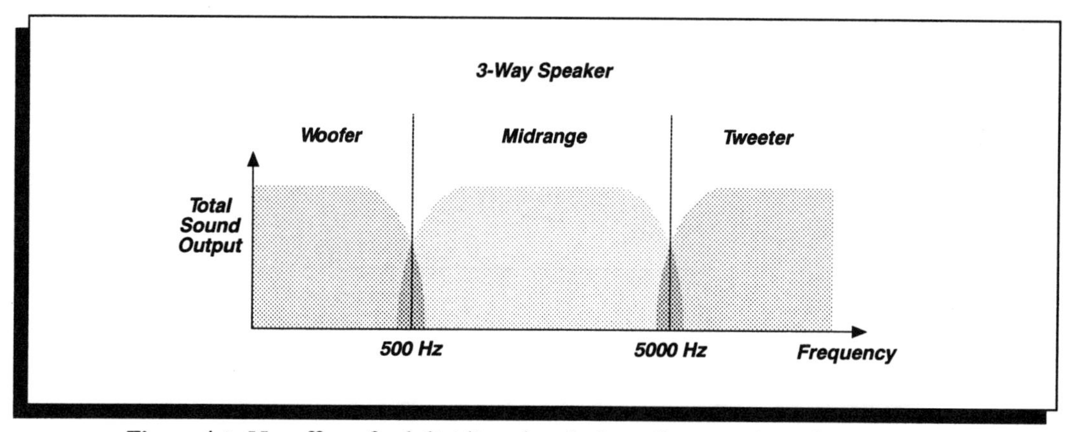

Figure 4.9 Net effect of subdividing the whole audio range into three sections.

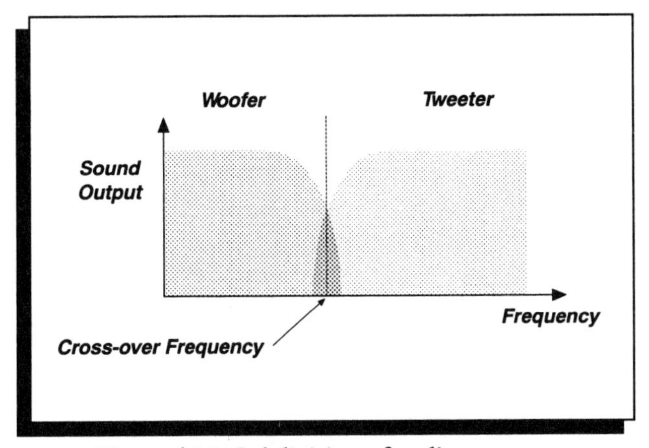

Figure 4.10 Subdivision of audio spectrum in a two-way system.

For a two–way system, there would be subdivision, as shown in Figure 4.10, into a woofer and a tweeter. The change–over from one driver to the next needs to be done smoothly, as the frequencies are changed. The frequency level at which one speaker decreases, while it increases in the next one, is called the cross–over frequency, as shown in Figure 4.10. The choice of this frequency depends on the drivers and the loudspeaker design. It has to be done in such a way that it is not noticeable. There should not be a void or intensity increase at the cross– over frequency.

So far the discussion has been on the subdivision of the audio frequency range into bands of frequencies that are sent to different drivers. This is achieved by a cross–over network, inside the loudspeaker and next to the drivers. It allows a certain range of frequencies to the woofer; then another network determines which frequencies will go to the midrange. The highest frequencies are then sent to the tweeter. The cross–over from one frequency range to the next must be done in a smooth and balanced manner which requires a special design.

■ VOLUME OF AIR MOVED

The amount of sound produced depends on the volume of air moved by the driver. Consider the driver shown in Figure 4.11. Assuming that the entire surface of the driver moves for case A as for case B, then the amount of sound produced will depend on the displacement of each driver. Hence, A will produce louder sound than B.

For two speakers of different sizes, the amount of sound produced will depend on the volume of

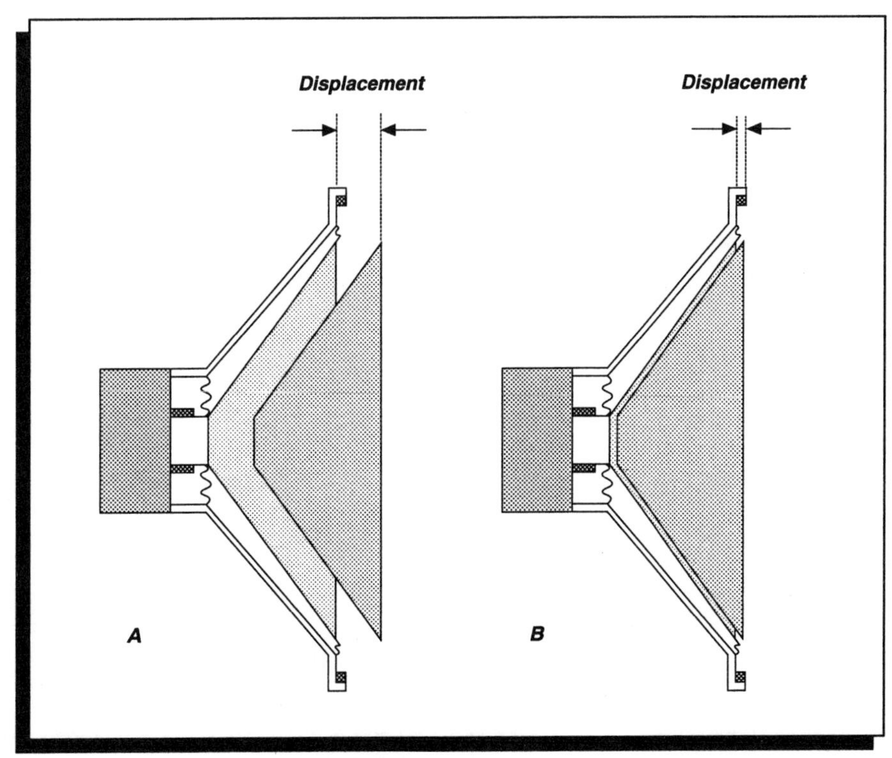

Figure 4.11 Amount of sound produced depends on volume displacement A is louder than B.

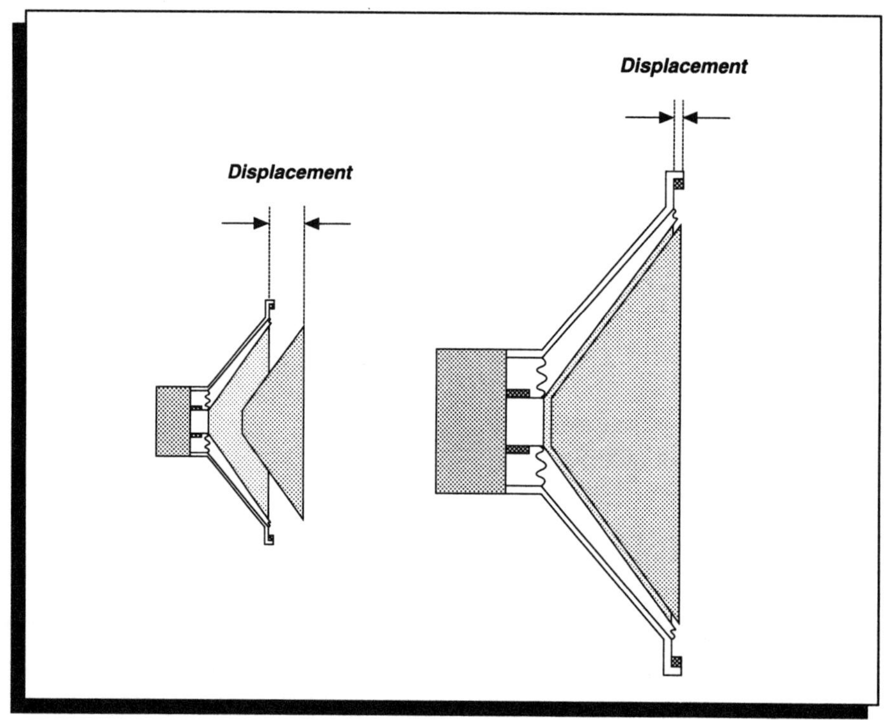

Figure 4.12 To produce the same amount of sound by both drivers at the same frequency the smaller one has to move through a larger distance than the bigger one.

air displaced. In Figure 4.12 the smaller speaker has to move through a larger distance than the larger one to produce the same amount of sound. For a driver whose radius is two times smaller than the larger one, it would have to move through a distance four times that of the larger one to produce the same amount of sound, at the same frequency.

The amount of sound produced by a speaker will depend very strongly on the frequency. Figure 4.13 shows the volume of air which needs to be

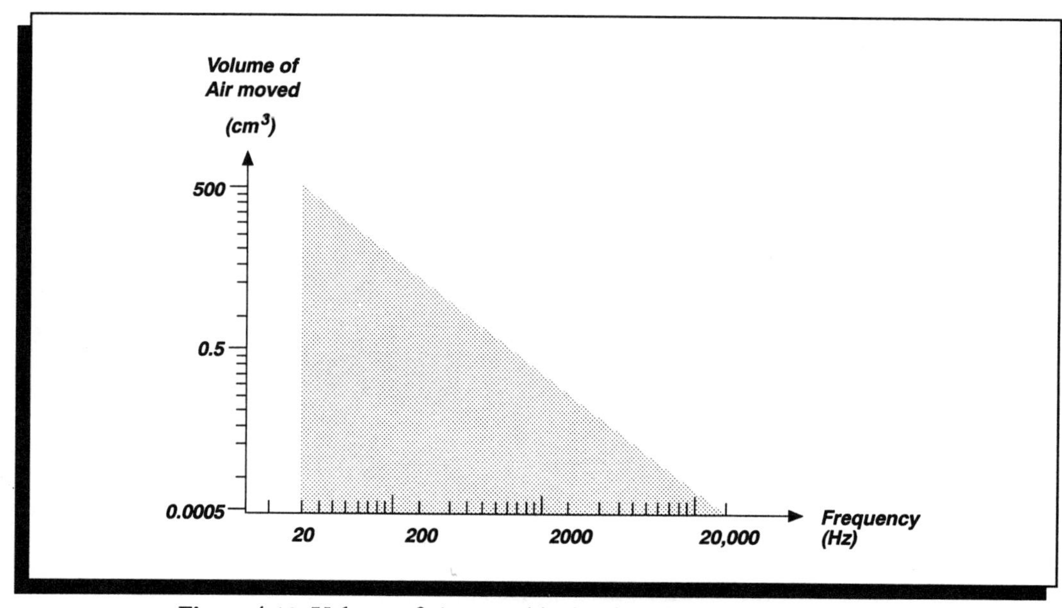

Figure 4.13 Volume of air moved by loudspeaker as a function of air and frequency, to produce the same level of sound.

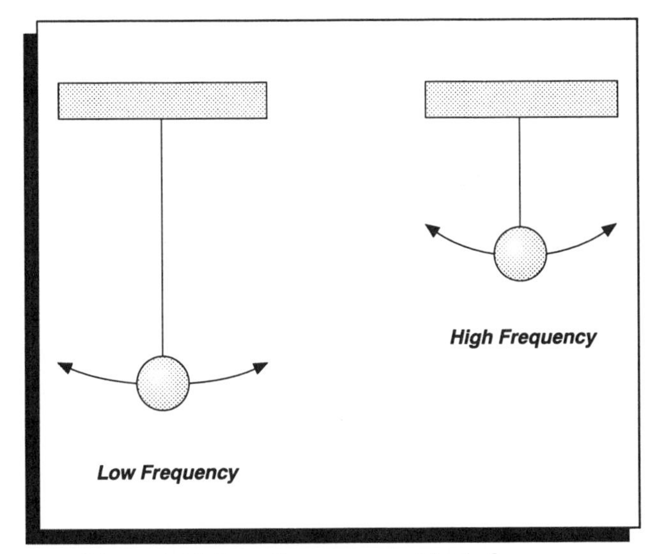

Figure 4.14 Low-frequency and high-frequency simple pendulums doing different amounts of work per second for same amplitude in displacement.

moved by a driver to produce sound of 0.01 watt at different frequencies. When a speaker moves 0.0005 cm^3 of air at 20 kHz, it must move 500 cm^3 of air at 20 Hz to produce the same amount of acoustic power. Hence, a low frequency speaker has to be large, while a high frequency speaker can be small and still produce a large amount of sound. The explanation for this can be illustrated by two simple pendulums. Figure 4.14 shows a low frequency pendulum (representing a low frequency driver) and a high frequency pendulum (representing a high frequency driver.) When both have the same amplitude, the high frequency one will do more work per second than the low frequency pendulum. If they are to do the same amount of work per second, the high frequency one does not have to move with such a large amplitude as it makes up for it by moving more times per second. Consequently low frequency speakers have to be large while the high frequency ones can be small (they also have to be small to move fast). This is an important question which tells us how effectively the speaker does its job.

EFFICIENCY

Since a driver converts audio electrical power to sound, its ability to do this will depend on the type of driver and how it is incorporated into a loudspeaker. Actually, the question is: how effective is the loudspeaker in converting power from an electrical source to acoustic waves? This question is important because it gives us an idea of where the input power goes and how effective different speakers are in converting power from the electrical to acoustic signals. In all drivers, a large fraction of the electrical power will be converted to heat dissipation. Fig. 4.15 illustrates the balance which will be maintained at all times in the energy conversion process by the speaker.

Its importance will become more evident when we compare the various speakers. An important definition which tells us how well the driver converts energy is:

Efficiency = **(acoustic power out / electrical power in) × 100**

It is expressed in percentage. Values of this quantity can typically vary from less than 1% to about 50%, depending on the type of speaker and its design. It is a quantity which is useful in the design of a hi–fi system, but it does not tell us about the quality of the speaker. A speaker can be of low efficiency, and yet it can produce sound of hi–fi quality. Many high–quality speakers are inefficient.

As an example, consider a loudspeaker receiving 80 watts of electrical power from a receiver and producing two watts of sound, as shown in Figure 4.16.

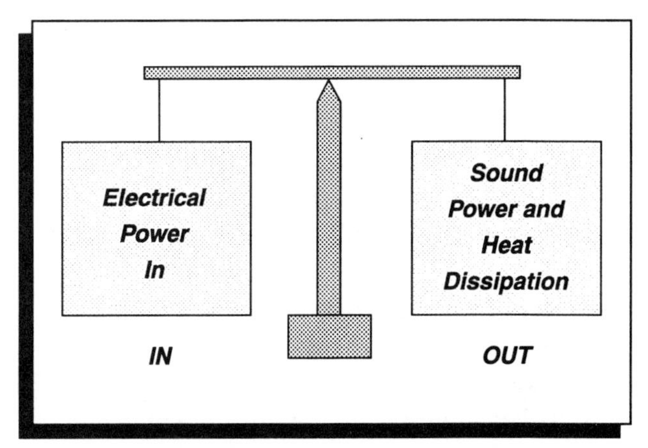

Figure 4.15 Balance between electrical power going to driver and the production of sound power and heat dissipation by driver.

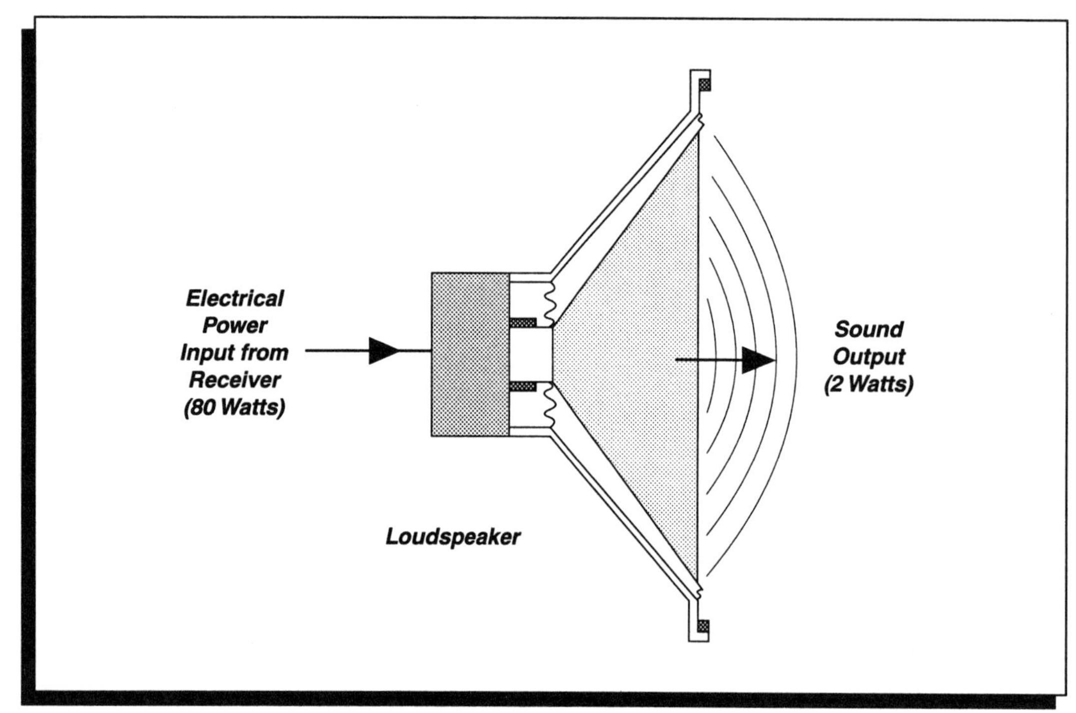

Figure 4.16 Example of a loudspeaker whose efficiency is less than 100%.

From the definition of efficiency:

$$\text{Efficiency} = \frac{2\text{watts}}{80\text{watts}} \times 100$$

$$= 2.5\%$$

What happens to the rest of the power (i.e. the 78 watts)? The answer is that it goes to heat dissipation (i.e. it is wasted). Typically the efficiency for a cone speaker is around 1%, while for horns it can be between 20–50%. The two types of speakers will be studied next.

4.2 CONE SPEAKERS

This type of speaker is sometimes referred to as a direct radiator. Its main parts are:

- — cone (diaphragm): creates the sound by a push–pull action on air pressure.
- — flexible edge or suspension.
- — basket: housing for supporting cone and the rest of driver.
- — voice coil: attached to cone and moves in a magnetic field of a permanent magnet. It receives electrical signals from the amplifier which drive the unit in push–pull action.
- — permanent magnet: clamped, it creates push–pull action on the voice coil.
- — spider: centers voice coil.

Figure 4.17 shows the basic cone speaker, referred to as a driver.

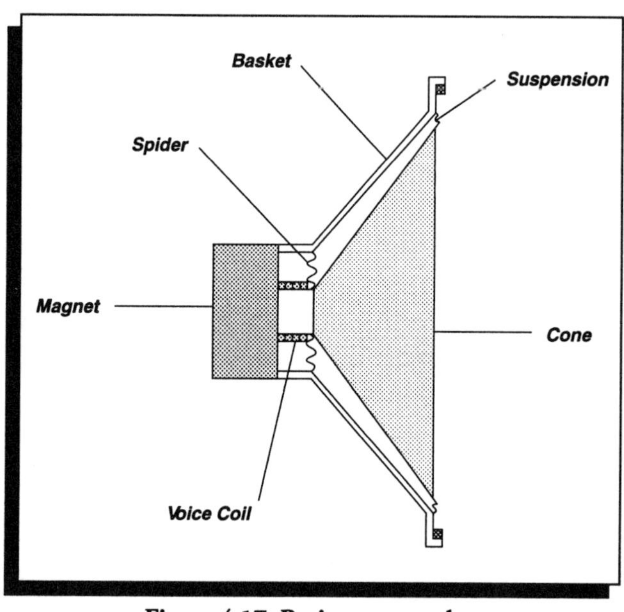

Figure 4.17 Basic cone speaker.

This type of speaker is popular because:
- — it is simple to construct and cheap to fabricate.
- — it is compact and hence it occupies a small space.
- — it is quite uniform in frequency response.
- — it is very useful for building a loudspeaker system.

Let us now study its mechanical behavior. Later on its electrical and magnetic performances will be considered.

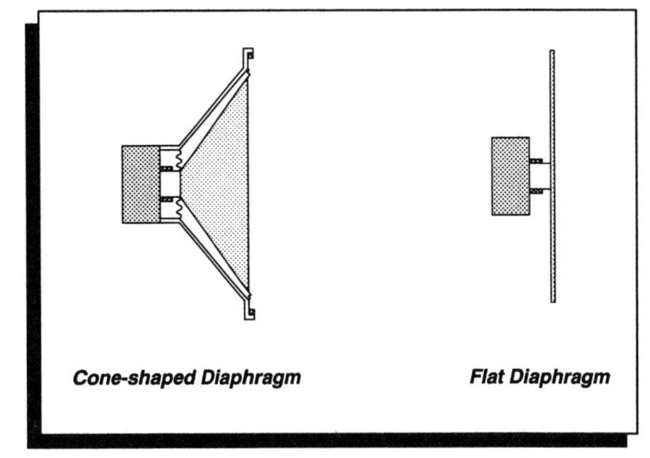

Figure 4.18 Comparison of cone-shape over flat shape for mechanical strength when thin material is used.

■ DIAPHRAGM BEHAVIOR

When the voice coil receives audio electrical signals it causes the diaphragm to move back and forth. As it vibrates back and forth, it forces the air in contact with it to vibrate in exactly the same manner, and thus sound waves are created. To be effective in doing this, the diaphragm must be strong and light.

Usually the diaphragm is formed into the shape of a cone; it will then exhibit the greatest strength when a force is applied to it by the voice coil. Figure 4.18 illustrates why the cone shape is preferred to a flat shape.

The diaphragm needs some sort of support which allows it to move back and forth, and this is provided by the flexible edge. At this point we have an oscillating system which, although complicated, can be modeled by a simple well–known system in physics, the mass–spring system. Figure 4.19 shows how cone behavior can be modeled by a spring–mass system.

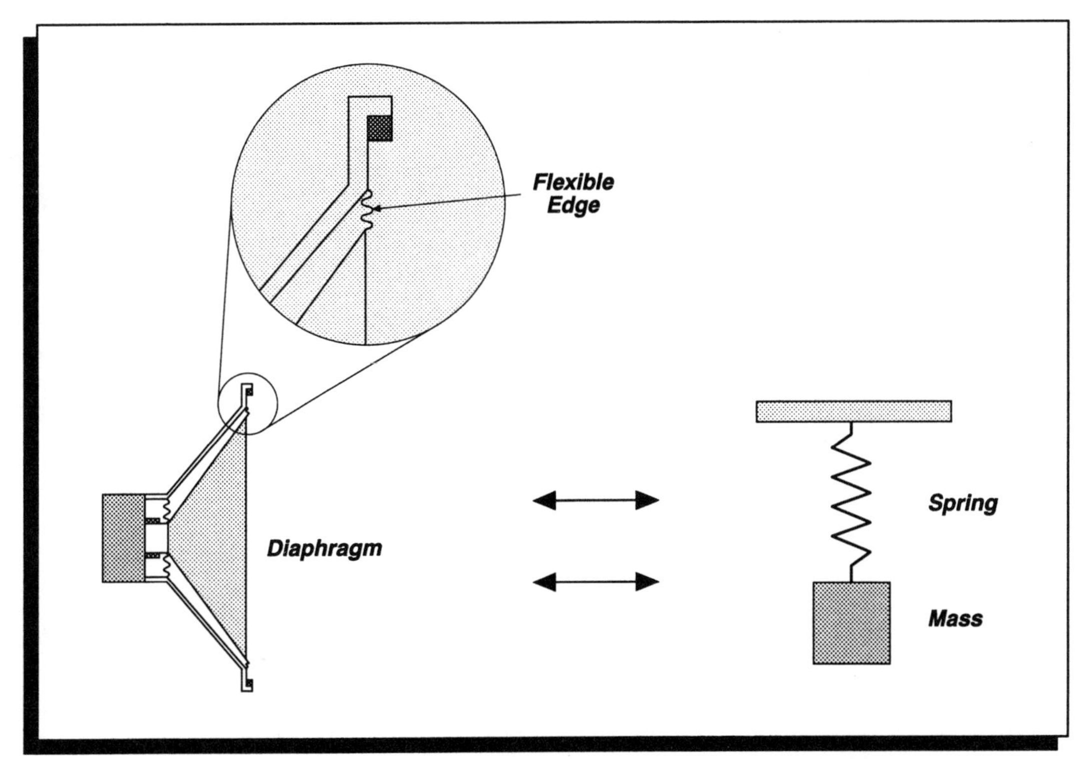

Figure 4.19 Modeling of diaphragm action by mass-spring oscillating system.

Let us study this model by comparing the two systems:

mass–spring system		**diaphragm of driver**
mass	↔	mass of diaphragm
spring	↔	flexible edge suspension

For simplicity only the mass of the diaphragm is considered as we have ignored the mass of air around it which also moves. There is a quantity which describes the spring action, the spring constant. It tells how much force is needed to produce a certain extension of the spring. The stiffer the spring, the larger the spring constant. For a diaphragm where the flexible edge suspension provides small stiffness but a lot of give, we define a new quantity which is the inverse of its stiffness (i.e. of the spring constant). This new quantity is:

***compliance* = displacement / force**

For speaker application, it is a very important quantity. It is a fixed quantity for each driver and it is given in units of cm/dyne, where dyne is a unit of force which is exerted by the coil on the diaphragm, and hence on the flexible edge. The resonant frequency can be determined quite simply with reasonable accuracy by following the modeling approach. For a mass–spring system the resonant frequency has been derived and it is given by:

$$\text{Resonant frequency} = \frac{1}{6.28}\sqrt{\frac{\text{spring constant}}{\text{mass}}}$$

where the mass is in grams and the spring constant in dynes/cm. Hence by comparison, the resonant frequency of the diaphragm in the driver will be:

$$\text{Resonant frequency} = \frac{1}{6.28}\sqrt{\frac{1}{\text{compliance} \times \text{mass}}}$$

where mass is the mass of the diaphragm in grams, and the compliance is in cm/dyne.

A large compliance (i.e. a loose suspension) will lead to a low resonant frequency. A small compliance will cause a high resonant frequency as the diaphragm will move quickly. Also, when the diaphragm is light its resonant frequency will be high, since it can move fast.

This is a useful expression, as it tells us approximately what the resonant frequency of the diaphragm will be and that is the main resonance of the driver.

Measurement of the frequency response will show the main resonance, but as the frequency is increased, there are other peaks and dips. Their origin is in the standing waves set up on the diaphragm itself. It is a deformed plate and standing waves can be set up on it as if they were on plates. Figure 4.20 and Figure 4.21 illustrate two of the many standing waves on a diaphragm (due to reflections off its edges). One shows the standing wave pattern for the up and down motion on the cone, while the other one shows a standing wave pattern around the circumference of the cone.

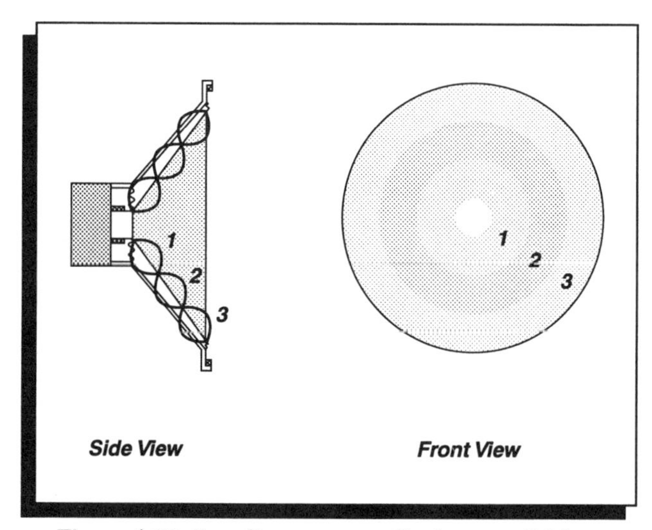

Figure 4.20 Standing wave on diaphragm of driver.

Of course many more patterns can be generated on the diaphragm. These standing waves are responsible for the small resonances in the frequency response shown in Figure 4.22. The standing waves on the cone are known as cone "break–up" because of the unpleasant noise they produce. These can be reduced by: the flexible edge which absorbs some of the energy, thus reducing the

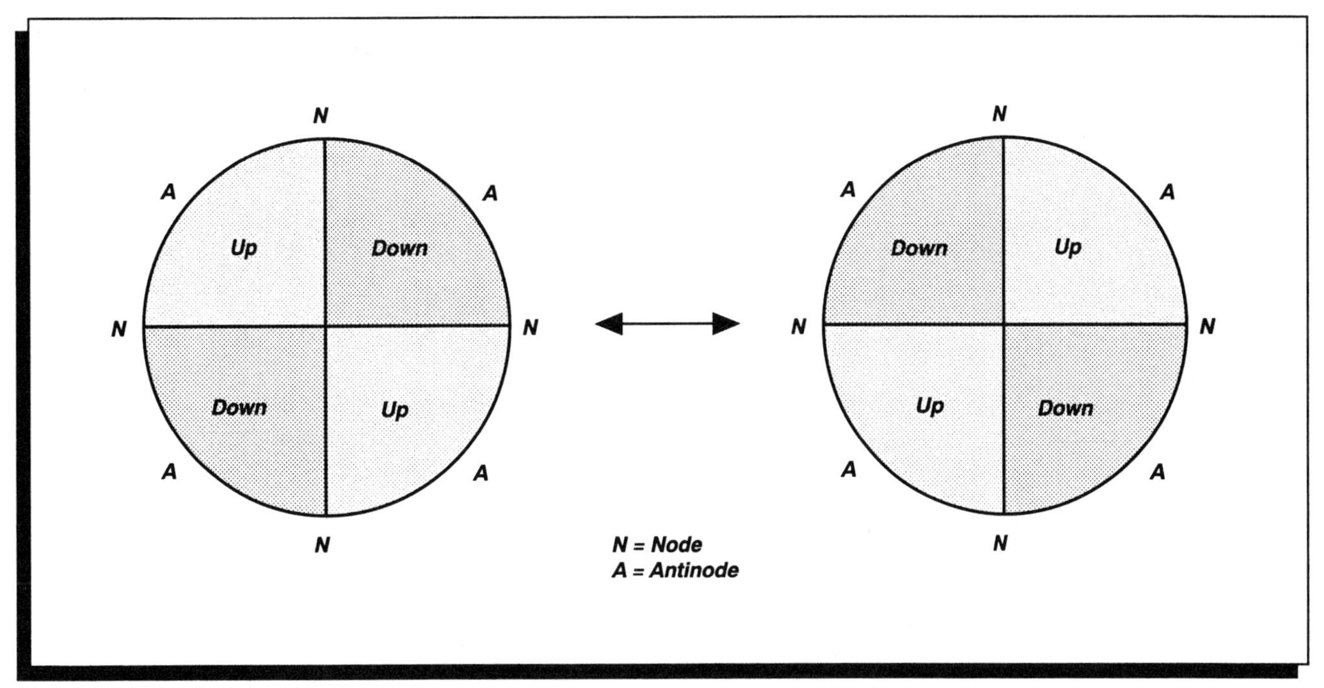

Figure 4.21 Standing wave around the rim of a diaphragm.

amount of reflected wave along the cone. Sometimes it is reached by placing ribs on the cone to stiffen it. In spite of such precautions, there will still be some standing waves remaining.

In analyzing the cone driver, there is another important characteristic that needs to be considered, the subject of baffles.

BAFFLE PROBLEM

When the motion of a cone speaker is analyzed its behavior can be strongly affected by how it is mounted. Consider the cone driver in Figure 4.23 producing sound at low frequencies. The front sound travels to the listener, and the sound from the rear sneaks around the driver and also goes to

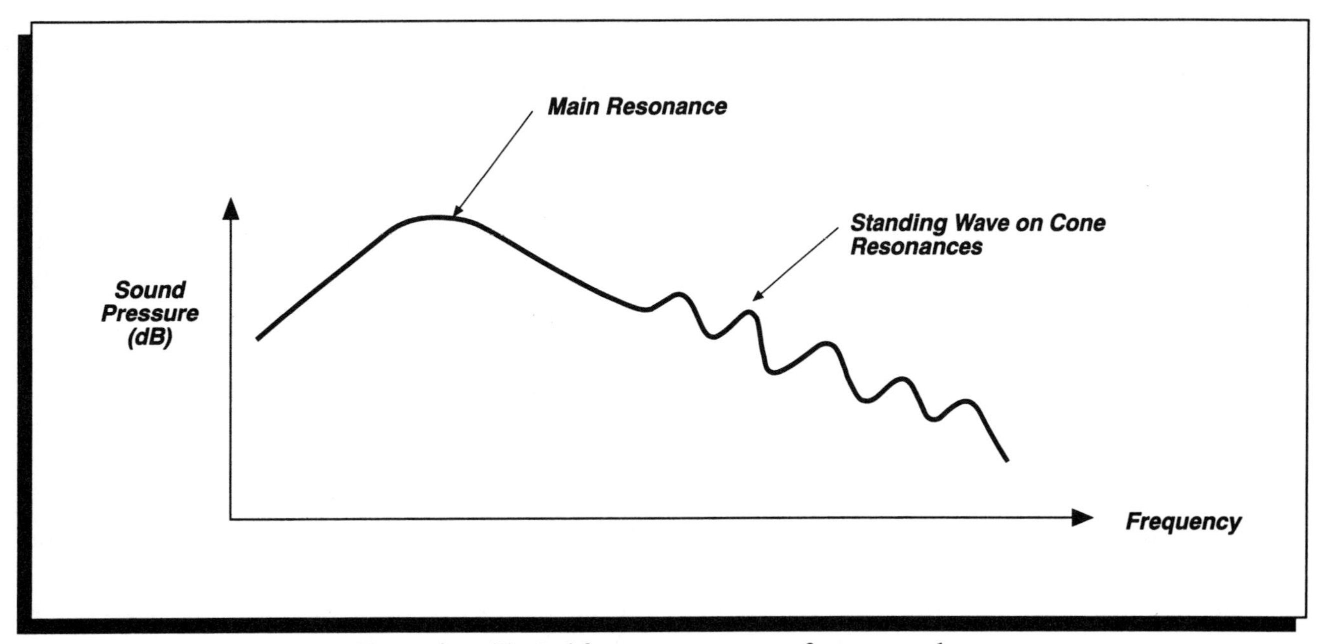

Figure 4.22 Typical frequency response of a cone speaker.

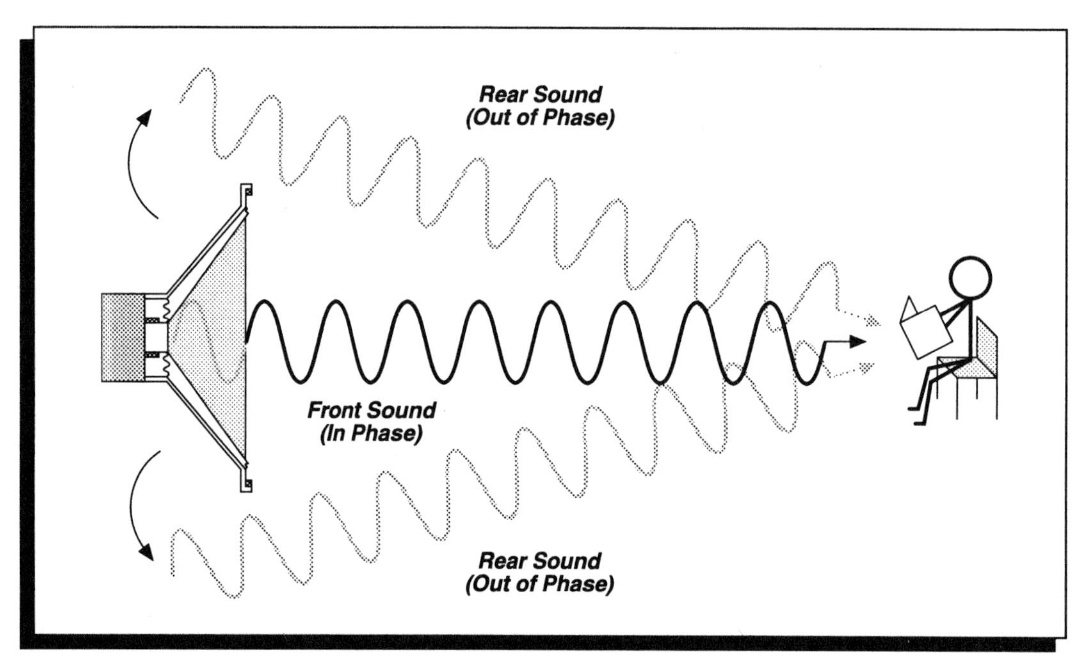

Figure 4.23 Baffle problem in cone driver.

the listener. Unfortunately, the front and the rear of the cone are 180° out of phase, Figure 4.24. At the time the cone is producing a condensation in front, the rear is producing a rarefaction, and vice–versa. They are always out of phase. This will cause interference around the speaker, leading to severe cancellation. This is just plain destructive interference. If you were to listen to an unmounted driver at low frequencies its output would be low due to this effect. There are two solutions to this problem:

- Separate front and rear sound so that they do not interfere destructively with each other.
- Delay rear wave long enough so that by the time it reaches the front it will be in phase with the next half wave of the front.

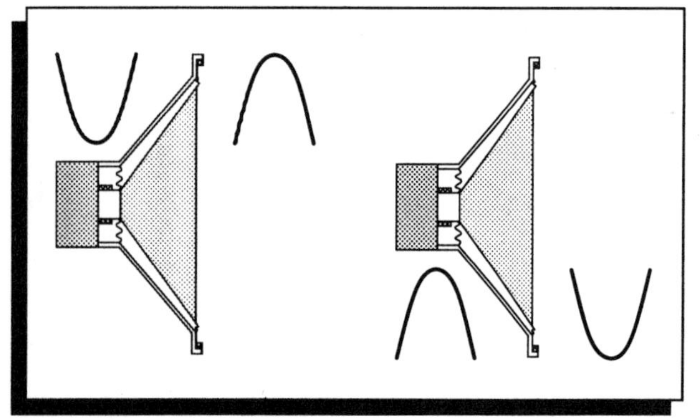

Figure 4.24 Front and rear of cone speaker are 180° out of phase.

The first solution leads to a variety of systems that use enclosures. The second solution leads to delays based on a baffle. Let us now address the second solution. Consider Figure 4.25. In this system a baffle is used around the cone driver to delay the rear wave when it arrives in front of the speaker. A delay by ½ wave will prevent destructive interference. In order for it to arrive ½ wave later and cause constructive interference, it must travel through a distance which is ½ wavelength long. The path around the baffle, from the rear to the front, must be ½ wave.

As an example, consider a cone driver producing 100 Hz. To prevent the rear from interfering destructively with the front, a baffle is used. The dimensions it should have are calculated:

$$\begin{aligned} \text{wavelength} &= (344 \text{ m/sec}) / 100 \text{ Hz} \\ &= 3.44 \text{ meters} \\ \tfrac{1}{2}\,\text{wavelength} &= 1.72 \text{ meters} \end{aligned}$$

The distance from the back of the driver around the baffle and down to the front must be 1.72 meters. Therefore, the baffle should be 3.44 meters tall, tip to tip, and wide to reduce the destructive

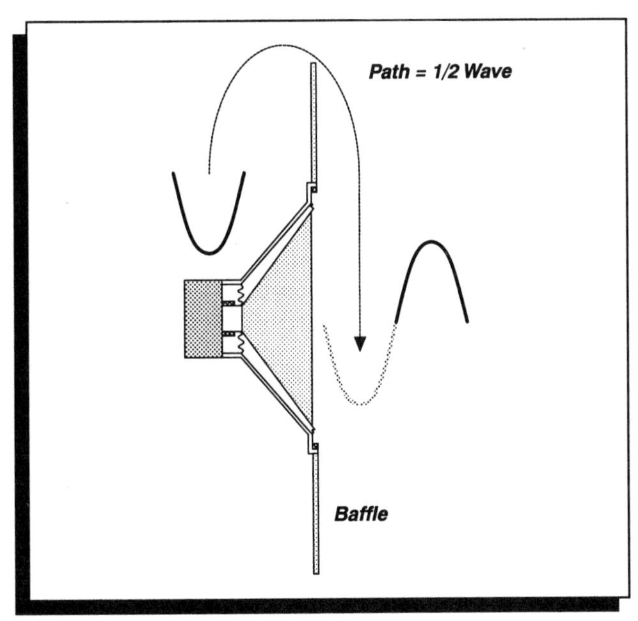

Figure 4.25 Baffle action.

interference cancellation for frequencies down to 100 Hz. For improvement at lower frequencies, larger baffles are necessary.

The discussion in this section has dealt with low frequency sound waves. For high frequencies the destructive interference effects are not as important. At high frequencies the waves are much shorter and they do not have sufficient time to reach the front of the driver. Low frequency solutions using baffles are not practical, and hence we will look into the enclosure approach.

4.3 SPEAKER ENCLOSURES

There are four basic enclosure types with large variations in their performance. Their main purpose is to separate the sound from the front of the speaker and the rear so that destructive interference does not occur. Before discussing them a definition has to be introduced as it often appears in speaker evaluation. Consider a tone at frequency f. Now, produce a tone at frequency $2f$. The ratio of the two frequencies is:

ratio of frequencies = $2f/f$
= 2 and this is known as a one octave interval

One octave = **ratio of 2 frequencies, which is equal to 2.**

Example: A frequency of 500 Hz when it goes up by one octave will be 1,000 Hz. Moreover, when one goes from 50 Hz to 100 Hz, that interval is also one octave. The interval is the same for both cases. This means that in going up or down in frequency in music one should not count how many Hertz it is up or down, but by how large a factor it is up or down. The basic factor or interval is the octave, a factor of 2. This interval is a building block in music.

Another example is the tone four octaves above 500 Hz. It will be:

500 Hz	→	1,000 Hz	= 1 octave
1,000 Hz	→	2,000 Hz	= 1 octave
2 000 Hz	→	4,000 Hz	= 1 octave
4,000 Hz	→	8,000 Hz	= 1 octave

Therefore, four octaves higher would be 8,000 Hz (and not 2,000 Hz, since an octave is not a harmonic).

■ INFINITE BAFFLE

The goal here is to separate the front and rear sound waves in a driver since they are out of phase. Other than mounting the speaker in a wall to satisfy this approach, one can use an enclosure. A box, known as an enclosure, is used to mount the driver and to trap the sound in the rear. What size of box should be used to trap the sound in the rear? Figure 4.26 shows two possible approaches, a large and a small enclosure.

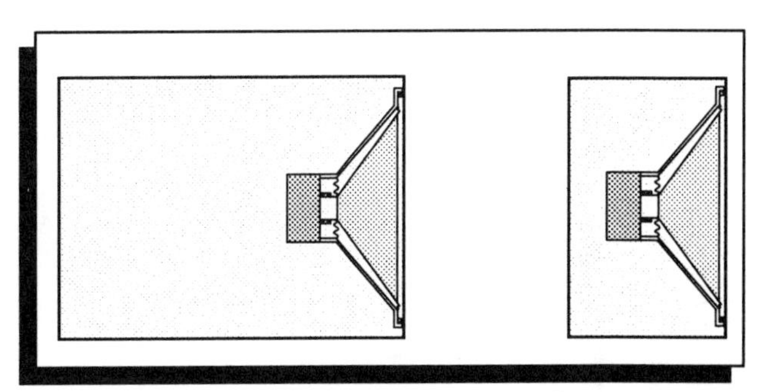

Figure 4.26 Two possible approaches for trapping rear sound in a speaker by means of an enclosure.

To find out which approach is better, let us look at the effect of the enclosures on the frequency response, Figure 4.27, of the whole system, driver and enclosure. The enclosure will affect the speaker performance because the air inside it acts as a spring, pushing against the driver. The trapped air inside the enclosure will change the compliance of the driver. Because the smaller enclosure will decrease the compliance of the driver more than the larger enclosure, the resonant frequency will then be shifted upward the most. This effect will be least for the large enclosure, as it does not introduce as much stiffness to the cone motion (i.e. the compliance is not reduced as much). In both cases the trapped air reduces the driver compliance. Such behavior dictates that the enclosure be large, as it will not push up as much the resonant frequency, and thus reduce sound output at low frequencies. Since all the rear sound is trapped in the enclosure, it is called an infinite baffle enclosure, and it acts like a baffle of infinite dimension with no rear sound reaching the front of the speaker.

As discussed earlier, standing waves could be set up inside the enclosure along the three major directions, very much like in a hall. Standing waves are reduced by lining the enclosure with wool, felt, glass wool, or a loose fabric. Absorption will reduce reflections and hence standing waves.

Characteristics of an infinite baffle speaker system are:

- — large volume enclosure which takes up space.
- — boomy sound.
- — not efficient because a large fraction of the speaker sound is dissipated in the enclosure.
- — not a popular approach due to its large size.

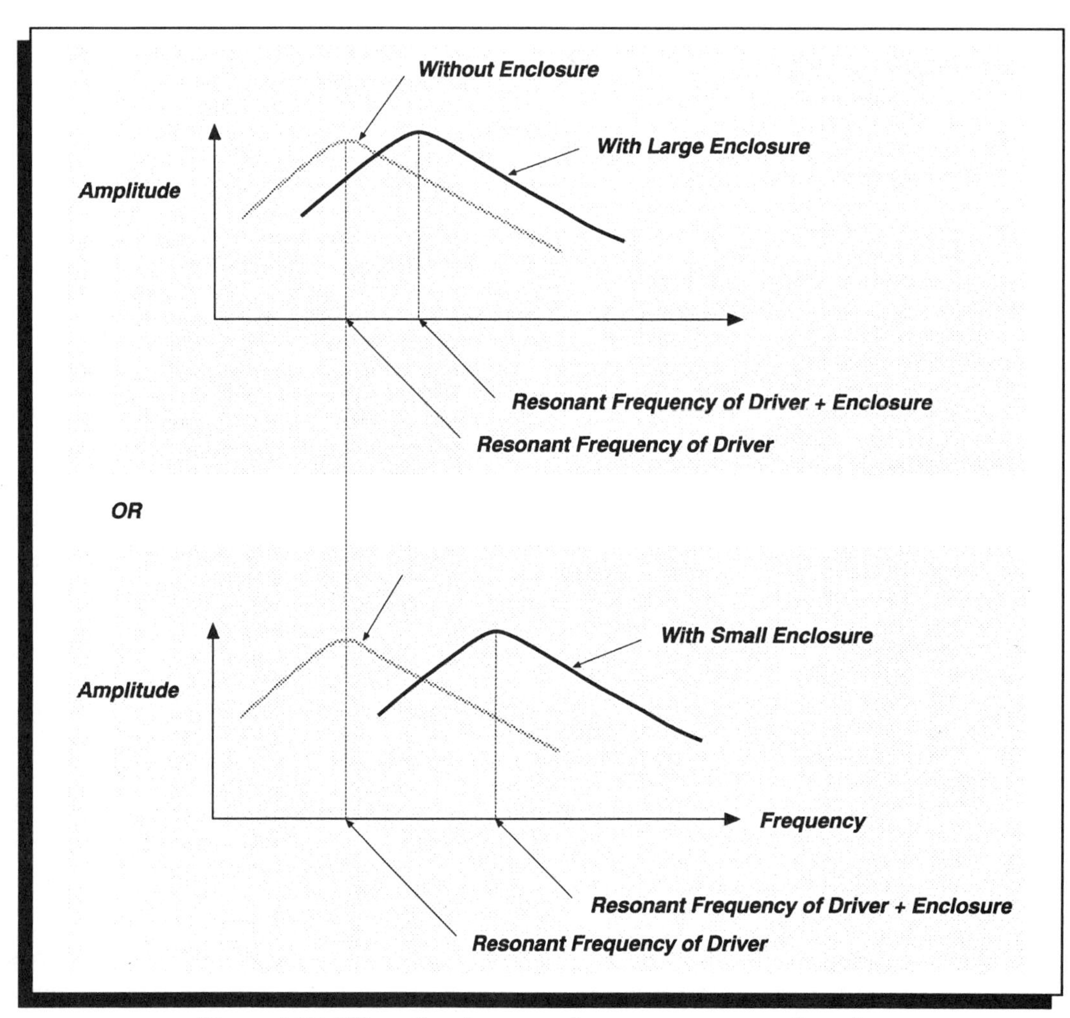

Figure 4.27 Effect of enclosure on frequency response of speaker.

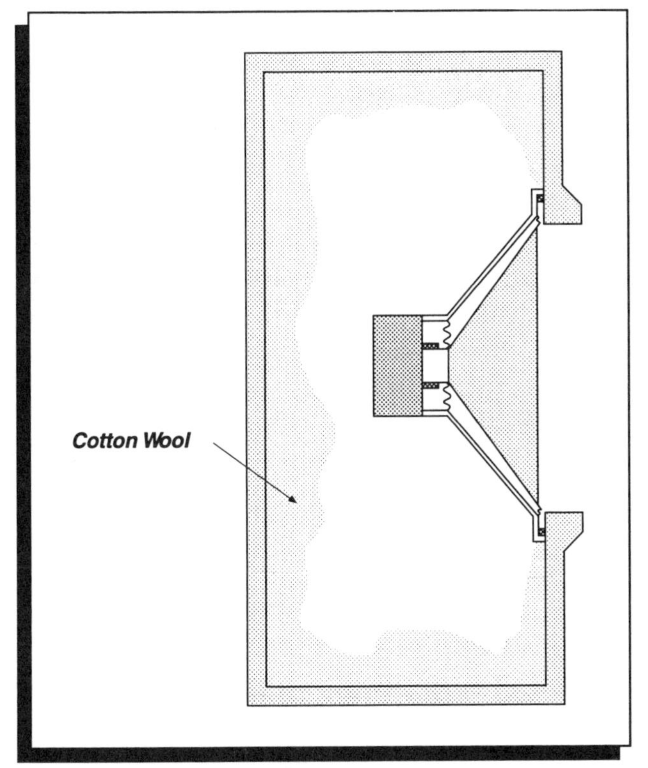

Figure 4.28 Reducing standing waves inside speaker enclosure.

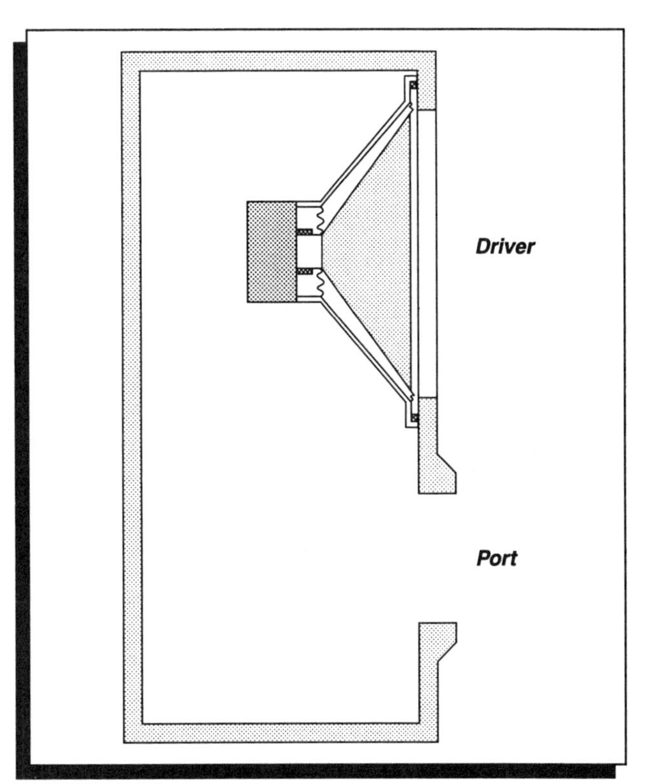

Figure 4.29 Basic bass reflect enclosure

Bass Reflex

This is a popular speaker system. It addresses an interesting and important question. Is it possible to use the sound from the rear to reinforce the sound in front, at least some of the time. If this were possible, our efficiency would improve. To use the sound from the rear it is necessary to have some sort of port in the enclosure and to delay the rear sound long enough that by the time it comes out it will be in phase with the front. Fig. 4.29 shows how the enclosure would be constructed so that low frequency sound from the rear could be used to reinforce the front sound. The mechanics of how the correct phasing will be achieved is subtle. Now our system has two oscillating components, the driver and the air from the port. They are coupled together in the enclosure. These two components are coupled and this can be demonstrated by pushing the driver in. Air comes out of the port.

The two components of a bass reflex speaker that are oscillating are the driver and the air in the enclosure through the port, Figure 4.30. Each will have its own resonant frequency. For proper action

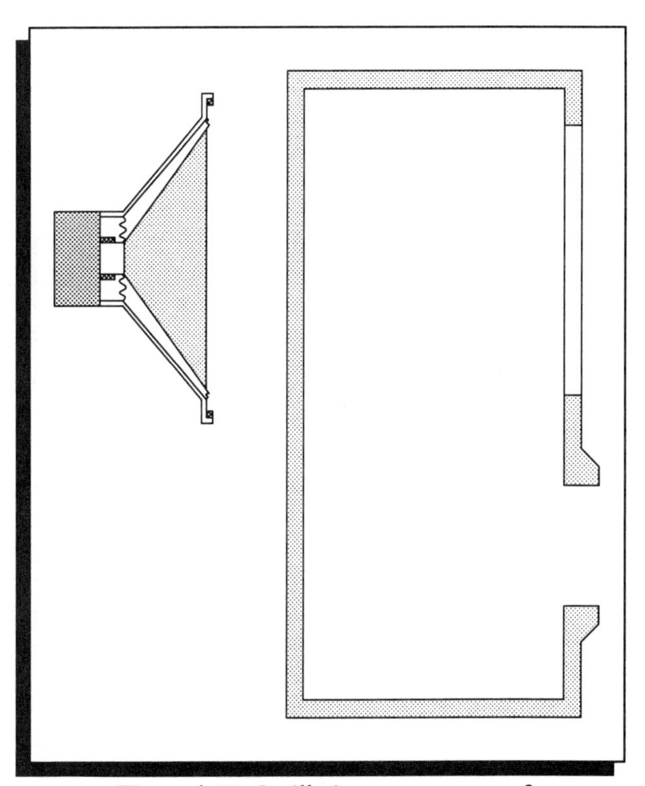

Figure 4.30 Oscillating components of bass-reflex speaker.

in the bass reflex, we select the resonant frequencies of the components so that:

resonant frequency of driver = **resonant frequency of enclosure**

When the two components are put together with the resonant frequencies tuned to each other, they become strongly coupled. Consequently, the system no longer shows a resonance at the original resonant frequency of the individual components. The original resonant frequency disappears entirely and two other resonant frequencies appear on each side of the original one, as shown in Figure 4.31. Such behavior is caused by the action of the two components, the driver and enclosure air, when coupled strongly. They will resonate at two frequencies corresponding to in–phase motion of one component relative to the other, and to out–of–phase motion of each component relative to the other, Figure 4.32.

The in–phase motion has a lower resonant frequency than the out–of–phase motion. These are the two resonances shown in the bottom of Figure 4.31. This result comes from a well–known analysis of a driven two–mass system when studying coupled vibrating systems.

There is one more step, and that is of relating the results above to the behavior of a bass–reflex speaker. The two coupled components forming the bass–reflex in fact appear as shown in Figure 4.33. The two resonant behaviors still exist; however the sound coming from the port and from the driver, both come out in front of the speaker. The in–phase motion causes the sound to come out of the port out of phase with the driver. The out–of–phase motion causes the sound to come out in phase with the driver. The net effect is:

— sound from driver is reinforced by sound from port at some frequencies (near the higher resonant frequency).
— sound from driver is reinforced at the lower resonance, because it is a resonance. Also the response is pushed down to lower frequencies.
— frequency response is flatter than for driver alone.

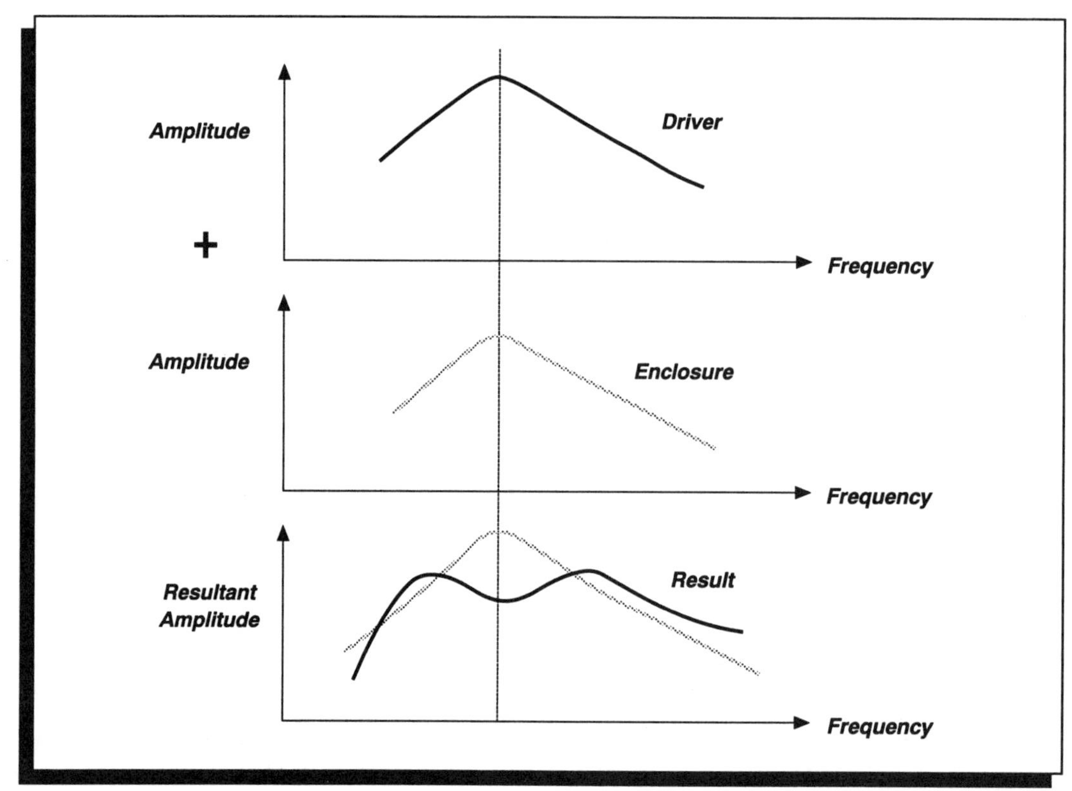

Figure 4.31 Splitting of original resonance into two new resonances in a bass-reflex system.

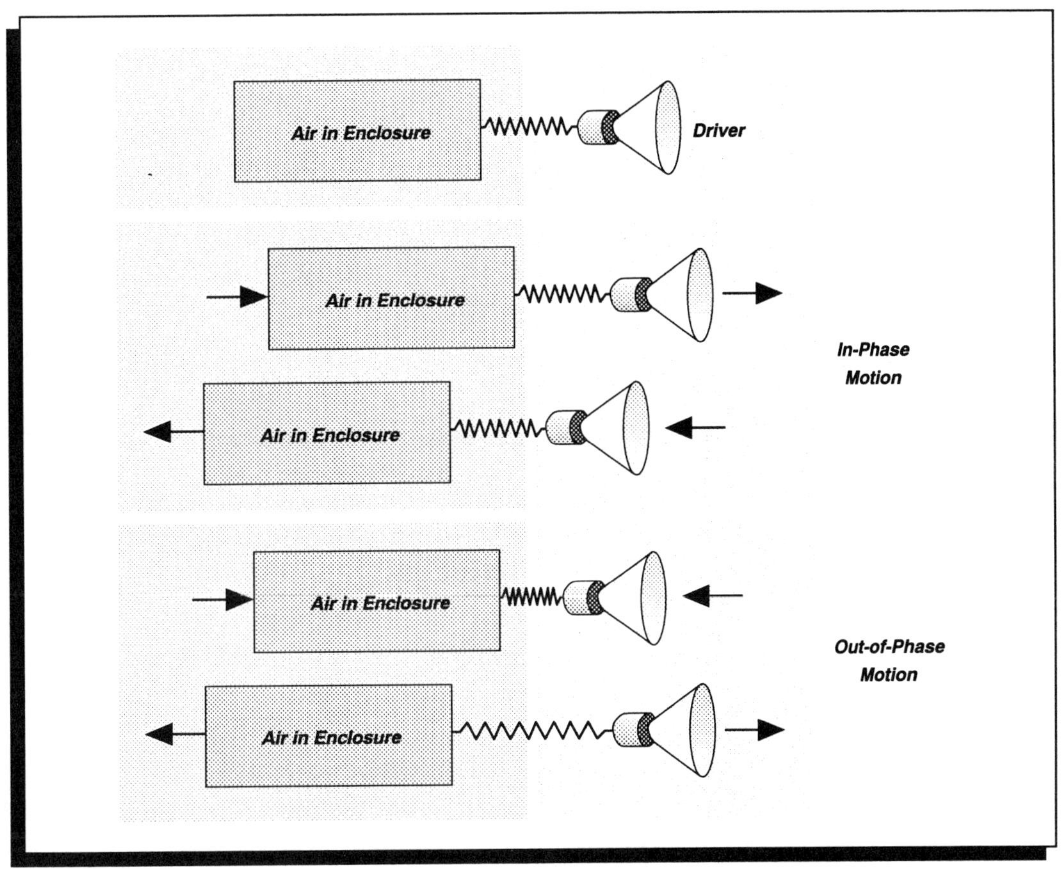

Figure 4.32 Resonant behavior, in-phase and out-of-phase, motion of strongly coupled components of bass reflex system.

— efficiency of the speaker is increased as more sound comes out of the speaker.

Some bass–reflex speakers use a passive radiator that covers the port. The action is still that of a bass–reflex except that possible turbulences around

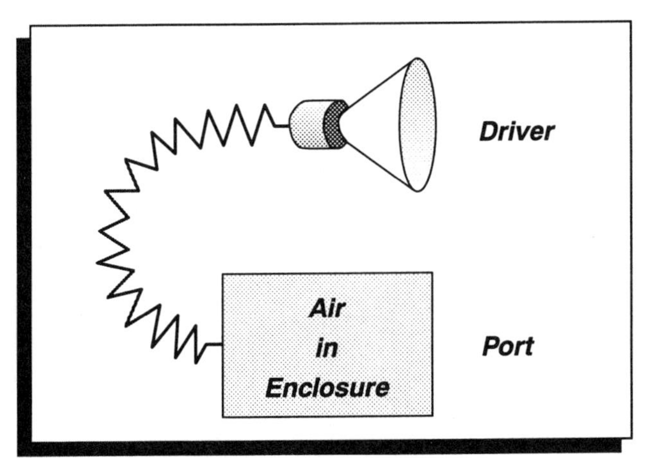

Figure 4.33 Coupled components of a bass-reflex speaker.

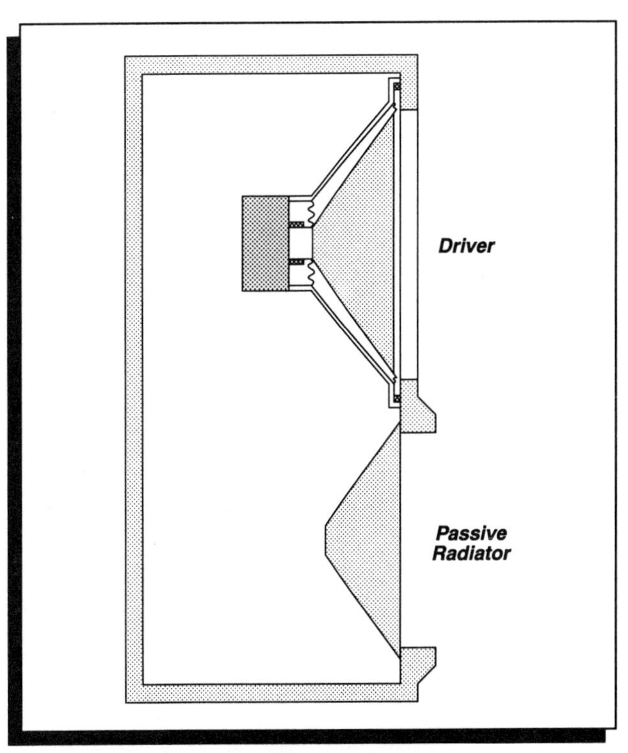

Figure 4.34 Bass-reflex speaker using a passive radiator over the port.

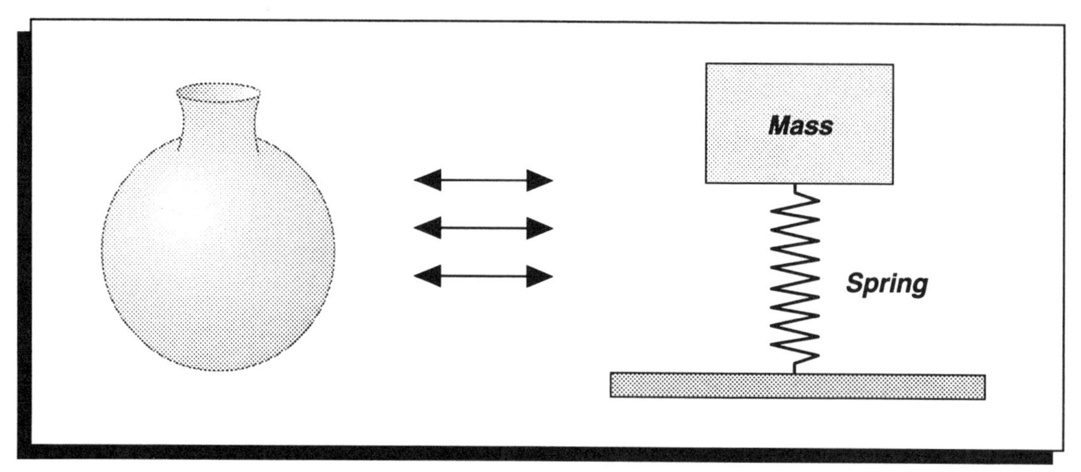

Figure 4.35 Helmholtz resonator behaves like mass-spring system.

the sharp edges of the port are reduced. The passive radiator has no electrical connections to its voice coil; only the sound in the enclosure from the rear of the driver provides the driving mechanism.

In the construction of the bass–reflex speaker it may not be that simple to satisfy the condition of the same resonance frequency for both the driver and the enclosure. The enclosure can be tuned to satisfy this condition. The tuning can be achieved by using the resonant properties of the enclosure, which is a cavity. In the 19th century, Helmholtz studied the resonant behavior of various acoustic cavities. One such cavity, now known as the Helmholtz resonator, Figure 4.35, has the following properties which were derived by comparing the resonator to a mass–spring system. The air inside the resonator behaves like a spring and the air in the neck or port of the resonator behaves like

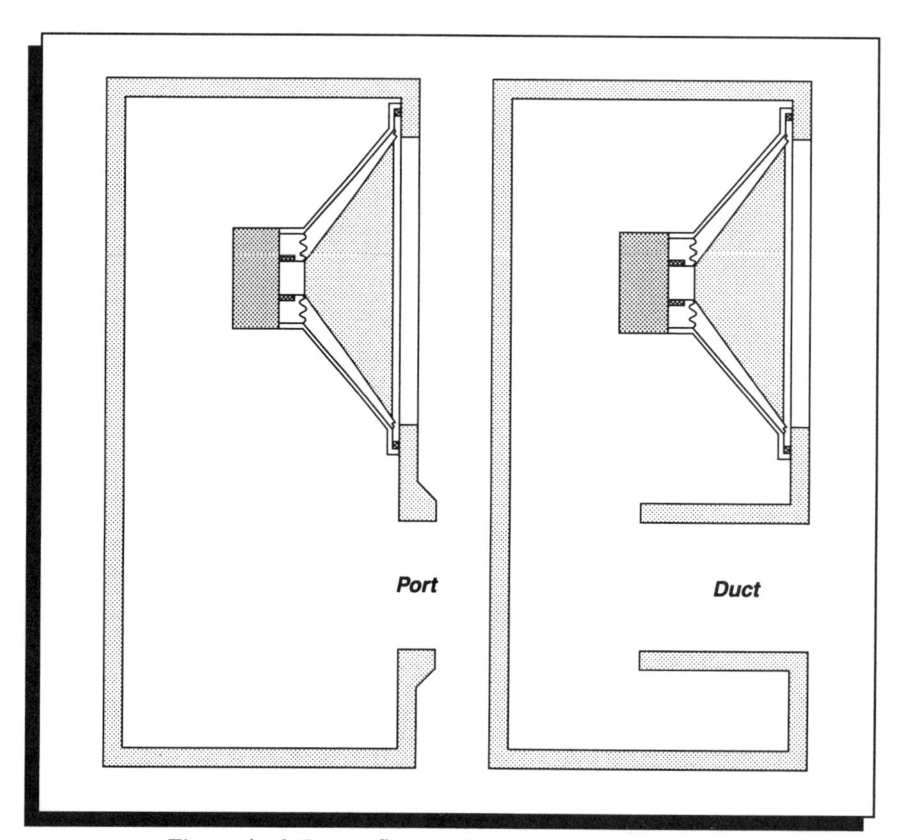

Figure 4.36 Bass-reflex speaker using a port or a duct.

an oscillating mass on top of a spring. Its resonant frequency has been derived to be:

$$\text{Resonant frequency} = \frac{\text{speed of sound within gas of resonator}}{6.28} \times \left(\frac{a}{Vl}\right)^{1/2}$$

where *a* is the cross–sectional area of neck; *l* is its length; and *V* the volume of the resonator. This tells us that a large Helmholtz resonator (large volume) will have a low resonant frequency. When the port size is increased, its resonant frequency goes up. Hence, by adjusting the port size it is possible to tune the resonator. Also, by changing the neck length, the resonator can also be tuned. A long neck leads to a low resonant frequency.

Since the speaker enclosure looks like the Helmholtz resonator , it is possible to tune the enclosure using the facts just presented. Indeed, enclosures are tuned by modifying the port size or by introducing a neck called a duct whose length can be adjusted, Figure 4.36.

The inside of the enclosure is coated with wool or other sound absorbing material to reduce standing wave effects.

The bass–reflex speaker design is very popular for the following reasons:

- — excellent frequency response; it is quite flat and it extends down to low frequencies.
- — efficient since sound from rear is used.
- — does not have to be large since tuning can be adjusted by port size or neck length.
- — distortion is reduced because of improved efficiency.

■ ACOUSTIC LABYRINTH

This speaker is also known as the transmission line speaker. Its design addresses the problem of what to do with the rear sound, when it is out of phase with the front. In theory, a transmission line or labyrinth enclosure can act like an infinite baffle enclosure when all the rear sound is absorbed. It essentially is a long pipe which has been folded a few times to conserve space, and hence it is called a labyrinth. The idea is to absorb all the sound in this volume without making it excessively large. Figure 4.37 shows a typical design approach for this enclosure. The folded transmission line is lined with sound–absorbing material.

It is possible to design this unit in such a way that a phase change is produced by the labyrinth. In this case, its internal path length would be ½ wave for frequencies below the driver resonance. Such design will cause an enhancement of the low frequency response of the speaker.

This type of speaker has good frequency response, but it is not efficient (especially if the rear sound is absorbed.)

■ ACOUSTIC SUSPENSION

The goal in this design is to have a small enclosure for space–limited situations. Quite often it is referred to as an air suspension or acoustic suspension speaker. It is usually used as a book–shelf speaker.

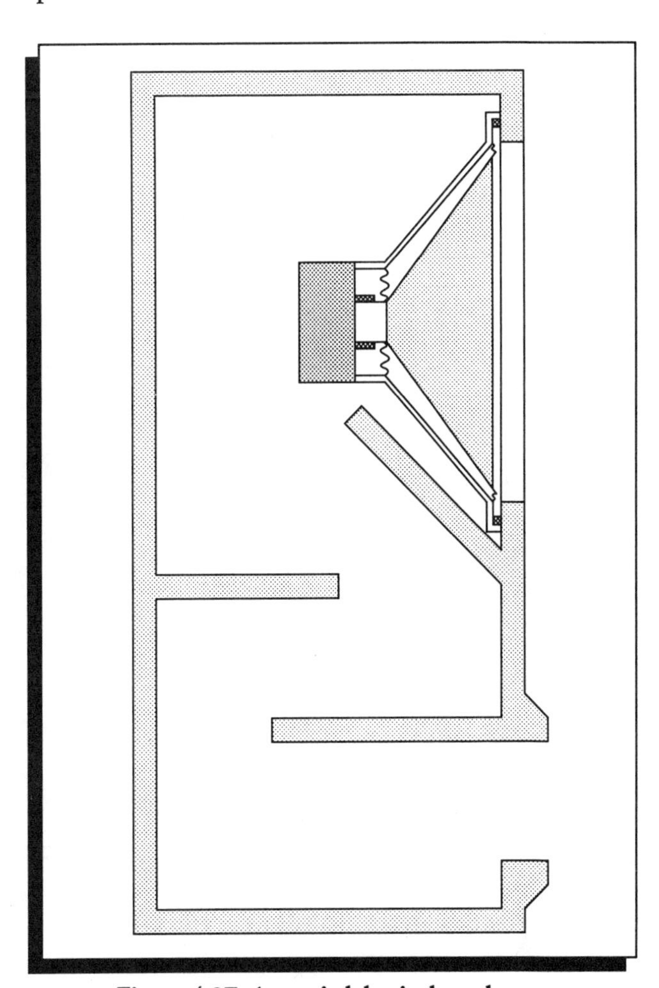

Figure 4.37 Acoustic labyrinth enclosure.

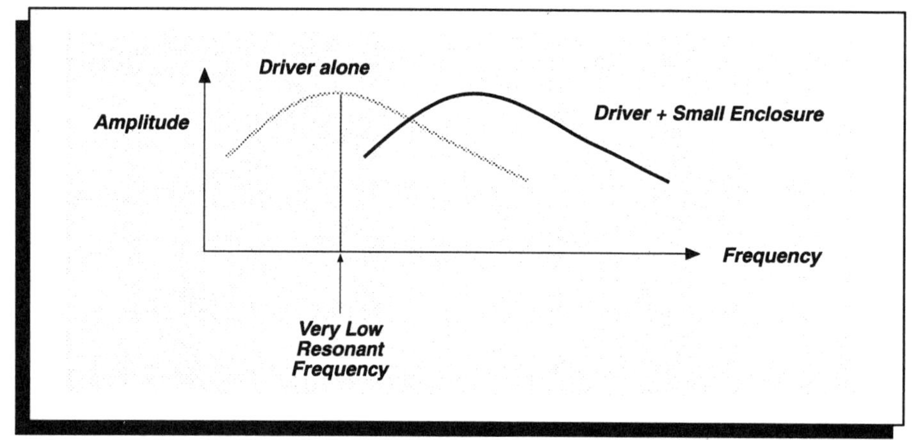

Figure 4.38 Change of frequency response of speaker when a small enclosure is used.

Our previous discussions have shown that the air inside an enclosure will reduce the compliance of a driver and hence move its resonant frequency up. The combination of a regular driver and a small volume enclosure would lead to a large increase in its resonant frequency, and hence to a large reduction of sound at lower frequencies. Figure 4.38 shows this for some arbitrary case. There is one way to overcome this loss and that is to start with a driver, which has a much lower resonant frequency. This will be possible in a driver with a very large compliance. Such a large compliance means that there is not much stiffness and it is the air in the enclosure that will provide this stiffness. The design approach would be as shown in Figure 4.39. In this, a driver of large compliance

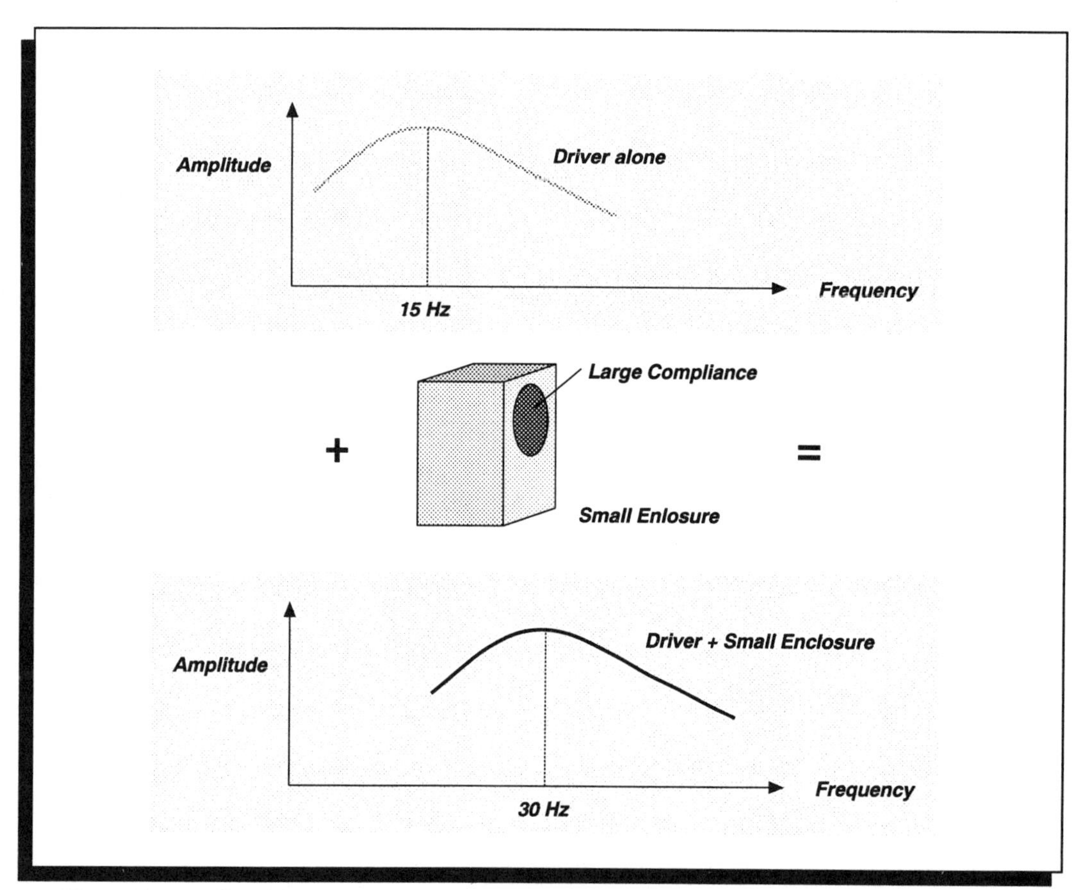

Figure 4.39 Effect of small enclosure on frequency response of acoustic suspension speaker.

has its resonance at around 15 Hz. When mounted in the enclosure, the resonance will shift to 30 Hz still providing a sufficient low frequency response. Important features of such speakers are:

— small size.
— very good frequency response.
— very low efficiency, since rear sound is trapped inside enclosure.
— efficiency is usually less than 1%; however this should not be confused with quality.
— popular since it is compact.

4.4 HORNS

This is probably one of the oldest loudspeakers used in acoustics. It is useful because it is efficient. The question of efficiency arises when one compares the performance of a horn to that of a cone driver. An understanding of why a cone driver is not efficient will provide an introduction to the subject of horns. A radiating surface will reproduce sound accurately when it can move a large mass of air and when it is small and light so that it can respond quickly. This is impossible to achieve with a cone driver, because in order to move a large mass of air it must be sturdy and large; to cause a large object to vibrate requires much power. The problem is a fundamental question that deals with the transfer of power from one body to another. We are trying to move air which is very light by using a relatively heavy cone. This is not an efficient method of power transfer. The problem is similar to the transfer of energy of an oscillating bob, depicted in Figure 4.40. When the bob on the left strikes the one on the right, all the energy will be transferred to the bob on the right since the masses of the two bobs are equal. However, when the bob on the left strikes a small bob, all of its energy will not be transferred to the small bob; this is an inefficient way of transferring energy as the mechanical characteristics of the two objects are quite different. In other words, the ideal way of

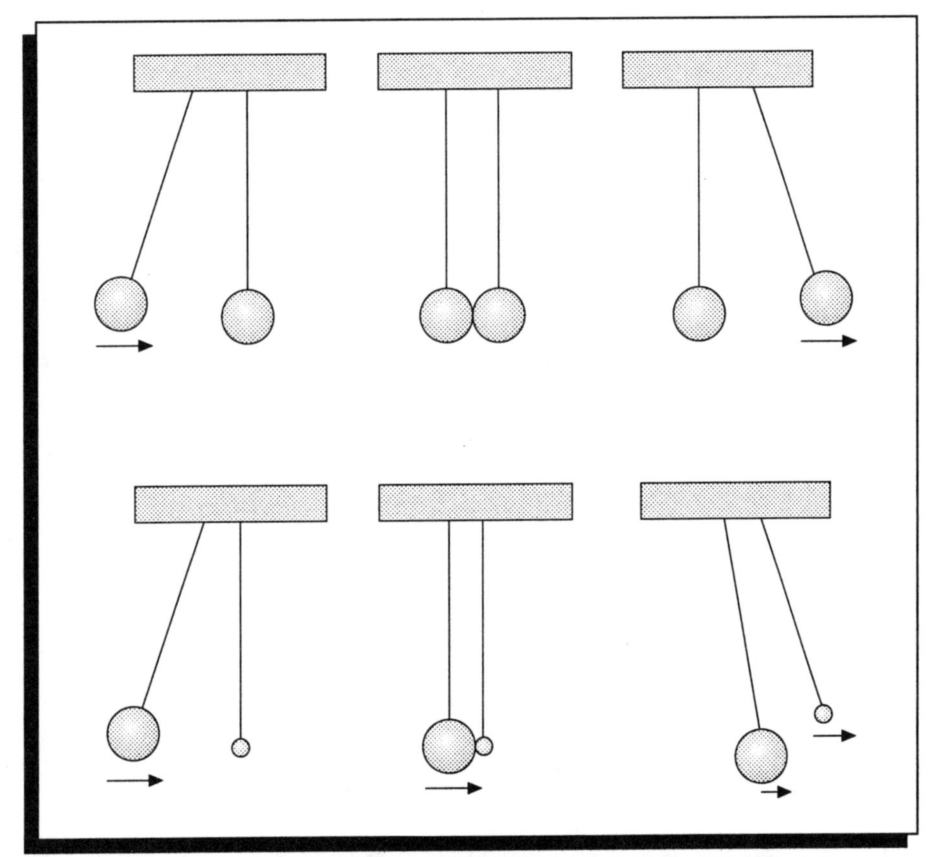

Figure 4.40 Transfer of energy from a bob on one of equal mass (A), and to one of different mass (b).

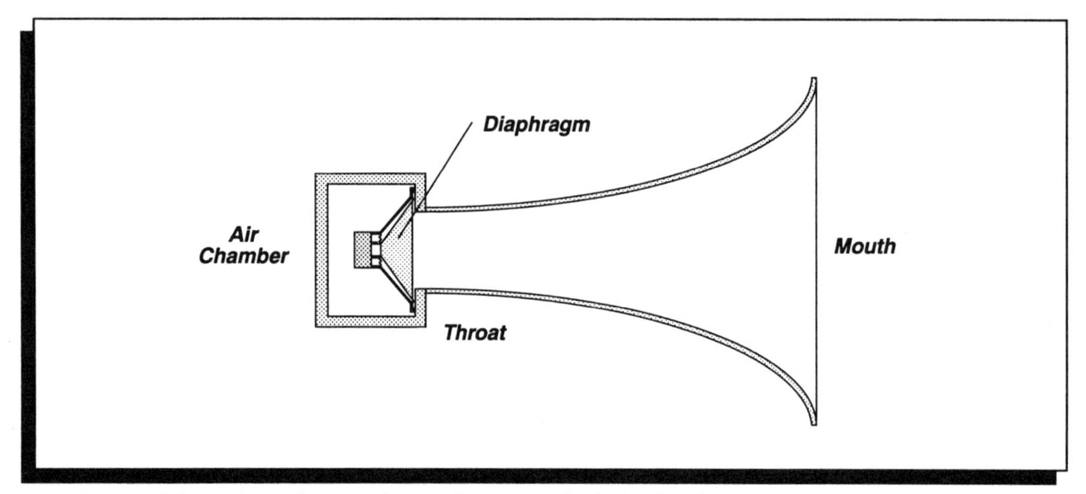

Figure 4.41 A horn for matching vibrations of a light diaphragm to a large volume of air.

transferring energy from one object to another is when their properties are matched. We have a similar situation with speakers when a relatively massive cone is trying to move air which is very light. This process is quite inefficient. The horn solves this problem by using the following:

— a small light diaphragm; it will not need much power to vibrate.
— a tapered matching structure which is coupled to a large volume of air.

This is shown in Figure 4.41. The vibrations of a light diaphragm are coupled to air at the horn throat and then they are matched gradually to a large volume of air at the horn mouth. An air chamber is placed behind the diaphragm to balance the diaphragm motion against the work that it must do on the horn when it moves forward. The horn is a match for the transfer of power between two acoustically dissimilar objects, a diaphragm and air. Because of such a match, the horn is much more efficient than the cone driver; its efficiency can be as high as 40 – 50%. The horn transforms small amplitude vibrations to large amplitude ones, and hence it is known as an acoustic transformer.

The horn has its limitations. The most severe is its response at low frequencies. Experiments and calculations show that below a certain frequency, called the cut–off frequency, there is almost no sound output from the horn. This characteristic frequency depends on the size of the horn's mouth. It can be determined from the diameter of the mouth as it corresponds to approximately $1/4$ wavelength of the wave of the cut–off frequency. Any longer waves are greatly attenuated by the horn. Figure 4.42 illustrates the low frequency response of a horn and its low frequency cut–off. To extend

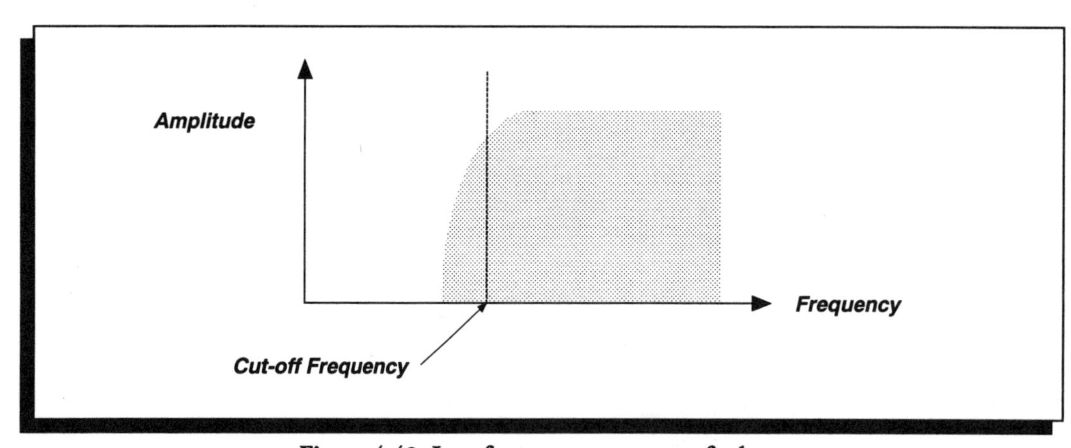

Figure 4.42 Low frequency response of a horn.

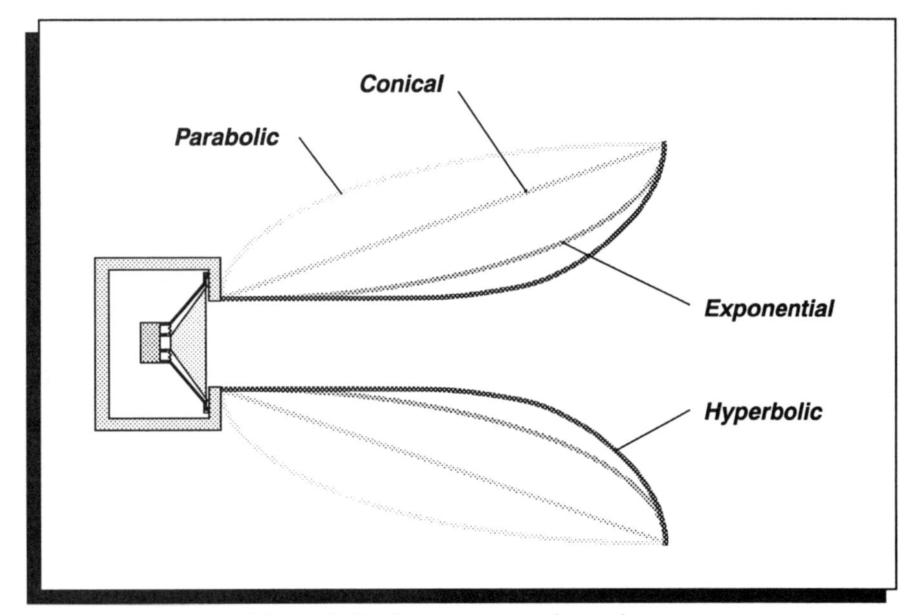

Figure 4.43 Some common horn shapes.

the response to lower frequencies, the cut–off frequency has to be lowered by using a horn with a larger diameter mouth. A calculation of this diameter shows that it becomes very large for low frequencies. For example, the cut–off frequency for a horn which has a mouth with a one–meter diameter would be:

since diameter	=	¼ wave
here, 1 meter	=	¼ wave
then 4 meters	=	1 wave
the cut–off frequency	=	344 m/sec/4 meters
and therefore	≈	86 Hz

This means that this horn will not reproduce well frequencies below 86 Hz unless the horn has a mouth larger than 1 meter in diameter.

Horns have all sorts of shapes and their application determines which is the most suitable shape. Figure 4.43 shows a few horn geometries and their characteristic flares. Important horn geometries are: the exponential horn; the hyperbolic horn; the parabolic horn; and the conical horn. Some horns are used in musical instruments where a standing wave is first set–up and then a small fraction of the sound is radiated. This requires a particular horn shape. A speaker has different goals: it projects the sound, as there is no need for standing waves on it. The best compromise between efficiency and low distortion for a loudspeaker is the exponential shape; it is the shape most commonly used for speakers.

For low frequencies, the horn has to be unreasonably large. This places a severe limit on the use of horns, though it is possible to compromise by folding the horn into a more compact unit known as the folded horn, Figure 4.44.

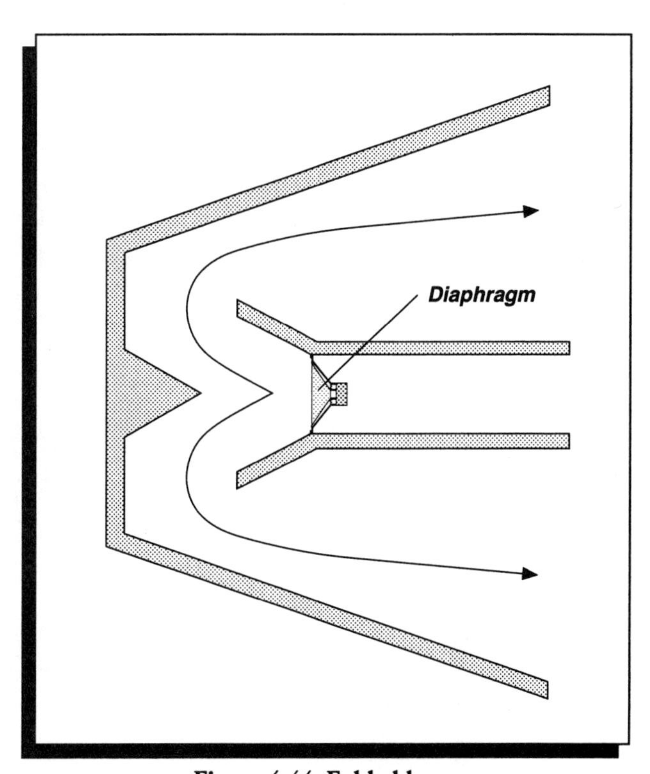

Figure 4.44 Folded horn.

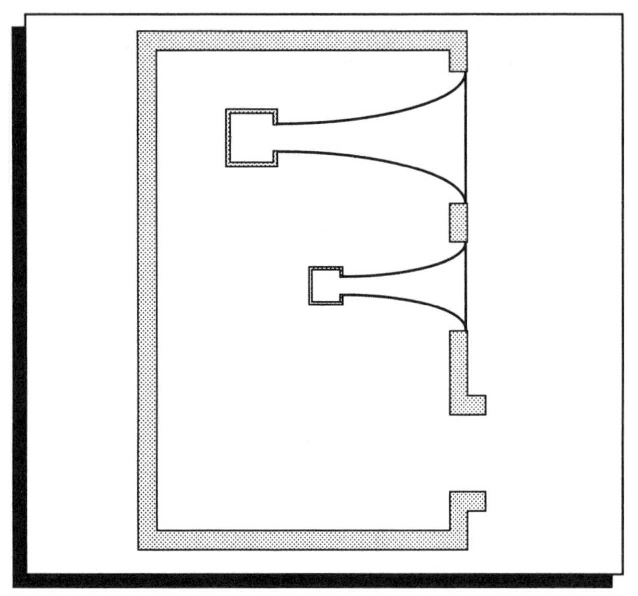

Figure 4.45 Two-way horn loudspeaker with bass-reflex enclosure.

The directional characteristics of a horn loudspeaker will depend on the frequency, shape, and the mouth. As the frequency is increased, the directivity becomes sharper and there is less dispersion. Due to directivity problems, some speakers use two or three horns. Figure 4.45 shows a two–horn system as used in a bass–reflex enclosure.

The advantage of very high efficiency, as found in a horn, can be illustrated by an example. If a folded horn speaker at 20% efficiency needs 5 watts of electrical power to produce an adequate level of sound, an acoustic suspension speaker at 1% efficiency would need 100 watts of electrical power to sound as loud as the horn. A horn loudspeaker does not need an amplifier with as much power as for a cone loudspeaker.

4.5 LOUDSPEAKER PLACEMENT

The acoustical performance of loudspeakers in a room can be varied, depending on the geometry of the room, its volume, and its sound absorption characteristics. Important points to recall here are:

— surfaces will reflect sound.
— loudspeaker radiation pattern will vary with frequency. In general, higher frequencies tend to be directed forward in a narrower beam than lower frequencies .
— sound can be projected in a room using a simple stereo arrangement or with a surround–sound set–up.

These important points will determine the sound distribution in a room. There are a few other factors that should be considered. First, a room is like a large enclosure and it can have several natural resonances determined by its dimensions. Second, a room with hard, smooth wall surfaces will reflect sound as mirrors reflect light. A well–known example is that of singing in a shower; the walls reflect the voice back and forth and cause resonances which enhance the singing. Similarly, reflections of sound in a room set up standing waves, enhancing sound where certain spots have high sound pressures, while other spots have minimal sound pressure. Figure 4.46 shows one such case at a room fundamental frequency. The position of the listener in such a room will effect the quality of sound that he hears.

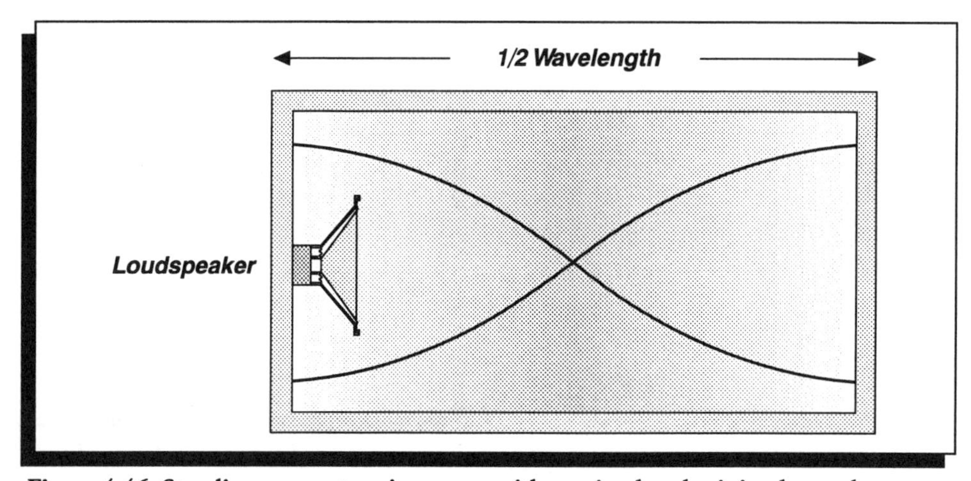

Figure 4.46 Standing wave set up in a room with maximal and minimal sound pressure.

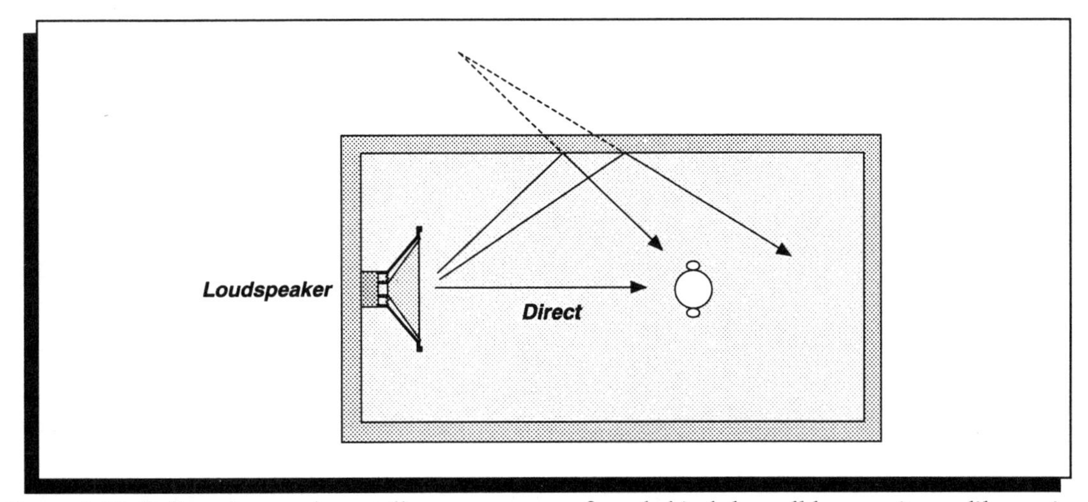

Figure 4.47 Reflected waves by a wall appear to come from behind the wall because it acts like a mirror.

The amount of reflected sound in a room will also depend on how much absorption there is by the walls and the contents of the room. Figure 4.47 shows a listener receiving reflected waves from a wall. If a comparison were made to that of light reflected by a mirror, the reflected waves appear to come from behind the wall. Hence, as a result of reflections the room appears larger and it gives a sensation of being a lively room. If the walls absorb most of the sound, there will be very little reflection and the room will appear "dead." Of course a happy compromise between these two extremes is needed to give the sound the correct quality which comes from the room reverberation time.

The placement of loudspeakers at specific locations in the room will determine the quality of the sound. When the loudspeaker is placed on the floor near a wall, there will be reflections by the wall and floor, creating an impression that part of the speaker came from behind. This is true for the low frequencies which diffract so easily. Hence, the low frequencies will be enhanced. The high frequencies will not be affected as much by the floor and the wall since the speaker beams them mainly forward. This effect is even stronger when the loudspeaker is in a corner.

When two loudspeakers are placed in a room for stereo coverage the dispersive properties of the speakers have to be taken into account in choosing the location in the room and their direction. A customary practice is to locate the speakers as in Figure 4.48.

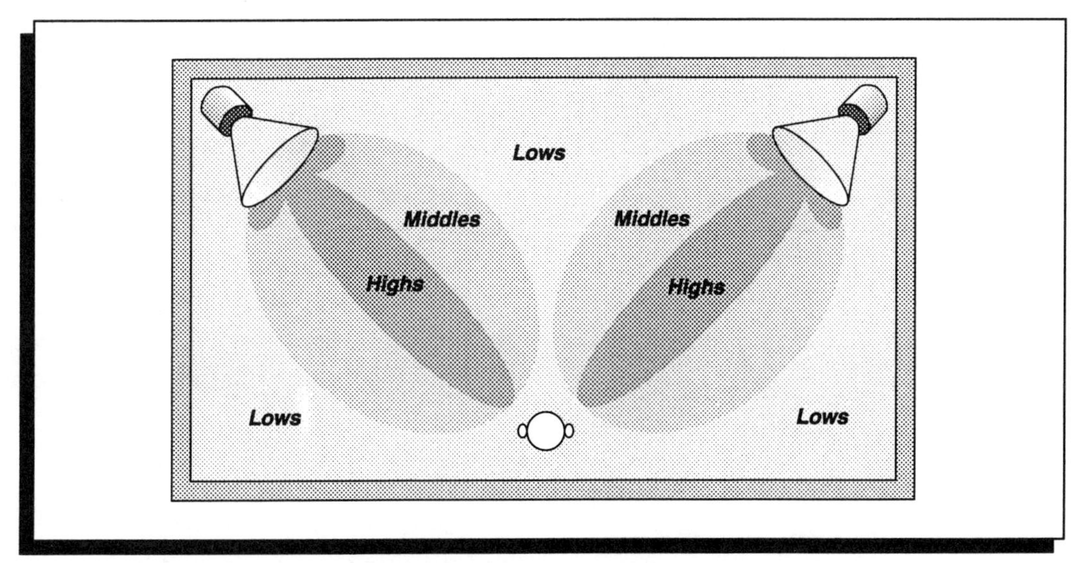

Figure 4.48 Stereo coverage in a room.

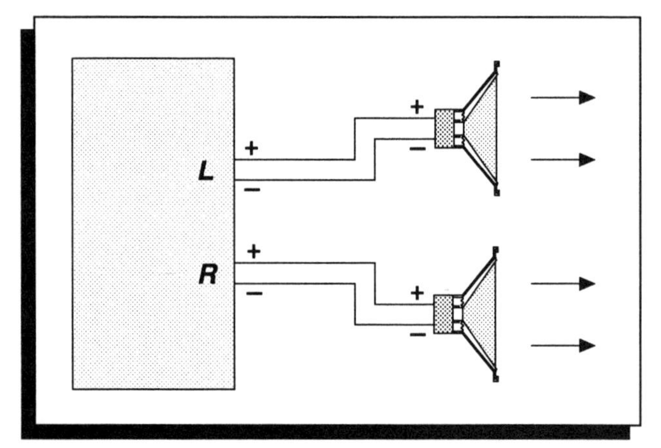

Figure 4.49 Speaker phasing: speakers are in phase.

In locating speakers in a room, it is important to consider the effect that one wishes to create and the dispersive properties of the speakers. A certain amount of experimentation is necessary to determine ideal conditions.

Finally, a few words should be mentioned about speaker phasing. Here there can be problems due to destructive interference especially at low frequencies when speakers are not properly wired to the amplifier. Figure 4.49 shows how the stereo speakers are attached to a receiver or amplifier, with speaker phasing being in phase.

When the polarity for one of the speakers is inverted, the speaker phasing would be incorrect since it would be out of phase, Figure 4.50. One can verify the correct speaker phasing by listening to the bass sound. For the out–of–phase connection due to one speaker being incorrectly wired to the amplifier, the bass sound will be quiet as a result of destructive interference between the left

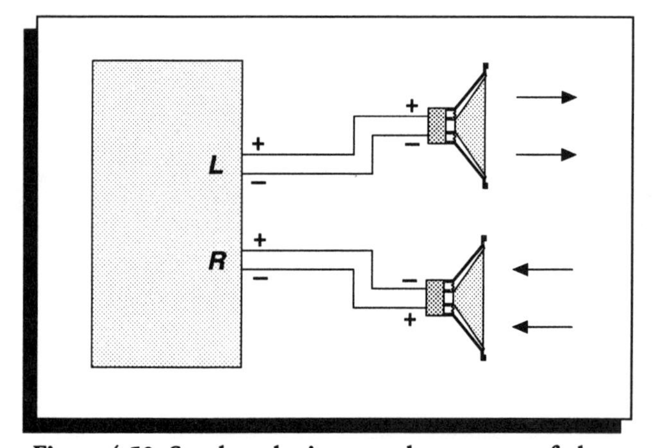

Figure 4.50 Speaker phasing: speakers are out of phase.

and right speakers. This can happen when the positive (+) of the amplifier goes to the negative (–) of the speaker, and the negative (–) of the amplifier goes to the positive (+) of the speaker, for one of the speakers.

4.6 LOUDSPEAKER SYSTEMS

So far we have introduced the two types of speakers, cones and horns, as well as the different types of enclosures. There are many speakers on the market fitting somewhere in one of the above categories. Since there are so many different speakers it is clear that each one is trying to do something better than the others. Some of the approaches are new and interesting, others are more traditional and follow popular tastes. To discuss them all would take too much space and time here. Therefore, a few speakers have been selected for presentation here , where the design principles and approaches are very different from the others. The speaker systems are the Bose 901, the Klipschorn, and the mini– and subwoofer system.

■ BOSE 901 SPEAKERS

This is a popular and high quality system which has distinctive features. The main ones are:

— direct radiator cones.

— spectrum of direct sound is different from that of reverberant sound.

— uses a total of nine equal drivers per speaker. One driver is in front and eight drivers are in the rear, Figure 4.51. Hence 11% of sound is direct radiation, while 89% of sound is reverberant radiation, aimed toward a wall.

— there are air columns in the enclosure to reduce distortion.

— there are no cross–over networks since each of the nine drivers per speaker receives the full spectrum of sound over all the audio frequencies.

— all the drivers are the same in size.

Because of anticipated difficulties in coping with the full sound spectrum handled by each driver,

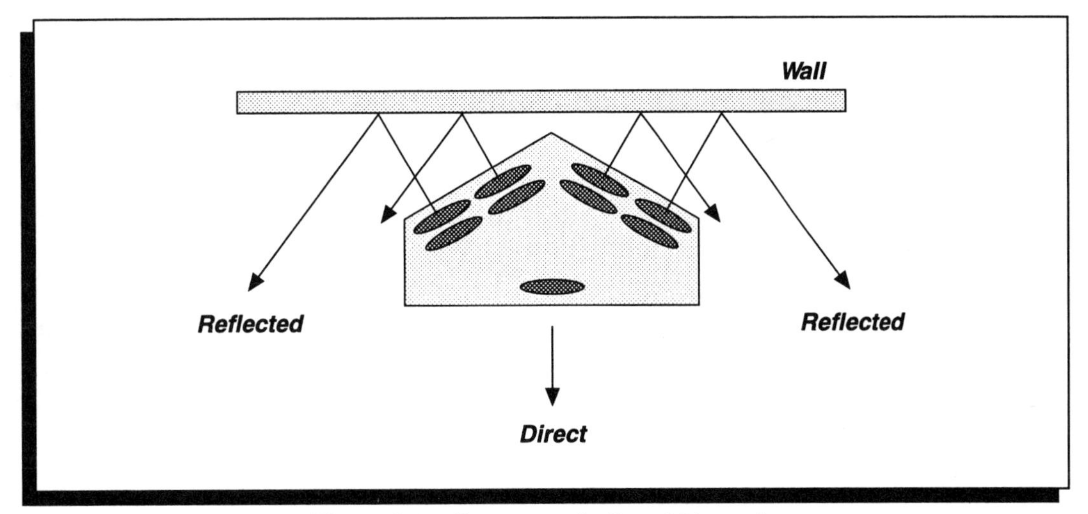

Figure 4.51 Geometry of a Bose 901 speaker.

this speaker requires its own special electronic unit which produces high fidelity response. It is the equalizer. The equalizer makes up for the deficiencies of the drivers. This is shown in Figure 4.52. The frequency response is poor at low and high frequencies because of the limitations of the drivers; they are all the same and there are no tweeters or woofers. To make up for this the special equalizer boosts the low and high frequencies by preselected amounts. This equalizer comes with the speakers.

Since each speaker deals with all the frequencies at the same time there can be problems with the

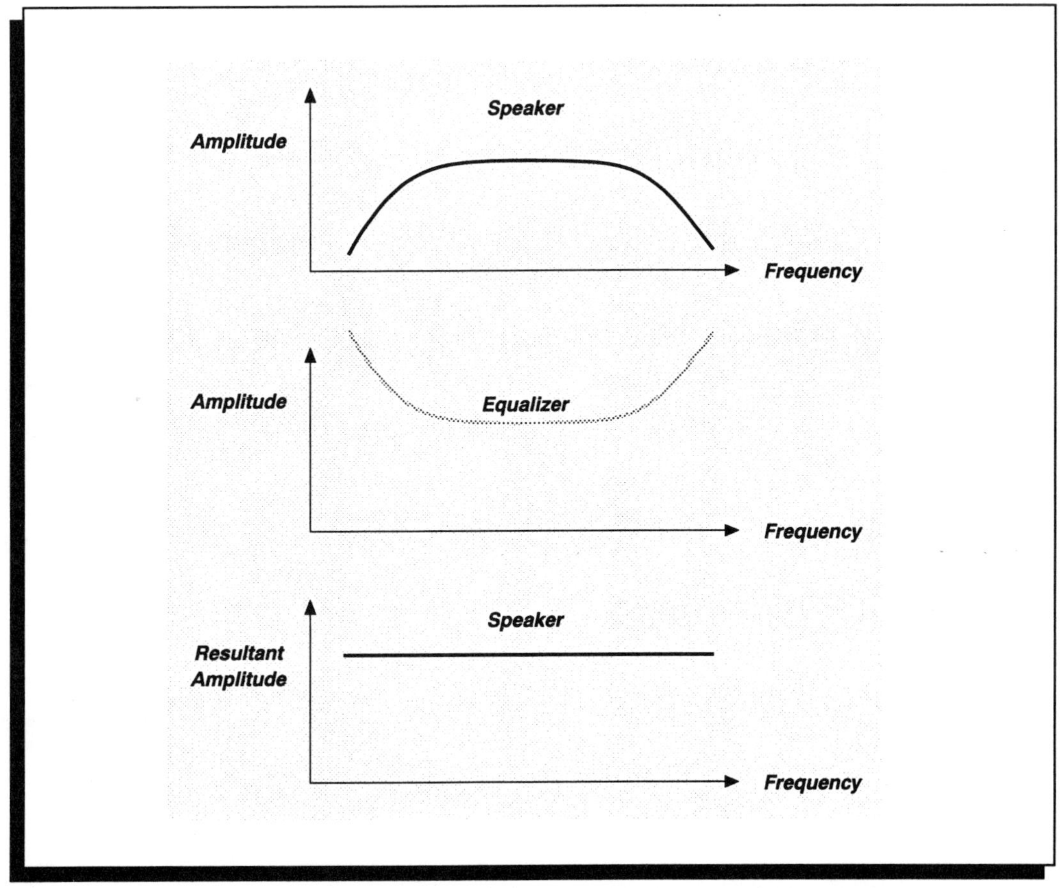

Figure 4.52 Effect of equalizer on frequency response of Bose speakers.

Doppler effect. This is reduced here because with 9 drivers/speaker the diaphragm does not have to move as much as with only one driver; hence each diaphragm has a small speed. Small relative speed between source and observer causes small Doppler Effect, and hence sound blurring is very small.

There can also be problems with distortion at low frequencies, as the speaker excursions can be large. It is reduced by using nine drivers per enclosure and they then have small diaphragm excursions.

KLIPSCH HORN SPEAKERS

This speaker uses all horns; the tweeter, the midrange, and the bass. The advantage of an all horn approach is that the efficiency is high and hence there will be low distortion, since the drivers do not need to have large diaphragm displacements. The most complicated part is that of the bass horn. From earlier discussions we know that the horn mouth has to be enormous if low frequencies are to be reproduced. This problem is solved by using corner reflections of the sound to create a mirror effect which makes the horn appear larger than it is. For such an approach to be effective folded horns are used at each corner of the room and the corners of the room become part of the loudspeaker. This is shown in Figure 4.53. Because of reflections at the corners, the mouth area is ¼ as great as would be required for a regular horn mouth.

SUBWOOFER AND MINISPEAKER SYSTEMS

Below 300 Hz, many authorities feel that pure tones are not perceived as directional. In this case, there is no need to have two speakers for this frequency range. Hence, a popular approach for stereo is:

— take LEFT and RIGHT bass speakers and place them in one unit, a subwoofer.
— use two small speakers, one for each channel, for the higher frequencies.

In order for the subwoofer to be a single unit, one of three approaches can be used:

1. two drivers, LEFT and RIGHT, are mounted together side by side in one enclosure.
2. a single driver with a double voice coil (one for the LEFT channel and one for the RIGHT channel), the dual voice coil.
3. a single driver which gets its power and information from a summing amplifier which causes the addition of the LEFT and RIGHT channels into a single channel.

Since the subwoofer is large, it can be used as a piece of furniture. The high frequency loudspeakers are small and they are located at an appropriate place to provide stereo effect. This system provides a popular approach that saves on room space.

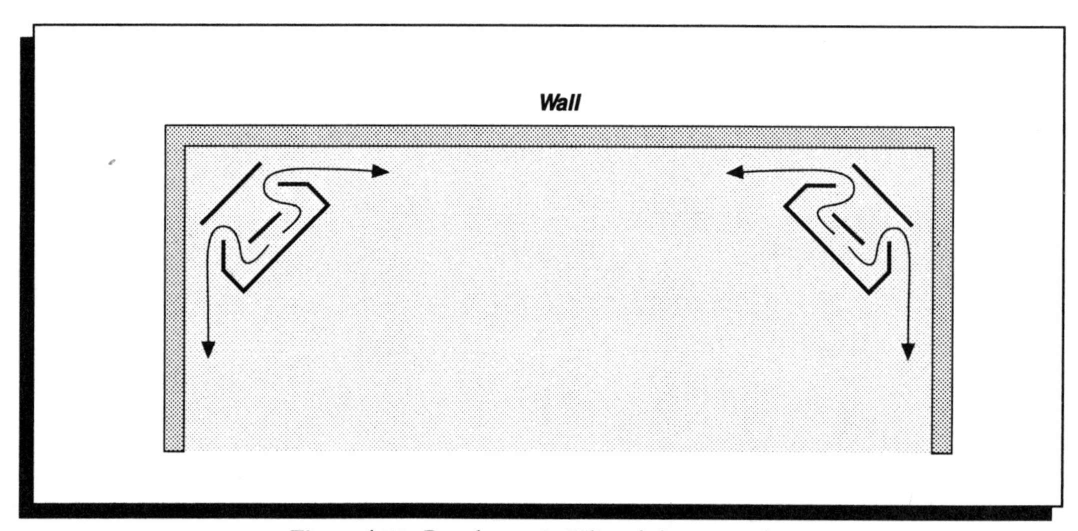

Figure 4.53 Bass horns in Klipsch horn speaker.

EQUALIZATION

It may be necessary to adjust the frequency response of a hi–fi in order to compensate for that of the room, speakers, or tape recordings, or for just exploring with sound. Usually it is necessary to create a more natural sound by compensating for room resonances. There is already in the pre–amp a control set, the BASS and TREBLE, which modifies the frequency response of the system but it is coarse and it would not be able to deal effectively in compensating for room resonances. The solution is an electronic unit, called the equalizer, which can selectively modify the frequency response of the system. An equalizer is defined as:

***Equalizer* = electronic unit which divides the audio frequency spectrum into several different bands, giving control over each by amplitude boosts or cuts with independent controls for the left and right channels.**

Usually there are eight to ten bands, each one octave wide. More expensive units are 1/3 octave wide. Figure 4.54 shows a simple equalizer, where levers can be moved to increase or to decrease the amplitude at frequencies around center frequency indicated with each lever.

There are two types of equalizers:

1. Graphic equalizer: most common. Levers form a rough graph of adjusted frequency response, and frequencies for each lever are fixed.
2. Parametric equalizer: more expensive, can vary center frequency continuously and its volume level. For example, a 1,000 Hz band can be shifted to 900 Hz or 1,200 Hz. On some models the frequency width of a band can also be varied.

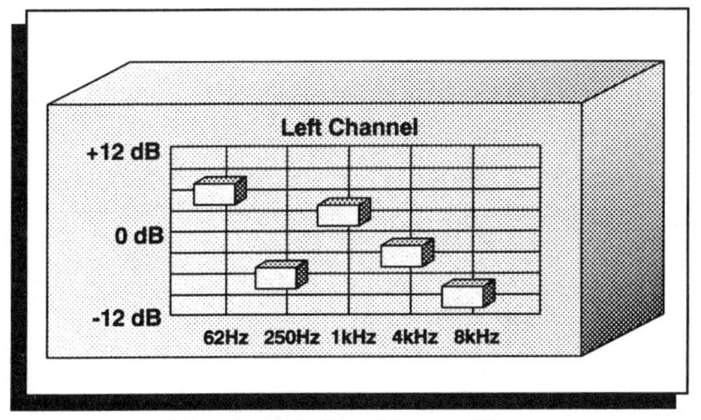

Figure 4.54 Graphic equalizer.

In operation, the equalizer is adjusted by ear or electronically. Ideally it is adjusted electronically by sending into the system a mixture of frequencies from 20 Hz to 20,000 Hz. A microphone is then used to measure the frequency response of the whole system, including the effects of the room. This frequency response is used to adjust the equalizer until the compensation is complete (i.e. when the total frequency response is flat). The mixture of frequencies injected into the system for testing is known as noise. But, how is the mixture of frequencies made? If we take the same amount of power at each frequency and add them all together, this is called white noise. There is a problem here in that this mixture does not give a fair test of the system. There are more frequencies at high frequencies than low frequencies. This would favor the high frequency end of the response of the system. To provide fairness, the spectrum from 20 Hz to 20,000 Hz is divided into equal bands in frequency, such as octaves, and then each band is given the same amount of power. The resultant mixture of test signals is a special type of noise called pink noise. Its definition is:

***Pink noise* = mixture of frequencies so that average power per octave or per frequency band is the same**

while white noise is:

***White noise* = mixture of frequencies so that average power per frequency is the same**

Usually an equalizer injects pink noise into the system and the resultant signal is picked up by a microphone. Its spectrum is then presented on a display. The levers are adjusted to equalize the frequency response.

This is a very useful unit which provides substantial changes and control in the frequency response of a system.

Sensitivity of Loudspeaker

In selecting a loudspeaker, the efficiency is of interest. This is specified as the speaker sensitivity. It tells us how much sound power is produced for a given amount of electrical power input. Its definition is:

> ***Loudspeaker sensitivity* = output of loudspeaker in dB measured 1 m away from speaker when 1 watt of electrical power goes into the speaker.**

A typical value of sensitivity is: 84dB / watt at 1 m. This means that a speaker will produce sound at 84dB measured 1 meter away from speaker when the loudspeaker receives 1 watt of electrical power.

The specifications on loudspeakers state the maximum power that the speaker can handle and the minimum power that is needed so that it can be heard. The maximum power specification tells us how much a speaker can handle safely. Higher power levels will damage it. The minimum power rating tells how much power is needed so that the speaker can be heard. In selecting an amplifier it is always prudent to choose one whose power output will not exceed the maximum permitted power to the speaker, otherwise the speaker could be damaged.

Summary of Terms

Baffle: partition to separate rear sound of driver from front sound to avoid destructive interference or to delay rear sound and avoid destructive interference.

Bass Reflex: enclosure which uses sound from rear of driver in enclosure to reinforce sound in front of speaker by means of a port or duct.

Compliance: ability of a mechanical system to move when a force is applied to it. It is the displacement per unit force.

Cross–over frequency: common frequency at which signal to one driver is reduced and to the next driver is increased and the two are equal.

Cross–over network: electric system which reduces signals going to one driver and routes them to the next one.

Diaphragm: vibrating membrane of driver which pushes air in and out.

Dispersion of Speaker: ability of speaker to radiate sound in all directions around it at low frequencies and to beam it mainly forward at high frequencies; it is caused by speaker diffraction.

Distortion: ability of system to modify original waveform and its spectrum.

Driver: speaker without enclosure

Efficiency: fraction of electrical power sent to speaker which becomes converted to sound, expressed in percent.

Enclosure: structure which separates rear sound from front sound of driver and adds quality to the sound radiated.

Helmholtz Resonator: resonant acoustic cavity whose oscillatory response is well– known and which can be used as a model for more complicated resonant structures, like a speaker enclosure.

Horn: structure which matches well the vibrations of a diaphragm to that of air. When combined with a driver it is capable of achieving very high efficiencies.

Infinite Baffle: enclosure which traps rear sound of driver so that it will not interfere destructively with front sound.

Loudspeaker: converts electrical signals into sound.

Octave: interval which consists of a factor of two in frequency.

Passive Radiator: driver without any electrical signal going to it and which is located in the port of a bass reflex. It is driven by the sound from the enclosure.

Spring Constant: force needed to extend or compress a spring by a certain distance.

Voice Coil: part of driver which carries electrical signals that make it move in the magnetic field of permanent a magnet. It is attached to the diaphragm and makes it vibrate.

NAME ____________________ DATE ____________

Questions for Review

1 Discuss two functions of a speaker enclosure.

2 Explain how standing waves on a cone can affect the driver's frequency response.

3 Why can one single driver not be used for the whole audio range?

4 Why is a driver not 100% efficient?

5 Discuss the four basic speaker enclosures.

6 Why is a horn speaker more efficient than a cone driver?

7 In designing a loudspeaker, what factors must be taken into account?

8 What is speaker dispersion?

9 Why are tweeters usually small and woofers are large?

10 What does a cross–over network do?

NAME ______________________________ DATE ____________

Please select the best answer.

1. Of the following types of speaker systems, the one considered to be the most efficient is the:
   ___A. Bass reflex
   ___B. infinite baffle
   ___C. acoustic suspension
   ___D. horn
   ___E. subwoofer

2. A 3–way speaker system means:
   ___A. there are 3 speaker enclosures
   ___B. there are 3 types of drivers in each enclosure, the woofer, midrange, and tweeter
   ___C. there are 3 surround speakers
   ___D. the speakers are connected to 3 separate amplifiers
   ___E. each driver has 3 parts, a diaphragm, a voice coil, and a magnet

3. A loudspeaker which contains a woofer, a midrange, a tweeter, and a supertweeter has _______ cross–over points.
   ___A. 1
   ___B. 2
   ___C. 3
   ___D. 4
   ___E. 0

4. When a speaker diaphragm is moving backward a _______ is created at the front surface.
   ___A. high pressure region
   ___B. rarefaction
   ___C. condensation
   ___D. node
   ___E. standing wave

5. As you increase the efficiency of a loudspeaker, the amplifier power needed to drive that speaker at a constant sound loudness must_______.
   ___A. remain the same
   ___B. increase
   ___C. decrease
   ___D. increase when the frequency increases
   ___E. remain the same since the amplifier determines the efficiency

6. 125 watts of electrical power are fed into a speaker whose efficiency is 5%. How much power is lost in heating the driver?
   ___A. 125 watts
   ___B. 5 watts
   ___C. 6.25 watts
   ___D. 118.75 watts
   ___E. 130 watts

7. If the minimum recommended power rating for a speaker is high, the speaker is:
   ___A. very efficient
   ___B. a low quality speaker
   ___C. very inefficient
   ___D. hard to damage
   ___E. easy to damage

8. The main purpose of all loudspeaker enclosures is to:
   ___A. separate the woofer from the tweeter
   ___B. separate the front–sound from the back–sound which has been produced by the diaphragm, especially at low frequencies
   ___C. hold the driver upright
   ___D. increase the compliance of the driver
   ___E. help set up standing waves behind the driver

9. The type of loudspeaker enclosure which necessarily uses the entrapped air to reduce the compliance of a speaker diaphragm is called:
___A. acoustic suspension
___B. infinite baffle
___C. acoustic labyrinth
___D. bass reflex
___E. bass reflex with passive radiator

10. Which one of the following systems would sound the loudest?

	Amplifier power to speaker	Speaker efficiency
___A.	110 watts	1.5%
___B.	80 watts	4.2%
___C.	9.5 watts	42%
___D.	9.0 watts	45%
___E.	225 watts	2.5%

11. Sound radiated from a speaker is most directional (spreads out least) at:
___A. the low frequencies
___B. the mid–frequencies
___C. the high frequencies
___D. sound spreads out equally well at all frequencies
___E. sound is a longitudinal wave and it does not spread out

12. Speaker X moves 2×10^{-2} cm when a force of 0.5 dyne is applied to it; speaker Y moves 0.8×10^{-2} cm when a force of 0.4 dyne is applied to it. Both have the same diameter. What force must be applied to Y so that it moves double the volume of air that X moves?
___A. the same as for X
___B. 2 times more than for X
___C. 2 times less than for X
___D. 4 times less than for X
___E. 4 times more than for X

13. As a general rule one can say that the air inside an enclosure (especially a closed one) will ______________ a cone driver's compliance.
___A. increase
___B. decrease
___C. not change
___D. double for every octave change
___E. double when the enclosure is halved in volume

14. Which type of loudspeaker needs to be unreasonably large in order to reproduce the lowest bass frequencies?
___A. the acoustic suspension
___B. the horn
___C. the bass reflex
___D. the infinite baffle
___E. the air suspension

15. The frequency response for a loudspeaker:
___A. tells how fast the diaphragm is moving at each frequency
___B. is equal to the efficiency of the loudspeaker
___C. tells how many driver are located within the enclosure
___D. tells how well the loudspeaker reproduces each frequency in relation to all others
___E. tells how well sound is dispersed in all directions

16. Cross–over networks are used in loudspeakers to:
___A. provide baffling
___B. separate the sound into different frequency bands, and then send each band to the appropriate driver
___C. increase the dispersion of the woofer
___D. separate the sound from left and right speakers
___E. increase the Doppler Effect

NAME ______________________________ DATE ____________

17. The speaker specification which tells the power needed to damage a speaker is called:
___A. cross–over frequency
___B. minimum recommended power
___C. maximum power rating
___D. frequency response
___E. dispersion

18. Of the following loudspeakers, the best sounding one will contain:
___A. 1 woofer and 1 tweeter speaker
___B. 2 woofers, 2 midrange, and 1 tweeter speaker
___C. 1 woofer, 1 midrange, and 1 tweeter speaker
___D. 2 woofers, 1 midrange, and 1 tweeter speaker
___E. none of the above will necessarily give the best sound

19. What is the frequency of a tone four octaves above 100 Hz?
___A. 1600 Hz
___B. 400 Hz
___C. 800 Hz
___D. 3,200 Hz
___E. 16,000 Hz

20. The speaker system which uses the back–radiated sound to reinforce the front–radiated sound is:
___A. a folded horn
___B. a bass reflex speaker
___C. an acoustic suspension speaker
___D. an infinite baffle speaker
___E. a cone driver

21. At what frequency will dispersion effects start to become important for a driver 0.25m in diameter?
___A. 1376 Hz
___B. 688 Hz
___C. 344 Hz
___D. 2752 Hz
___E. 172 Hz

22. When the cross–over frequency for a two–way loudspeaker is 500 Hz, at that frequency:
___A. a small amount of sound will be sent to the woofer and a very small amount of sound will be sent to the tweeter
___B. a small amount of sound will be sent to the woofer and a large amount of sound will be sent to the tweeter
___C. a large amount of sound will be sent to the woofer and a small amount of sound will be sent to the tweeter
___D. equal amounts of sound will be sent to the woofer and to the tweeter
___E. no sound will be sent to the woofer and all the sound will be sent to the tweeter

23. For a horn which has a mouth 0.5 m in diameter, the lowest sound which can still be produced by the driver is:
___A. 688 Hz
___B. 172 Hz
___C. 1376 Hz
___D. 86 Hz
___E. 20 Hz

24. When the compliance of a driver is decreased, its natural resonant frequency will:
___A. increase
___B. decrease
___C. remain the same
___D. become zero
___E. remain the same since it only depends on the cone mass

25. An equalizer is used to:
___A. improve the frequency response of loudspeaker
___B. reduce standing waves in a room
___C. improve the frequency response of amplifier
___D. make up for deficiencies of our ears
___E. all of the above are correct

26. When the port area in a Helmholtz resonator is decreased, its resonant frequency will_______ .
___A. increase
___B. decrease
___C. remain the same
___D. be almost zero
___E. depend only on the volume of air inside the resonator

27. Pink noise is:
___A. mixture of many frequencies such that the power is the same for each frequency
___B. mixture of many frequencies such that the power is the same for each octave
___C. special music in an equalizer which is added to the music being played to make it brighter
___D. all the odd harmonics of a complex wave
___E. noise at one frequency

28. A horn is more efficient than a cone driver because:
___A. it has a small diaphragm
___B. it is long
___C. it is usually made out of metal
___D. it matches well the vibrations of a small diaphragm to those of a large volume of air
___E. it has standing waves

29. A loudspeaker with only a single driver would have problems of:
___A. efficiency
___B. dispersion
___C. sensitivity
___D. dynamic range
___E. standing waves in the loudspeaker enclosure

30. A speaker receives 12 watts of electrical power and it wastes 10.5 watts in the form of heat. What is the efficiency of this speaker?
___A. 1.5%
___B. 12.5%
___C. 87.5%
___D. 12%
___D. 0.125%

CHAPTER 5 ELECTRICITY

The Hi–Fi system deals with electrical phenomena for acoustic information storage, broadcasting and play back. Electrical processes are at the center of the complicated chain which deals with the processing of audio signals. Electricity made possible the hi–fi and the spectacular electronic advances that we are witnessing. In this chapter the subject of electricity will be introduced and then applied to various parts of the hi–fi system. This chapter is important as an introduction to subsequent chapters where electrical phenomena control the functioning of hi–fi systems. For the sake of completeness the subject of electricity is introduced at a fundamental level and then it is developed for the sake of understanding how a hi–fi system works.

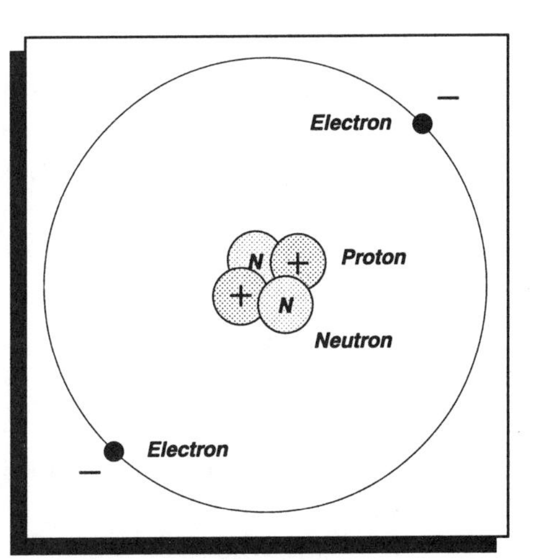

Figure 5.1 Example of an atom, a Helium atom.

5.1 INTRODUCTION: ELECTRICITY AT REST

The fundamental quantity that is the basis of all electrical phenomena is the electric charge. Since the charge comes from an atom, let us consider an atom. Every atom is made up of a nucleus which has neutrons and protons, and of electrons which are around it. As an example, Figure 5.1 shows a helium atom and atom's basic parts. The protons are positively charged and the electrons are negatively charged.

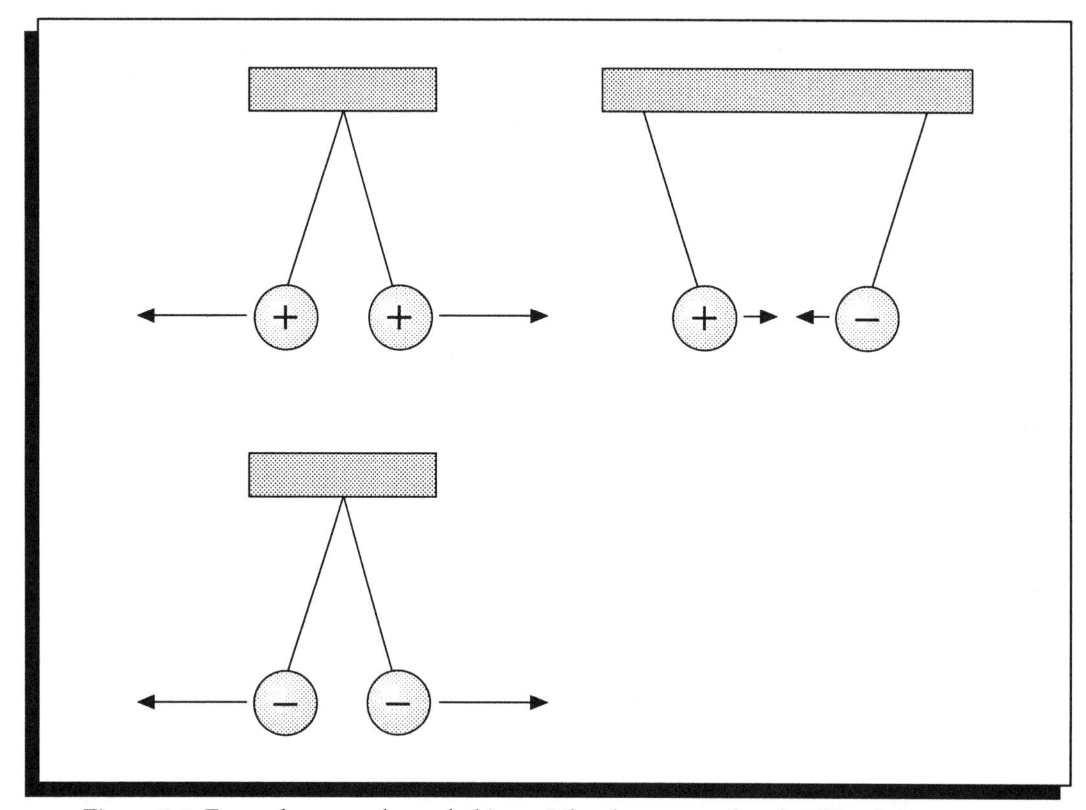

Figure 5.2 Forces between charged objects. Like charges repel and unlike charges attract.

In a neutral atom the number of electrons will balance the number of protons. The atoms of one element differ from another element by the number of electrons and of protons. Neutral particles, neutrons help to keep the protons together inside the nucleus. Hydrogen, for example, has one electron and one proton, and carbon has six electrons and six protons. An atom can acquire one or more extra electrons, and hence it becomes negatively charged. An atom can lose one or more electrons, and then it will become positively charged. Note that it is always a transfer of electrons. One can extend these arguments to objects which can be positively charged or negatively charged, or remain neutral.

Suppose that two positively charged objects are placed next to each other, as in Figure 5.2. There is a force between them causing them to repel each other. When a positively charged object is placed near a negatively charged object, there is a force of attraction between them. Also, two negatively charged objects will repel each other. Our conclusion is:

the force between two charged objects will cause:

— like charges to repel
— unlike charges to attract.

The nature of the force is electrical. Investigations in the 18th century by Coulomb into this phenomenon led to a formulation of the observed facts known as Coulomb's law. It states how the force between charged objects depends on the distance between them and on the charges.

The force between two charged particles depends on:

(i) charges on each particle
(ii) $1/(\text{separation between particles})^2$

when the size of the particles is much smaller than the separation between them.

According to this law, the force between two charged objects increases as they are brought closer together, and it decreases when they are separated further apart.

The unit of charge = 1 Coulomb and it consists of the charge of 6.28×10^{18} electrons.

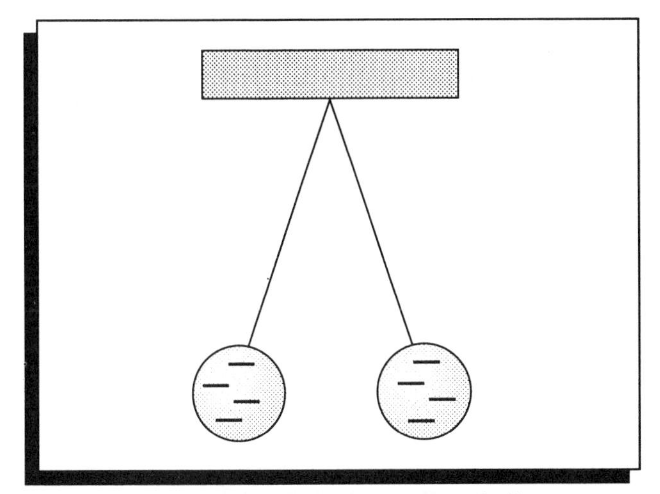

Figure 5.3 Charged ping-pong balls repelling each other.

Every charged object exerts a force on every other charged object even though they do not touch. An experiment can be performed on two suspended ping–pong balls, Figure 5.3. When they are rubbed with fur, some of the electrons from the fur are transferred to the ping–pong balls. They become negatively charged and hence they repel each other. They do not touch, and yet they know that they must repel each other. This ability of knowing how the force will be exerted resides in the concept of an electric field existing around each charged object. We define it as:

electric field **= energetic region of space surrounding a charged object.**

This can be verified by bringing a test charge near the charged object. The presence of a force on that test charge, whether attractive or repulsive, by the charged object indicates that there is an electric field around that object. Figure 5.4 indicates this situation. We represent the electric field by imaginary lines. The lines show the direction of the force

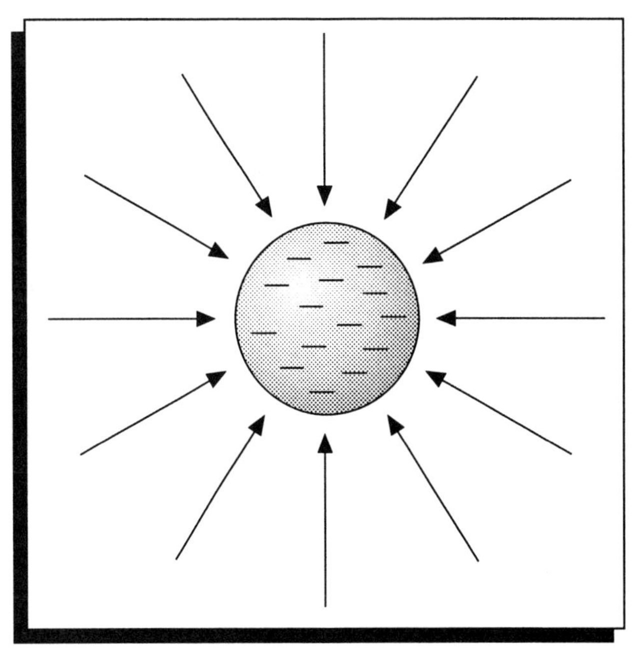

Figure 5.4 Electric field produced by a charged object.

on a small positive test charge. The density of lines indicates the strength of the electric field. Hence, an electric field can also be defined as:

electric field **= region of space where electric force is felt by test charge.**

As an example, consider two plates shown in Figure 5.5, where charges are placed on each plate. When a small positive test charge is located between the plates, it will move toward the negative plate. This is indicated by an arrow; imaginary arrows represent an electric field.

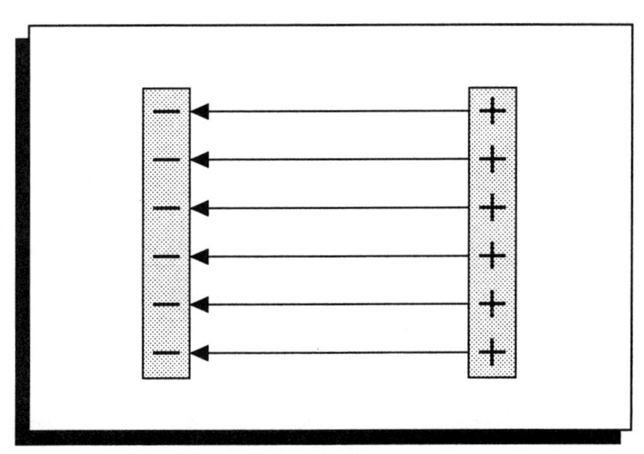

Figure 5.5 Electric field between two charged plates.

If free to move, a positive test charge at the positive plate will go from the positive plate to the negative plate. The test charge moves from a point of high energy to one of low energy, just like a ball rolling down a hill. A certain amount of energy was needed to move that test charge. If there were two test charges then twice as much energy would have been used to move the two charges. Rather than keep track of the total energy used, it is simpler to consider the energy lost (or gained) for every charge (i.e., the energy difference per charge between the two plates). This is known as the voltage difference. Its definition is:

voltage difference **= energy difference/charge**

Since it is the energy difference that is important, one plate can be considered as the reference, while the other one would be at a certain energy per electron above it (i.e. at a certain voltage). The unit of voltage is the volt.

In some cases the voltage will be small and hence, a few simple definitions are introduced here. Terms that will come up are:

0.001 volt = 1 millivolt (1mV)
0.000001 volt = 1 microvolt (1μ V)

An example of voltage is the energy difference between the terminals of a battery, where one terminal is above the other by so many volts. Another example is the energy difference across the output terminals of an amplifier, there are so many volts. Figure 5.6 illustrates this. In the case of a battery (here the energy is stored chemically) one electrode can be considered as a reference at 0 volt, while the other one is (+) 12 volts, for example. This voltage can make a positive test charge move from (+) 12 volts to 0 volts. For an amplifier, the voltage at the output terminals can be used to push electrons through the speaker attached to its terminals.

Before discussing the flow of electrons in circuits, there are some interesting applications of the static electricity concepts to the hi–fi which illustrate the principles presented here. Such applications will now be discussed.

5.2 ELECTROSTATIC AND PIEZOELECTRIC SPEAKERS

The electrostatic speaker makes use of the force between two charged objects to create motion of a diaphragm at audio frequencies. Electrical audio signals are transformed into mechanical motion through electrostatic forces of attraction or repulsion. When a voltage is applied across two fixed plates, an electric field is created between them, as in Figure 5.7. If a positive charge is placed anywhere in this space, there will be a force on it causing it to move to the right. If the polarity of the voltage across the plates is reversed, the force on the charge will be to the left. Hence, an alternating voltage will cause the force on the charged object to vary in direction, leading to an oscillation of the object.

A simplified approach to an electrostatic speaker is shown in Figure 5.8. Here a thin sheet of some light–weight material, such as a conducting mylar,

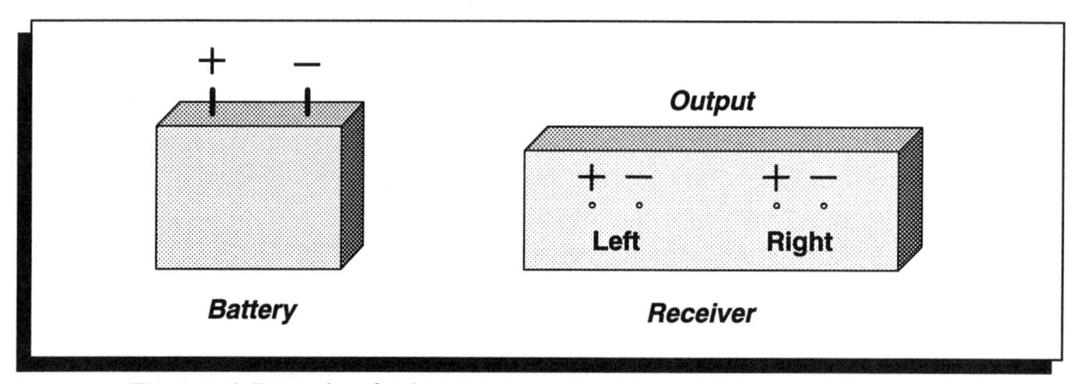

Figure 5.6 Examples of voltage sources; a battery, and the output of a receiver.

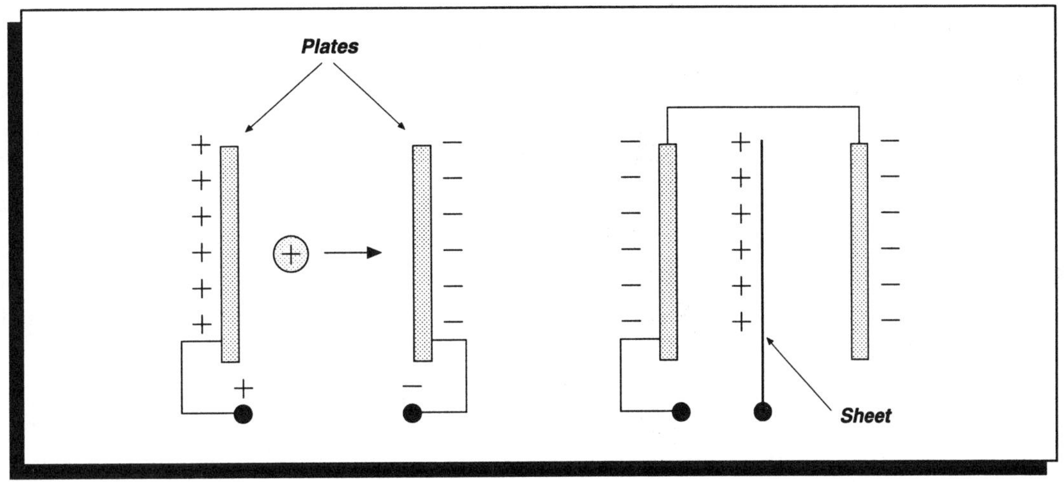

Figure 5.7 Electrostatic speaker: basic principle and actual speaker.

is mounted between two conducting plates. When appropriate voltages are placed between the sheet and the plates, the sheet can be made to vibrate by electrostatic forces acting on it, thus producing sound. The sound is transmitted through perforated plates. A high voltage power supply produces a constant electrical charge on the sheet and at the same time it establishes an electric field between the sheet and the plates. This is a static arrangement, where the sheet is in equilibrium between the two plates. The voltage producing such a state is known as the bias voltage. When an audio signal voltage is applied on top of the existing bias voltage and between the two plates (how this can be achieved will be discussed in a later chapter), the static arrangement will be disrupted and there will be an extra force acting on the sheet causing it to vibrate. At some instant, one plate will be more positive than the other one, Figure 5.9, and since a positive object will be attracted to a negative one but repelled by a positive one, the sheet will move to the right. Likewise, when the signal causes the voltage across the plates to change polarity, the

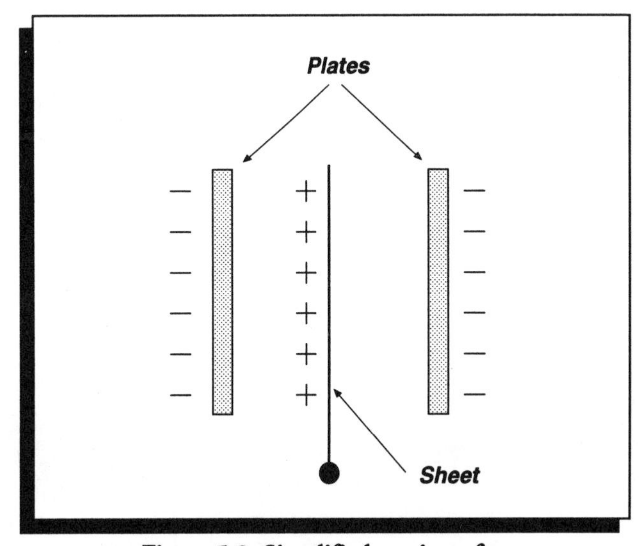

Figure 5.8 Simplified version of an electrostatic speaker at equilibrium.

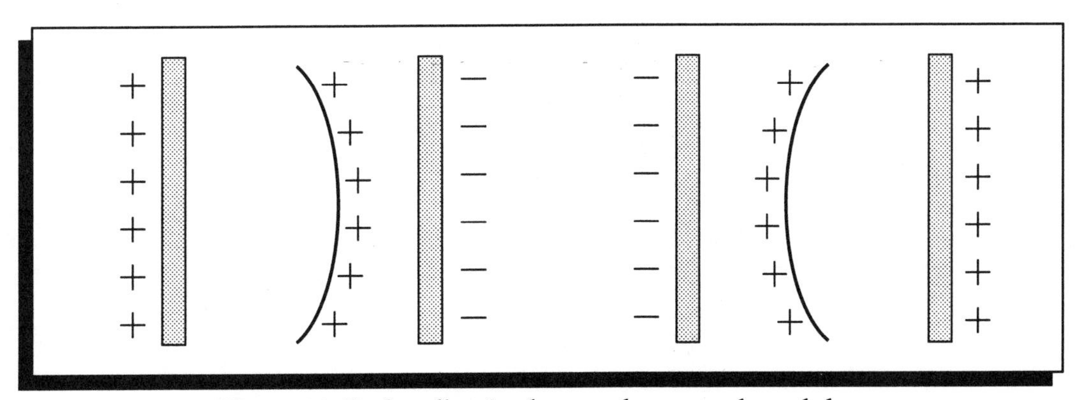

Figure 5.9 Push-pull action by two plates on a charged sheet.

sheet will move to the left. Such pushpull action by the signal voltage on the sheet will cause it to vibrate at the frequency of the signal and hence sound of that frequency will be produced. This action requires the simultaneous presence of two voltages between the plates:

— a bias voltage, between the sheet and the plates, to establish an electric field
— a signal voltage between the two plates to cause vibrations

Figure 5.10 shows an actual electrostatic speaker. To produce high intensity sound, a large sheet is used and it is placed in between two closely–spaced plates. A regular amplifier or receiver needs special equipment to be adapted to such speakers. In particular, a high voltage power supply is necessary for the bias voltage, and appropriate connections have to be made of the amplifier signal to the plates in the presence of the bias voltage.

Usually this type of loudspeaker does not use an enclosure. The low frequency cancellations between front and rear are reduced by making the electrostatic speaker large; it acts like its own baffle between front and rear. Some speakers are as large as 8 feet high and 3 feet wide. The high voltages in this speaker can cause break–down of the air between the sheet and plate, with a resulting cracking sound, especially when the speaker is driven hard. To reduce this problem, some manufacturers place an inert gas in the space between the sheet and the plates.

This type of loudspeaker has been popular for many years with some listeners. The driving action of the diaphragm is designed so that standing waves on it are minimized. Distortion is low since the sheet does not have to move with a large amplitude to produce loud sound. In some speakers, only the tweeter and mid–range are electrostatic, and the woofer is a cone driver. The efficiency of an electrostatic speaker is higher than that of a cone because the diaphragm is thin and light and is a good match to the mechanical characteristics of air, which is also light.

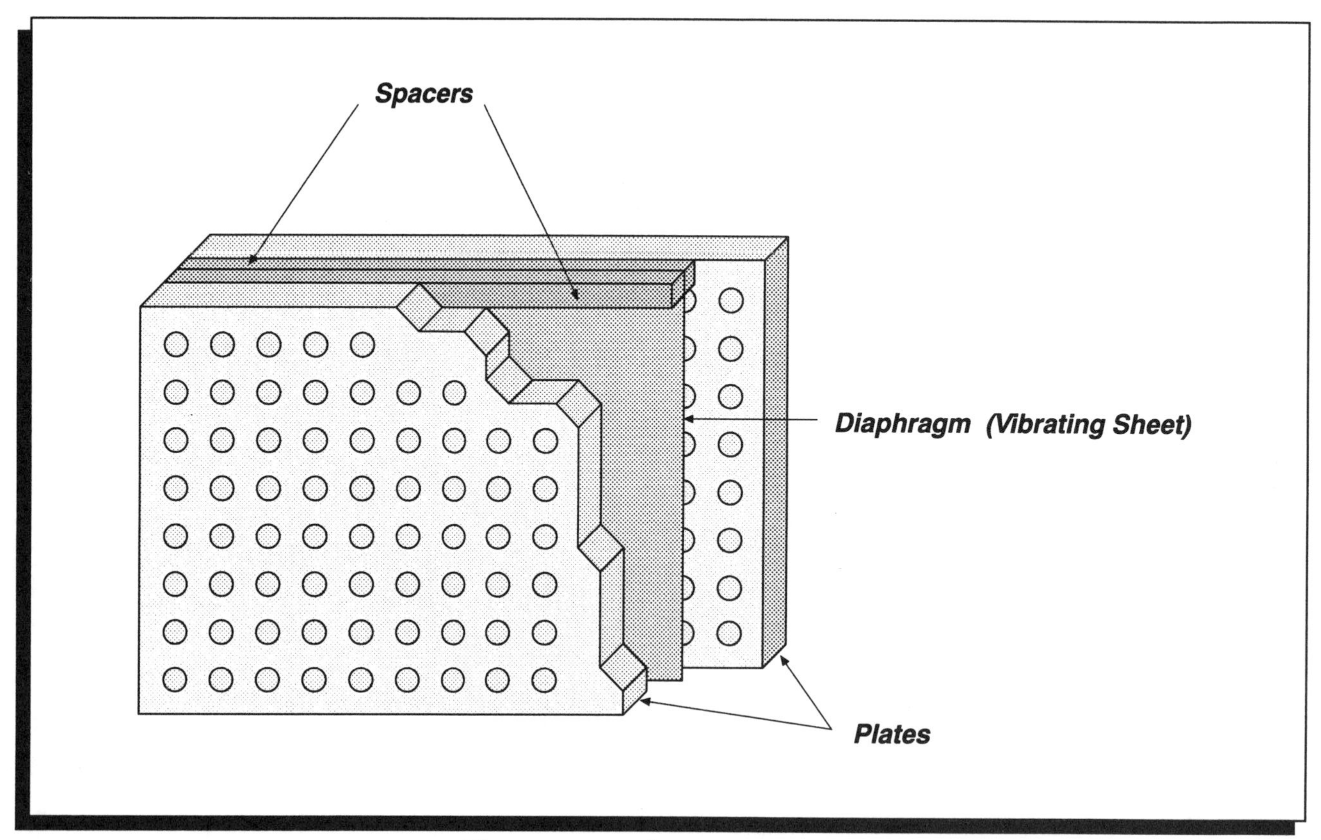

Figure 5.10 An electrostatic speaker.

PIEZOELECTRIC DRIVER

Electrostatic forces are also used in some drivers, but with an entirely different approach than in the electrostatic speakers. The mechanical properties of this driver make it attractive only for high frequency speakers, such as a tweeter driver. When pressure or stress is applied to certain crystals or ceramics, electric charges are produced at the surface. This is known as the piezoelectric effect, and it was discovered in 1880 by Jacques and Pierre Curie. The name comes from the Greek word *piezo,* meaning pressure. It occurs in solids which are not symmetric at the molecular level. Consider one such material, quartz crystal as shown in Figure 5.11. It consists of silicon oxide, SiO_2. The ions of Silicon and Oxygen are held together by electric forces between them. When squeezed, the ions are displaced, and due to the lack of symmetry in these molecules, the negative ions of oxygen are pushed closer to one electrode in contact with the crystal, while the positive ions of Silicon are moved closer to the other surface. This creates a positive charge on one side of the crystal, and a negative charge on the other side. When the pressure is released, the surface charges redistribute themselves, making the crystal neutral again.

Soon after this discovery, it was realized that this phenomenon is reversible. The application of voltage across the material will cause a deformation due to the electric interaction between the electric field and the charge distribution in the material. These two features of piezoelectricity lend themselves to a variety of applications. Two well–known applications are:

— ***microphone:*** Stress due to sound waves will cause a voltage across the material. This provides a simple and effective detector of sound.

— ***loudspeaker:*** Voltage across the material will cause deformation. This is a convenient method of producing sound.

New man–made ceramic materials have been processed and treated thermally to achieve a high electric–mechanical conversion efficiency. One such material is PZT, which consists of oxides of lead (P for Pb), Zirconium (Z), and Titanium (T). Its behavior in the presence of a voltage is illustrated in Figure 5.12.

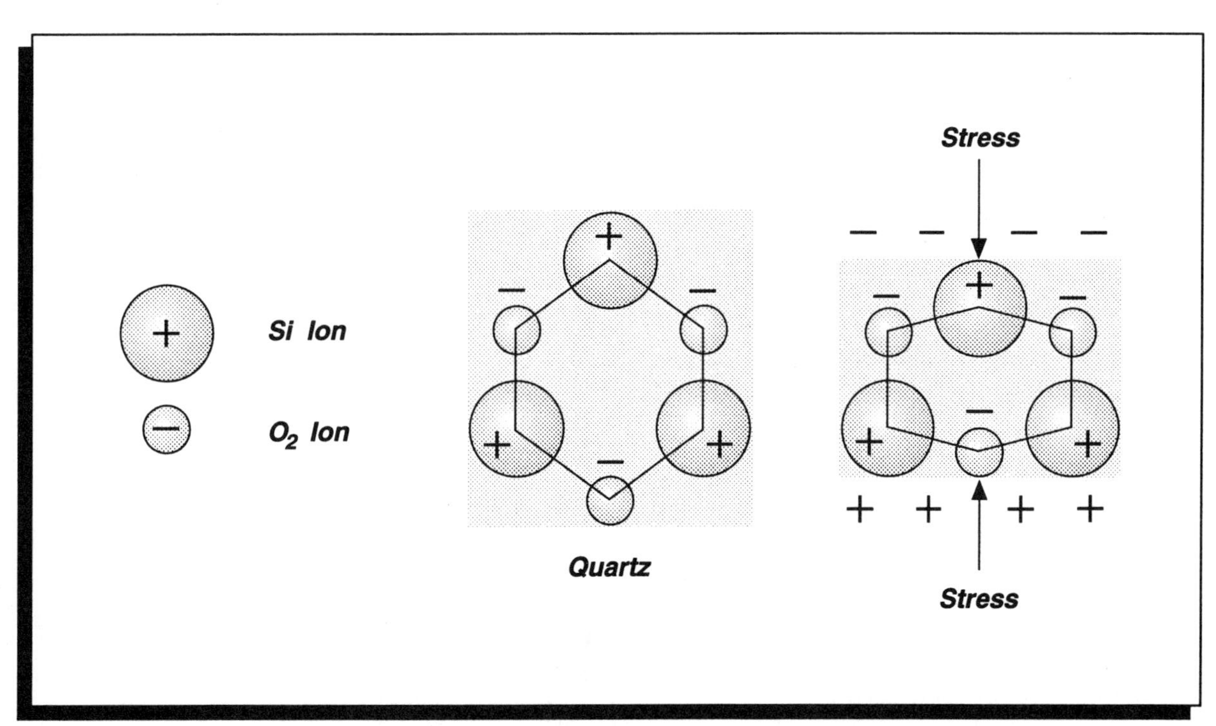

Figure 5.11 Some crystals under pressure produce positive and negative charges on surface.

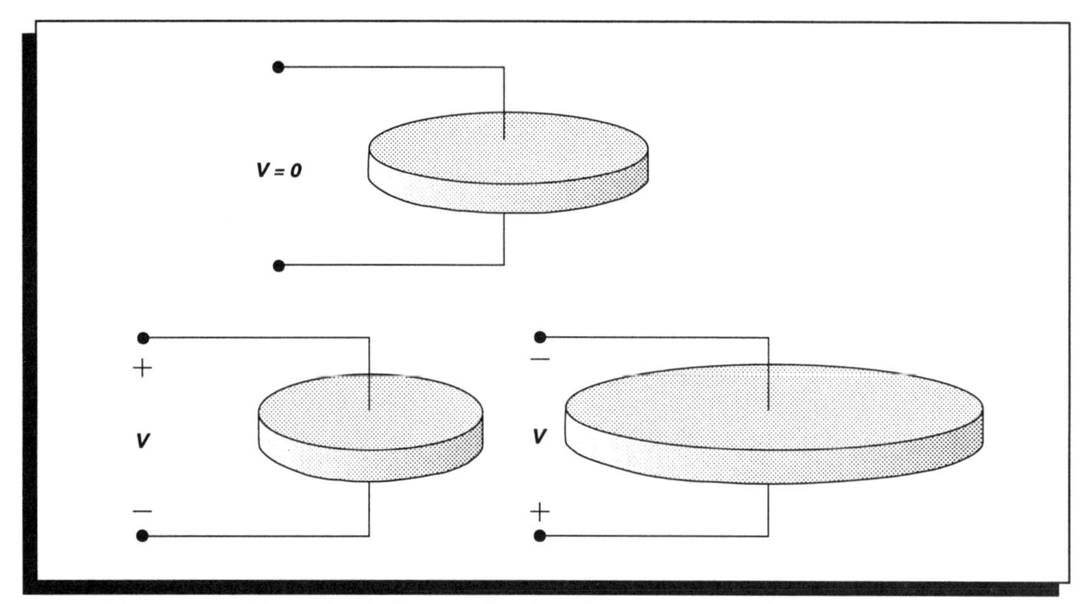

Figure 5.12 Dimensional changes of a piezoelectric ceramic when voltage is applied.

The deformations shown in Figure 5.12 appear to be large, but in reality they are quite small (10^{-6} cm when 1 volt is applied across the sample). For applications such as loudspeakers, much larger displacements are needed. This can be achieved by using a lever action where two piezoelectric plates are attached together as in Figure 5.13. The result is a push–pull action which causes a large displacement of the plates. The piezoelectric plates are sandwiched face to face between a thin metallic sheet in such a way that when one expands, the other contracts, and vice–versa. This action of one plate working against the other causes a larger bending of the plate, resulting in loud sound. A combination of two piezoelectric elements is called a **bimorph.** Based on these principles, Motorola developed a high frequency speaker, the piezo tweeter, which is capable of delivering high intensity sound with a sensitivity of approximately 92 dB/1 watt/1m. Such high efficiency was achieved by:

— attaching a small cone to the center of the bimorph, Figure 5.14, clamped at its edges. Vibrations of the bimorph at the frequency of the applied voltage cause a pumping action of the cone at the same frequency.

— matching the cone vibrations to air by means of an exponential horn, Figure 5.15.

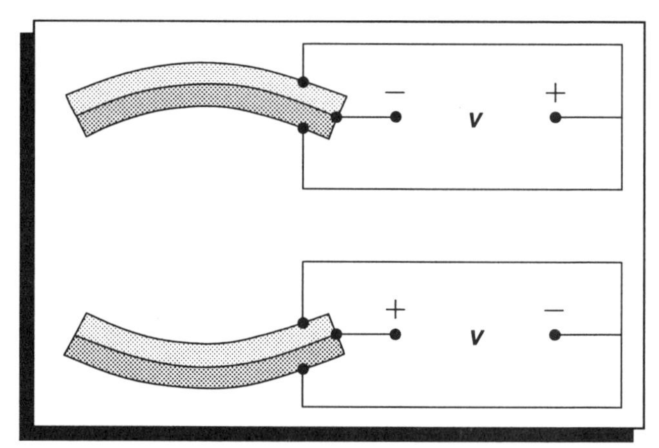

Figure 5.13 Bending action of a double piezoelectric driver

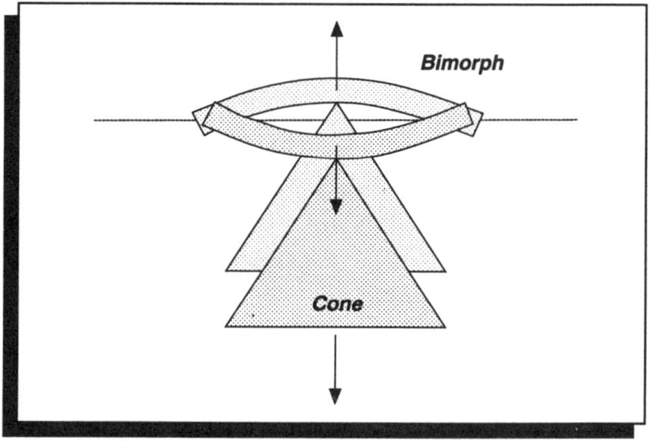

Figure 5.14 Pumping action of cone caused by bending of bimorph.

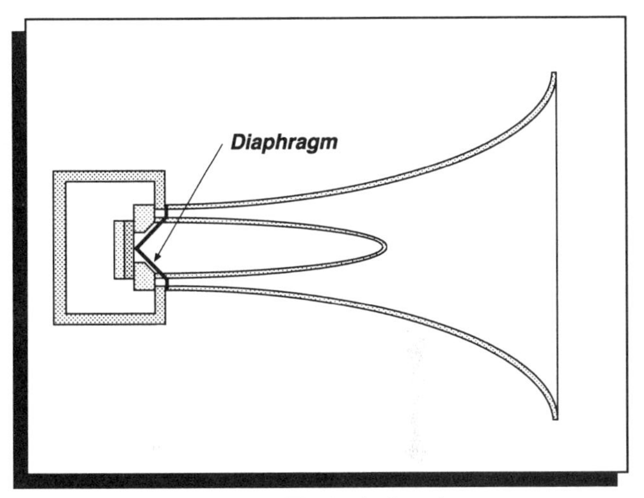

Figure 5.15 Typical piezo horn.

Typically, the frequency response ranges from about 3 kHz to 29 kHz without any cross–over network. It is a popular driver because it is light and efficient.

5.3 ELECTRICITY IN MOTION: CURRENTS AND CIRCUITS

In the previous section, static phenomena of electricity were introduced. The concept of voltage described the ability of an electric system to store energy and do work. Associated with work is the notion of motion, which can be applied to electric charges. This was implied in some of the examples of the previous section, when charges were moved from one system to another. Consider two charged objects as shown in Figure 5.16. When a wire is connected between them the

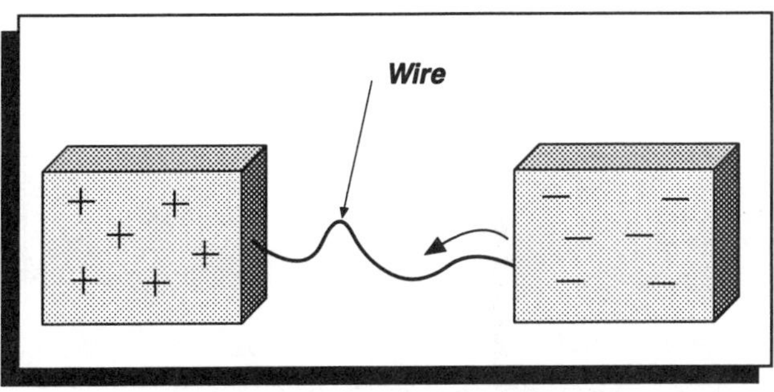

Figure 5.16 Wire connected between two charged objects allows charges to be transferred.

charges will be able to move, or to be transferred. Since there is an excess of electrons on one body and because like charges repel, there will be a tendency for electrons to move away from that body. The wire provides a means for charges to be transferred from one body to the other one. In this case, there is a flow of electrons along the wire, and this is called a current. We define it as:

> ***Current* = flow of electric charge (i.e. the number of electric charges passing a point per second).**

It is measured in amperes. One ampere is one Coulomb of electric charge passing a point per second. If a wire carries 5 amperes then another wire passing 10 amperes will be passing twice as much charge per second as the first wire. The concept of electric current should not be confused with voltage. A voltage gives electrons enough energy to flow, as in the wire of the example given here. Because electrons flow in a wire, it does not mean that the wire has a net charge; it is neutral, since there are an equal number of positive and negative charges.

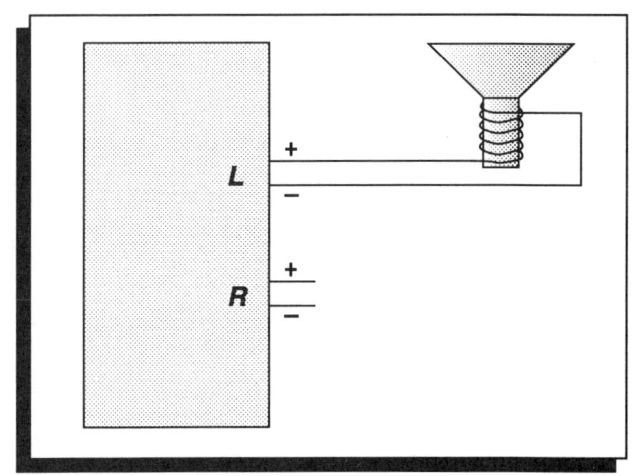

Figure 5.17 Flow of electric current from amplifier to speaker.

The concept of current can be extended to an amplifier and speakers, as in Figure 5.17. The output voltage of the amplifier causes a flow of electric current to a speaker through a set of wires. Because of the voltage at the output of the amplifier, electrons are moved along in a circuit. Although the dynamics of a circuit will be discussed in detail later in this chapter, for now, you need to know they exist. Therefore, electrons are forced to leave one terminal end of the amplifier, and they are conducted by the wire through the speaker voice coil, ultimately returning to the other terminal end of the amplifier.

The reason why current does not flow when there are no wires, even though there is air between the two bodies, can be explained by a phenomenon known as electric resistance. Electric resistance creates a friction–like effect in opposition to electrons as they flow. It is defined:

***Electric Resistance* = opposition to flow of electron current due to friction.**

The unit of resistance is called ohm. To explain why electrons flow between two bodies when a wire is attached to them, and why they do not flow when there is no wire can be explained when you consider the levels of resistance. Wire is much less resistant than air, and hence it allows a current flow. Air offers such a large resistance that there is no current flow. Electrical resistance will depend on:

— material
— geometry
— temperature.

ELECTRICAL RESISTANCE OF DIFFERENT MATERIALS

The electrical resistance can vary by enormous factors depending on the class of materials. At one end of the range of materials there are insulators and at the other end we have conductors, in particular metallic conductors.

Insulators: This is a class of materials where electrons are bound so tightly to their respective atoms or molecules that when an electric field is placed across the sample, electrons will not move, and hence they will not form a current in the samples. Some examples of insulators are: helium gas, air, plastics, wood, etc. Figure 5.18 shows a solid where the electrons are tightly bound to the atom, forming an insulator. No

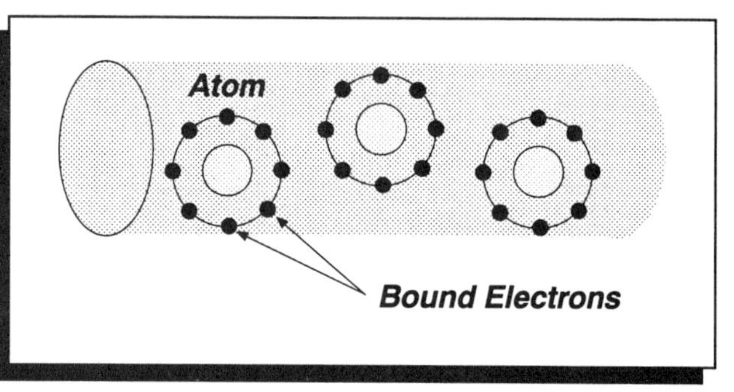

Figure 5.18 Solid with atoms where electrons are tightly bound and do not conduct electricity under normal circumstances.

current flows when a voltage is applied. An insulator has enormous resistance, in many cases its value is infinite.

Metallic Conductors: In a metal, atoms have one or more electrons in the outer orbit and these electrons are loosely attached to the individual atoms. These electrons are free to wander throughout the material and then form a sea of electrons called conduction electrons. The conduction electrons will respond to an applied electric field. These are the ones that will carry an electric current, in contrast to an insulator where the electrons are not free, but are bound tightly to each atom. The scattering of the conduction electrons by defects and vibrating ions in the material will lead to the electrical resistance of the material. The specific resistance will depend on the composition of the material. For example, lead has a higher electrical resistance than copper, and that is one reason why we do not have wires made out of lead going from the amplifier to the speaker. Figure 5.19 illustrates the motion of an electron

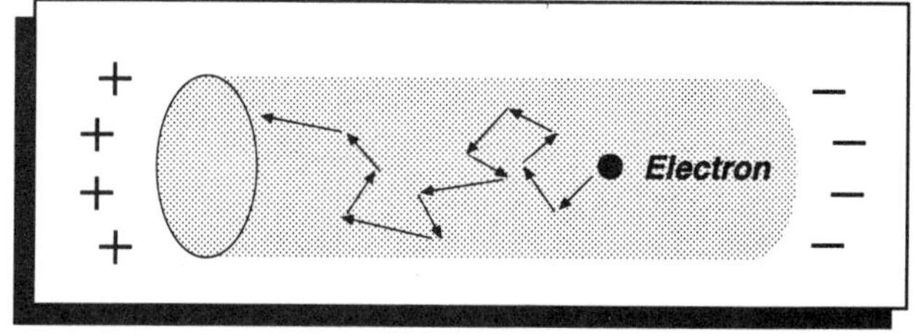

Figure 5.19 Motion of one electron in a conductor in the presence of an electric field. Changes of direction are due to scattering.

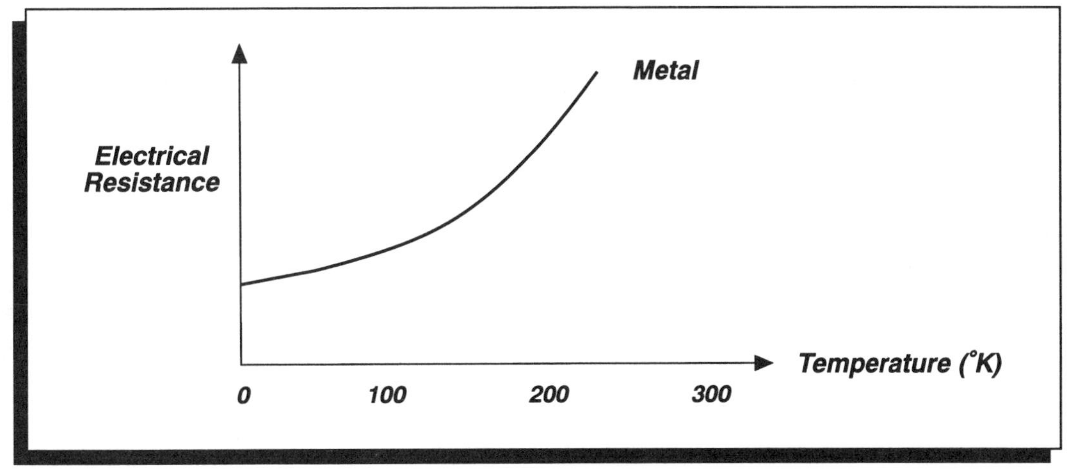

Figure 5.20 Temperature dependence of the electrical resistance in a metal conductor.

in a material when an electric field is applied to it. Every time it gets scattered by atoms, impurities or defects it changes direction. Overall, it goes in the direction imposed by the electric field, due to the voltage across the sample. These collisions reduce the current for a particular voltage.

An important question arises as to what will happen to the electrical resistance of a conductor when its temperature is reduced. In this case the vibrating atoms will have less thermal energy and they will not vibrate as vigorously, causing less scattering of the conduction electrons. Consequently, the electrical resistance will decrease as shown in Figure 5.20. The resistance change can be quite large. In copper between room temperature and liquid helium temperature (4K, which is 4 degrees Kelvin above absolute zero), the resistance changes by a factor of about one hundred or more.

There is a class of metallic materials which show at low temperatures strong departures from the behavior shown in Figure 5.20. It was discovered in 1911 that some materials when cooled to sufficiently low temperatures exhibit no resistance; their electric resistance is R=0, and they are known as superconductors. Below a critical temperature (T_c) the electrical resistance vanishes entirely. This is shown in Figure 5.21 for a sample of Yttrium Barium Copper Oxide ($Ba_2Cu_3O_7$) whose superconducting properties were discovered in the late 1980s. Its T_c = 93K. Although there are many superconductors with low T_c, this is one of the highest critical temperatures, and its discovery rep-

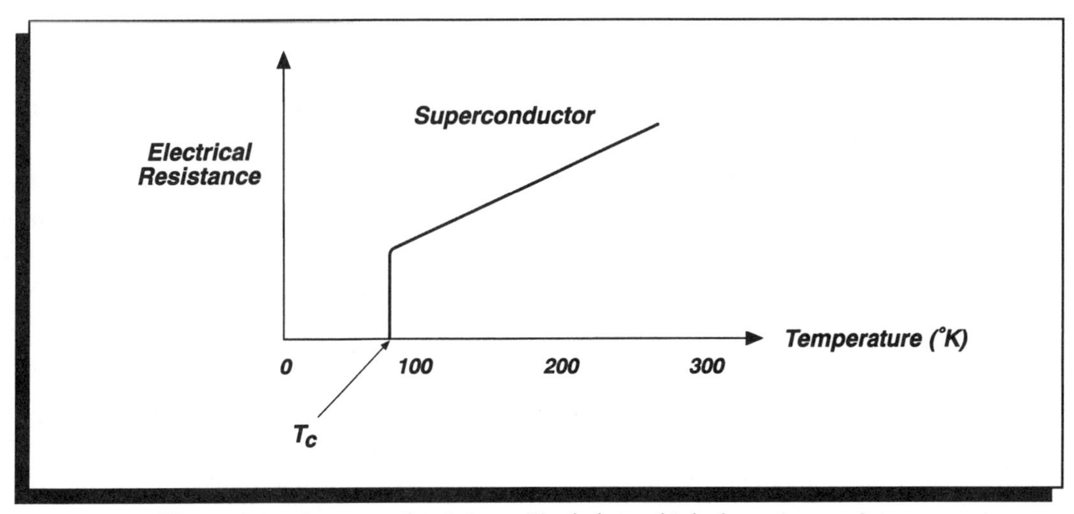

Figure 5.21 Superconductivity at T_c, below which the resistance is zero.

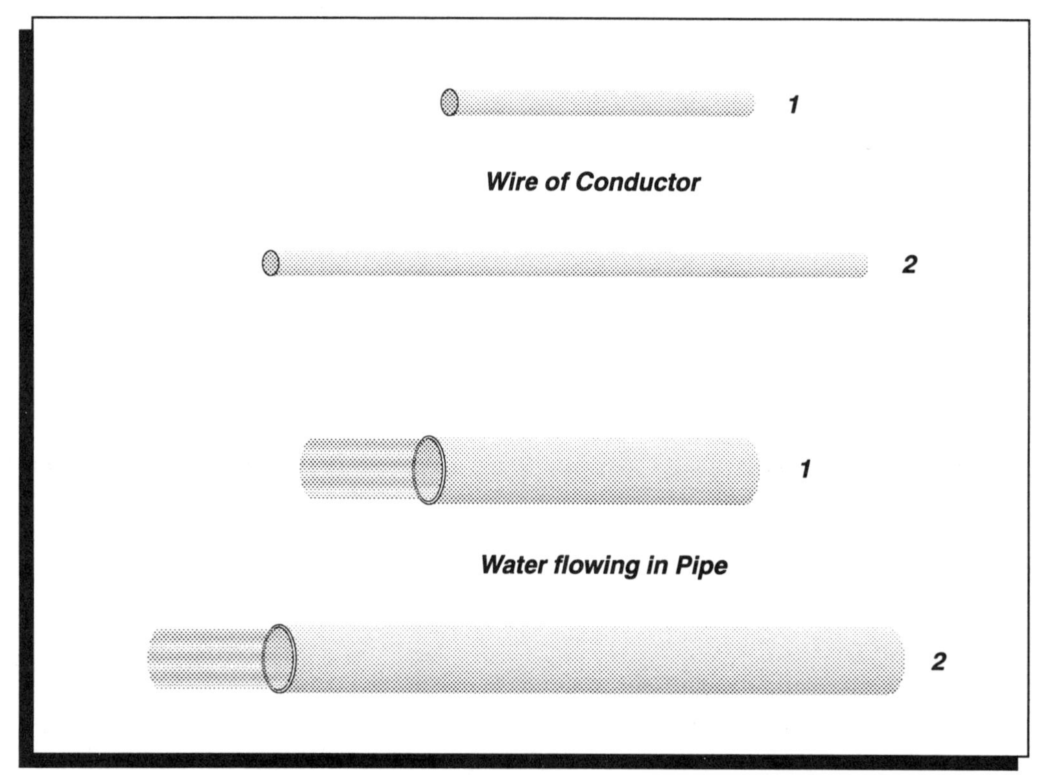

Figure 5.22 The resistance to current or to water flow increases as the length of a conductor or pipe increases. Resistance of 2 is double that of 1.

resents a major breakthrough. For any application based on superconductivity the materials must be refrigerated from room temperature (300K) down to below T_c. Presently the highest critical temperature superconductor has a T_c of around 133K. Although superconductors have no application to the hi–fi at present, it is interesting to introduce the effect as its importance will grow.

ELECTRICAL RESISTANCE AND GEOMETRIC FACTORS

For a given material how does its resistance depend on its shape? Geometric factors will determine what the resistance is. This can be inferred by analogy to the resistance of pipes with water flow. The geometric factors which affect the electrical resistance of an object are its length and its cross–sectional area.

— ***length:*** The longer the conductor, the larger its resistance. By doubling the length of a conductor, its electrical resistance doubles. The dependence of the resistance of a material on its length can be compared to water flow in a pipe. By doubling the length of a pipe, the resistance to liquid flow doubles. This is shown in Figure 5.22. For a given thickness of material, the longer the wire, the more resistance it will offer to the flow of electric current.

— ***cross–sectional area:***
The resistance of a material to an electric current varies inversely with its cross–sectional area. By doubling the cross–sectional area of a conductor, its resistance drops by half. Just as large diameter pipes will offer less resistance to the flow of water, thick conductors will offer less resistance to the current flow. This is shown in Figure 5.23.

For wires connecting the speaker to the amplifier, the resistance must not be large, so the wire must not be thin nor very long. If long wires have to be used then their cross–sectional area must be made large. The decrease in the resistance of the

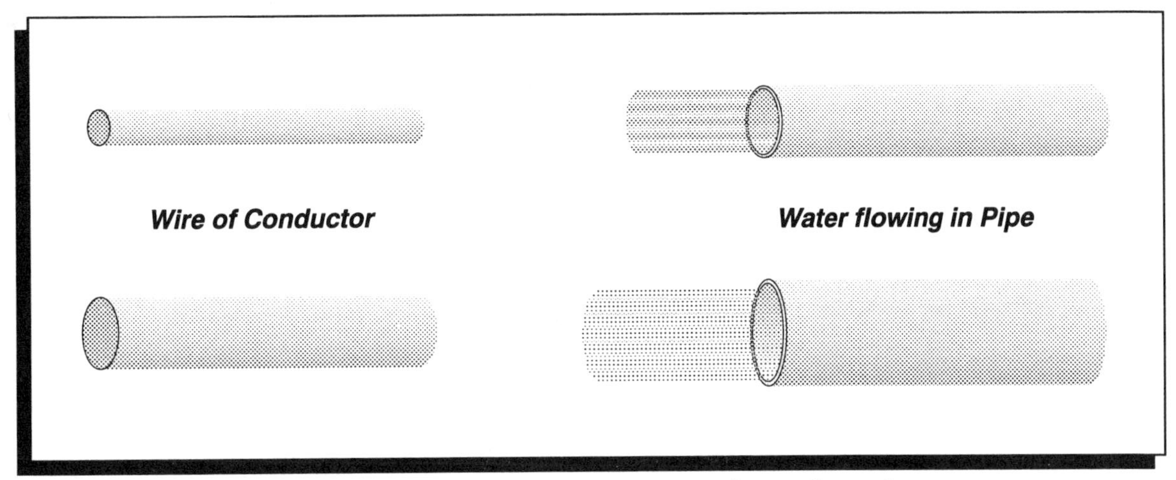

Figure 5.23 By increasing the cross-sectional area of a conductor, resistance to current or water flow decreases.

wire will compensate for the resistance increase due to its long length.

In a hi–fi there are many applications of resistors. One usually sees a variety of them inside the amplifier. They are encapsulated in plastic and they have color bands to specify the value of the resistance. Figure 5.24 shows a resistor with its color bands.

SEMICONDUCTORS

Semiconductors are a class of materials that have resistance levels between insulators and conductors. Usually these materials are good insulators when they are very pure. However, when some of the atoms are replaced with certain atoms known as impurities, then they have a tremendous decrease in resistance. The control of the resistance by the amount of impurity doping can lead to a variety of interesting phenomena and applications.

Consider pure silicon. The outer orbit of each atom has four electrons. When a solid is formed by bringing together many silicon atoms, the four outer electrons of each atom are shared by the nearest neighboring atoms; this forms the bond which holds the atoms together in this solid. Since there are no free electrons for conducting electricity, its electrical resistance is very large. When impurities are added to this solid, significant changes in the resistance will occur. Let the impurities be arsenic atoms. The arsenic atom has five electrons in its outer orbit. Four of the electrons will be shared with the neighboring silicon atoms, but one electron will be free, available for conducting electricity. This type of semiconductor is called n–type, since a negative charge per impurity atom is available for electrical conduction. In this case the impurity atom is known as a donor impurity, as it provides a free electron for conduction. It is also possible to dope a semiconductor with other impurities. When gallium is added to the silicon, bonding will also occur. However, since a gallium atom has three electrons in its outer orbit, one electron is missing for bonding. This deficiency of one electron per impurity is known as a hole, and the semiconductor is then referred to as p–type, since each hole acts like a positive charge. Electrical conduction will take place via the holes, where a nearby electron can fill the hole pushing it effectively somewhere else. The impurity atom in this case is known as acceptor impurity. Figure 5.25 illustrates the pure silicon, the arsenic–doped silicon, and the gallium–doped silicon. Although the semiconduc-

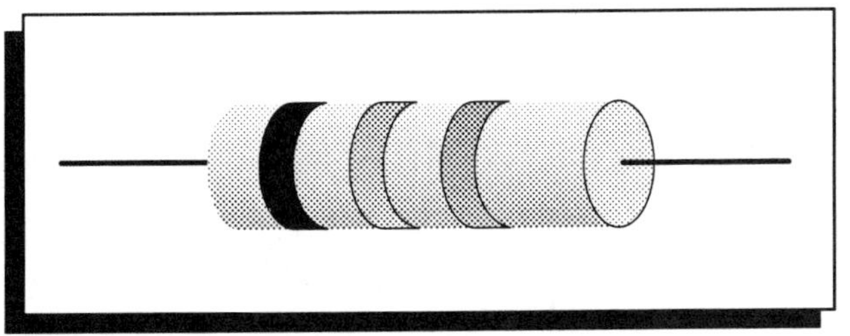

Figure 5.24 Resistor with colored bands to specify its resistance value.

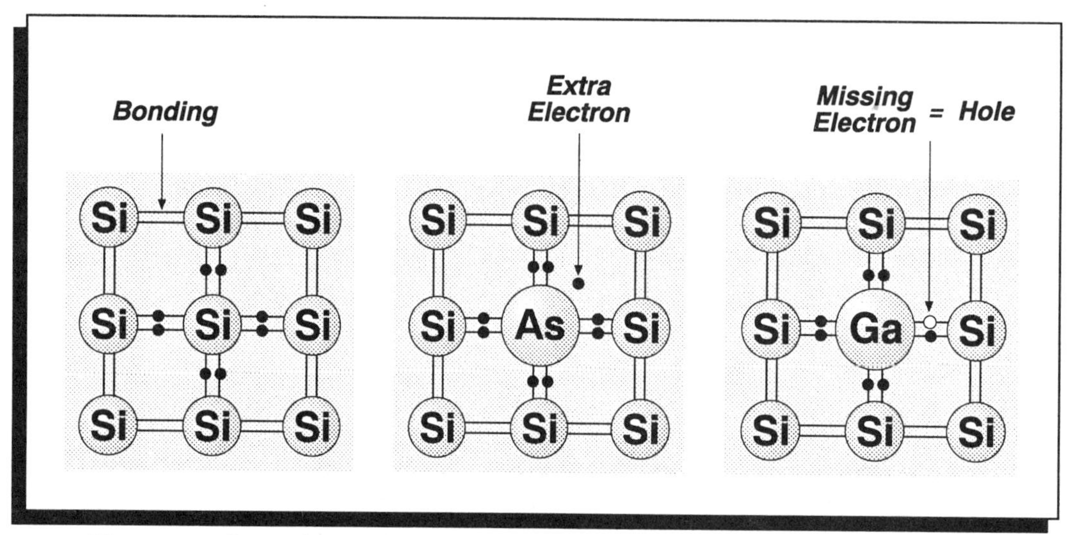

Figure 5.25 Pure silicon, silicon doped with arsenic, and silicon doped with gallium.

tor is doped, it is still electrically neutral, since there are as many negative charges as positive ones. The doping has made free electrons available or free holes for conduction of electricity. When an electric field is applied to the doped semi– conductor the principal carriers of current will be electrons for the n–type of semiconductor and holes for the p–type of semiconductor. This class of materials has led to major developments in electronics and devices. It has advanced the field of hi–fi to very high levels of excellence.

■ CIRCUITS

Having introduced the basic concepts of electricity we will now apply them to some of the components in the hi–fi. There will be a transfer of electric power from one place to another and the electric power will be used for specific functions. Starting with the definition of voltage, current, and resistance, these are all combined together to form a circuit, Figure 5.26. Here a resistor is connected to a voltage source, and the resistor is a speaker. The symbol for a resistor is shown in Fig. 5.26. How much current will flow through the speaker? In order to present a clear picture of what is going on a comparison is made to water being pumped through a hose, as in Figure 5.27. Such a comparison will make it easier to understand how much current will flow in the circuit of Figure 5.26. Water will flow in the system when there is pressure that pushes it. In an electric circuit a voltage difference causes electric current flow, but this current flow

Water System	Analogy	Electric Circuit
pressure	↔	voltage
water flow	↔	current
resistance (friction)	↔	resistance

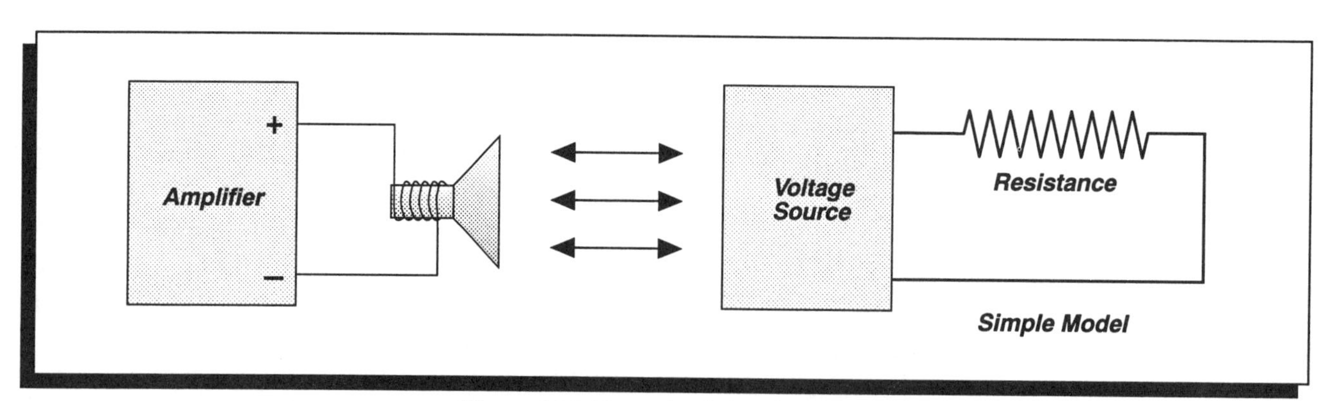

Figure 5.26 Example of simple circuit.

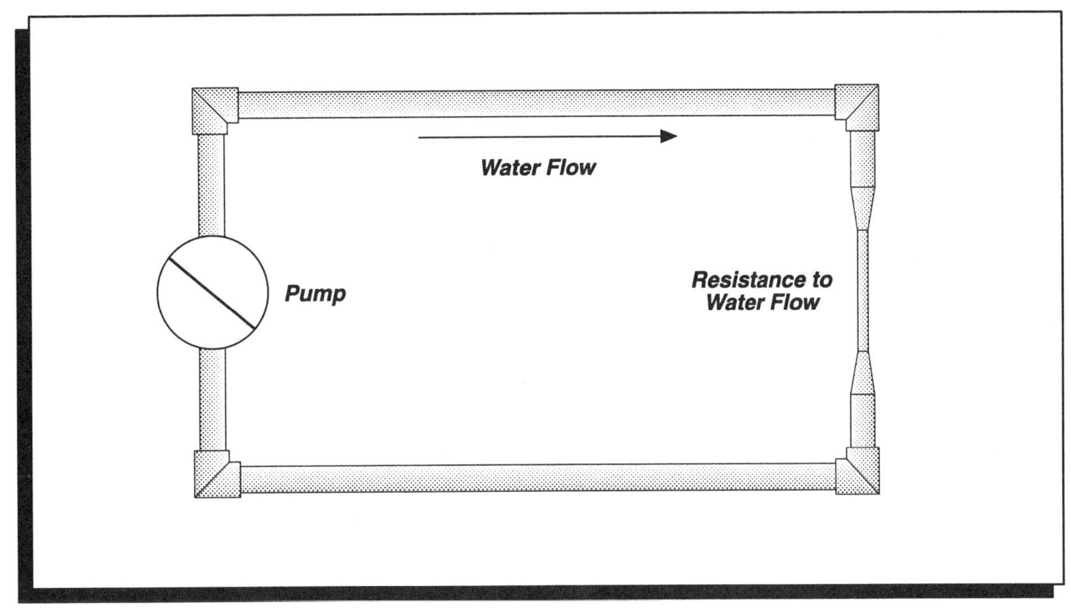

Figure 5.27 Model using water for electric circuit.

will be limited by the resistance. Hence, these factors can be put in a quantitative form as:

current in circuit = **voltage / resistance**

This is known as Ohm's law, and it can be applied directly to the circuit in Figure 5.26 or to any other circuit. It makes sense that in order to have a current, it is necessary to have voltage and the resistance will determine how much current will flow. Knowing the voltage output of the amplifier and the resistance of the speaker, we can calculate how much current will flow through the speaker. We are ignoring here the resistance of the wires connecting the speakers to the amplifier.

An example will illustrate how to use Ohm's law. For an amplifier voltage of 24 volts, how much current will flow through an 8–ohm speaker?

According to Ohm's law,

current = voltage / resistance
= 24 volts / 8 ohms
= 3 amperes

DIRECT CURRENT AND ALTERNATING CURRENT

When a battery is attached to a circuit, the current as a function of time will probably be like in Figure 5.28. It does not change with time, and hence it is called direct current, **dc.** When the current varies with time, as in Figure 5.28, it is known as alternating current, **ac.** Since the origin of the currents just discussed is voltage, the voltage can also be dc or ac. The ac part is important for audio signals since an audio signal can be represented by an ac signal, Figure 5.29.

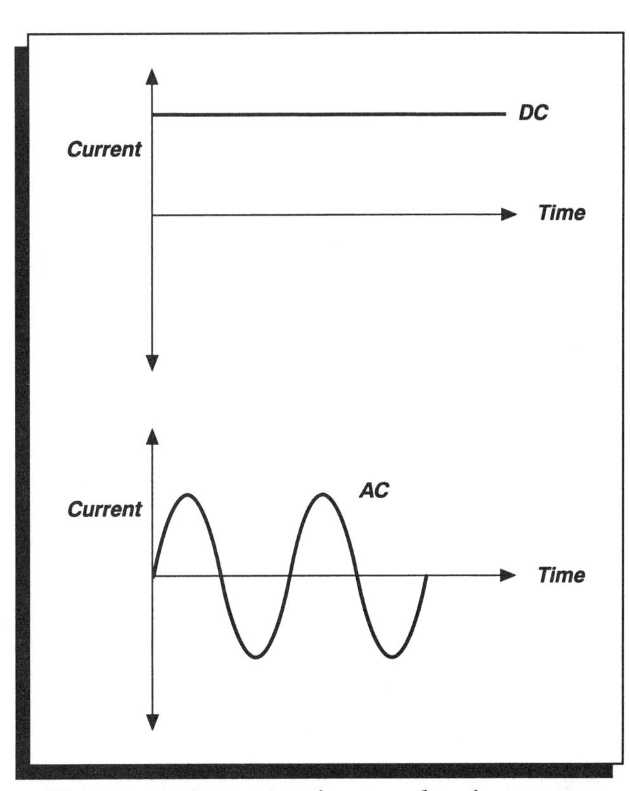

Figure 5.28 Comparison between *dc* and *ac* current.

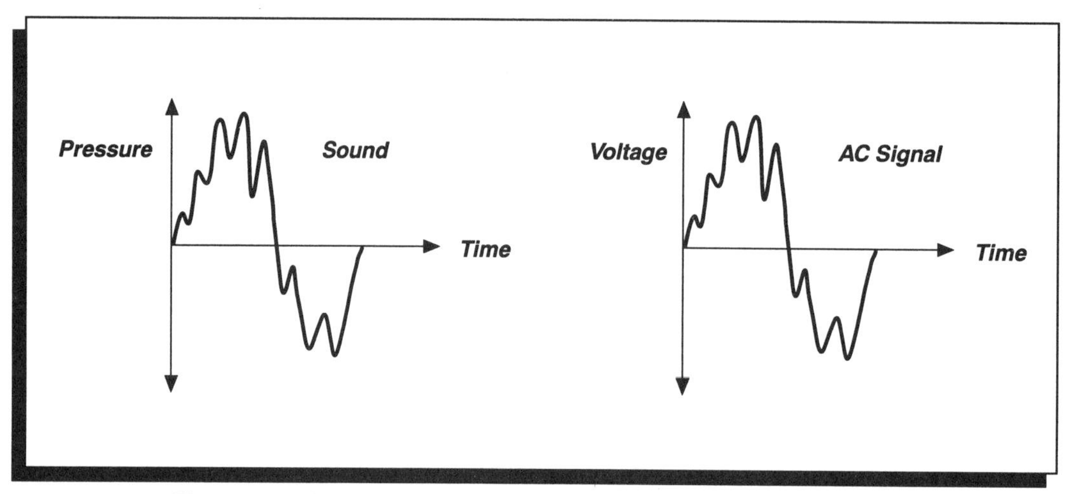

Figure 5.29 Representation of a sound wave by an ac electrical signal.

VARIABLE RESISTANCE

According to Ohm's law, the current in a circuit depends on the voltage and the resistance; by varying the resistance the current will change. This is a practical way of changing the current. It can be achieved by a variable resistor with a sliding contact across it, Figure 5.30. The resistor is wire XZ and the sliding contact is placed at selected positions. For example at the position shown, the resistance of XY is half that of XZ. Similarly, by sliding the contact across XZ the resistance between XY can be varied. This works because, as shown earlier, the resistance of a wire will depend on its length. Here we are changing the effective length of the resistor using the sliding contact, and hence the resistance is varied.

ELECTRIC POWER

The concept of power has not been defined yet, even though it was used extensively when discussing speakers and amplifiers. Power relates to work, energy and their time dependence, and also to the rate at which work and energy are used or produced. Therefore,

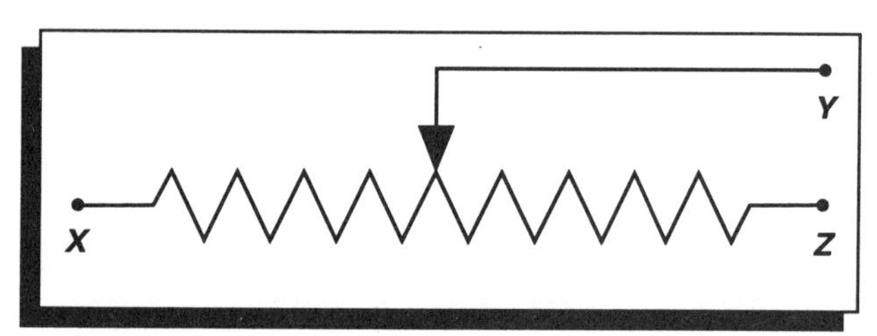

Figure 5.30 Variable resistance between XY.

Power **= rate of doing work**

This means that for electrical power it is necessary to have voltage and current. Thus,

Power **= voltage × current**

The unit of power is the watt.

The definition of power makes sense since:

Power = (energy / charge) × (charge / time)
= energy / time (watts)

Although power is defined as

$$\mathbf{P = V \times C}$$

this formula can be rewritten in different forms using Ohm's law.

Since	C = V / R
and	P = V × C
we get that	P = V × V / R
	$\mathbf{P = V^2 / R}$
Also since	V = C × R
	P = C × R × C
	$\mathbf{P = C^2 \times R}$

There are three ways of writing the basic formula for power; the particular application will dictate which is the most useful form.

Example: How much power will go to an 8–ohm speaker that by mistake gets hooked up to the electrical outlet at 120 volts?

We select the formula which has the quantities of interest to this problem. Hence, we choose here power, voltage, and resistance and we use:

$$
\begin{aligned}
P &= V^2 / R \\
&= 120 \text{ volts} \times 120 \text{ volts} / 8 \text{ ohms} \\
&= 1{,}800 \text{ watts}
\end{aligned}
$$

This is a very large amount power which will blow up the speakers.

FUSE

In the example above there is a simple way to prevent the speaker from being damaged by excess power. Usually, the damage occurs in the voice coil where the wire melts at one spot causing a discontinuity in the wire. This is called an open circuit, since the circuit is not complete and no current can flow through it. The preventive measure that can be taken is to allow a preselected piece of wire to melt before damage is done to the speaker voice coil. This preselected wire is called a fuse and it comes in specified currents above which it will melt. Figure 5.31 shows a fuse. It is connected in series with the speaker. This is a very simple way to protect speakers. The value for the fuse, specified in amperes, can be calculated quite simply from:

$$\textit{Power} = \textbf{current}^2 \times \textbf{resistance}$$

As an example, for a speaker of 4 ohms capable of handling a maximum power of 144 watts, the fuse size would be:

$$
\begin{aligned}
144 \text{ watts} &= \text{current}^2 \times 4 \text{ ohms} \\
\therefore \text{current}^2 &= 36 \text{ amperes}^2 \\
\text{current} &= 6 \text{ amperes}
\end{aligned}
$$

A fuse rated at 6 amperes or less would protect the speaker.

SERIES AND PARALLEL CIRCUITS

In the electric circuit just discussed, the basic elements were the voltage source, resistors, connecting wires and perhaps a switch. Here we will discuss how to connect speakers to an amplifier, and in particular when there is more than one speaker per channel. There are two ways of connecting two or more speakers to an amplifier, in series or in parallel. In a series connection, the current flows from the voltage source through each speaker consecutively and back to the source. In a parallel connection, each speaker is in branches, and each has its own current path from the source. Both parallel and series connections have special characteristics which we will study now.

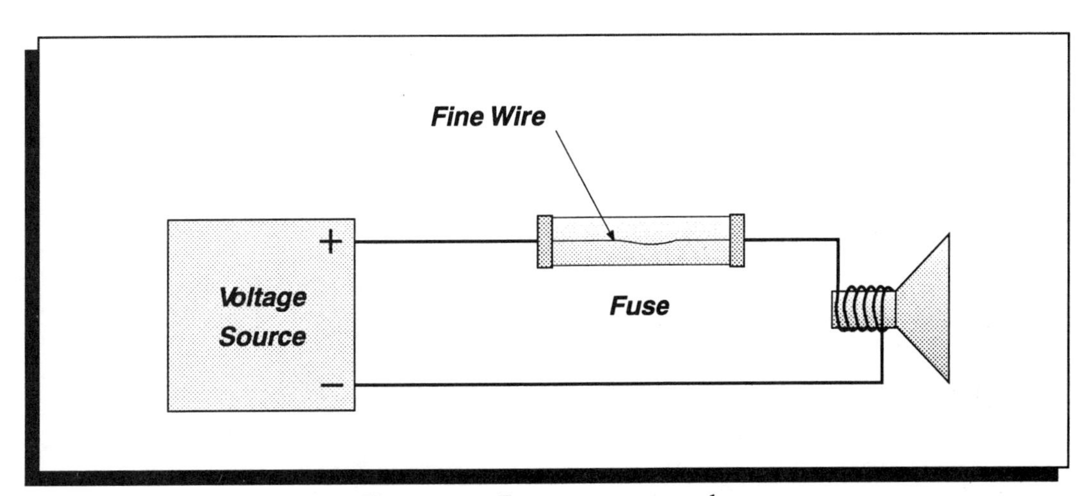

Figure 5.31 Fuse to protect speaker.

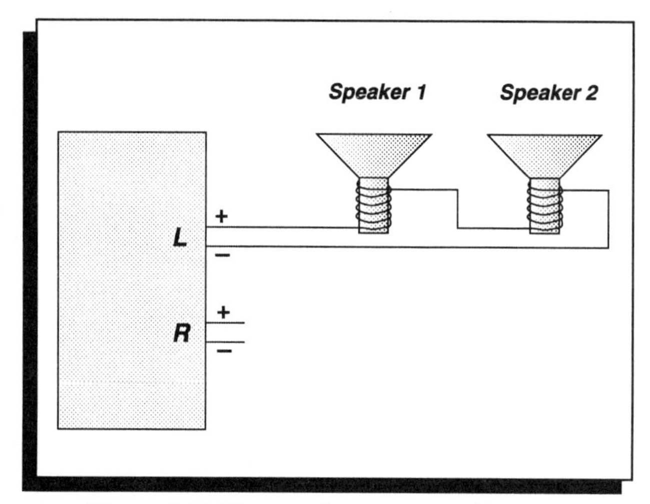

Figure 5.32 Two speakers connected in series to one channel of amplifier.

— Series Connection

A simple series circuit is shown in Figure 5.32, where two speakers are connected in series, one after the other, to an output of one channel of an amplifier. Current leaves the negative terminal of the amplifier and goes through each speaker, without piling up in any speaker, and continues to the positive terminal of the amplifier. Because there is no current pile–up, the same current flows through speaker 2 as through speaker 1. The current path shown above is the only path that current can take. It is possible to calculate the current in this circuit from Ohm's law. Hence,

$$\text{current} = \text{voltage} / \text{resistance} = \frac{\text{voltage at amplifier terminals}}{\text{total resistance}}$$

where total resistance = resistance of speaker 1 + resistance of speaker 2.

Does it make sense? Since the amount of current in the circuit is determined by the resistance, the presence of the second speaker adds extra resistance to the circuit. This reduces the circuit current. This case can be simplified by drawing a model of our circuit, where we have a voltage source, wire connections, and a single effective resistor, as in Figure 5.33. But what is the effective resistance? It represents the combined resistances of the two speakers; they are simply added together. In this case, it is simple to calculate the current through the circuit using Ohm's law.

Example: Two 8–ohm speakers are connected in series to the output of an amplifier which puts out 32 volts. What is the current flowing through each speaker?

The current is calculated from Ohm's law:

current = 32 volts/(8 ohms + 8 ohms)
= 2 amperes

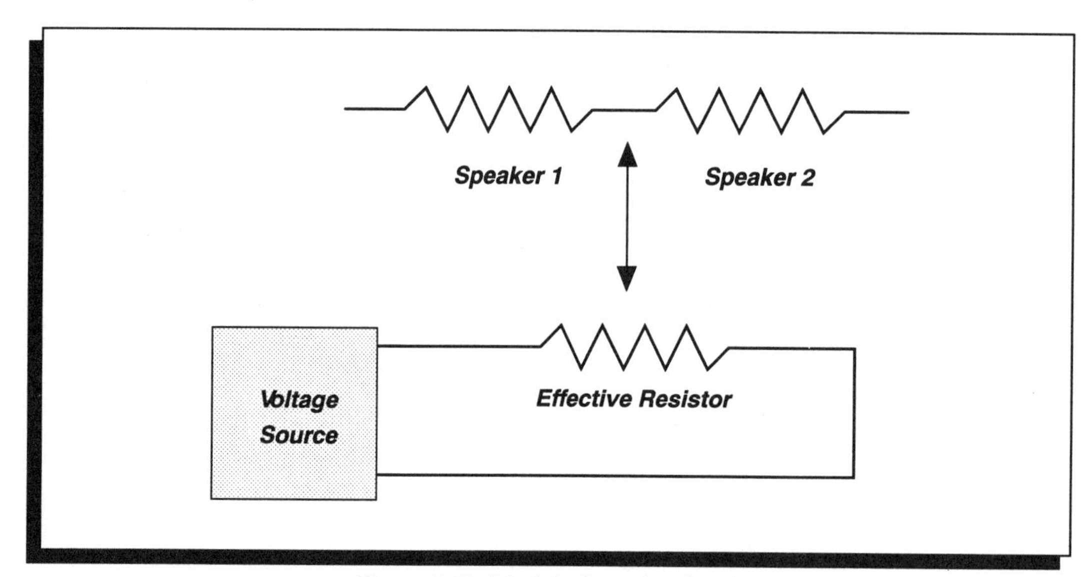

Figure 5.33 Model of a series circuit.

Two amperes flow through one speaker and then through the other one. The same current flows through each speaker, there is nowhere else for it to go.

A few more questions can be asked about this circuit.

— ***How many volts are there across each speaker?***
Because the amplifier supplies 32 volts of energy per charge, this voltage will be divided equally between the two speakers, and hence there are 16 volts across each speaker.

— ***What is the total power delivered by the amplifier?***
Remembering that one of the power equations is

$$\begin{aligned} \text{Power} &= \text{current}^2 \times \text{resistance} \\ &= (2 \text{ amperes})^2 \times 16 \text{ ohms} \\ &= 64 \text{ watts} \end{aligned}$$

— ***How much power is used up by each speaker?***
Since they both have the same resistance, each one uses half of the total power put out by the amplifier, which is 32 watts.

— ***Show that each speaker uses 32 watts of power. Power into one speaker is:***

$$\begin{aligned} \text{Power} &= \text{current}^2 \times \text{resistance} \\ &= (2 \text{ amperes})^2 \times 8 \\ &= 32 \text{ watts} \end{aligned}$$

Characteristics of series connections are:

— Should one speaker blow up, the other will stop producing sound. The circuit will then be interrupted and no current flows (i.e. this is known as a open circuit). The speaker blowing up has opened the circuit.

— By adding speakers in series the current coming out of the amplifier is reduced because the effective resistance has increased. Hence, there is less sound per speaker.

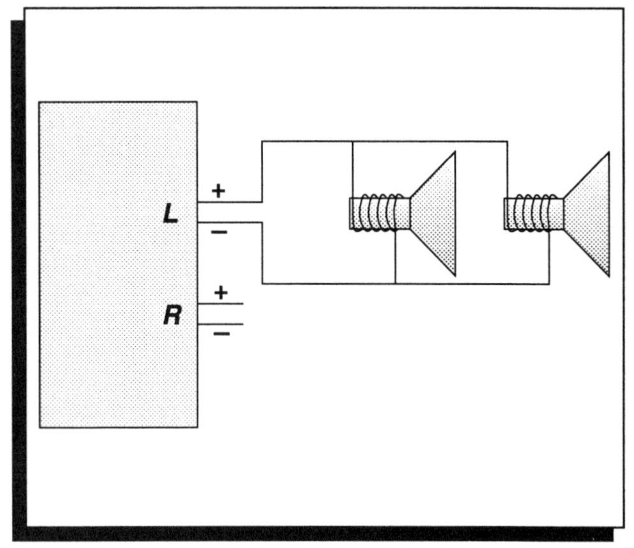

Figure 5.34 Parallel connection of two speakers to an amplifier.

— The equivalent resistance representing the speakers is equal to the sum of the speaker resistances.

— The voltage across each speaker is proportional to the resistance of each speaker.

PARALLEL CONNECTION

There is another way of connecting two or more speakers to an amplifier. Consider Figure 5.34, again for the same two speakers connected to one channel of the amplifier. Current leaves the negative terminal of the amplifier and some of it goes through speaker 1, while some goes through speaker 2. The amount of current through each speaker depends on the resistance of each speaker. Since in this example each speaker has the same resistance, the currents will divide equally between the two speakers and then return to the positive terminal of the amplifier. An interesting but important fact is that should one speaker blow up the other one would not be affected, and the same current would continue flowing through it. Also, the presence of the second speaker drains extra current from the amplifier. The current supplied by the amplifier is divided into the loop of speaker 1 and into the loop of speaker 2. The currents in each speaker add up to the total current delivered by the amplifier. Here as well, the circuit can be modeled by a sim-

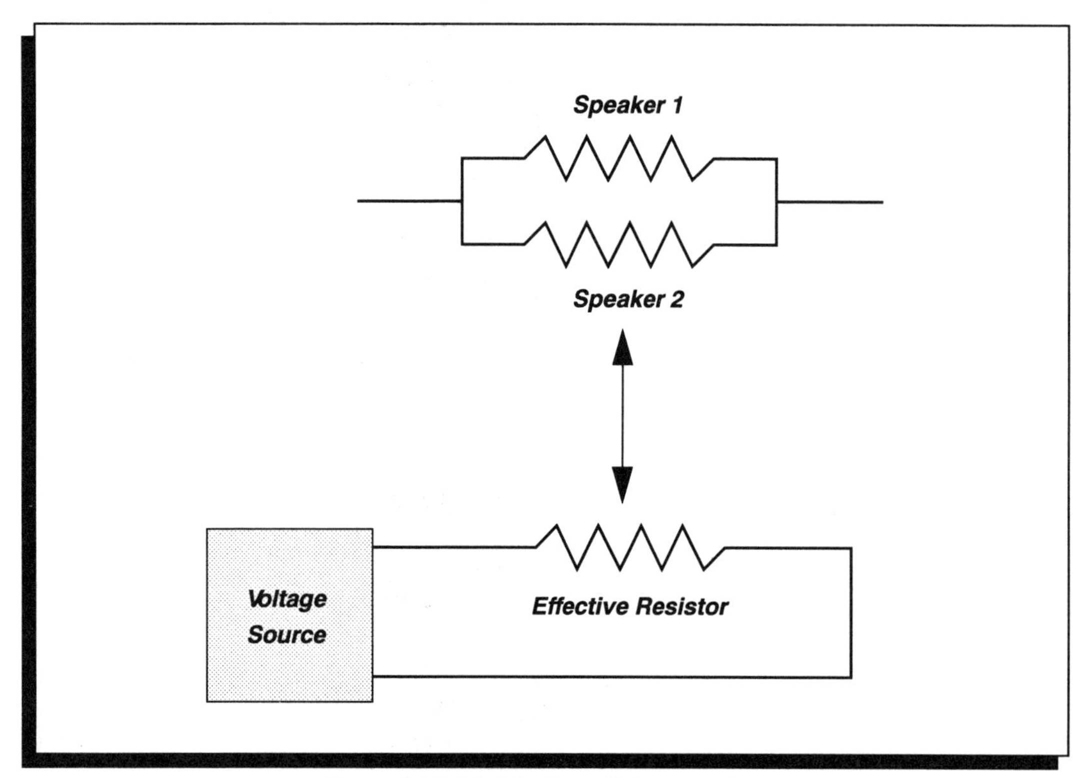

Figure 5.35 Model of parallel connections.

ple circuit represented in Figure 5.35. The equivalent resistance now represents the combined speaker resistances in parallel. Before trying to calculate the equivalent resistance some simple reasoning can help lead to the correct answer. The presence of the second speaker will require more current from the amplifier. But from Ohm's law, there will be more current when the resistance is reduced, for a fixed voltage. Thus, the equivalent resistance must be smaller than any of the other two resistances. The presence of two speakers with the same resistance doubles the current from the amplifier, meaning that the equivalent resistance is half that of each speaker. If the two speakers were 8 ohms each, the equivalent resistance would be 4 ohms.

There is a formula for calculating the equivalent resistance for parallel combination of resistors of any value. Let us do a simple example of two resistors, R1 and R2, in parallel.

The current through resistor R_1 is:

$$C_1 = \text{voltage} / R_1$$

and the current through resistor R_2 is:

$$C_2 = \text{voltage} / R_2$$

But according to Figure 5.35, the total current is:

$$C = \text{voltage} / R_{eq}$$

where R_{eq} is the equivalent resistance.

The total current C from the amplifier is:

$$\begin{aligned} C &= C_1 + C_2 \\ &= (\text{voltage} / R_1) + (\text{voltage} / R_2) \\ &= \text{voltage}\,(1/R_1 + 1/R_2) \end{aligned}$$

Hence, $$1 / R_{eq} = 1/R_1 + 1/R_2$$

and $$R_{eq} = \frac{R_1 \bullet R_2}{R_1 + R_2}$$

Simply stated, to get the equivalent resistance, the rule is:

— divide each resistor into one
— add all these fractions
— flip them over and this is the equivalent resistance

An example of such a calculation deals with two 8–ohms resistors and one 4–ohm resistor , all added in parallel. To get the equivalent resistance the rule gives:

$$1/8 + 1/8 + 1/4 \text{ becomes } 2/8 + 1/4 = 1/4 + 1/4 = 2/4$$

The equivalent resistance is then 4/2 ohms, which is 2 ohms. It is smaller than any of the resistors, as expected.

In the case of the two 8–ohm speakers, the concepts can be extended to another example of parallel connections. Consider two identical resistors connected in parallel to one channel of an amplifier putting out 36 volts. What is the total current from the amplifier?

$$\text{Total current} = \text{voltage} / R_{eq}$$

The equivalent resistance is calculated as shown above:

$$1/8 + 1/8 \text{ becomes } 2/8$$

which gives an equivalent resistance of 8/2 ohms (i.e. 4 ohms).

Now the total current can be calculated as:

$$\text{current} = 36 \text{ volts} / 4 \text{ ohms} = 9 \text{ amperes}$$

However, the current through speaker 1 is half of the total current. Therefore:

$$\text{current}_1 = 36 \text{ volts} / 8 \text{ ohms} = 4.5 \text{ amperes}$$

and likewise for speaker 2. The two currents do add up to a total current of 9 amperes.

A third example is:

— What is the total power supplied by the amplifier in the previous example?

$$\text{Power} = \text{current}^2 \times \text{Equivalent resistance} = (9 \text{ amperes})^2 \times 4 \text{ ohms} = 324 \text{ watts}$$

This is considerably more than for the series case.

— What is the power delivered to speaker one?

$$\text{Power} = \text{current}_1^{\,2} \times R_1 = 4.5^2 \times 8 = 162 \text{ watts}$$

Characteristics of parallel connections are:

— should one speaker blow up, the others will continue to produce the same sound. Only the circuit with the one speaker has been interrupted.

— by adding speakers in parallel, more current is drawn from the amplifier.

— the equivalent resistance representing the speakers is always less than any of the speakers.

— the voltage across each speaker is the same, but the current through each depends on its resistance.

Usually if one is to connect more than two speakers to an amplifier, the parallel connection is used. Of course, there is a limit as to how many speakers can be connected to the amplifier and this is determined from the amplifier specifications. By placing speakers in parallel, the equivalent resistance decreases. There is a limit to how many speakers can be placed in parallel. This limit is dictated by the smallest allowed resistance across the amplifier, as given in the specifications.

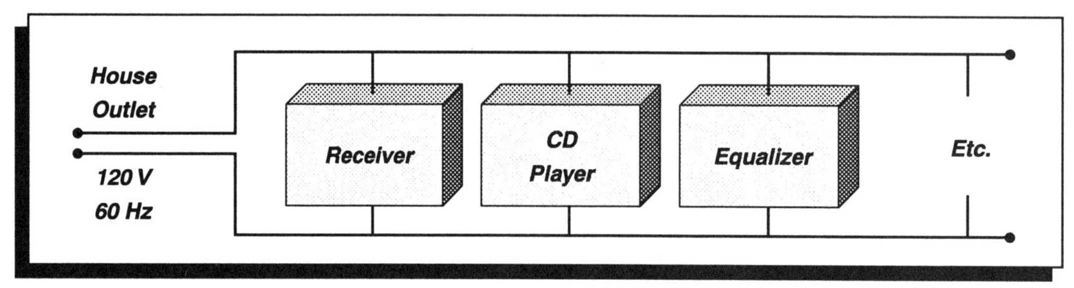

Figure 5.36 Parallel connections of hi-fi components to house electrical outlet.

While on the subject of parallel connections, it is interesting to look at the connections of the amplifier and other hi–fi components to the house electrical outlet at 120 volts. Figure 5.36 shows one example.

5.4 IMPEDANCE

When circuits were introduced with speakers as examples of electrical resistance, the concept of resistance was related to friction which causes losses in the form of heat. A light bulb is a clear demonstration of this as the filament glows red (and even white) hot due to its resistance in the flow of current. However, heating is not the only function in a circuit. Electricity must be used for other things.

Let us look at a cone speaker from a mechanical point of view. Its motion, resulting from a force acting on it, is determined by:

— mass of cone
— compliance of suspension
— friction

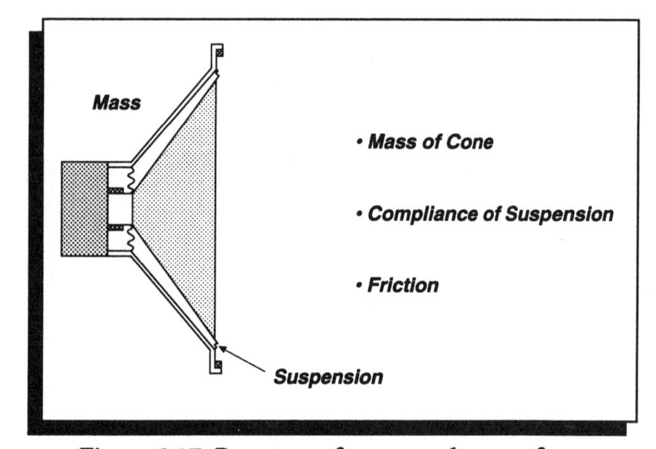

Figure 5.37 Response of cone speaker to a force.

Figure 5.37 shows how the speaker motion is affected by its mass, compliance, and friction.

The friction part of the response causes heating. But there is one part of the response which depends on the mass, and another part which depends on the springiness of the cone suspension. Hence, the force applied to the cone is not used up only in overcoming friction. The three elements, mass, compliance, and friction, all of which contribute to the response of the system to a force, can in fact impede the motion of the cone. Hence, these three elements contribute to the impedance of the cone when a force acts on it; they impede the action of a force.

In electrical circuits we have a similar situation. The voltage across a circuit does not have to be used only to overcome friction losses, the electrical resistance. There must be other elements like the mass and compliance for the mechanical system. These elements are the inductance and the capacitance, and they contribute to the electrical impedance of the circuit.

Hence, the current through a circuit will be, according to a general Ohm's law:

current = **voltage / Impedance**

Contributions to the impedance will come from:

— resistance
— inductance
— capacitance

of the circuit. Let us discuss each contribution.

RESISTANCE

The effect of friction on the flow of electrons causes heating. This effect is very useful for light bulb filaments and stove elements. In the audio frequencies of interest here, the resistance does not depend on the frequency. Heating is produced equally at all audio frequencies. It is wasted energy.

INDUCTANCE

It is possible that in a circuit there can be a coil (as shown in Figure 5.38) for creating a magnetic field (this will be discussed in a later chapter) when a current flows through it. Its ability to create a magnetic field is characterized by its inductance and a bar magnet near this coil will feel a force. As alternating currents of 20 to 20,000 Hz are extensively used, how will the coil respond to such currents? Current flows into the coil and it creates a magnetic field. It must flow out, then flow in the opposite direction, creating a magnetic field also in the opposite direction. How fast can this coil keep on doing this? As the frequency of the current is increased will the coil keep on performing its job of creating magnetic fields at these frequencies? It is possible to imagine that as the frequency of the current increases, the coil will not have enough time to do all this magnetic field creation and reversing. In that case, as the frequency increases, less current will be flowing into the coil and one can say that at high frequencies the inductance of the coil impedes the flow of current. This is shown schematically in Figure 5.39. This figure shows that much less current will flow through this coil at high frequencies than at low frequencies. This fact is very useful and it will be applied to hi–fi components.

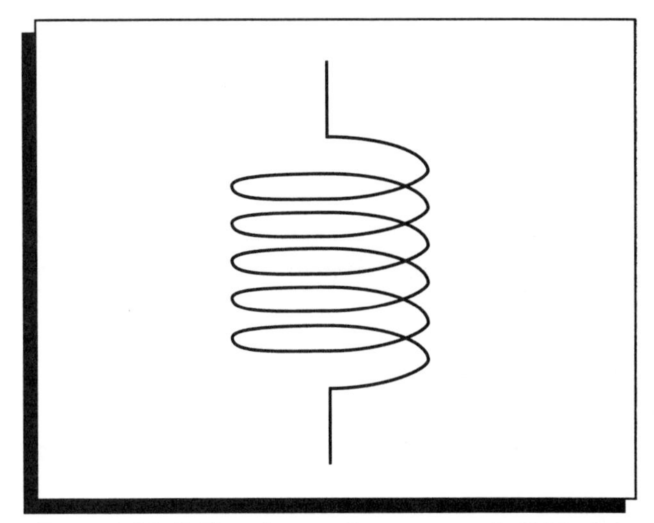

Figure 5.38 Coil used to produce a magnetic field when a current flows through it. It has inductance.

CAPACITANCE

An electrical circuit can have two plates where electrical charges are stored. The property dealing with the plates' ability to store electrical charge is called capacitance. Figure 5.40 shows two such plates connected to a voltage source. Electrons will

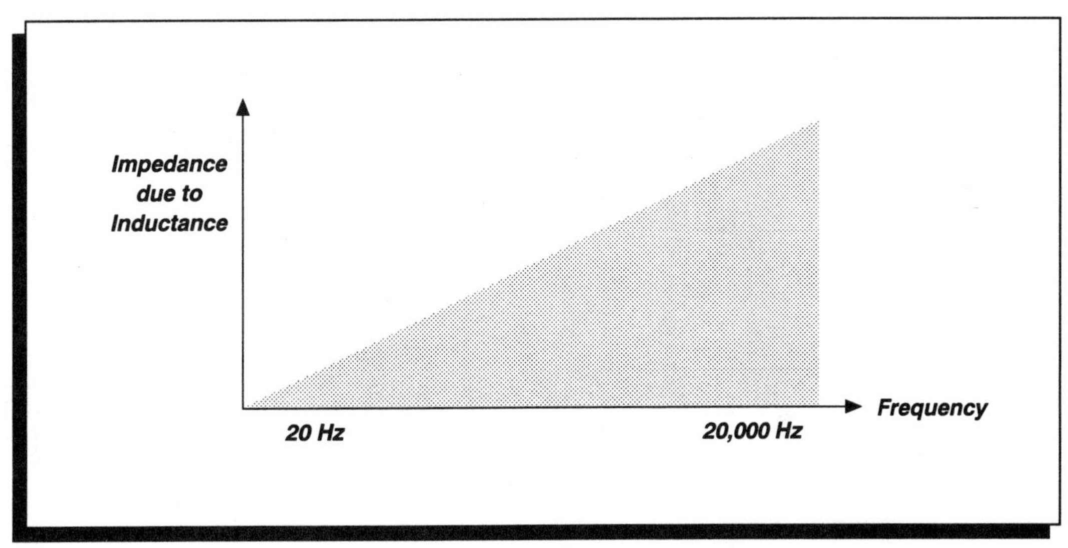

Figure 5.39 Frequency dependence of impedance associated with inductance.

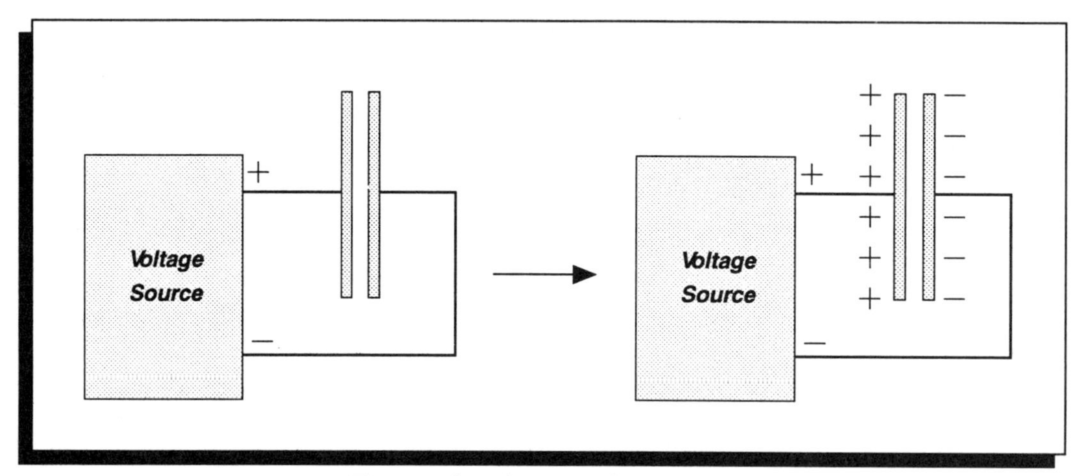

Figure 5.40 Charging of a capacitor.

flow from the negative terminal to the plate connected to it. In the other plate, the electrons from that plate will be attracted to the positive terminal leaving the plate positively charged. After a certain time the plates will be full of charges and then no more current will flow toward the plates. The plates act like a bucket storing water, except they store electrical charges. If the voltage source polarity were reversed we would have the situation in Figure 5.41. Here as well, after a certain amount of time the plates get filled up with charge and the current flow stops. This can be rephrased by saying that after a while the impedance due to this capacitance is so high that no current flows. In fact, at dc no current flows once the capacitance is charged up. However, if the voltage source were alternating in polarity at a high rate, the capacitance would never have enough time to be charged completely and hence, current would always be flowing, toward and away from the plates. In that case the impedance would be low. All this is presented as a graph, in Figure 5.41. The impedance of a capacitance electric circuit behaves totally opposite to one with an inductance as the frequency is varied.

■ APPLICATION: CROSS–OVER NETWORK

In the loudspeaker a special electric circuit is used to divide the sound spectrum into specific sections, which are sent only to certain drivers. This is achieved by cross–over networks located inside the speaker enclosures.

There are two circuit elements which could be applied to such frequency discrimination in the speaker. Consider the woofer; it should handle only the low frequencies. To prevent high frequencies from entering it, a circuit with large impedance

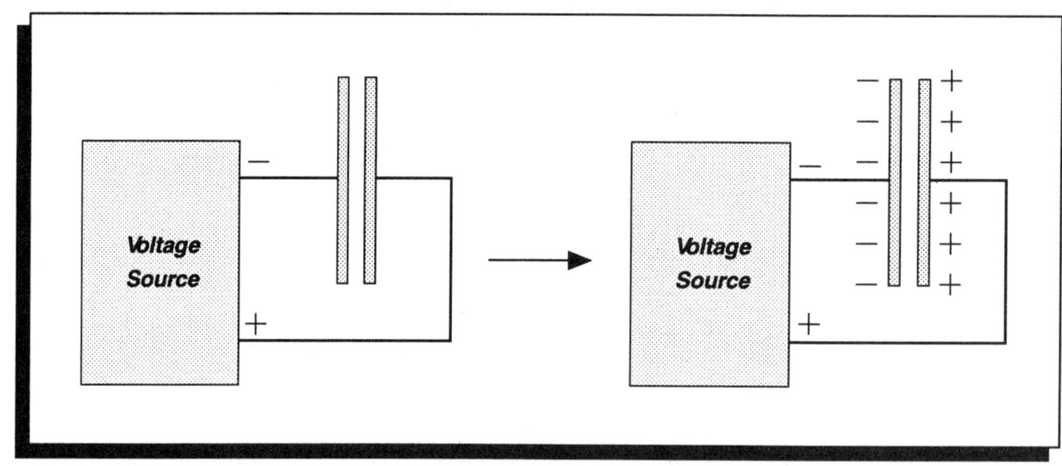

Figure 5.41 Charging of a capacitor when polarity of voltage source is reversed.

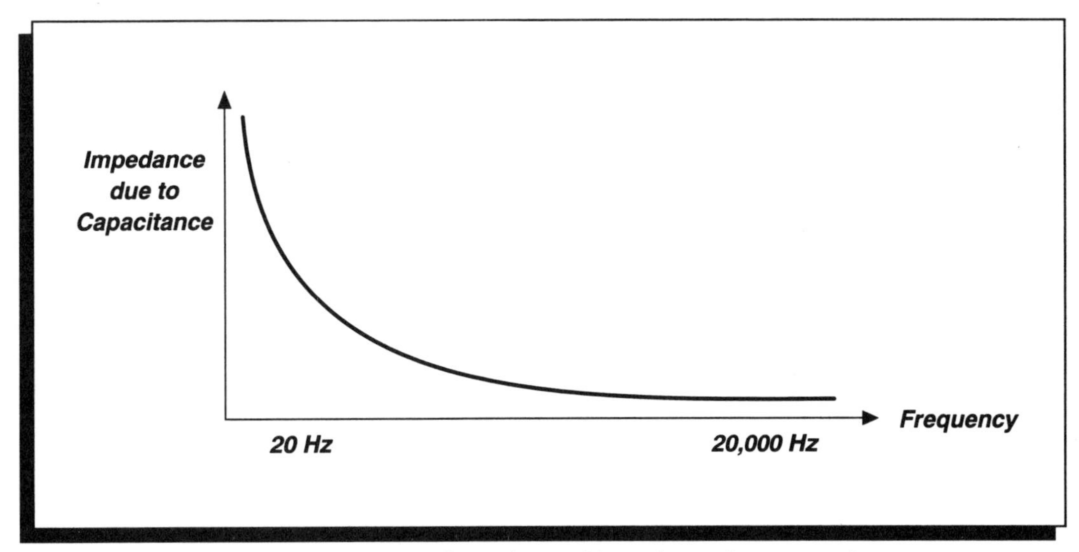

Figure 5.42 Frequency dependence of impedance due to capacitance.

at high frequencies should be connected to it. Hence, the simplest approach to this problem is as shown in Figure 5.43; an inductance is placed in series with the woofer.

For the tweeter we have the opposite problem; low frequencies should be prevented from getting to it. Figure 5.42 shows that the impedance due to capacitance is large at low frequencies. Hence, it is connected to the tweeter, Figure 5.44.

For the mid–range, a combination of the two elements, a capacitor and an inductor in series will provide a high impedance for low and high fre-

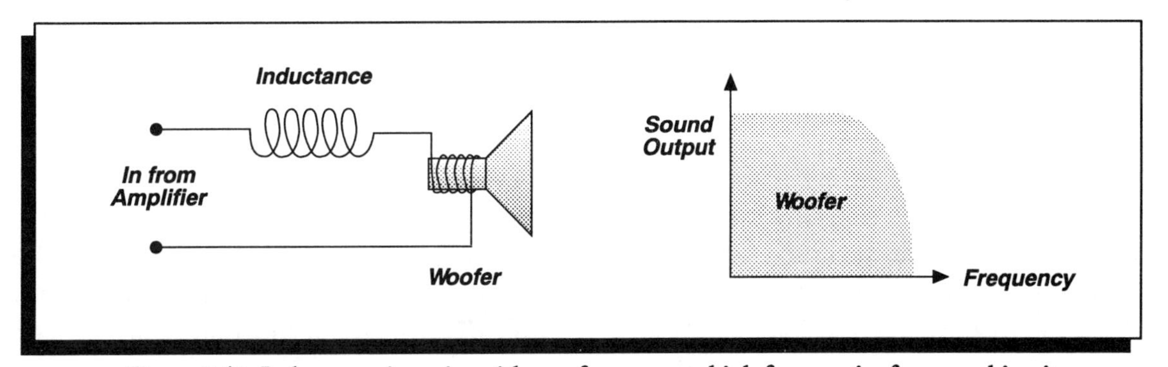

Figure 5.43 Inductance in series with woofer prevents high frequencies from reaching it.

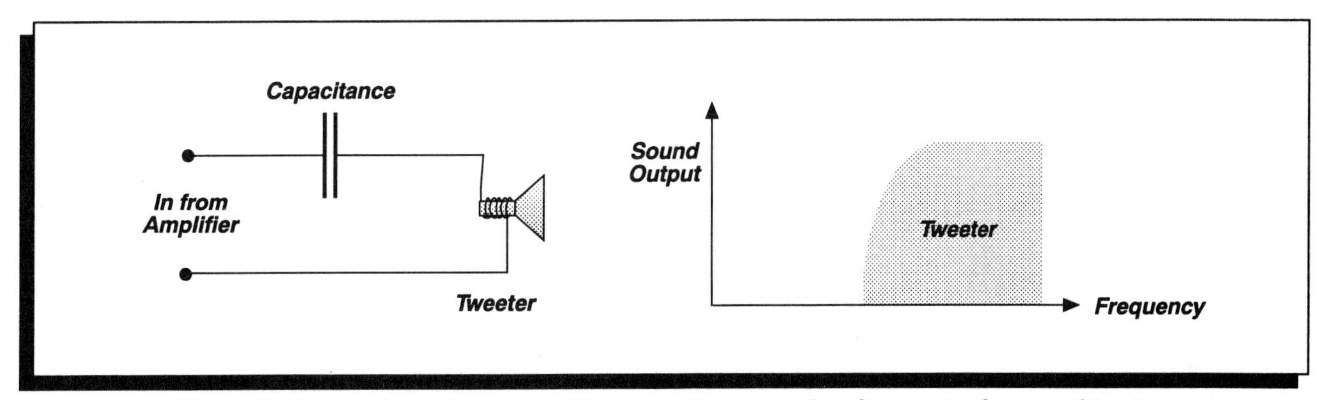

Figure 5.44 Capacitance in series with tweeter. It prevents low frequencies from reaching it.

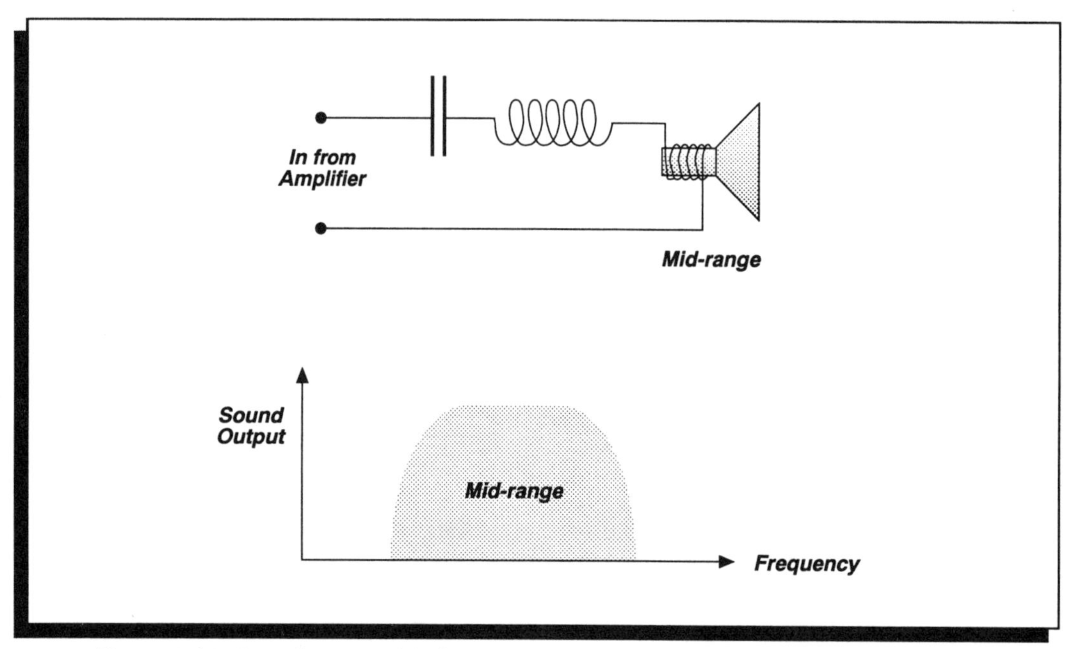

Figure 5.45 Capacitance and inductance in series with mid-range speaker to prevent the high and low frequencies from reaching it.

quencies. Figure 5.45 shows this.

This application of impedance of capacitance and an inductance to cross–over networks has been presented for the simplest example. There are more sophisticated cross– over networks in use, but all are based on similar principles.

IMPEDANCE OF SPEAKER

A speaker is a complex device which has both mechanical and electrical characteristics. Its voice coil has inductance, and the wires in the coil have some resistance as they heat up. The mass and compliance of the cone determine its resonant frequency. With all this, how does the total impedance of a speaker vary with frequency? Figure 5.46 shows the impedance of a driver as a function of frequency. The impedance reaches a maximum at the speaker's resonant frequency. The minimum below the resonance bump is usually the nominal value of impedance as specified for the driver (example: 8–ohm speaker). At high frequencies the impedance increases due to the inductance of the voice coil. Each driver has its own particular impedance curve. This curve is useful in enclosure and system design.

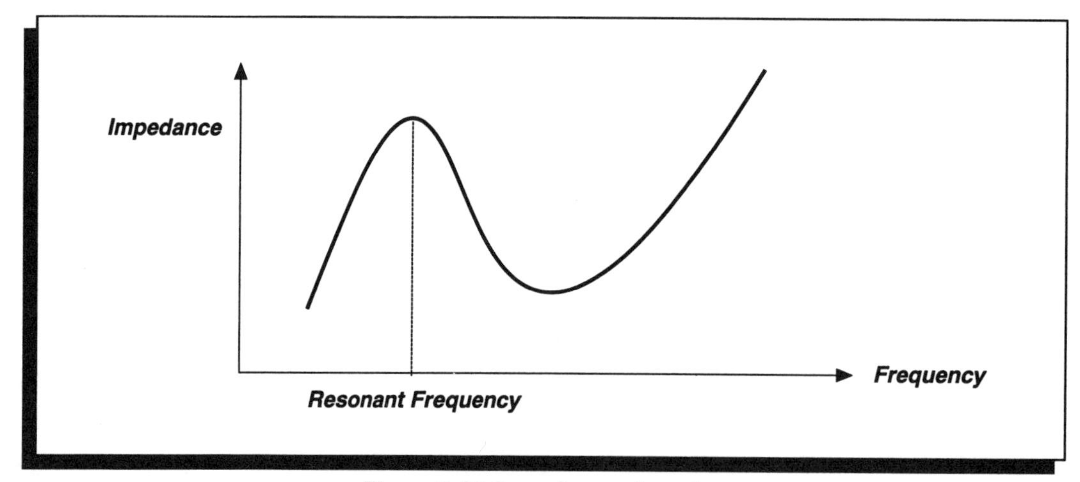

Figure 5.46 Impedance of speaker.

SUMMARY OF TERMS

Ampere: Unit of electrical current specifying how much charge flows per second.
Bimorph: Two piezoelectric discs connected in parallel in order to produce a push–pull action when a voltage is applied across them.
Capacitance: Property of a device for storing electric charge.
Electric field: Energetic region in space around an electric charge.
Electrostatic speaker: A loudspeaker which uses the principle that like charges repel and unlike charges attract.
Impedance: Response of a system to a force or a voltage. It determines how much current will flow in an electrical circuit.
Inductance: Ability of a circuit to store magnetic energy.
Insulator: Material composed of atoms where the electrons are tightly bound to the atoms, and they do not conduct electricity. It has infinite resistance.
Ohm's law: Determines how much current will flow in a circuit of resistance R, when a voltage is applied to it.
Piezoelectricity: Production of electricity by certain objects when stressed or twisted. The converse is true.
Resistance: Property of a body that determines how poorly the electrons respond to an electric field. It causes heating when current flows.
Superconductor: Material whose resistance drops to zero below a critical temperature, Tc.
Voltage: Difference in energy per charge across two points of a circuit.

NAME ______________________ DATE __________

QUESTIONS FOR REVIEW

1 Explain how an insulator differs from a conductor.

2 What is the difference between voltage and electric power?

3 How does a p–type semiconductor differ from an n–type semiconductor?

4 Why is ac voltage useful for processing acoustic information?

5 What factors determine the resistance of a metal wire?

6 Explain how you would select the length and thickness of wires for connecting speakers to an amplifier.

7 How does voltage differ from current?

8 When insulated cables are connected between speakers and an amplifier, why does current flow in the copper metal of the cable and not through the plastic around the copper metal?

9 Discuss the principles of an electrostatic speaker.

10 Explain what a cross–over network does in a speaker and how it works.

NAME________________________________ DATE ____________

EXERCISES

1. If you make the cross–sectional area of a wire 5 times as great, its resistance will be _______________ as large.
   ___A. 5 times
   ___B. 1/5 times
   ___C. 25 times
   ___D. 5/5 times
   ___E. 1/25 times

2. A current of 2 amperes is passing through a speaker of 8–ohms resistance. The voltage across the speaker is:
   ___A. 2 volts
   ___B. 8 volts
   ___C. 4 volts
   ___D. 16 volts
   ___E. 1/4 volt

3. When two resistors are added together in parallel, then the effective resistance:
   ___A. increases
   ___B. decreases
   ___C. remains the same as the resistors
   ___D. one cannot say what will happen since the values of the resistors are not given
   ___E. will be that of the smallest resistance.

4. If two speakers are connected in series and one of them burns out, then the sound from the other speaker will:
   ___A. become 1/2 as loud
   ___B. become twice as loud
   ___C. not change in loudness
   ___D. be turned off
   ___E. become slightly louder.

5. When three 8 ohm speakers are wired in series, the total resistance becomes:
   ___A. 24 ohms
   ___B. 8/3 ohms
   ___C. 3/8 ohms
   ___D. 8 ohms
   ___E. 3 ohms

6. Two objects, one of which has a deficiency of electrons, and the other one has an excess of electrons, will:
   ___A. attract each other
   ___B. repel each other
   ___C. do nothing
   ___D. repel each other first and then become attracted
   ___E. repel each other only when they move

7. Three speakers are connected in parallel. Their resistances are 8–ohms, 8–ohms, and 4 ohms, respectively. The effective resistance is:
   ___A. 2 ohms
   ___B. 16 ohms
   ___C. 6 ohms
   ___D. 8 ohms
   ___E. 4 ohms

8. When will a current flow from one end of a wire to another?
   ___A. when both ends have the same positive charge
   ___B. when there is a voltage between the two ends of the wire
   ___C. when the resistance of the wire is less than 8 ohms
   ___D. when both ends have the same negative charge
   ___E. when the wire is an insulator

9. A speaker is connected to an amplifier. Doubling the voltage output by the amplifier __________ the power received by the speaker.
___A. quadruples
___B. doubles
___C. cuts in half
___D. does not change
___E. halves

10. An 8–ohm speaker can accept a maximum of 50 watts before it burns out. What size fuse must be used to protect the speaker from damage?
___A. 6.25 amperes
___B. 2.5 amperes
___C. 400 amperes
___D. 50 amperes
___E. 8 amperes

11. When eight speakers, each 16 ohms, are connected in parallel, the effective resistance will be:
___A. 128 ohms
___B. 2 ohms
___C. 8/16 ohms
___D. 8/3 ohms
___E. 1 ohm

12. Two speakers, each of 8 ohms, are connected in series across the output terminals of an amplifier. If the amplifier is delivering a total of 52 watts of power, how much power does each speaker receive?
___A. 26 watts
___B. 52 watts
___C. 52/8 watts
___D. 416 watts
___E. 104 watts

13. Two speakers, whose resistances are 8 ohms and 16 ohms are wired in series. Which one of the following statements is correct?
___A. the current in each speaker is different and therefore they each receive the same amount of electrical power
___B. the current flowing through each speaker is not the same, but each receives a different amount of electrical power
___C. the current through each speaker is the same, but the 16–ohm speaker receives double the power that the 8–ohm speaker receives
___D. the current through each speaker is the same but the 8–ohm speaker receives double the power that the 16–ohm speaker receives
___E. the current through each speaker is different and therefore they each receive different amounts of electrical power

14. A current of 0.3 ampere is flowing through an 8–ohm speaker. What power is being delivered to the speaker?
___A. 0.24 watt
___B. 0.72 watt
___C. 26.7 watt
___D. 2.4 watt
___E. 19.2 watt

15. A 4–ohm speaker and an 8–ohm speaker are wired in parallel. What is the effective resistance of the combination?
___A. 8/3 ohms
___B. 3/8 ohm
___C. 12 ohms
___D. 1/32 ohm
___E. 3/32 ohm

16. In a Bose 901 speaker there are nine drivers per speaker, each driver being 8 ohms. What combination of these drivers should there be to produce an effective resistance of 8 ohms?
___A. they are all in series
___B. 8 in series and in parallel with one
___C. they are all in parallel
___D. 8 in parallel and in series with one
___E. there should be three groups in series, with each group having three in parallel

17. The impedance offered by a capacitor is large when:
___A. the frequency of the current is low
___B. the frequency of the current is high
___C. the amplitude of the AC current is large
___D. the amplitude of the AC current is low
___E. the period of the AC current is short

NAME ______________________ DATE ____________

18. An electric current which reverses directions is called:
    ___A. alternating current
    ___B. direct current
    ___C. 60 Hz current
    ___D. impedance
    ___E. standing wave current

19. Why are thick wires rather than thin wires normally used to carry the current?
    ___A. thick wires offer less resistance to the current
    ___B. more watts can be carried by thick wires
    ___C. thick wires offer more resistance to the current
    ___D. more volts can be carried by thick wires
    ___E. thick wires are stronger

20. Which wave properties does an alternating current have?
    ___A. frequency
    ___B. amplitude
    ___C. speed
    ___D. period
    ___E. all of them

21. When an inductor is connected in series with a speaker:
    ___A. low frequency currents will reach the speaker without being reduced in amplitude
    ___B. high frequency currents will reach the speaker without being reduced in amplitude
    ___C. currents at all frequencies will reach the speaker without being reduced in amplitude
    ___D. midrange frequencies will reach the speaker without being reduced in amplitude
    ___E. no current for all frequencies will reach the speaker

22. When a wire is made 4 times shorter and its cross–sectional area is made 3 times smaller, its resistance will be:
    ___A. 4 times larger
    ___B. 3 times larger
    ___C. 12 times larger
    ___D. 3/4 times larger
    ___E. 12 times smaller

23. An 8–ohm speaker is connected to an amplifier by wires whose total resistance is 0.1 ohm. When a current of 3 amperes is sent to the speaker, it dissipates ________ watts of power; the connecting wires dissipate ________watts.
    ___A. 72, 0.9
    ___B. 24, 0.3
    ___C. 8, 0.1
    ___D. 24, 0.1
    ___E. 72.9, 72.9

24. Why are thick wires rather than thin wires used to carry the current to speakers?
    ___A. thick wires offer more resistance to the current
    ___B. thick wires offer less resistance to the current
    ___C. alternating currents from amplifier go only in thick wires
    ___D. thick wires carry the volts well
    ___E. thick wires dissipate more heat than thin wires

25. If an electric current makes 100 complete cycles in 0.2 second, then it is called a ________________ .
    ___A. 100 Hz AC current
    ___B. 100 Hz DC current
    ___C. 500 Hz AC current
    ___D. 20 Hz AC current
    ___E. 500 Hz DC current

26. A neutral atom has:
    ___A. the same number of neutrons and protons
    ___B. the same number of neutrons and electrons
    ___C. the same number of electrons and protons
    ___D. only neutrons
    ___E. only neutrons and protons

27. The fundamental law upon which an electrostatic speaker is based is:
    ___A. like charges repel, unlike charges attract
    ___B. like charges repel, unlike charges repel
    ___C. like charge attract, unlike charges repel
    ___D. there is no force between charged objects
    ___E. Ohm's law

28. When an inductor and capacitor are connected in series, the impedance of this combination is greatest at the:
    ___A. midrange frequencies
    ___B. midrange and treble frequencies
    ___C. midrange and bass frequencies
    ___D. bass and treble frequencies
    ___E. the impedance is the same for all frequencies

29. A capacitor and resistor are placed in parallel across the output of an amplifier. When the frequency of the output is increased, the current through the capacitor:
    ___A. decreases
    ___B. increases
    ___C. does not change
    ___D. is zero
    ___E. will always be less than through the resistor

30. Each of two objects is electrically charged and they exert a force on one another. As the two objects are moved further apart, the force between them:
    ___A. becomes stronger
    ___B. does not change
    ___C. becomes weaker
    ___D. changes from attraction to repulsion
    ___E. changes from repulsion to attraction

CHAPTER 6 AMPLIFIERS

Some of the concepts of electricity discussed earlier will be applied here to the component that deals entirely with electrical signals, the amplifier. The main goal is to take weak electrical signals which represent the audio information and to increase their level to the point that they can be used to drive speakers. This chapter will deal with how this is done, how well it is done, and some of the limitations. The performance of amplifiers and their limitations will also be presented in the amplifier specifications.

6.1 Basics

In the hi–fi system, sources of audio signals such as the CD player, the tape deck, the microphone, etc. produce weak electrical signals which have to be raised in level (i.e. amplified) to be able to drive the speakers. Figure 6.1 illustrates this. It is important to stress that in this part of the system, sound is not involved. The system deals only with electrical signals that represent the sound. What would be the basic design of such a unit? A

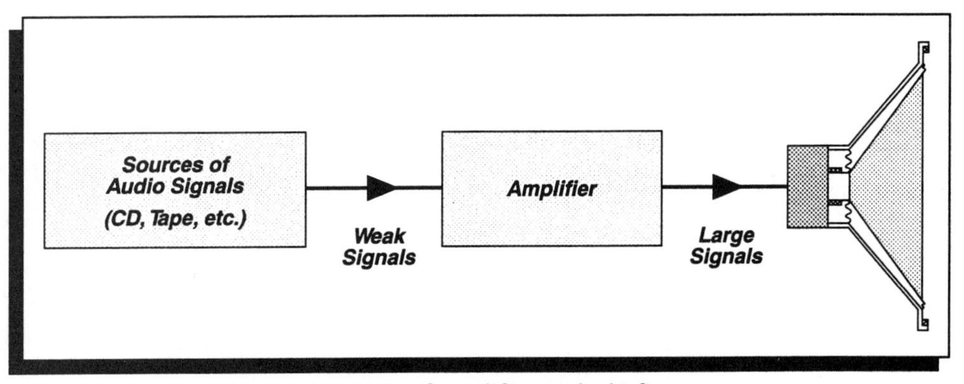

Figure 6.1 Role of amplifier in the hi-fi system.

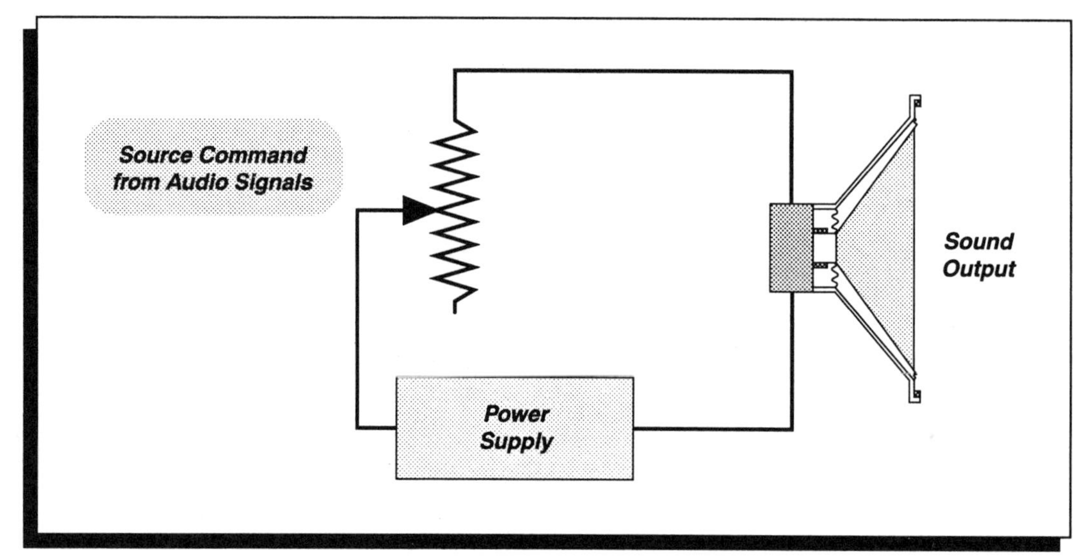

Figure 6.2 Basic amplifier.

simple approach consists of the conceptual unit shown as in Figure 6.2. A power supply is used as the source of power that will then be delivered in format suitable to the speaker. The amount of power delivered to the speaker is controlled by some sort of device that behaves as a variable resistor. This variable resistor determines the amplitude of the signal which is sent to the speaker, while the rate at which it changes the amplitude determines the signal frequency. Commands to the variable resistor determine how much and how often current should be sent to the speaker (Figures 6.3 and 6.4). Such commands will come from audio sources, which by their nature produce weak electrical signals. Hence, the basic approach in developing an amplifier is to design it to control large signals by weak ones. Of course, the variable resistor will have to move at rates determined by frequencies in the range of 20 Hz to 20,000 Hz. Its speed has to be so fast that at this point, the model has reached its limit and a very fast device, acting like a variable resistor, has to be implemented. Such a device is the transistor. In fact, it behaves as a very fast variable resistor but with no moving parts; it is the basic element of an amplifier. A simple definition of an amplifier is:

***amplifier* = device or unit where small input signals control large output signals.**

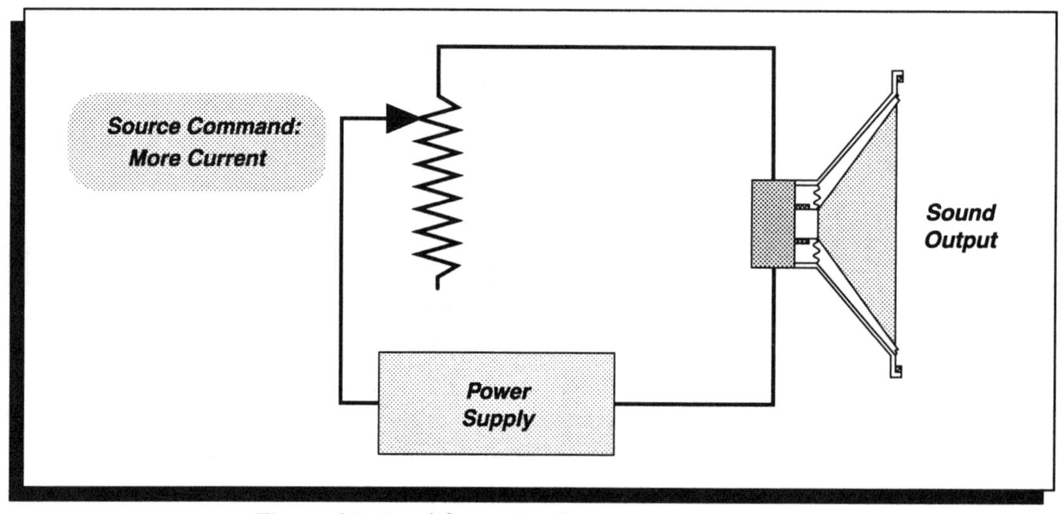

Figure 6.3 Amplifier action for more output current.

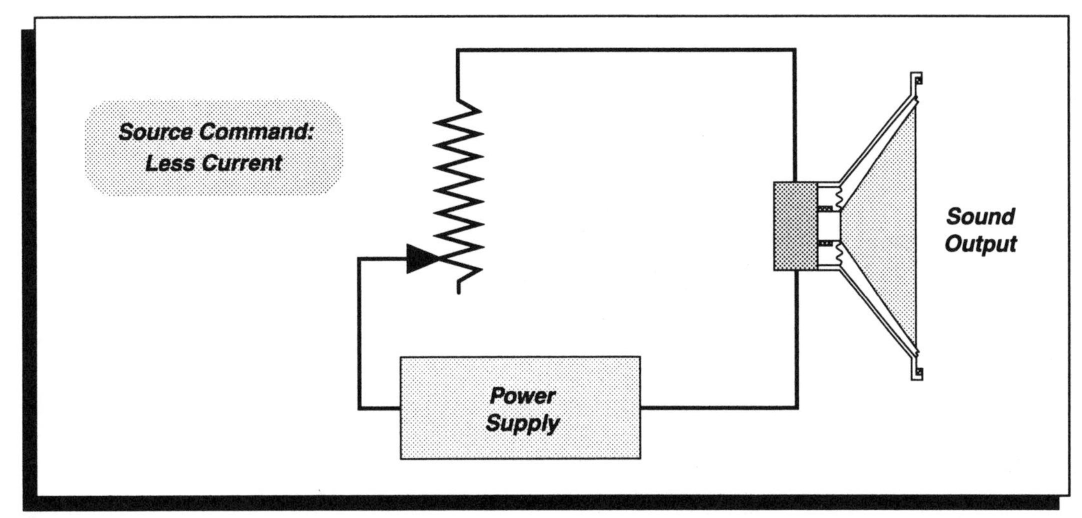

Figure 6.4 Amplifier action for achieving less output current.

To understand what a transistor does and how it does it, we need to go back to semi– conductors, which were introduced in the previous chapter.

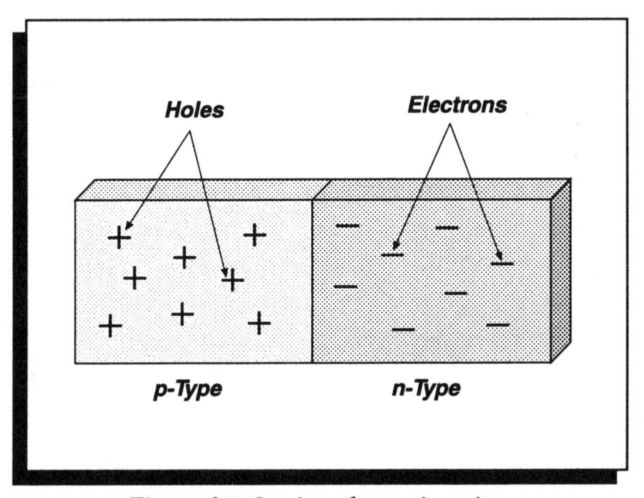

Figure 6.5 Semiconductor junction.

SEMICONDUCTOR JUNCTION

The transistor is based on the behavior of junctions between dissimilar semiconductors, one is a p–type, and the other n–type. Semiconducting junctions are the basis for a variety of devices. Consider a pure semiconductor crystal that is doped at one end with n–type of impurities and at the other end with p–type impurities, with a junction as an interface between the two, as in Figure 6.5. The crystal is neutral. In order to investigate its electrical behavior a battery is attached across this device, Figure 6.6. In this case, the device is biased. The electric field set up by the battery causes the free electrons to be attracted to the positive terminal of the battery and on the other side of the junction, the negative battery terminal exerts a force on the holes. The result is that there are no carriers left on both sides near the junction, and

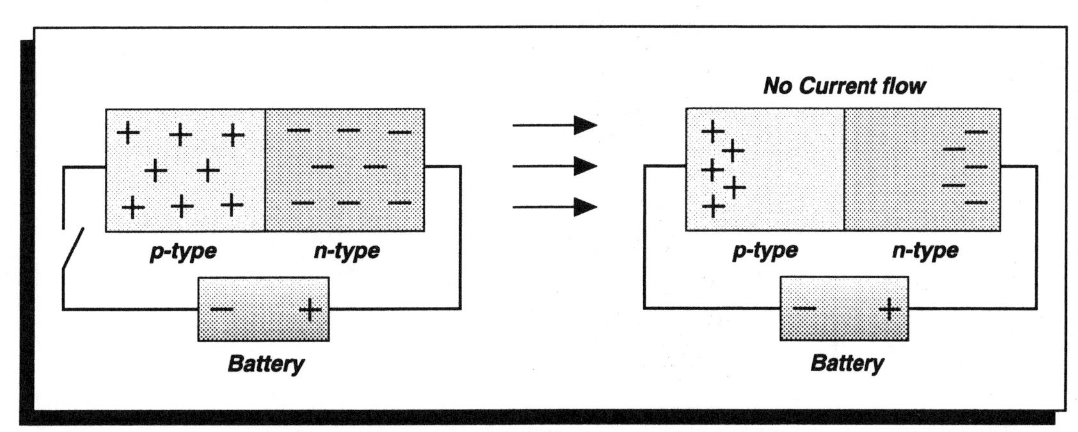

Figure 6.6 Reverse-biased semiconductor junction.

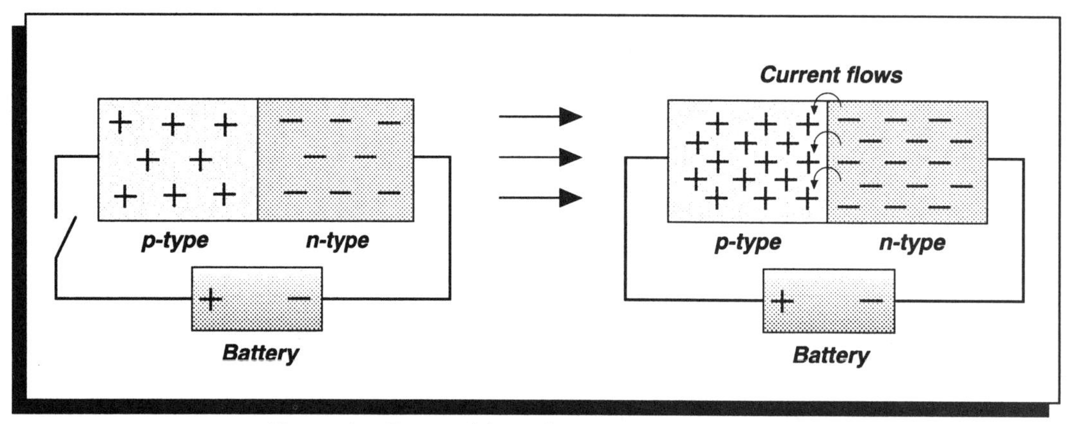

Figure 6.7 Forward-biased semiconductor junction.

this means that no current flows through the junction. This is known as reverse–biased.

When the polarity of the battery is changed to the configuration presented in Figure 6.7, the electric field produced by the battery repels the electrons from the negative terminal, and it repels the holes from the positive terminal causing them to move to the junction area. The electrons are attracted to the holes as they cross the junction. The process continues because the battery re–supplies electrons. This means that current now flows through the junction. It is forward–biased. A device doped differently on each side of a junction is known as a diode. A diode conducts current when it is forward–biased, and it essentially does not conduct when it is reverse–biased. The circuit symbol for the diode (as you will find in the amplifier schematic) is shown in Figure 6.8, as well as the current–voltage characteristics. The characteristics show that in one voltage direction there is a large current, but in the other voltage direction there is almost no current.

One application of the diode is in a rectifier. Consider a diode as shown in Figure 6.9 where instead of a battery, an alternating voltage is applied to it. A resistor is attached so that the circuit is complete and current can flow. The result is that during the positive part of the input signal, the diode is forward–biased and it conducts current. During the negative part of the input signal, the diode is reverse–biased and therefore no current flows. The net result is that only part of the waveform gets through the circuit. Such uni–directional action is referred to as the rectifying property of a diode. This effect has an important application: it can convert an ac voltage to a dc voltage, as in the power supply of an amplifier, or to dc. The rectified waveform in Figure 6.9 is not dc, but a capacitor across the resistor or filter can help to make it become dc.

■ TRANSISTOR

Consider a double junction as in Figure 6.10. There are two possibilities with a double junction,

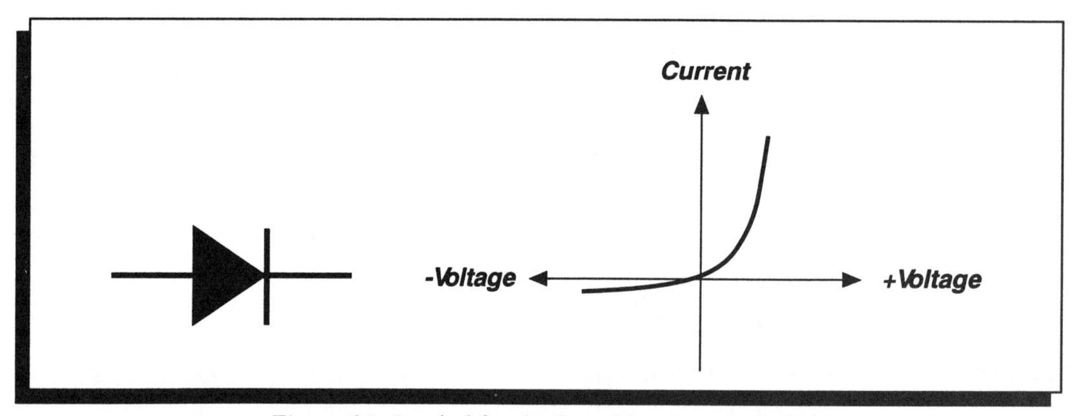

Figure 6.8 Symbol for diode and its characteristics.

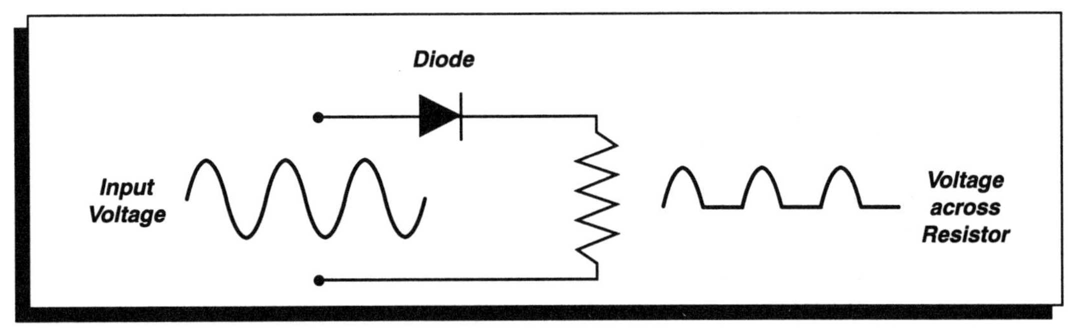

Figure 6.9 Rectifier action of a diode when an ac voltage is applied to it.

an n–type of semiconductor between two p–type materials or a p–type between two n–type materials. Consequently, we can have an n–p–n device or a p–n–p device. Such devices are known as transistors.

Because a transistor has three elements, there are three connections to it: electrical connections made at the base, the emitter, and the collector. There are different ways of connecting this device in a circuit, depending on the application. However, consider the configuration shown in Figure 6.11. It has some similarity with the basic model of the amplifier discussed earlier. The role of the transistor is that

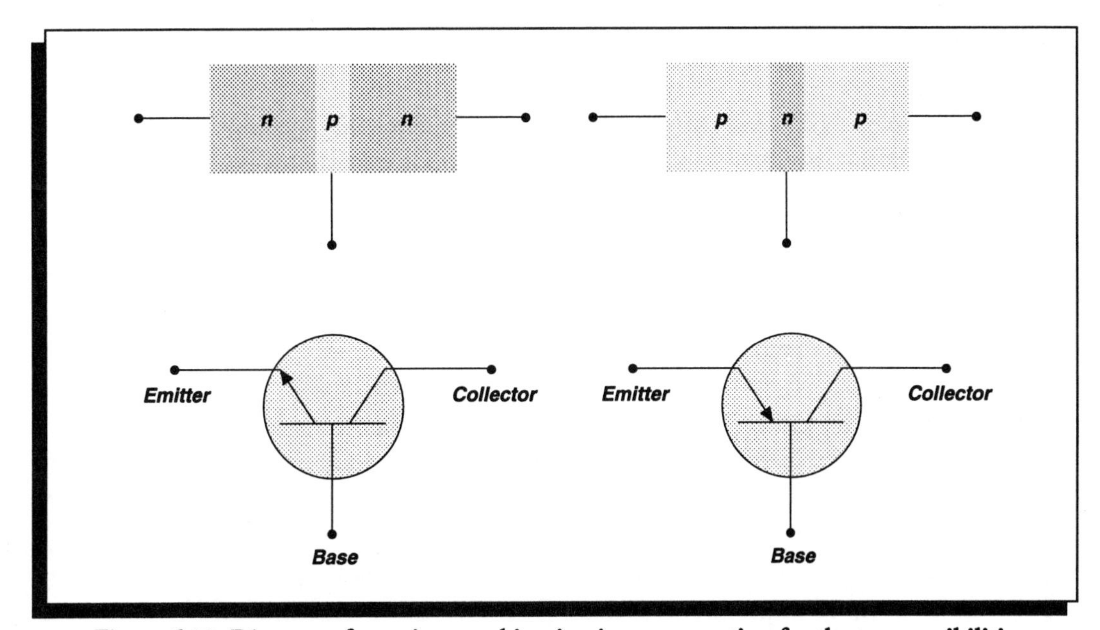

Figure 6.10 Diagram of transistor and its circuit representation for the two possibilities.

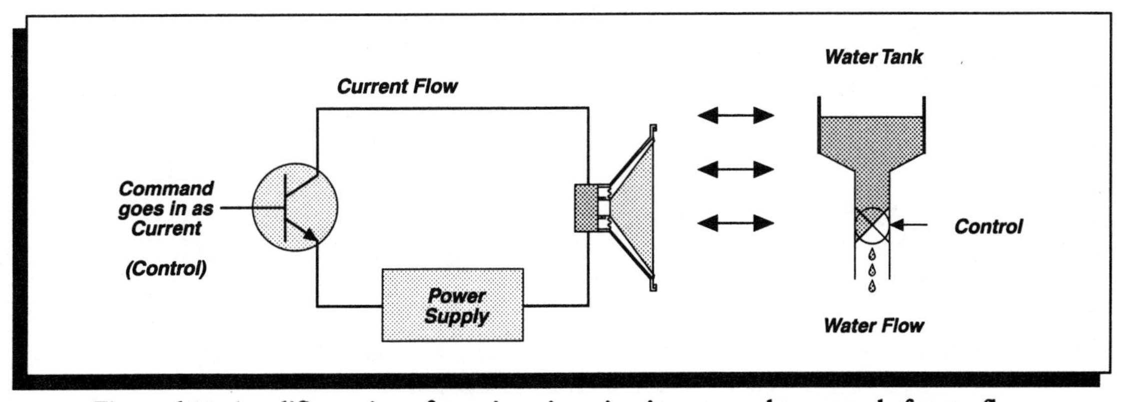

Figure 6.11 Amplifier action of transistor in a circuit compared to control of water flow.

of emanating a very fast variable resistance controlling the flow of large currents. Current flow between the emitter and the collector is further controlled by signals applied to the base. This circuit can be compared to water flowing from a container where an adjustment by a small valve can control large quantities of water flow.

The transistor can be defined as:

Transistor = **semiconducting device which can behave like a very fast variable resistor controlling large currents by means of a small input signal current.**

Because in this device a small input can control large outputs, it amplifies (makes larger) the input signal causing an output signal.

A transistor can also amplify voltage signals or current signals along with providing power amplification. It is the basic element of an amplifier. It can be used in a variety of configurations, but these applications are beyond the scope of this book.

Since the main feature of an amplifier is to increase the level of a signal, its amplification ability is represented by the gain, which is defined as:

gain = **voltage out / voltage input**

Gain is a number indicating how many times the input signal is amplified by the amplifier, as in Figure 6.12.

Example: What is the gain of an amplifier when the input signal is 1,500 μvolts and the output is 0.3 volt?

By definition,

gain = output signal/ input signal

Here, gain = 0.3 volt / 0.0015 volt

= 200

Note that the answer is a number, and that it has no units. It is a factor. It means that the input signal is 200 times larger. That is why we have an amplifier, to make input signals larger.

The representation of an amplifier as in Figure 6.12 is a simplification of the actual unit. Inside this box there are transistors, resistors, capacitors, inductors, and diodes. The design of this type of unit is interesting, but it is a major development project. Is it possible to get a general unit where one can just get a specific gain without having to design the whole amplifier? There is such a unit, it is an operational amplifier. This device is the basic building block of most amplifiers.

OPERATIONAL AMPLIFIER

Consider an amplifier which consists of many devices (transistors, resistors, capacitors, etc.) on a semiconductor chip. Its characteristics (such as gain, frequency response, etc.) would depend on the components inside it, the design, and perhaps even the manufacturer. It would be very useful to arrange this device in a circuit so that:

— it has stable gain, even in varying environmental conditions, like change in temperature

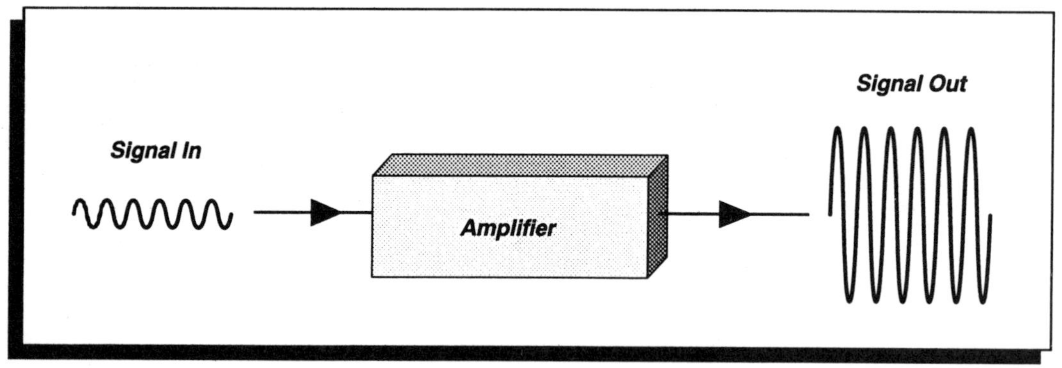

Figure 6.12 Function of an amplifier.

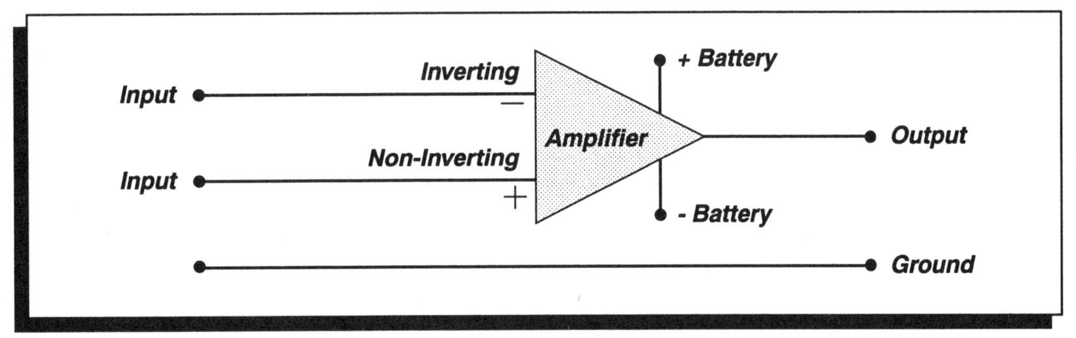

Figure 6.13 Amplifier integrated on a chip.

— the gain will be of a specific value that we can select
— its gain is independent of the components in it, the manufacturer, etc.

The ordinary amplifier on a chip does not usually have these features. The basic circuit of a integrated amplifier on a chip is shown in Figure 6.13. It has all sorts of features in order to be useful for various applications. It has:

— two inputs, one which does not invert the signal (+) and one which inverts it (–).
— an output.
— leads which attach to a power supply, usually a battery.
— a high gain which depends on the circuit and the manufacturer's design.

Although this device has a high gain and will amplify a signal, it lacks stability.

Its performance can be modified to achieve high stability and gain, which is independent of internal parameters by adding a few simple components in a special way:

— Part of the output signal is fed back to the inverting input. To ensure that only part of the signal is fed back, a resistor is placed in this path. The larger the resistor, the less signal will be fed back. This resistor is called the feedback resistor, R_f.
— Since the output is 180° out of phase with the inverting input, the signal at the input will be reduced. It is called negative feedback. Its definition is:

***negative feedback* = feeding part of output signal back to the input, but 180° out of phase with it**

— A resistor is placed at the input and it is called the input resistor R_{input}

By analysis of this circuit, Figure 6.14, the gain is simply given as:

$$\textbf{gain} = R_f \,/\, R_{input}$$

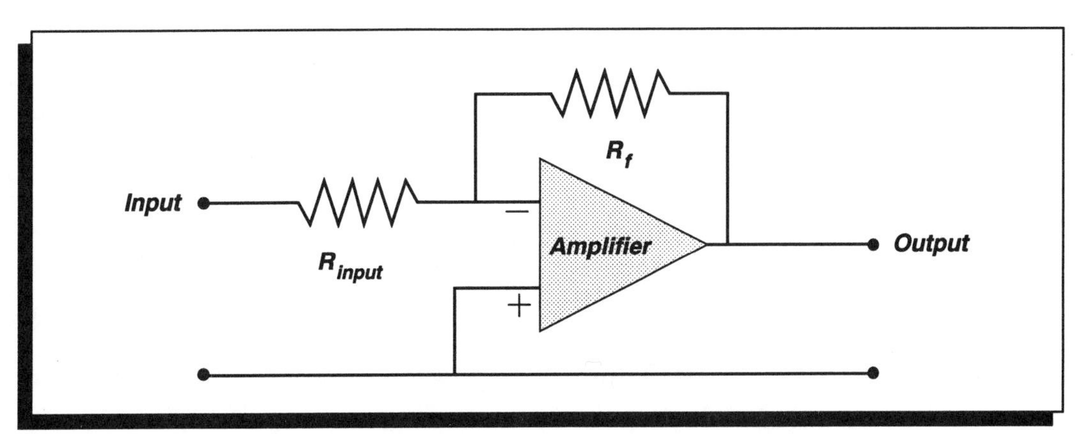

Figure 6.14 Operational amplifier with negative feedback.

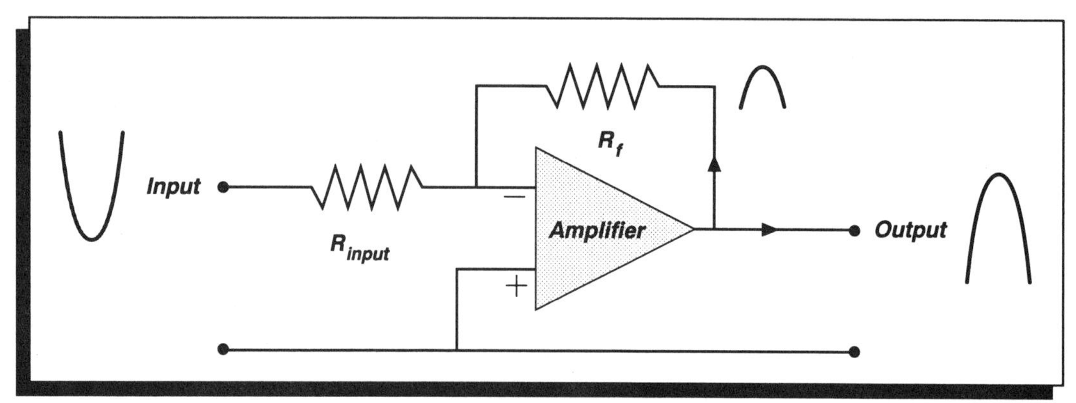

Figure 6.15 Negative feedback corrects fluctuations in gain.

This amplifier is called operational amplifier or "op–amp." We have achieved these goals:

— Its gain is simply given by two resistors that we select. The gain is independent of what is inside the amplifier or of who made it.
— The gain of the amplifier has been reduced because part of the signal is fed back, but the amplifier becomes very stable, which is an important feature.
— We can select any gain that we need just by the choice of two resistors.

Negative feedback is a very important method of stabilizing the response of a system. We even use it in our everyday life. For example, while driving a car, if we get too close to the sidewalk, we correct and steer the car away from the sidewalk. Had we not corrected this situation, there would have been an instability as the car hit the sidewalk. Let us examine how negative feedback stabilizes an amplifier. Consider the op–amp in Figure 6.15 with the input and output waveforms as shown. At some instant assume that the gain increases perhaps due to heating inside the amplifier, and the output waveform appears as the one shown. Consequently, the feedback signal also increases and there is a larger signal to subtract (since it is 180° out of phase with input) from the input, reducing the total input signal. Almost instantly, the output signal will decrease. Thus negative feedback has stabilized the gain against fluctuation and instability and it has prevented the output from following the fluctuation.

What would happen if part of the output signal were fed back in phase with the input? This is known as positive feedback.

positive feedback **= part of output signal is fed back to the input signal in phase with it.**

The consequences of this action would be that the resultant input signal would get larger and the output would also get larger. There would then be an even larger signal to feedback and the input signal would become even larger. Very soon this would become unstable. For the car approaching the sidewalk, positive feedback would cause the wheels to be steered even more toward the sidewalk, resulting in a collision, a highly unstable situation!

There is a well–known example of positive feedback presented when using a microphone in a convention hall and a loudspeaker system. This is shown in Figure 6.16.

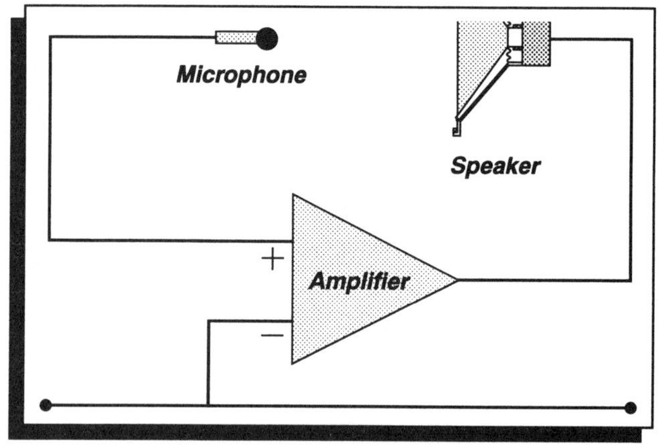

Figure 6.16 Positive feedback in large hall with a microphone picking up the sound and driving the loudspeaker system.

Sound is picked up by the microphone and its level is increased by an amplifier which drives the speaker. When the signal is picked up by microphone in phase with the speaker and amplified by a non–inverting amplifier, the sound will become increasingly louder, leading to an instability known as oscillation; this produces a loud piercing sound.

Although positive feedback is not useful for maintaining stability, it can be used as a method of producing an oscillator.

In the pre–amp and the power amplifier there are many circuits which have op–amps with negative feedback. This maintains the high performance level of the components to produce high–fidelity sound.

6.2 CONTROLS IN PRE–AMP

The pre–amp has a variety of buttons, switches, and controls. We will explore them now.

■ VOLUME CONTROL

This controls the amplitude of the signal coming out of the pre–amp. There is a simple way of doing this. A variable resistor is modified so that it has four leads, Figure 6.17, and it is located just before the output. Consider one channel. When there is an input signal, current flows into the circuit, goes through the resistor and returns to the lower terminal. The lower terminal is called "the ground," and this means that it is a relative point of reference for the voltage. As the current flows through the resistor, electrons lose energy, and the voltage, which is energy/charge along the resistor relative to the ground, decreases. At the point where the sliding contact touches the resistance wire as shown in the figure, the electrons have lost half of their energy, i.e. the voltage is one half of what it was at the top. The voltage between the sliding contact and the ground lead indicates that at this position only half the voltage is left. This means that the input has been reduced by half in amplitude between the output leads as shown in Figure 6.17. The sliding contact can be pushed down and consequently the output voltage will be further reduced. Thus, this presents a simple way of controlling the amplitude of a signal, the volume. The resistor with its four leads is known as a potentiometer, or simply called a "pot." Such a device is used in the volume control of the pre–amp, one per channel. Usually the two potentiometers are attached together so that the volume of both channels can be controlled simultaneously.

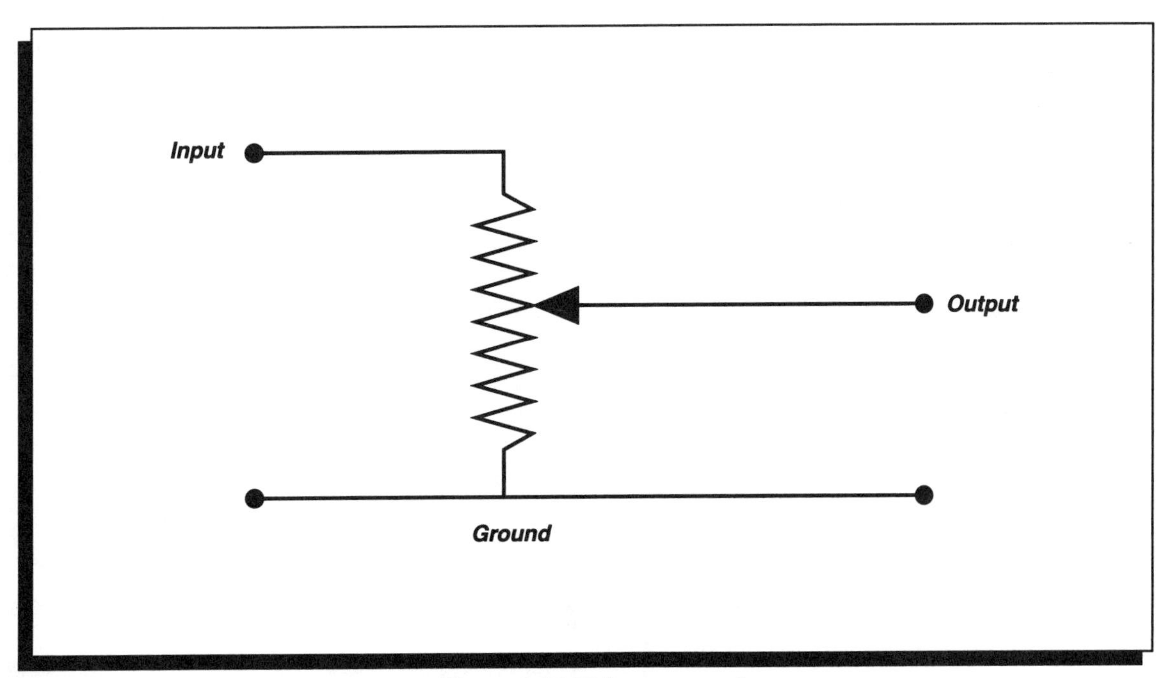

Figure 6.17 Volume control.

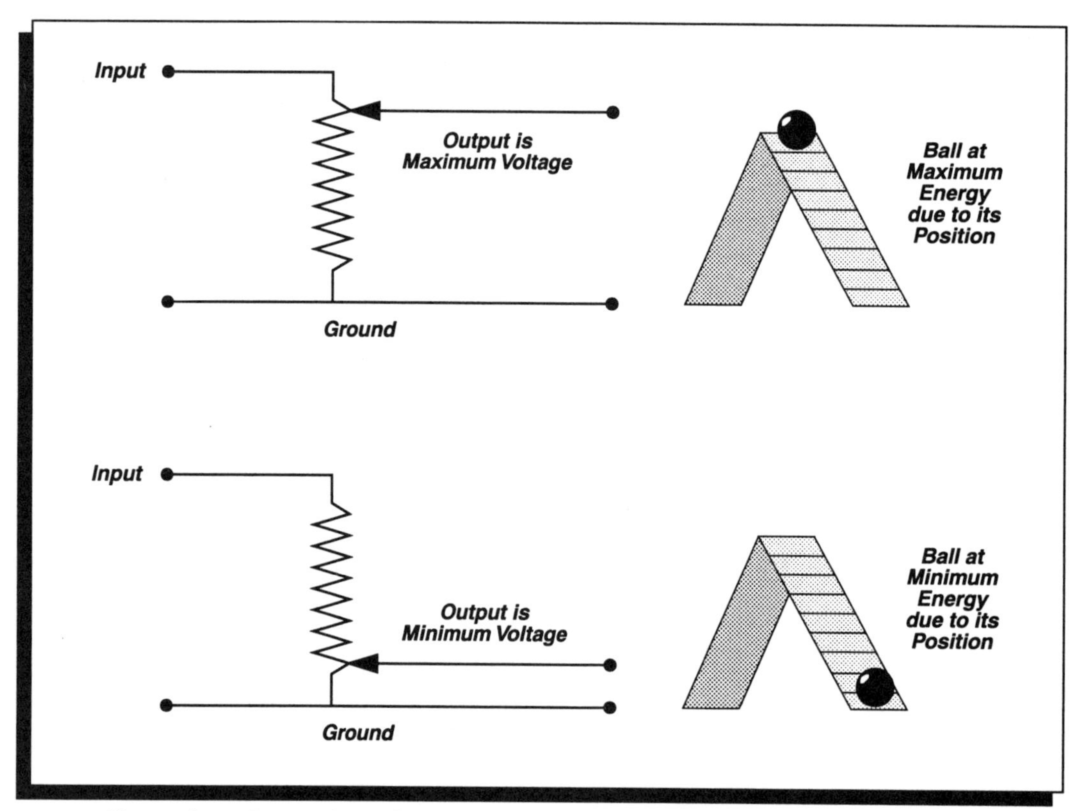

Figure 6.18 Comparison of potentiometer action with energy of a ball on a ladder.

The action of the potentiometer can be compared to the energy of a ball on a ladder, Figure 6.18. The position of the ball relative to the ground determines its energy. When at the top of the ladder, the ball's position energy is maximum. This corresponds to the sliding contact on the resistor when it is at the top giving maximum signal. Since the sliding action of the contact against the resistor is essentially continuous, the output level is also continuous with position of contact. Sometimes the potentiometer is referred to as a continuous voltage divider, since the voltage is reduced depending on the slider contact setting.

■ BASS AND TREBLE CONTROL

These are the two knobs on the front of the pre–amp and they can be rotated clockwise and counter–clockwise, with a middle position. Each knob controls the two channels, LEFT and RIGHT, at the same time.

For the case shown in Figure 6.19, clockwise rotation of the bass increases the basses up to a maximum amplitude while a counterclockwise rotation decreases the basses to a minimum amplitude. The middle position is in between these two extremes. The effect of rotating the knobs on the spectrum of the signals coming out at the output is presented in Figure 6.20. The mid–range frequencies are unaffected. These controls should be used discretely. Reasons to use them are:

— Treble control can reduce noise from a phono record or hiss noise from a tuner.
— Bass control can compensate for problems due to room acoustics (standing waves) or speaker deficiencies.

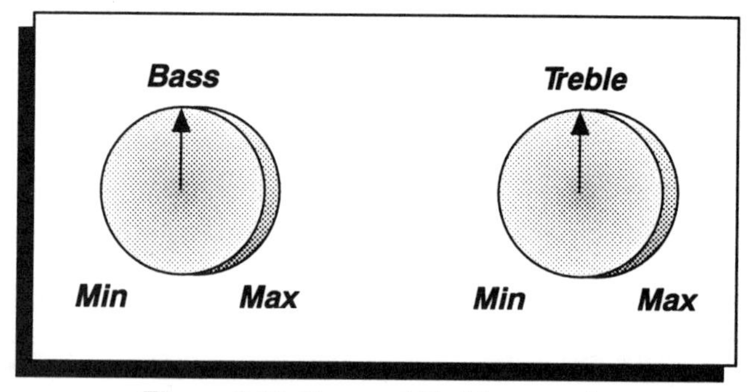

Figure 6.19 Bass and Treble controls.

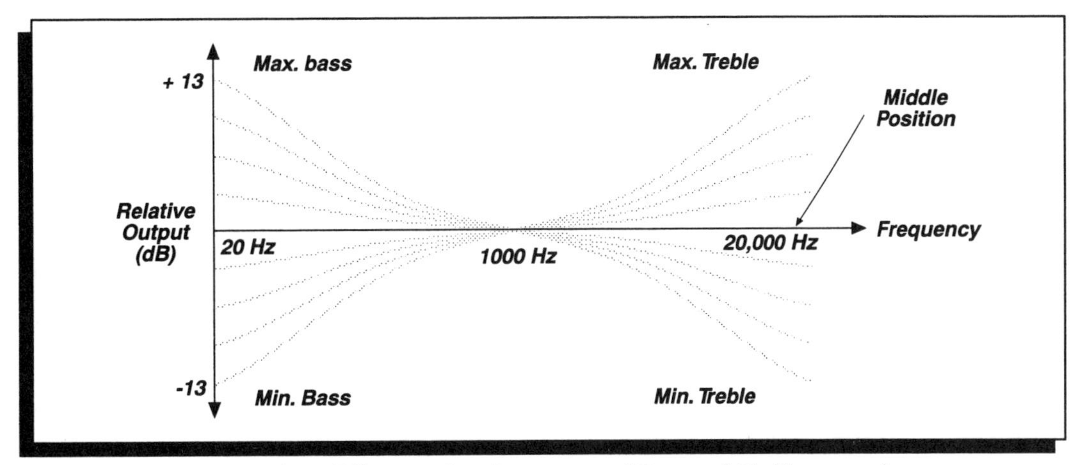

Figure 6.20 Effect on signal spectrum of Bass and Treble controls.

FILTERS

These controls and their characteristics will depend on the unit and the manufacturer's specifications. Here is what is usually available:

— Infrasonic filter: if the frequency response of an amplifier goes well below 20 Hz this filter is very important. It will cut out noise below 20 Hz, which could damage the speaker even though it would not be heard.

— HIGH and LOW filters: the HIGH attenuates the signal at high frequencies above a fixed predetermined frequency, and the LOW attenuates the low frequencies below a certain frequency. The amount that it attenuates will depends on the unit, but one of the following options usually exist:

— attenuation of the signals by 18 dB/octave
— attenuation of the signals by 12 dB/octave
— attenuation of the signals by 6 dB/octave.

The filter effect is shown in Figure 6.21 for both ends of the audio signal spectrum. When the filters are in, their action cancels part of the musical spectrum. This may be desirable for cases where there is too much noise at low frequencies or at high frequencies. The 18 dB/octave filter is more effective than the 6 dB/octave filter.

Example: If a 12 dB/octave LOW filter is used and it starts being effective at 100 Hz when switched in, at what frequency will it attenuate the signal by 24 dB?

Since it attenuates at 12 dB for every octave range in frequency down, the signal will be cut by 12 dB at 50 Hz, and a total of 24 dB at 25 Hz.

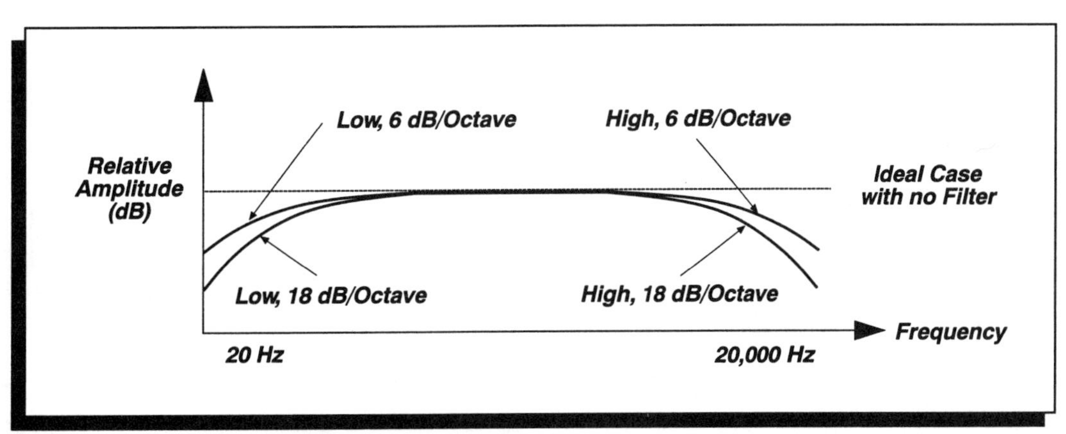

Figure 6.21 Action of LOW and HIGH filters with 6 dB/octave attenuation, and also with 18 dB/octave attenuation.

LOUDNESS

This is a clever circuit which makes up for the usual hearing deficiencies at low VOLUME levels. Recall the Fletcher–Munson curve of the response of the human ear. At low volume levels, we do not hear well the low and high frequencies. When this switch is in, it makes up for these deficiencies by boosting the low and high frequencies by a predetermined amount at low volume levels. As the level gets louder the frequency response curves become flatter and the LOUDNESS switch has a progressively weaker effect. Essentially, it automatically follows and compensates the Fletcher–Munson curve when it is in. Note: it has nothing to do with the VOLUME control.

BALANCE

It varies the power between the LEFT speaker and the RIGHT speaker. This may be useful when one speaker is against a corner reflecting much of the bass frequencies, while the other speaker is by only one wall, reflecting less of the bass frequencies. The BALANCE control can remedy this situation.

6.3 AMPLIFIER PERFORMANCE

An amplifier is expected to raise the level of input signals in the frequency range of 20 Hz to 20,000 Hz, without any preference for any frequencies. This means that it must have a flat frequency response. However, the performance of an amplifier may have certain deficiencies and imperfections or even peculiarities leading to various forms of distortion and noise. We will now cover these deficiencies and their origins.

TOTAL HARMONIC DISTORTION

Total Harmonic Distortion is usually referred to as THD. Although musical instruments produce sound with many harmonics, amplifiers should reproduce the incoming signals without adding or creating any new harmonics. The amount of new harmonics introduced by the amplifier is known as THD. Figure 6.22 shows how an amplifier can produce THD. Let us look at a signal of 1 kHz at the input. One would expect the output to be at 1 kHz, but of larger amplitude due to the gain of the

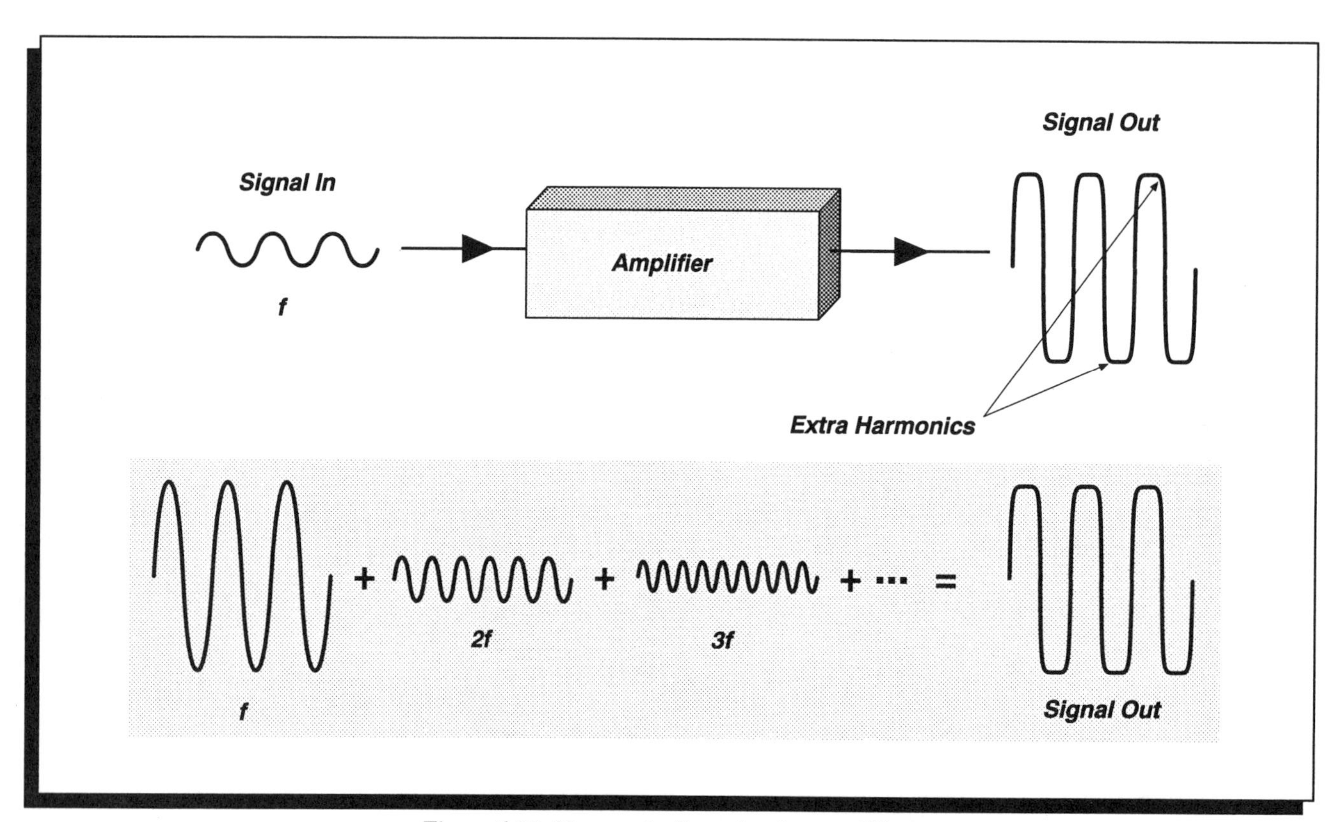

Figure 6.22 Harmonic distortion by amplifier.

amplifier. A careful look at the output shows that the wave has been made more complex, so extra harmonics have been added by the amplifier, such as: 2 kHz, 3 kHz, 4 kHz, etc. These extra frequencies are unwanted.

To specify how much distortion there is, a simple formula can be used:

$$\text{Total Harmonic Distortion} = \frac{\text{voltage of unwanted harmonics}}{\text{voltage of wanted fundamental frequencies}} \times 100$$

The human ear can detect distortions down to a level of about 1%. In practice, a reproduced sound never sounds like the original one, even with ideal amplifiers and speakers. There is always some THD. The reasons why an amplifier causes THD are:

- — Limited frequency response; for example, when an amplifier works only up to some frequency lower than 20,000 Hz.
- — Non–linear gain: the amplifier does not amplify by a fixed factor. Usually at high input levels the output is not made proportionally larger. An example is shown in Figure 6.23.

INTERMODULATION DISTORTION

Intermodulation distortion is known as IM. In this, the amplifier also introduces tones of different frequencies, but they are not multiples of the signal frequency. This type of distortion occurs when two or more signals enter the amplifier, and results in

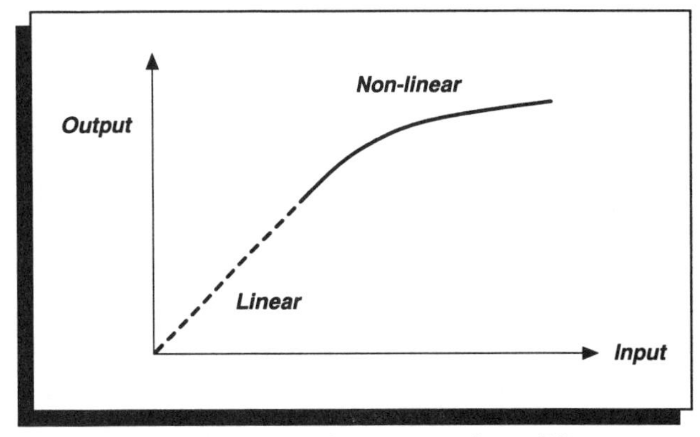

Figure 6.23 Non linear gain of amplifier.

distortion of the output, which now has new frequencies on top of the original ones. The new frequencies are the sum and difference of input frequencies, with all sorts of combinations of sum and difference.

Intermodulation distortion **= creation of new frequencies at the output and they are the sum and difference of input frequencies.**

This is shown in Figure 6.24 for two input frequencies, but it applies to any combination of input frequencies. This distortion is also expressed in a percentage, as the amplitude of unwanted frequencies is divided by the amplitude of input frequencies times one hundred. It is caused by non–linear behavior of amplifier. This effect has been known for a long time, particularly for the response of the human ear When two loud sounds are played together the ear will hear extra frequencies at the sum and difference of the original frequencies. Here also, it is due to non–linear response.

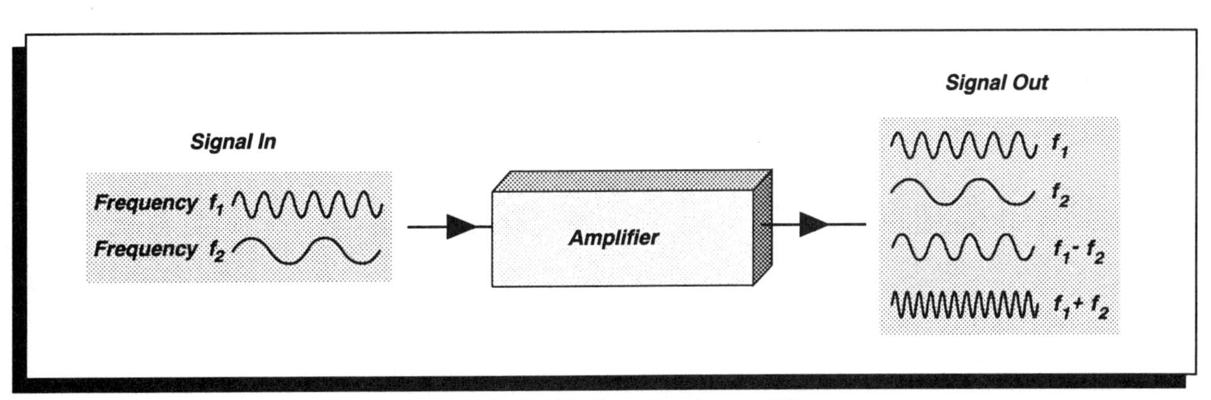

Figure 6.24 IM distortion in amplifier.

Both distortions, IM and THD, increase significantly when the maximum power output of the amplifier is approached and exceeded, Figure 6.25. Beyond the maximum power rating the amplifier starts clipping because of the limited power available in the amplifier. The waveforms become highly distorted then, as shown in Figure 6.26.

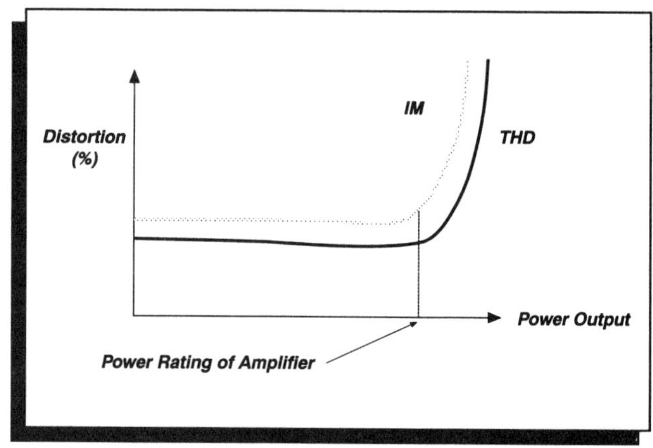

Figure 6.25 Distortions, THD and IM increase sharply above power rating of amplifier.

■ SIGNAL–TO–NOISE RATIO (S / N)

This is a very important characteristic associated with an amplifier. When a signal is amplified it will pick up small extra signals from the amplifier due to fluctuations of electrons in various components of the amplifier. The situation is shown in Figure 6.27. The fluctuations are called noise.

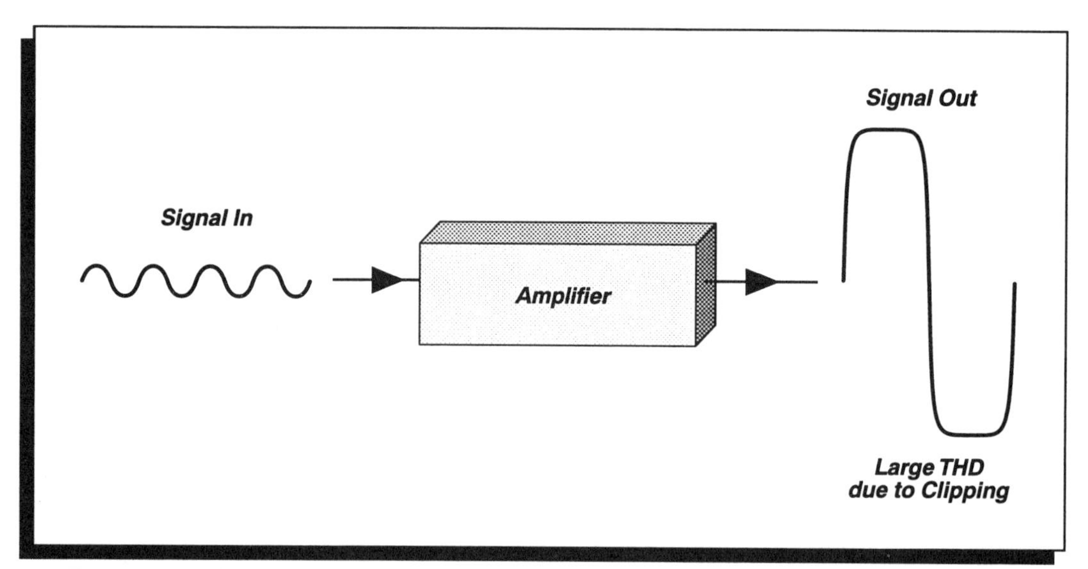

Figure 6.26 Clipping of waveform by amplifier at high input levels beyond the rated value.

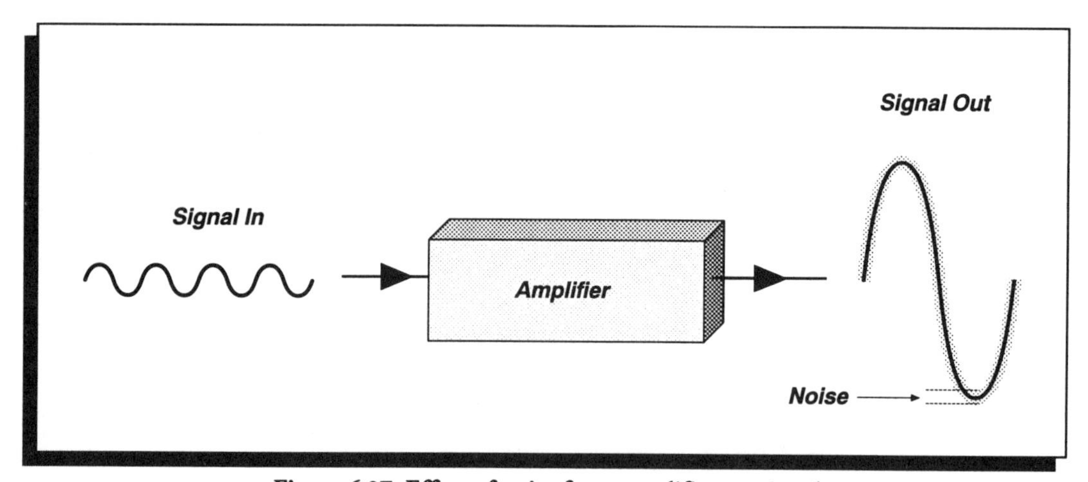

Figure 6.27 Effect of noise from amplifier on signal.

Hence the output of the amplifier contains:

signal	=	**wanted**
noise	=	**unwanted**

Since the volume of the amplifier can be set to any value, it is best to make a comparison of signal to noise. This is known as:

SIGNAL / NOISE (S / N)

It is expressed in decibels. A S/N of 60 dB means that the power of the signal is 10^6 times that of the noise. One should look for large S/N. The larger it is, the better it is.

There is a special kind of noise known as hum and it originates in the power line. It is annoying because it can be heard. Its definition is;

hum **= noise from power line at 60 Hz or harmonics of 60 Hz**

6.4 Amplifier Specifications

These specifications should be considered and understood when buying an amplifier. They usually are included with the amplifier. Consult them to see what sort of specifications your amplifier has.

■ Power

This is the maximum output power that an amplifier can deliver continuously and it is given in watts per channel with both channels simultaneously driven. It is the maximum continuous power that the amplifier can deliver without exceeding its distortion ratings. Furthermore the resistance of speakers should be stated since power depends on resistance.

The specifications should have:

— [*] watts continuous / channel
— into [*] ohms speakers, both channels simultaneously driven
— over frequency range 20 – 20,000 Hz
— distortion of [*] % (should be less than 1%) THD at that power.

*These amounts will vary according to manufacturing design.

The power requirements vary widely depending on speaker sensitivity, room(s) size, and the type of music. Remember that 10 times more power is needed to get sound which is two times louder. If speakers of 80dB/1W/1m sensitivity rating are driven by a 200 watt amplifier, then speakers of 90dB/1W/1m sensitivity would sound just as loud with a 20 watt amplifier. A 92dB/1W/1m sensitivity speaker driven by a 20 watt amplifier would sound louder than when a 200 watt amplifier is driving speakers of 80dB/1W/1m sensitivity. Most listening in a room corresponds to the power output by amplifiers of a few watts. Since music has many transients, once in a while the power output is higher and this is where it is good to have some power in reserve. Distortion in this case will be kept low.

Terms like "music power," "peak power" or "instantaneous power" are interesting specifications, but they only relate to very short bursts of sound and hence they are not the main specifications which would be very useful.

■ Dynamic Headroom

Dynamic headroom tells how much additional power, above the normal continuous power rating, can be delivered for a short time, 0.02 second. This power is expressed relative to the rated continuous power in decibel. If a 50 watt amplifier can deliver 100 watts for a burst of 0.02 second, the dynamic headroom is 3dB. Because music and speech are transient in nature, it is useful to have extra power for short bursts once in a while. This specification is particularly useful for CD music which has a high dynamic range.

■ Total Harmonic Distortion (THD) and Intermediate Distortions (IM)

These two specifications usually come together and they should be less than 0.5%. The THD refers to a 1 kHz signal, and the IM refers to a signal at 60 Hz and one at 6 kHz, in the testing. Usually it is hard to detect by ear distortions of less than 0.1%.

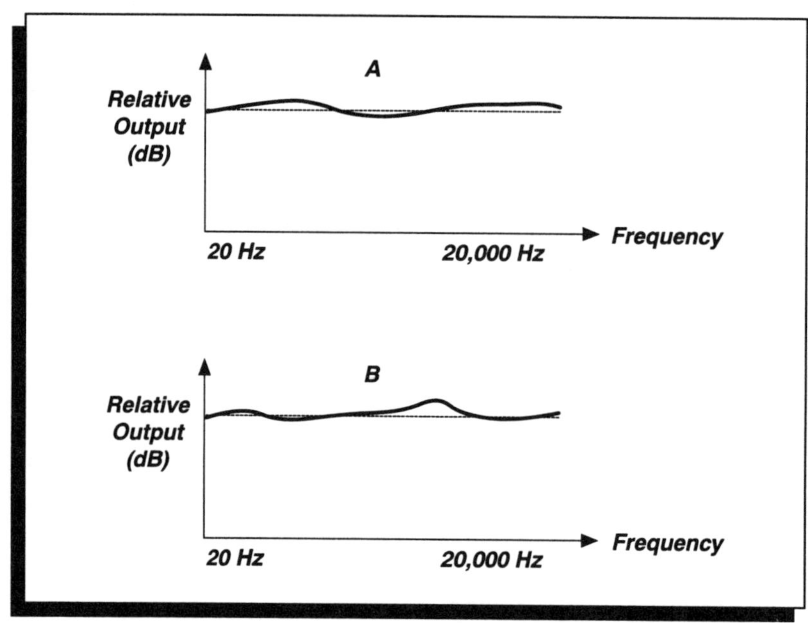

Figure 6.28 Comparing two amplifiers with the same specifications. Even though their specifications are the same, the amplifiers will sound different.

FREQUENCY RESPONSE

Ideally, the frequency response of an amplifier should be flat from 20 Hz to 20,000 Hz. This means that at a constant amplitude input signal the amplifier delivers the same level for each frequency over the frequencies of 20–20,000 Hz. In that case the frequency response will be stated as 20 Hz–20,000 Hz ± [*] dB.

Two amplifiers may sound different because there could be a variation in their frequency response. Consider amplifiers A and B and their frequencies response.

Amplifier A: 20 – 20,000 Hz ± 0.5 dB
Amplifier B: 20 – 20,000 Hz ± 0.5dB

The resonance of one of them may be at a different frequency than the other. This is shown in Figure 6.28.

SIGNAL–TO–NOISE RATIO (S/N)

This specification is determined by taking the ratio of the output signal from a test signal to the noise introduced with this signal; the ratio is expressed in decibels. Recently, this specification has been measured by using an "A–weighted" method, and this refers to a method of measuring noise which is correlated to the response of human ears at low levels (since noise will be at low levels). This is shown in Fig. 6.29. The A–weighted specification gives impressive values for the S/N ratio

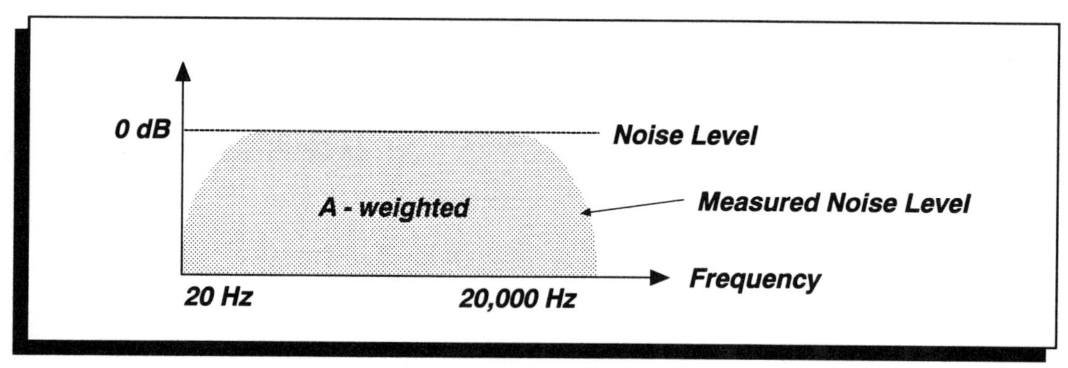

Figure 6.29 A–weighted noise.

since there is less noise in this measurement than when all the noise was measured.

Good values for the S/N are:

S/N = 80dB or higher for AUX, TAPE, TUNER
S/N = 70dB or higher for PHONO

■ CROSS–TALK

The basic hi–fi system deals with two channels, the LEFT and the RIGHT. To what degree does one channel leak into the other? This specification is expressed as a difference in decibels between LEFT and RIGHT channel leakage.

If the LEFT channel puts out 90dB and the RIGHT channel puts out 60dB with no signal going to that channel, then the cross–talk is 90 – 60 = 30dB. This number should be large. Cross–talk is caused by the sharing of the same power supply in the amplifier and by wires for each channel being near each other.

Every component has it own cross–talk (or stereo separation). This can be easily demonstrated by:

— removing the connection for one channel, LEFT for example, from any of the sources one is listening to.
— rotating BALANCE button to LEFT channel only
— listening to music sent out by LEFT speaker, but received by RIGHT channel. It "sneaked" into the LEFT channel.

■ DAMPING FACTOR

It is a specification which deals with a mechanical problem but it has an electrical solution. When electric signals stop going to the speaker, the speaker continues to vibrate, and hence it produces sound for a while. This is because the speaker is a mechanical system. The vibration blurrs the sound coming from the speaker. To solve this problem, it is necessary to use some sort of "shock–absorbers" in the speakers. The "shock– absorber" is provided by the amplifier's internal resistance; it damps the speaker motion. The damping factor is defined as:

***Damping factor* = speaker impedance / amplifier internal resistance**

Typically good values for this number range from 20–100.

■ INPUT SENSITIVITY

This is an important specification for compatibility with other equipment that is hooked into the amplifier. The definition of sensitivity is:

***Sensitivity* = input voltage required to produce a 1–watt output signal at full VOLUME.**

This specification deals with the input for PHONO, but there is also one for the TUNER and other inputs. It tells you how small a signal you can have at the input of the PHONO (for example) to get a certain level of output signal.

SUMMARY OF TERMS

Amplifier: A device where input signal controls the output signal. It increases the voltage or current or power of a signal.
Cross–talk: How much signal from one channel gets mixed up in the other channel.
Damping factor: The ability of an amplifier to stop oscillations of a speaker when output signal has decreased to zero. It is given by impedance of speaker divided by the internal resistance of amplifier.
Gain: The factor by which an amplifier increases the level of a signal. It is the amplitude of the output signal divided by the amplitude of the input signal.
Hum: Noise picked up in amplifier and other components from power lines and it is at 60 Hz and its harmonics.
Intermodulation Distortion: Amplifier specification which tells how much new frequencies which are the sum and difference frequencies of the input have been added.
Negative Feedback: Feeding back to input a signal 180° out of phase with it.
Operational Amplifier: A high gain amplifier with negative feedback.
Positive Feedback: Feeding back to input a signal in phase with it.

Rectifier: A diode which conducts only a portion of an ac signal.
Signal–to–Noise Ratio (S/N): A signal level divided by noise level, expressed in dB.
Total Harmonic Distortion: An amplifier specification which tells how much undesired harmonics have been added to a fundamental frequency. It is expressed in percent as the ratio of amplitudes of all unwanted frequencies relative to amplitude of fundamental signal.
Transistor: A solid state device made of three different semiconductor types of material in a sandwich arrangement. Electrical connections are made to the base, emitter, and collector. It is a building block of amplifiers.

NAME ____________________ DATE ____________

QUESTIONS FOR REVIEW

1 What is the function of an amplifier?

2 Why is it necessary to have amplifiers in a hi–fi system?

3 What is the importance of ac signals in a hi–fi system?

4 Distinguish between total harmonic distortion and intermodulation distortion.

5 What is meant by a forward–biased diode?

6 What is an operational amplifier?

7 Why does an amplifier introduce noise?

8 What is the purpose of damping in an amplifier and what does damping factor mean?

9 What would happen in an operational amplifier if the feedback resistor were replaced by a capacitor?

10 Explain how negative feedback maintains stability in an amplifier.

NAME____ DATE ____________

Please select the best answer.

1. All amplifiers tend to produce unwanted output signals whose frequencies are the sum and difference frequencies of the input signals. This undesirable behavior is measured by the amplifier's:
   ___A. S/N ratio
   ___B. THD rating
   ___C. IM rating
   ___D. frequency response
   ___E. cross–talk

2. Suppose that amplifier X has a S/N ratio of 70dB, and amplifier Y has a S/N ratio of 40dB. The noise generated by Y appears to be about ___________ than the noise produced by X.
   ___A. 30 times softer
   ___B. 30 times louder
   ___C. 8 times louder
   ___D. 8 times quieter
   ___E. 40 times louder

3. Which of the following specifications tells how an amplifier can control the natural tendency of a speaker to vibrate after the music has stopped?
   ___A. Intermodulation distortion
   ___B. Damping factor
   ___C. (S/N) ratio
   ___D. Frequency response
   ___E. Compliance

4. The control on a preamplifier which compensates for the ear's response at low levels for low and high frequencies is called:
   ___A. the LOUDNESS control
   ___B. the LOW filter
   ___C. the HIGH filter
   ___D. the BALANCE control
   ___E. the BASS and TREBLE control

5. An 800 Hz signal is being sent to an amplifier. The output signal contains the following frequencies: 800 Hz, 2400 Hz, 3200 Hz. The amplifier is producing:
   ___A. intermodulation distortion
   ___B. total harmonic distortion
   ___C. S/N distortion
   ___D. cross–talk distortion
   ___E. hum noise

6. When an amplifier amplifies all audio frequencies by exactly the same amount it is said to have:
   ___A. a flat response
   ___B. a large S/N ratio
   ___C. a large damping factor
   ___D. resonances
   ___E. a large N/S ratio

7. Which one of the following is the best frequency response for an amplifier?
   ___A. 20 → 20,000 Hz ± 2.5dB
   ___B. 20 → 20,000 Hz ± 1dB
   ___C. 15 → 25,000 Hz ± 0.4dB
   ___D. 15 → 25,000 Hz ± 2.5dB
   ___E. 15 → 25,000 Hz ± 5dB

8. What is the gain of an operational amplifier, when the input resistor is 800 ohms and the feedback resistor is 24,000 ohms?
___A. 30
___B. 300
___C. 0.00125
___D. 19,200,000
___E. 1/30

9. If the power output from an amplifier is increased beyond its rated value, the THD and IM will:
___A. increase
___B. decrease
___C. remain the same
___D. differ from each other more and more
___E. stop changing

10. An amplifier is delivering 75 watts of electrical power to the right speaker. Due to imperfections of the amplifier, 0.075 watt of electrical power containing right channel music is sent to the left speaker. What is the stereo separation of this amplifier?
___A. 20dB
___B. 30dB
___C. 40dB
___D. 10dB
___E. 5.62dB

11. In an operational amplifier, of input resistance of 1.22 **k** Ω, the output is 1.55 volts when the input is 1.55 mvolts. In that case the feedback resistor is:
___A. 1,000 ohms
___B. 1.22 k ohms
___C. 1,220 k ohms
___D. 1.22 ohms
___E. 1.29 k ohms

12. In negative feedback in an operational amplifier, part of the output signal is added to the input,
___A. 180° out of phase with it
___B. 0° out of phase with it
___C. in such a way that frequencies match
___D. in such a way that phases match
___E. in such a way that periods match

13. In an operational amplifier, as the feedback resistance increases, the gain of the amplifier:
___A. increases
___B. decreases
___C. remains the same
___D. depends on the number of transistors
___E. becomes larger than for an amplifier with positive feedback

14. What is the damping factor for an amplifier whose internal resistance is 0.1 ohm and when an 8–ohm speaker is attached to it?
___A. 8
___B. 80
___C. 800
___D. 0.8
___E. 0.0125

15. With the same speaker, the damping factor of amplifier X is 10 times that of Y. In that case the internal resistance of X is ________ than that of Y.
___A. 10 times larger
___B. 10 times smaller
___C. 80 times larger
___D. 80 times smaller
___E. one cannot tell because the speaker resistance is not given

16. Amplifier X has a S/N of 60dB and amplifier Y has a S/N of 40dB. In that case, considering noise only, amplifier X:
___A. is better that Y
___B. is worse than Y
___C. has quieter sound than Y
___D. has louder sound than Y
___E. has noise which is 20 times louder than Y

17. What is the gain of an amplifier which puts out 7.5 volts when the input signal is 15μvolts?
___A. 500,000
___B. 50,000
___C. 0.5
___D. 112.5
___E. 500

NAME ______________________________ **DATE** ____________

18. When a sine wave at the input to an amplifier becomes a square wave at its output, the distortion is:
    ___A. Intermodulation distortion
    ___B. Total Harmonic distortion
    ___C. unknown because the frequency of the fundamental is not given
    ___D. cross–talk
    ___E. related to the damping factor

19. A diode does not conduct current when it is:
    ___A. reverse–biased
    ___B. forward–biased
    ___C. attached to an ac source
    ___D. attached to a dc source
    ___E. attached to another diode

20. A power rating of an amplifier (i.e. 100 watts) is meaningless by itself. The rating must also include:
    ___A. the frequency range over which power is delivered
    ___B. the THD produced at the power rating
    ___C. the speaker resistance into which power is delivered
    ___D. the number of channels driven
    ___E. all of the above must be included

21. An amplifier has a gain of 250. This means that when the input is 125μvolts, the output will be __________ volts.
    ___A. 250
    ___B. 0.03125
    ___C. 3,125
    ___D. 2
    ___E. 0.5

CHAPTER

7 ELECTROMAGNETISM

This chapter deals with the connection between electricity and magnetism, and aspects of electromagnetism which are relevant to the hi–fi. After an introduction to the basics of magnetism, electromagnets are developed as the basic elements in cone loudspeakers, in magnetic recording, and transformers. Faraday's law of induction provides a fundamental connection between electricity and magnetism. There are many applications of this law in the hi–fi, ranging from microphones to magnetic playback heads, and many others. Finally, electromagnetism is applied to speakers which rely on magnetic forces on current–carrying wires in a magnetic field.

7.1 BASICS

The beginning of magnetism goes back centuries, when it was discovered that certain stones, known as lodestones, could attract pieces of iron. One of the most important early applications of magnetism was to navigation, where a compass needle would always point in the direction north–south because the earth itself is magnetic. In 1820

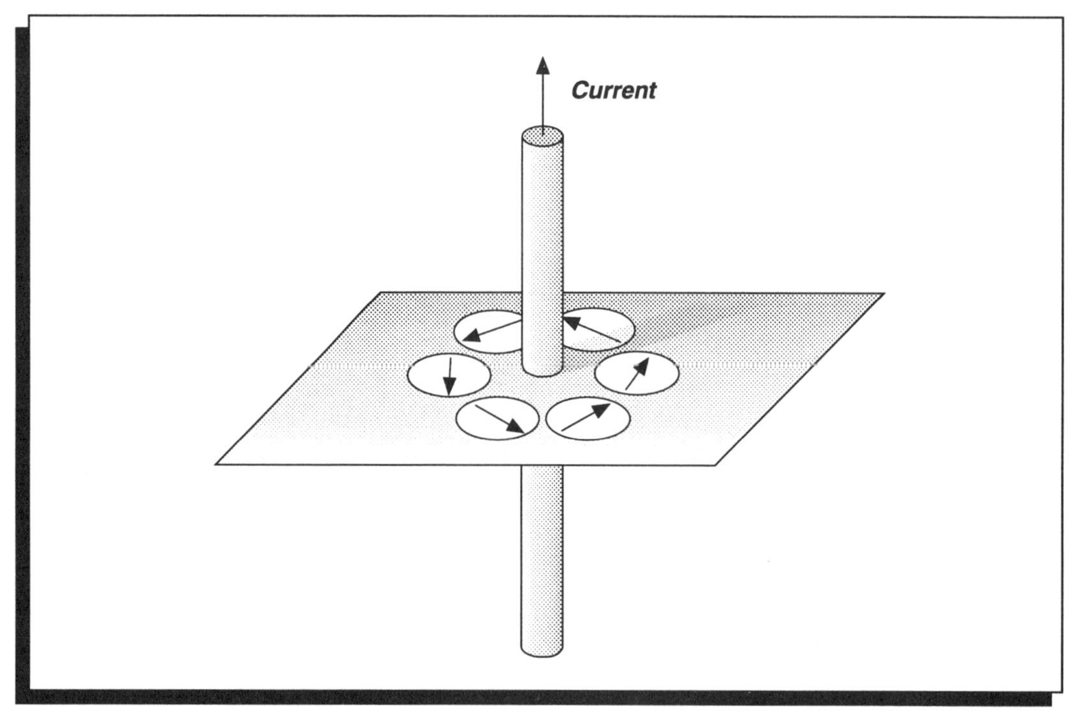

Figure 7.1 Effect of current in a wire on compasses around it.

Oersted discovered that electric currents affect a compass and that consequently, magnetism was related to electricity. This led to the observation that electric currents are the source of magnetism. Hence:

> ***magnetism* is caused by moving electric charges**

since a current is a flow of electrons. Figure 7.1 shows a wire carrying a current and its effect on a set of compasses around it. The current in the wire forces the compass to make a circular alignment around the wire. Since there is an effect on compasses, which are small needle–shaped bar magnets, the current–carrying wire produces a magnetic field in the region around it. This wire has a similar effect on the compass as the magnetism of the earth. Hence, the definition of magnetic field is:

> ***magnetic field* = The energetic region of space surrounding a magnetic system. There is magnetic force on any magnetic test object brought within this space.**

Tests on a bar magnet will show that one end aligns itself in the earth's magnetic field and points north, while the other end points south. Therefore, we label the bar magnet as having a north pole and a south pole, Figure 7.2. This bar magnet has a magnetic field around it and when another bar magnet is brought close to it, there will be a force of attraction when the north and south poles are close to each other. There will be a force of repulsion when two similar poles are next to each other, north–north or south–south. The law governing the forces between magnets is:

Like poles repel each other
and
unlike poles attract each other.

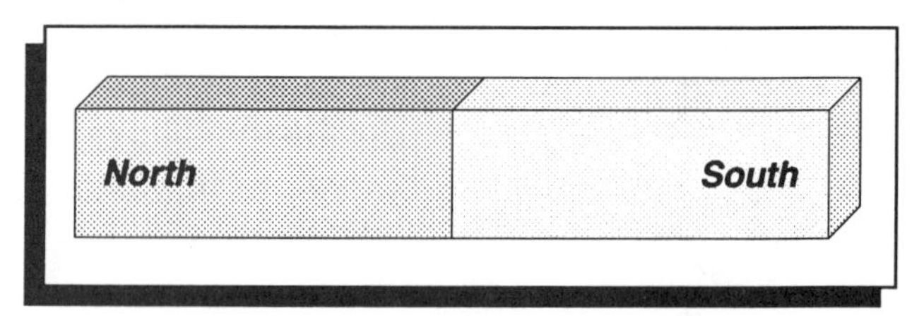

Figure 7.2 Bar magnet has a north pole and a south pole.

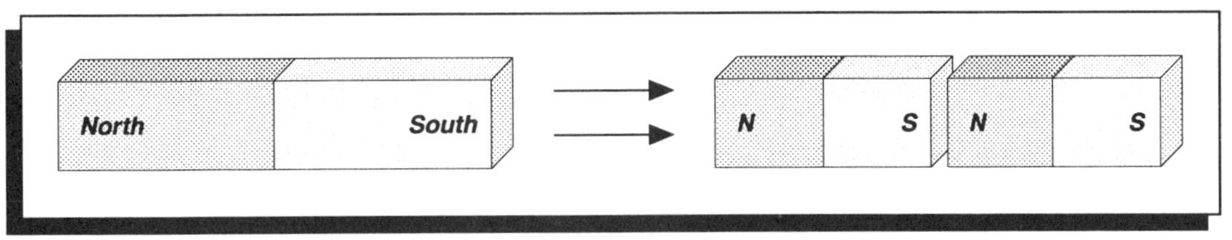

Figure 7.3 Cutting a bar magnet produces shorter magnets each, with its own respective north and south poles.

If one bar magnet were to be cut in two, each would have a north and south pole, Figure 7.3. The bar magnet behavior is not like that of electrically charged objects, where positive and negative charges can be separated. In magnetism, the north and south poles always come in pairs; a single north or south pole has not been isolated.

The origins of magnetic forces are not magnetic charges, since they do not exist, but they are electric charges in motion. The basic unit of magnetism is a ring of electric current as shown in Figure 7.4. It is called a magnetic dipole and it behaves like a tiny bar magnet. There is a north–seeking end and a south–seeking end. The connection between a ring of current and a bar magnet seems puzzling, especially since the origin of magnetism is moving electric charges. Where are the moving electric charges in a bar magnet? This question can be answered at the atomic level. In an atom moving charges are:

— electrons in orbits around the atom
— electrons with a spin around their axis just like a top
— nuclei moving around in the nucleus with characteristic motion

Indeed, there will be magnetism as a result of all those electrical charges moving. Nuclear magnetism due to moving nuclei is very weak and is not useful for the applications in hi–fi (however, it has very important applications in medical imaging, known as Magnetic Resonance Imaging, or MRI). The electron spinning contribution is the strongest in a solid and so it has applications to the hi–fi. The electron orbital motion for atoms in a solid tends to produce only a small contribution to the overall magnetism of the material. For materials like iron, cobalt, nickel, it is the spin contribution which is the most significant.

Since moving electric charges produce magnetism in atoms, then why are all atoms not magnetic? The electrons in an atom are arranged in specific orbits around the nucleus with definite energies as described by the laws of quantum mechanics. In most atoms the magnetism due to one electron is cancelled by a neighbor electron in the same orbit. The net result causes the atom to be non–magnetic

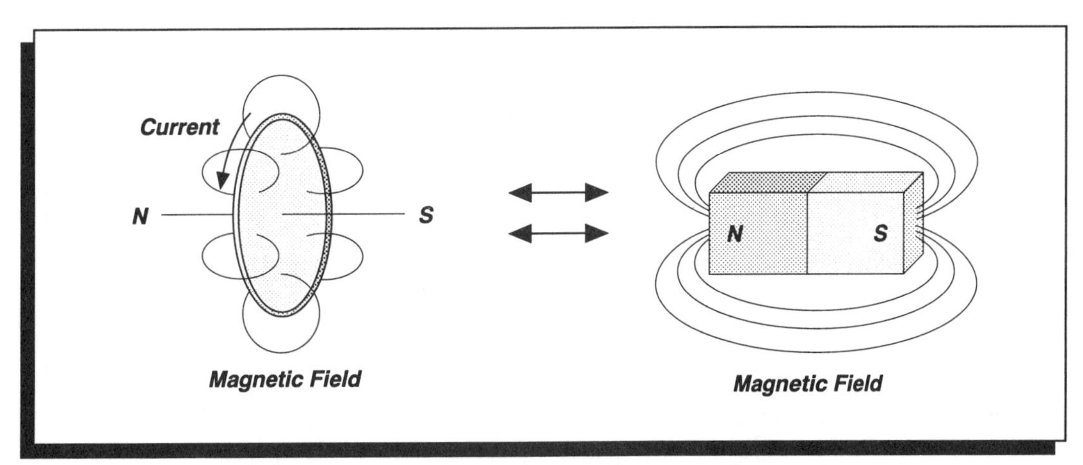

Figure 7.4 Magnetic dipole is the basic unit of magnetism.

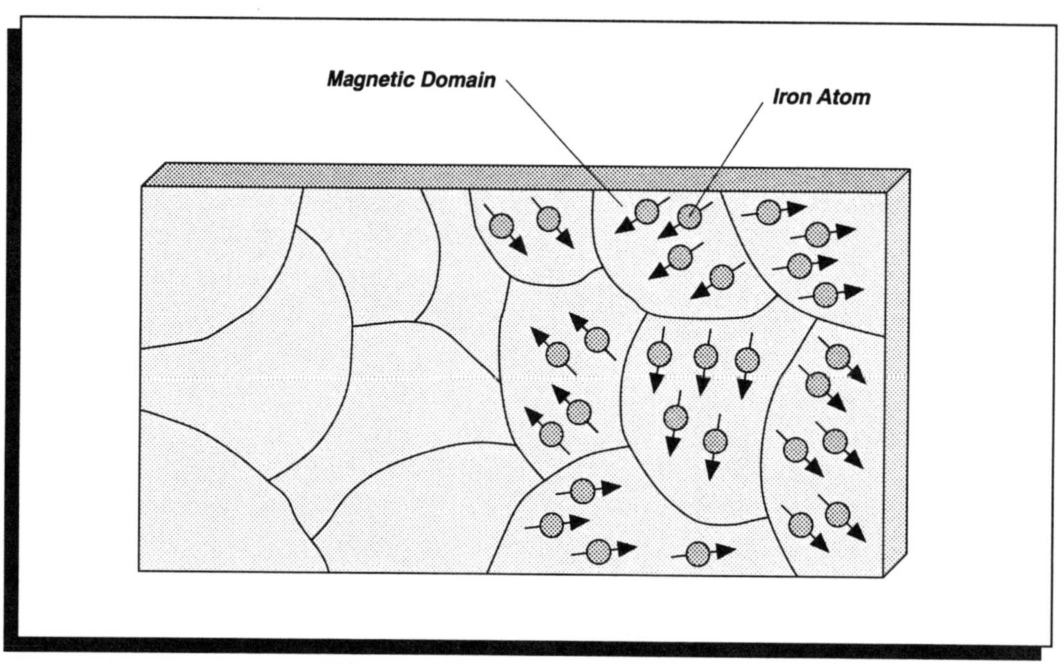

Figure 7.5 Unmagnetized piece of iron.

then. In some atoms, like iron, electrons combine their magnetism constructively resulting in an atom which behaves like a tiny bar magnet i.e. it has a magnetic dipole.

A solid is made up of these magnetic atoms, each one interacting strongly with its neighbors. This causes them to form very small clusters where each cluster has all the atomic magnets pointing in the same direction. Each cluster is called a magnetic domain. Figure 7.5 shows a piece of iron where all the atoms are aligned within each domain, but the domains are randomly aligned. Hence, this iron is in an unmagnetized (unaligned) state and it is not a magnet. However, by bringing a magnet near the piece of iron, the north pole of the magnet will attract south poles and repel the north poles of the iron. The is described in Figure 7.6 where the magnet forces alignment of domains and hence the iron becomes magnetized. In fact, a strong enough magnetic field will cause an alignment of most of the magnetic domains. If a piece of iron is forced into this state, it becomes a bar magnet.

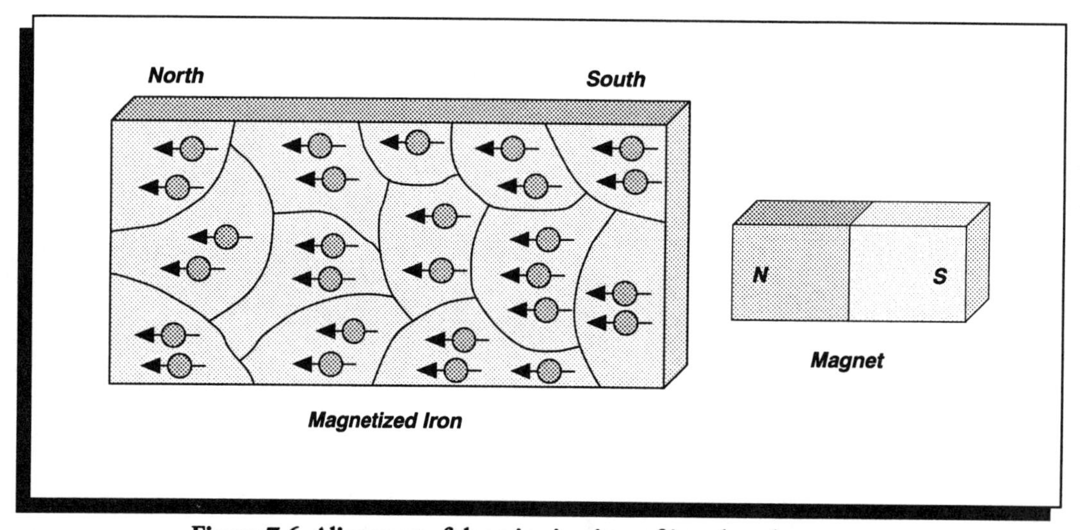

Figure 7.6 Alignment of domains in piece of iron by a bar magnet. Iron becomes magnetized.

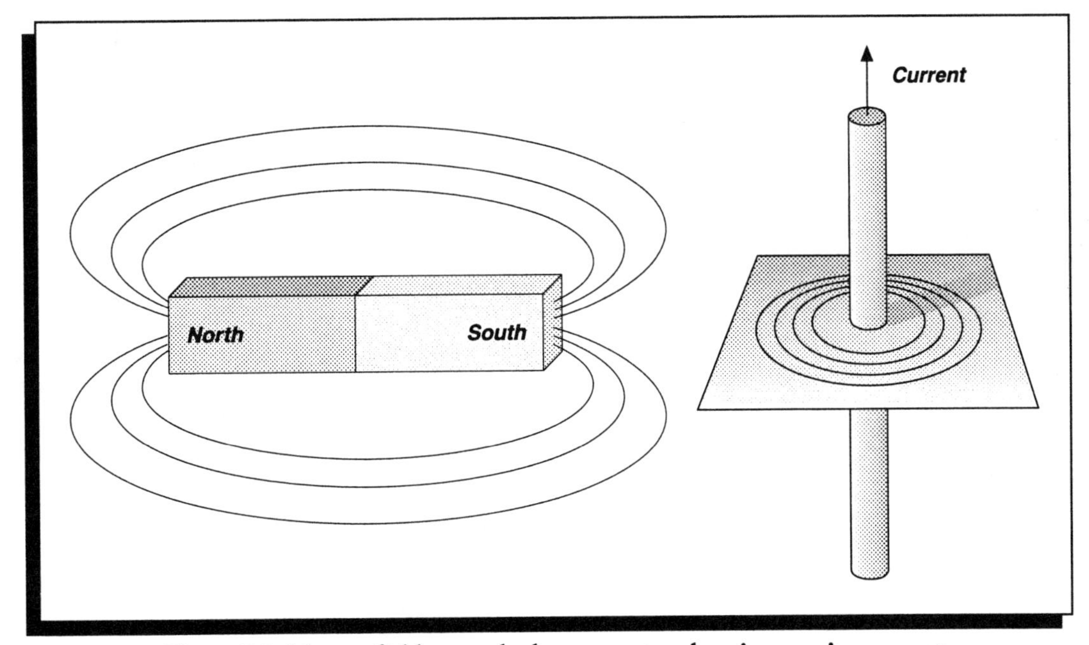

Figure 7.7 Magnet field around a bar magnet and a wire carrying current.

7.2 ELECTROMAGNETS

It is possible to visualize the presence of a magnetic field by sprinkling fine powder of a magnetic material like iron. Figure 7.7 shows the pattern that will be observed around a bar magnet and around a wire carrying an electric current. This means that a wire carrying a current can be used to magnetize a piece of iron if its magnetic field is strong enough. The magnetic field created by a wire carrying a current can be increased by concentrating the magnetic field. This is achieved by bending the wire into a loop where the magnetic field will be bunched up inside the loop. Placing many such loops next to each other leads to a proportional increase in the resulting magnetic field. Figure 7.8 shows a single loop of wire with the magnetic field concentrated at its center and a coil with many loops increasing the strength of the resultant magnetic field. Even stronger magnetic

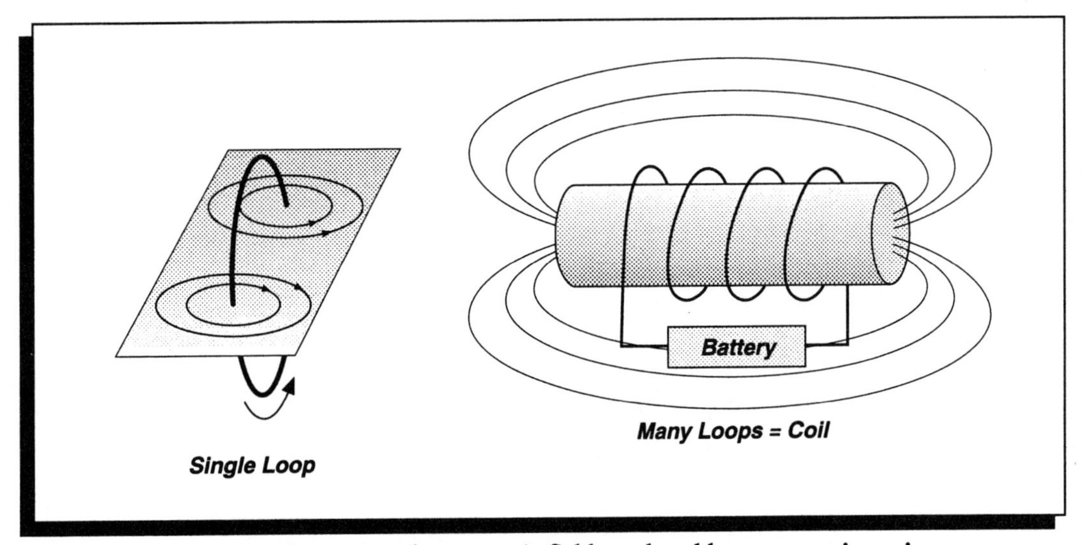

Figure 7.8 Increasing the magnetic field produced by a current in a wire: by forming a loop, and by using many loops.

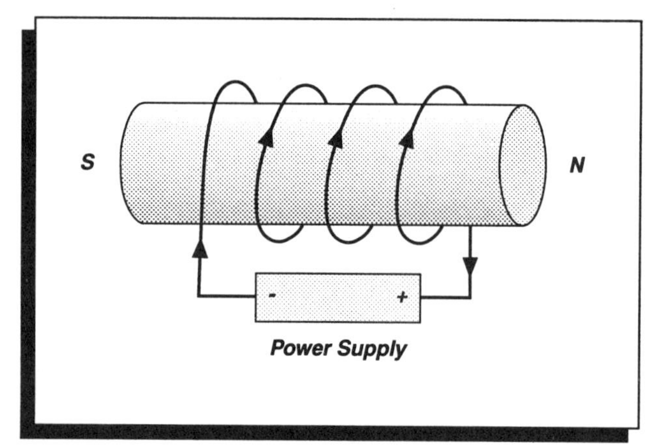

Figure 7.9 An electromagnet.

fields can be produced by placing a piece of magnetic material, like iron, inside the coil. The material is known as the core. A coil is shown in Figure 7.9, and it is known as an electromagnet. Its characteristics are:

- An increase in the amount of current will increase the magnetic field in the coil.
- An increase in the number of turns in the coil will increase the magnetic field.
- A change in the direction of the current, will change the direction of the resulting magnetic field.

This last characteristic can be used to set down a rule, which tells us the direction of a magnetic field created by a coil, when current flows in a particular sense. This rule is known as the first Left–Hand Rule.

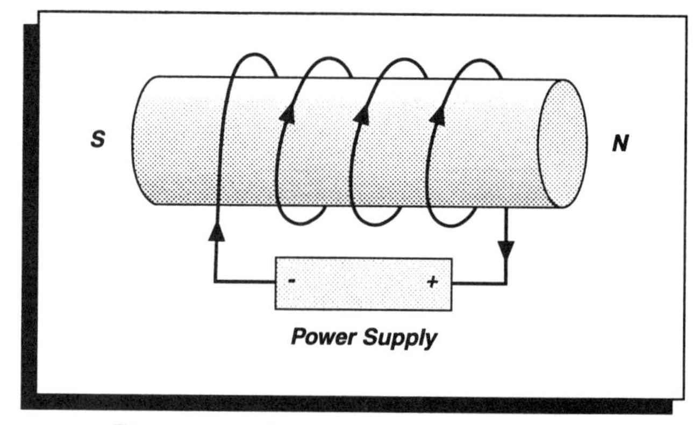

Figure 7.10 Determination of direction of magnetic field using First Left-Hand Rule.

FIRST LEFT–HAND RULE

Consider the electromagnet in Figure 7.10 with current flowing as indicated. What will be the pole on the right side? To determine this, the rule simply tells how to do it.

Rule:
- indicate the direction of the electric current (we will use the convention that it flows from negative to positive).
- let fingers of left hand point in direction of current.
- the thumb of left hand will indicate the north pole.

Figure 7.11 shows the steps in this rule. The electromagnet will have all sorts of applications in the hi–fi. It provides a controlled way of producing a specific magnetic field at a particular point, so that it can exert a force of a given magnitude (amount) and direction.

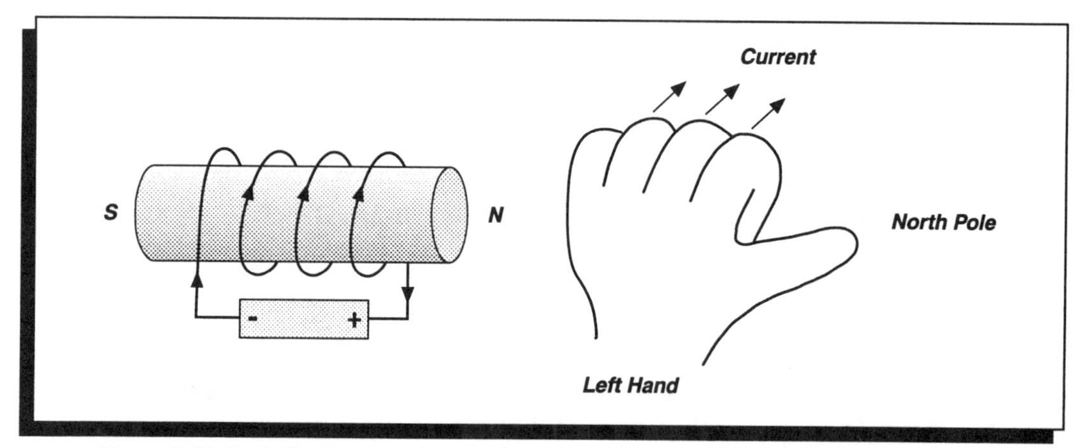

Figure 7.11 Rule for determining the direction of magnetic field in an electromagnet.

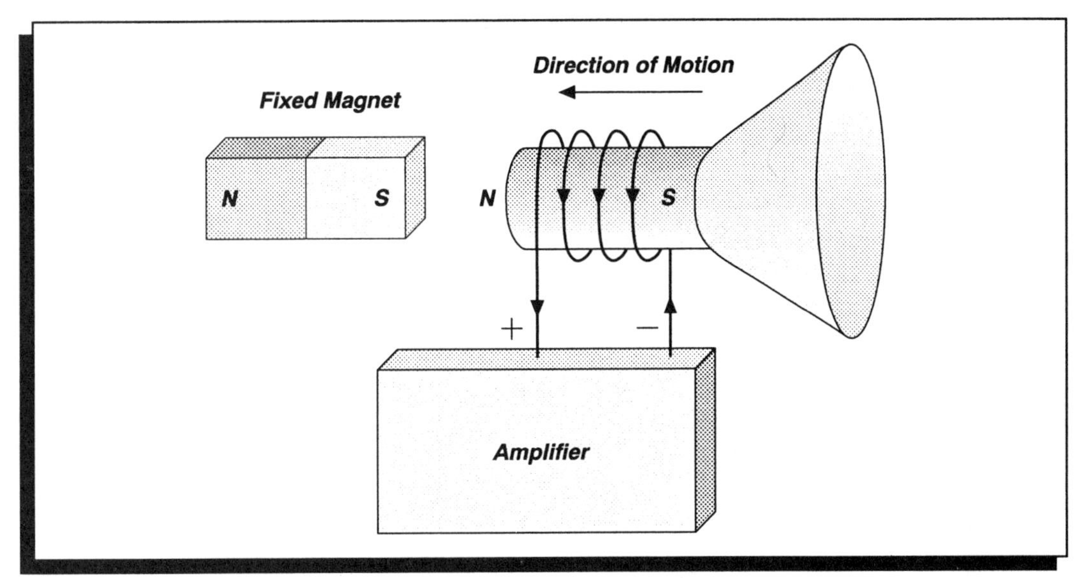

Figure 7.12 First Left-Hand Rule and how a cone speaker works.

EXAMPLE IN HI–FI: ACTION OF LOUDSPEAKER

— ***Cone loudspeaker:*** It consists of a voice coil attached to a cone and located near a magnet. When current flows in the voice coil in a given direction, a magnetic field will be created which will cause an attraction or repulsion from the changed permanent magnet. Alternating the direction of current in the voice coil leads to an alternating force on the cone, thus producing sound. This is shown in Figure 7.12 where the voice coil is connected to the output of the amplifier. At a particular instant as shown in Figure 7.12, the current goes from negative to positive. Applying the First Left–Hand Rule indicates that the north pole will face toward the left. That magnetic field polarity in the presence of the magnet will lead to a force of attraction, and hence, motion to the left. An alternating current coming from the amplifier leads to an alternating force on the cone, and hence, to cone vibrations and sound.

MAGNETIC FORCES ON CURRENTS

A wire carrying a current produces its own magnetic field. When placed in an external magnetic field there will be a force on this wire. Here we will discuss this force, its direction, and its applications to the hi–fi.

Consider a wire carrying a current and placed between the poles of a horseshoe magnet, as in Figure 7.13. The magnetic field around the wire interacts with the magnetic field of the magnet so

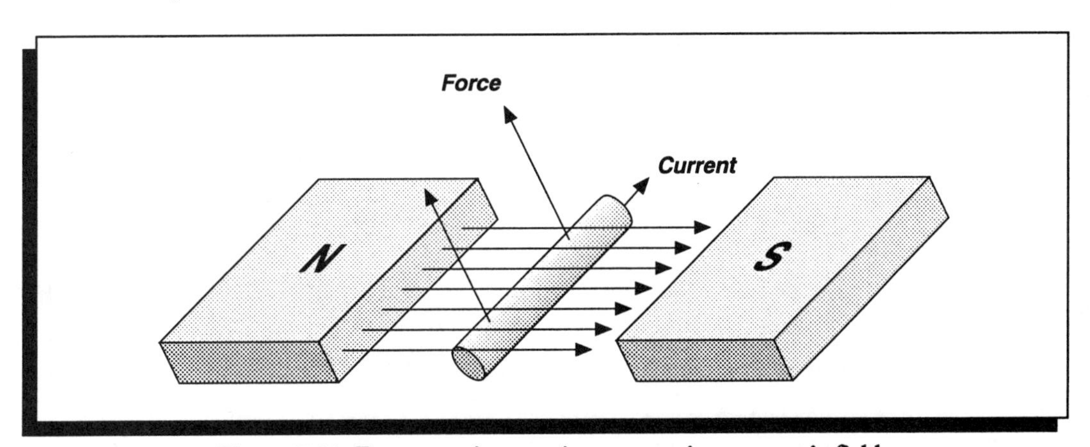

Figure 7.13 Force on wire carrying current in a magnetic field.

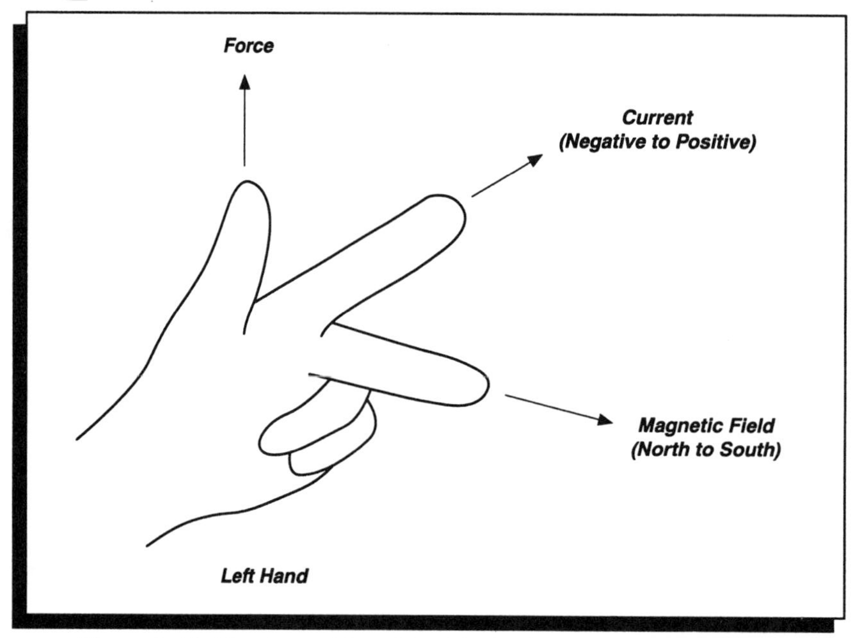

Figure 7.14 The Second Left-Hand Rule showing direction of force on wire carrying a current in a magnetic field.

that there will be a force on the wire lifting it up. For current flowing in the opposite direction, the force pushes the wire down. The direction of the force can be obtained by taking into account the direction of the magnetic fields. There is a rule, known as the second Left–Hand Rule, which will tell us the direction of the force.

Second Left–Hand Rule:

— let index of left hand point in the direction of current, from negative to positive.

— let second finger of left hand point in the direction of magnetic field, from north to south.

— the thumb of left hand when at 90° to other two fingers indicates the direction of the force.

Figure 7.14 illustrates this. It is important to note that each finger must be 90° to the other two. In this case, the force is maximum for a given current and magnetic field. Figure 7.15 shows how the force will vary with the orientation of the wire in

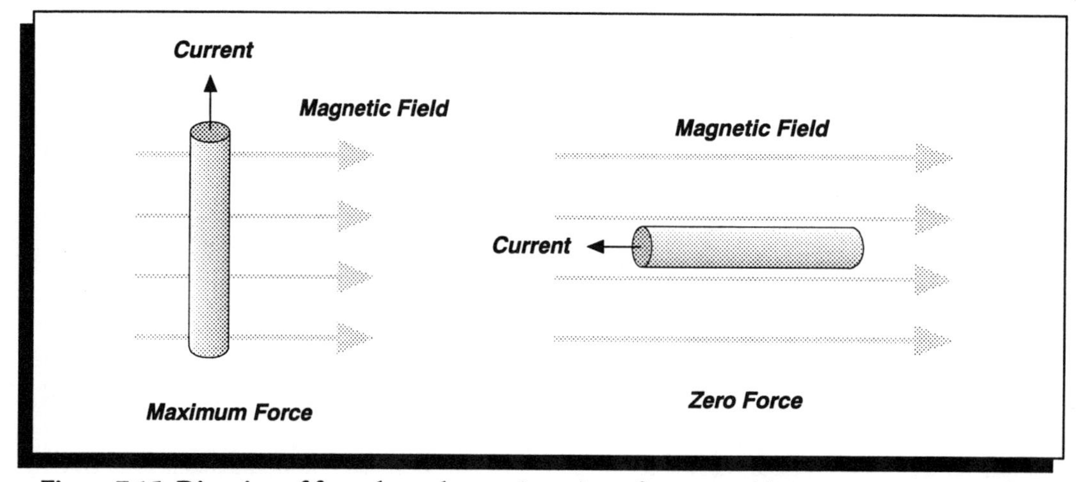

Figure 7.15 Direction of force depends on orientation of current with respect to magnetic field.

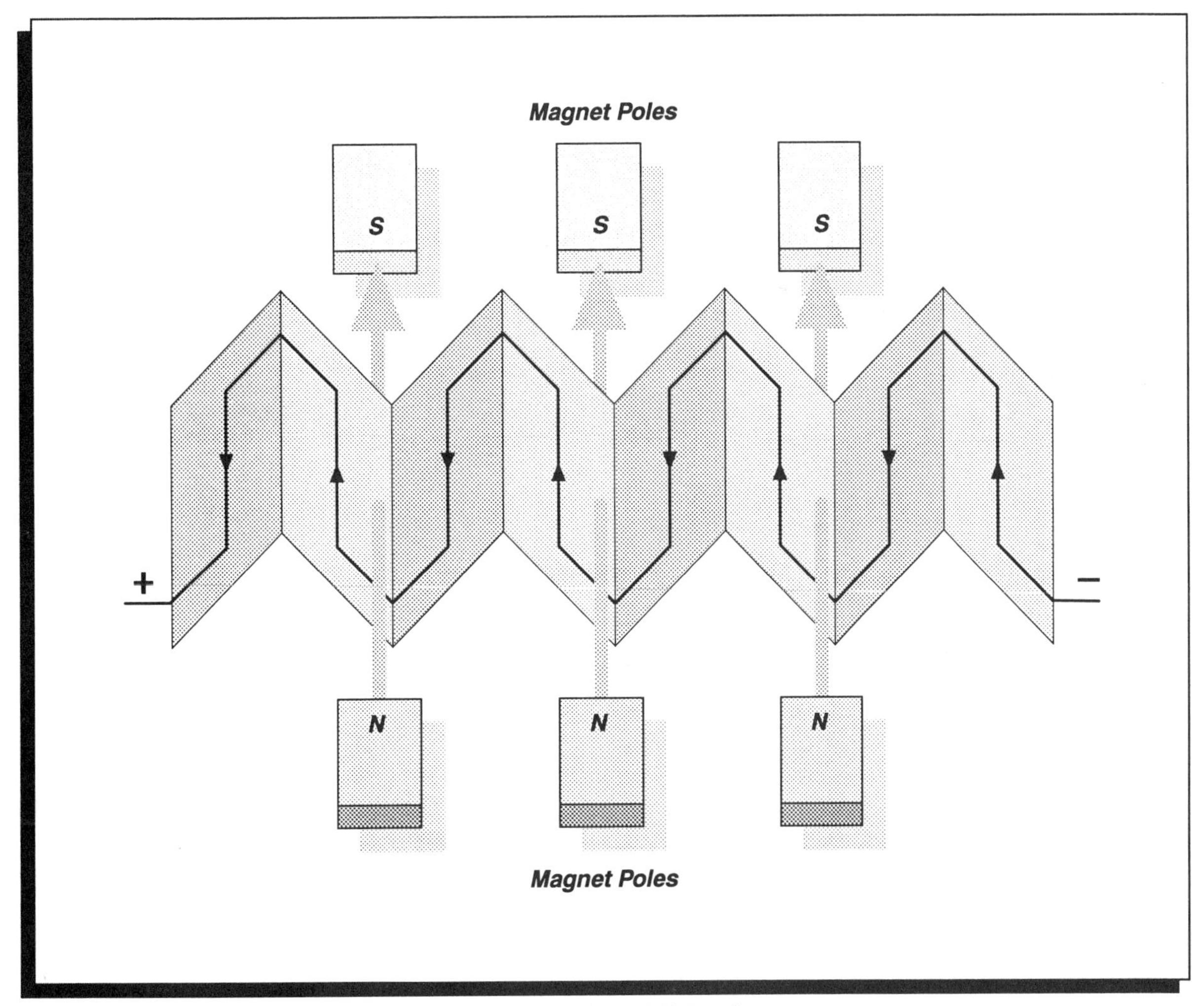

Figure 7.16 A Heil Speaker.

the field. The second Left–Hand Rule has many applications. A very important one is in the mechanism of motors. We will now consider two applications in loudspeakers.

Applications of Second Left–Hand Rule to Loudspeakers

There are two systems of loudspeakers which use this principle, the Heil speaker and the magnetic planar speaker. They both have wires or ribbons carrying a current in a magnetic field; the wires are attached to large sheets of plastic. Although both speakers have some characteristics in common with the electrostatic speaker, they offer some advantages. High voltages are not needed, thus avoiding arcing problems. Also, the speakers appear to the amplifier as resistors, whereas for the electrostatic speakers the speakers appear as capacitors, which complicates the amplifier performance.

Heil Speaker

A corrugated sheet of plastic has a thin ribbon of aluminum glued to it. The ribbon runs up and down each fold. The ends of the ribbon are connected to the output of an amplifier. In the front and back of this structure there is a magnet with the north pole on one side and a south pole on the other side. Figure 7.16 shows this basic arrangement. The fundamental principle governing this speaker is the second Left–Hand Rule. As current leaves the negative terminal of the amplifier, it goes through the ribbons, then returns to the amplifier at the positive terminal. The forces acting on each fold can be visualized by considering a pair of folds,

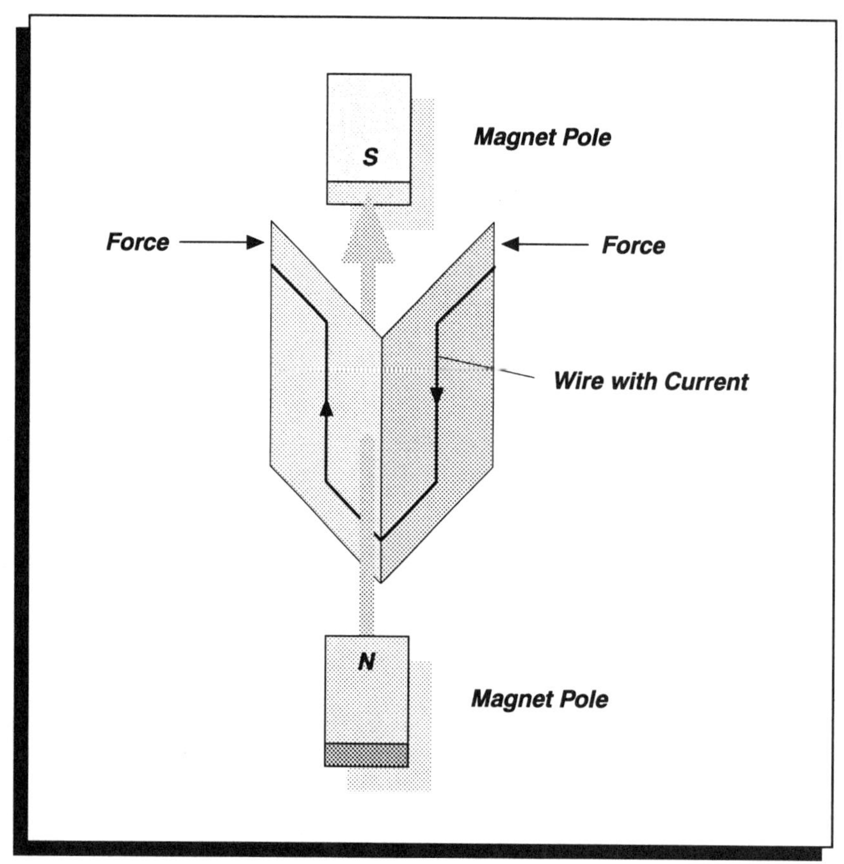

Figure 7.17 One set of folds in Heil speaker.

as in Figure 7.17. Assume the current goes as indicated at some period of time. Using the left hand, the index points in the direction of current, and the second finger points in the direction of the magnetic field from north to south. Here, the force on the right ribbon is to its left. For the left ribbon, since the current is going the opposite direction, the force on it will be to the right. Hence, during this cycle of the current from the amplifier, the force on the folds tends to close them in. During the next half cycle of the current, when its polarity is reversed, the folds will feel a force pulling them out. This motion continues as the polarity of the current from the amplifier keeps on alternating. Air is squeezed out and squeezed in by the accordion action of the folds. Such squeezing action imparts a high speed to the air producing a loud sound. Some features of this type of speaker are:

— it usually is used for high frequencies. At low frequencies it tends to rattle.

— it has low distortion since large displacements of the folds are not necessary. The speaker area is large, and hence it produces high intensity sound for small displacements.

— it is efficient. The voice coil moves rather than the voice coil moving a cone as in regular cone speakers.

— the sheet is light, hence it has a good high frequency response

— no cone "break up." Each set of folds moves independently of the others.

— no harmonic distortion. Displacements are small.

— no resonances as each fold moves independently of the other. The resonances are averaged out.

— since front and rear of this speaker are out of phase, there can be destructive interference effects at low frequencies

This speaker is used mainly as a tweeter and sometimes as midrange.

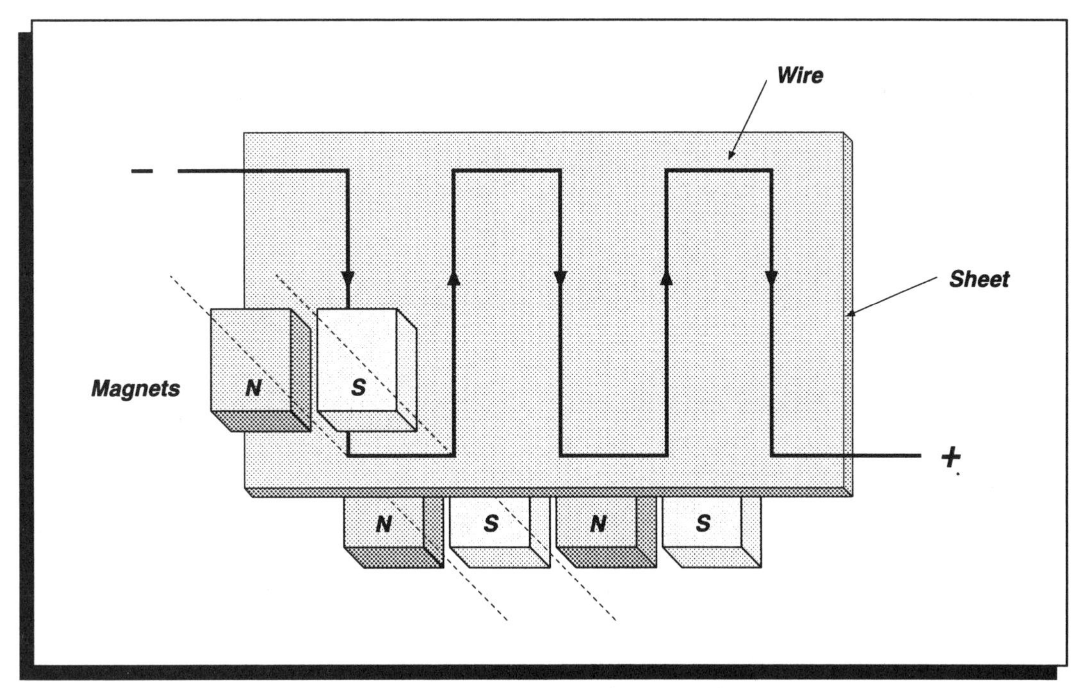

Figure 7.18 Magnetic Planar Speaker.

■ MAGNETIC PLANAR SPEAKER

The basic approach here is similar to the previous speaker in that ribbons or fine wires are glued to a flat piece of thin plastic, which is the diaphragm. The wires run up and down the diaphragm and its ends are connected to the output of an amplifier. On each side of the diaphragm are a set of small magnets located at positions between adjacent wires, as shown in Figure 7.18. Its action can be analyzed by looking down on the diaphragm. In this case the current comes up from the plane of this paper and down into it. When the current comes out of the plane, a special symbol is used to indicate this: ⊙ . This symbol resembles the tip of an arrow coming out of the plane of this paper. When the current goes into the plane the symbol is: ⊗ . This symbol corresponds to the feathers of an arrow, and represents the direction of current flow. Figure 7.19 shows how using the second left-hand rule the forces act on the diaphragm during one half cycle of current from the amplifier. On the next half cycle, when the current changes

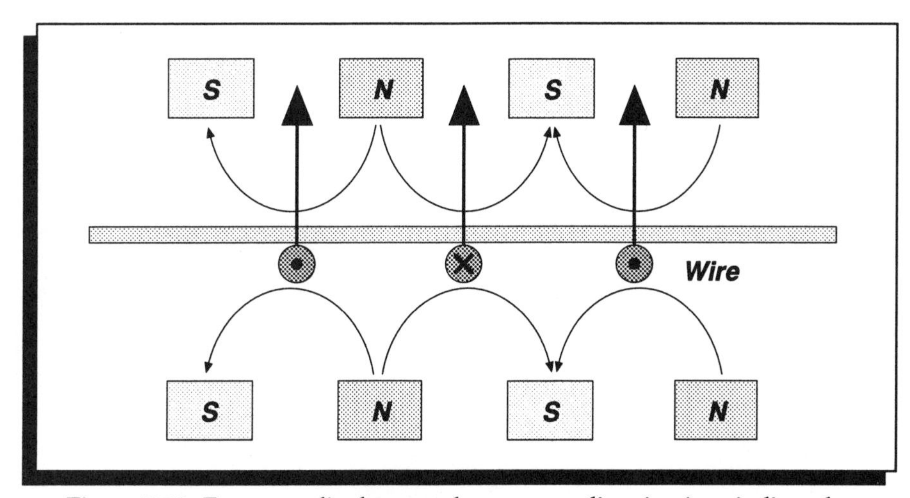

Figure 7.19 Forces on diaphragm when current direction is as indicated.

direction, the direction of the force on the diaphragm is reversed. As the cycles repeat themselves the diaphragm vibrates back and forth in response, thus producing sound. The diaphragm can be quite large — a few feet in width and length. Special features of this speaker are:

— No cabinet resonances as there is no enclosure.
— Low distortion because the displacement is small.
— No cone "break up" since each section is independent of its neighbors.
— Large radiating surface.
— It can be used for decoration.
— Tricky to place correctly in room especially for low frequencies where cancellation effects between front and rear can be significant.
— the diaphragm can be subdivided into sections. There can be a tweeter section, a midrange, and a woofer section.

7.3 FARADAY'S LAW OF INDUCTION

This is a fundamental law of electromagnetism which has been central to the development of electromagnetic theory and has a variety of applications in our everyday lives and in the hi-fi.

As an introduction to the subject, it is useful to recall that moving electrical charges produce magnetism. This leads to the question:

— if moving electrical charges produce a magnetic field, can magnetism produce an electric field and hence current?

That was what Faraday wanted to address experimentally. His experiment was similar in concept to the one shown in Figure 7.20 where a bar magnet (a source of magnetism) was brought near a coil attached to a meter. All sorts of interesting effects occur.

— When the magnet moves toward the coil a current is induced in the coil due to an induced voltage across the coil
— When the magnet moves away from the coil, a current of opposite polarity is induced.
— When the bar magnet remains stationary inside the oil, no current is induced in the coil.
— When the number of turns in the coil is increased, the induced voltage becomes larger.
— When the magnet moves in faster the induced current is larger than when it moves slowly
— The magnet can be stationary but the coil can be moved toward or away from the magnet thus inducing a current in the coil. It is the relative motion between the coil and the magnet which causes the voltage.

All these facts can be summarized as a law:

$$\textit{Induced voltage} = \frac{\textbf{\# of turns in coil} \times \textbf{change of magnetic field}}{\textbf{time}}$$

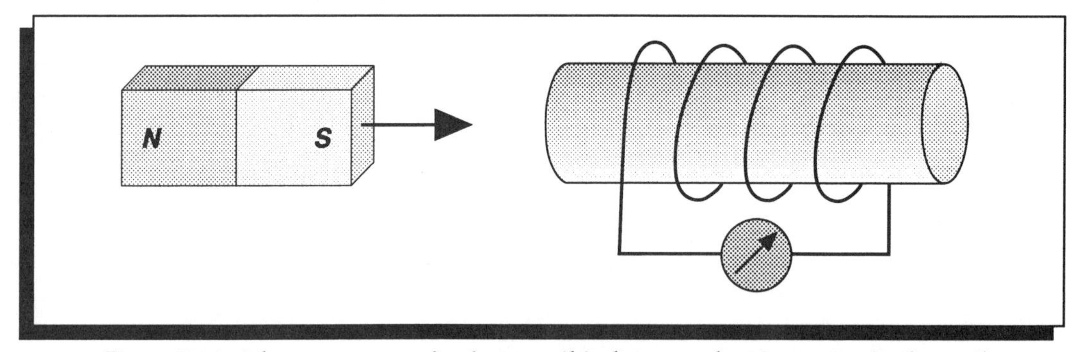

Figure 7.20 A bar magnet moving into a coil induces an electric current in that coil.

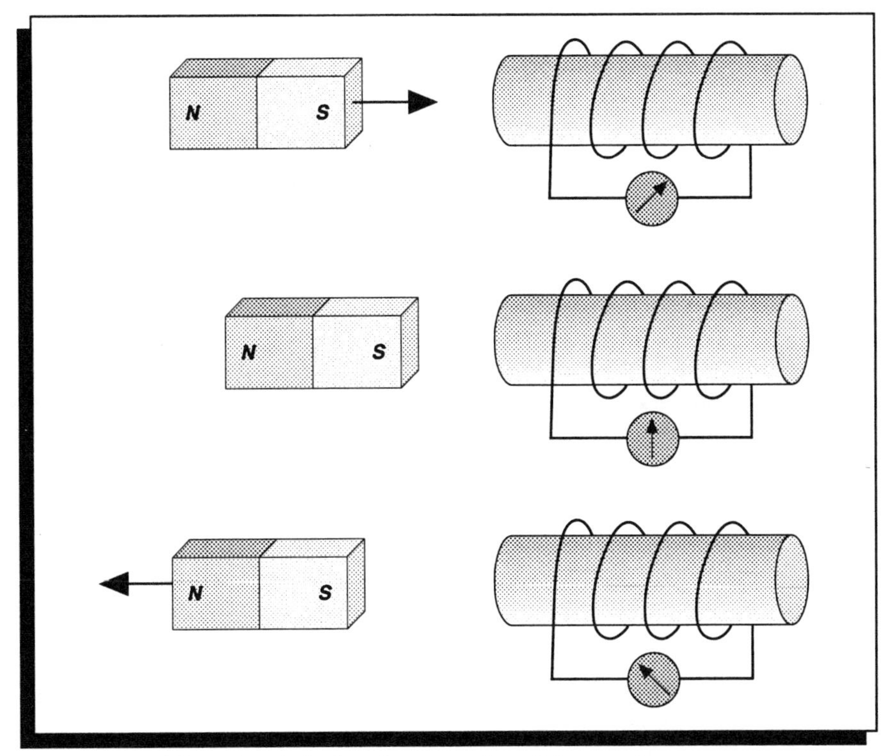

Figure 7.21 Induced current in coil by moving magnet.

This is known as Faraday's law of induction. It is represented in Figure 7.21 for a bar magnet moving into or away from a coil. A meter indicates the induced current. A large voltage will be induced only when there is a large change of magnetic field in the coil for a given amount of time. Hence a rapid change will lead to a large induced voltage.

Since it is the relative motion between coil and magnet that is important, a few other examples can illustrate this situation, as in Figure 7.22. The question could arise as to why an electric current is induced according to Faraday's Law. In simple words, this effect can be explained by recalling that a moving electric charge in a magnetic field will be subject to a force. Here we have the opposite effect, where charges (electrons in the coil) are

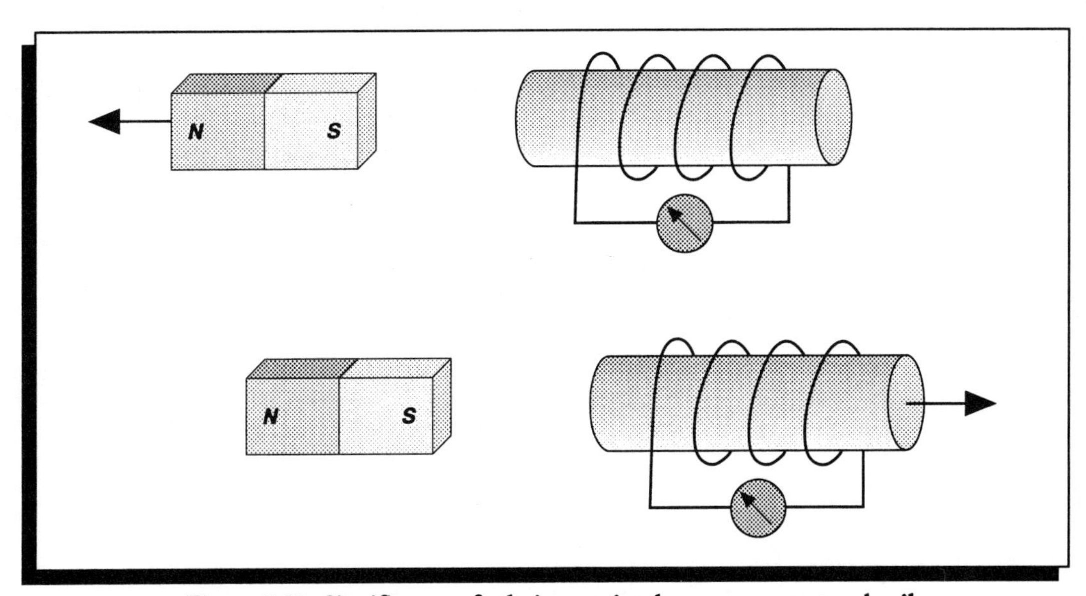

Figure 7.22 Significance of relative motion between magnet and coil.

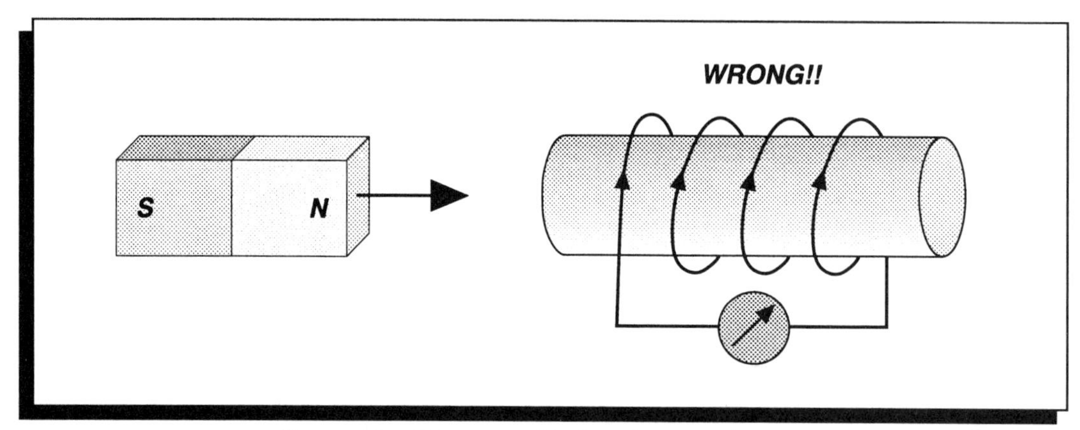

Figure 7.23 Direction of induced current (wrong).

forced to move in a magnetic field and this will cause the electrons to move under this force, leading to an induced current.

In the statement of Faraday's law of induction, a minus sign was introduced. What is its significance? By considering the details of the induced current, the minus sign will become clearer. The first question is: which way will the induced current flow? Since there are two possibilities it is instructive to explore them both. In Figure 7.23 a bar magnet is moving toward a coil which has a meter to show the current direction. Let us assume just for a brief moment that the induced current will flow as in Figure 7.23. Is it possible? The induced current in the coil will cause the left side of the coil to be a south pole. This will make the magnet with its north pole move in faster (since unlike poles attract!) and faster, creating a larger current, etc...

Obviously, it seems wrong because when one lets go of the magnet, it will move faster, creating even more electricity. We would start getting something for nothing and hence we are violating the first law of thermodynamics.

Let us now see if an induced current in the opposite direction will make sense. The induced current in the coil will cause the left side to be a north pole, which will then repel the north pole of the incoming magnet. This means that if we induced electricity we had to work hard for it. The Principle of Conservation of Energy will not be violated here. Hence the direction in Figure 7.24 must be the correct direction for the induced current. The minus sign tells us that:

The induced current's direction will be in opposition to the change of magnetic field.

This is referred to as Henry's Law since he was the first to emphasize this point, although the information is given in Faraday's law of induction. Opposition to a change in magnetic field is required because of the fundamental principle of the conservation of energy.

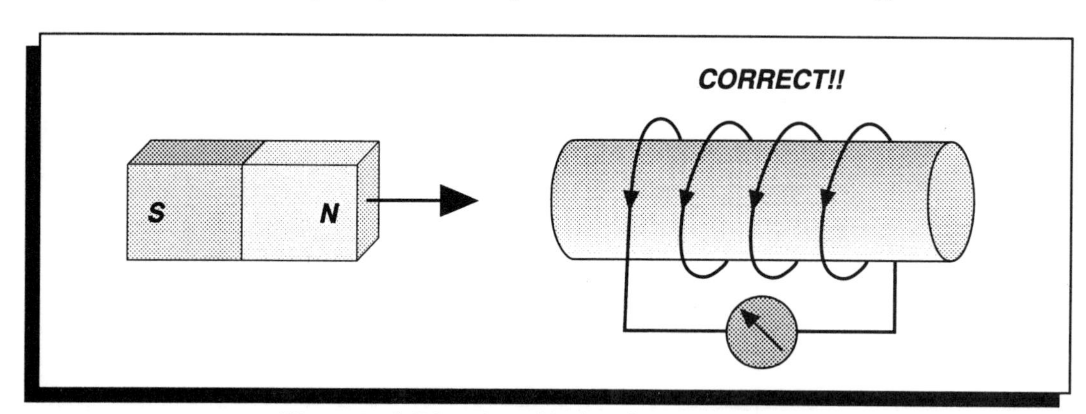

Figure 7.24 Direction of induced current (correct).

7.4 APPLICATIONS OF FARADAY'S LAW OF INDUCTION IN THE HI–FI

Faraday's law of induction has a wide range of applications, from the production of electricity to the detection of magnetic signals recorded on a magnetic tape. Here, a few selected examples will be presented to illustrate this law and its importance in the hi–fi. Other examples will be presented later.

■ TRANSFORMERS

In presenting Faraday's law of induction, a moving bar magnet induced a voltage in a coil. Is it possible to avoid the magnet? The answer is yes, in that a coil or electromagnet could be used instead. Another question is whether the coil has to be moved. The coil magnetic field can be changed by keeping the coil fixed in space but using an alternating current in it. These possibilities led to the development of a device known as a transformer. There are many applications of transformers in the hi–fi; every amplifier has one in the circuit for the power supply.

Figure 7.25 shows the principles involved in the transformer. The primary coil has alternating current producing the changing magnetic field at the secondary coil. In order to increase the magnetic field produced by the primary coil and to help concentrate the changing magnetic field at the secondary coil, a core of magnetic material is used in a typical shape shown in Figure 7.25. This is the basis of a principle known as the transformer principle. It is defined as:

Transformer Principle **= changing magnetic field produced by a primary coil will induce a voltage across the secondary coil.**

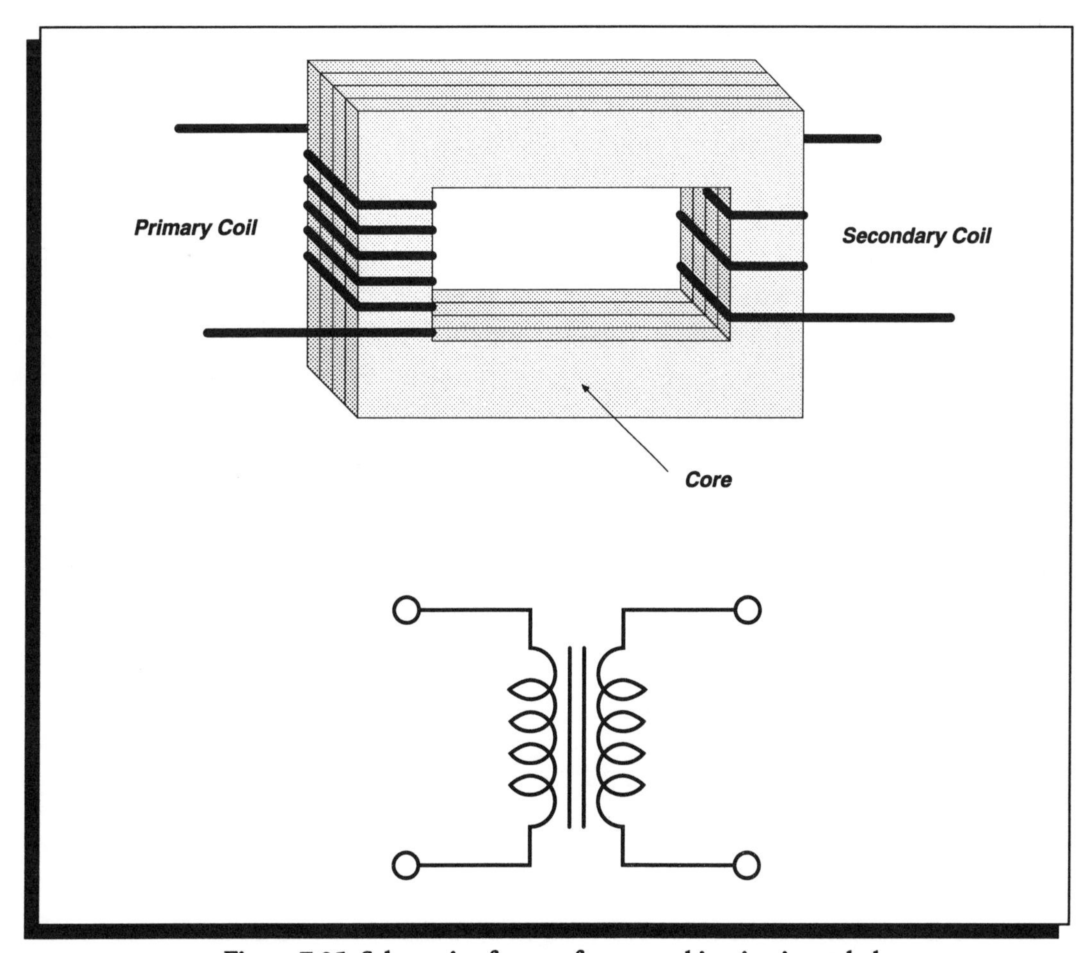

Figure 7.25 Schematic of a transformer and its circuit symbol.

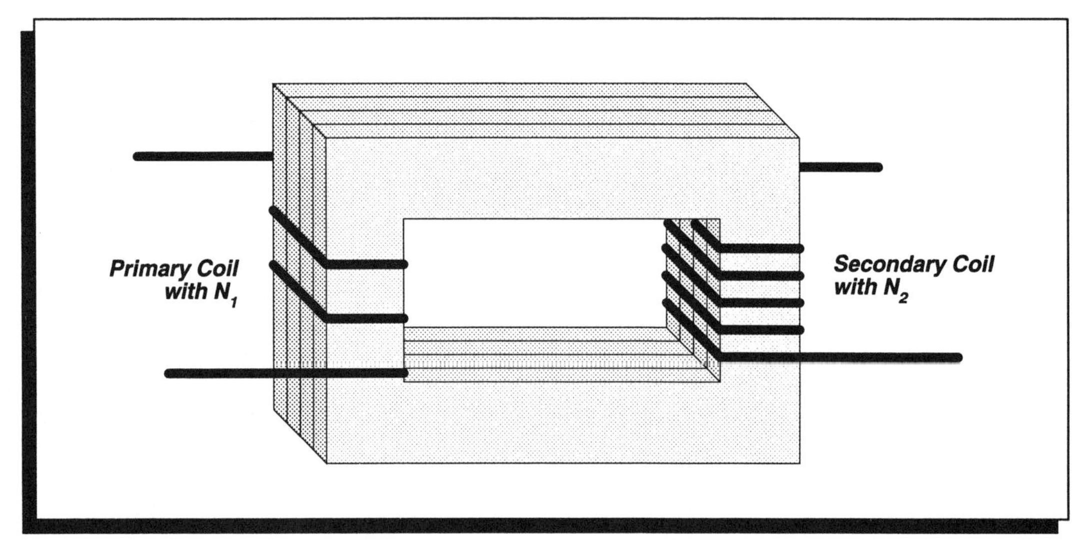

Figure 7.26 Step-up transformer.

The transformer is based on Faraday's law of induction and it works only with ac currents. The number of turns in each coil is important as it will indicate the particular application of the transformer.

Consider the case where the secondary coil has N_2 turns. The larger the number of turns N_2, the larger the induced voltage in that coil; to get a quantitative result it is important to compare the number of turns N_2 at the secondary to the number N_1 in the primary. From Faraday's law of induction, the voltage V_2 at the secondary can be calculated, depending on how much voltage there is at the primary and on the turns ratio. Hence,

$$\mathbf{V_2 = (N_2 / N_1)\, V_1}$$

If N_2 is larger than N_1, the voltage at the secondary will be larger than that at the primary by the ratio of turns N_2 / N_1. This transformer is called a step–up transformer as the secondary voltage is higher than the primary one. Figure 7.26 shows this case. One application of this transformer is for the high voltage in a TV, where 120 volts are stepped up to about 25,000 volts.

Let us now consider a case in which there are less turns N_2 at the secondary than at the primary. This means that the secondary voltage will be less than the primary voltage, as shown in Figure 7.27. The factor by which the voltage is reduced depends on the ratio N_2 / N_1. Such a transformer has application in situations where the voltage has to

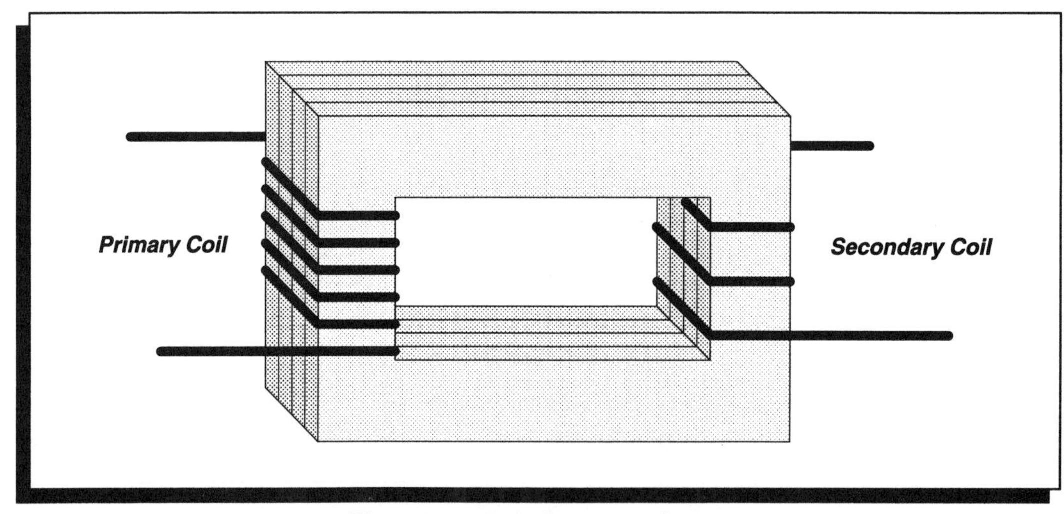

Figure 7.27 Step-down transformer.

be reduced. An example in the hi–fi is the transformer which steps down the outlet voltage of 120 volts to a lower value for the power supply circuits in the amplifier.

When dealing with transformers, the question of power has to be addressed. Ignoring losses, the power coming in must equal the power coming out, regardless of the type of transformer. This means that since:

$$\textbf{Power in} = \textbf{Power out}$$
$$\textbf{then } V_1 c_1 = V_2 c_2$$

Thus, if the voltage at the secondary V_2 is stepped–down, the current available at the secondary c_2 will be stepped up. This makes sense because in an amplifier the power supply needs a lot of current for running the speakers and this is achieved by stepping the voltage down. Similarly in a step–up transformer, an increase in voltage at the secondary is achieved at the expense of having less current available at the primary. Hence,

a transformer transforms voltage or current

but the power remains the same at each side when the losses are ignored.

■ MICROPHONE

This device converts sound into electrical signals. Faraday's law of induction provides a method for doing this. A simple approach is shown in Figure 7.28 where a funnel–like light structure collects the sound and its mechanical vibrations cause a magnet to vibrate in a coil inducing a voltage in the coil. The induced voltage is proportional to the sound. There are other very important applications of Faraday's law of induction that will be developed in later chapters.

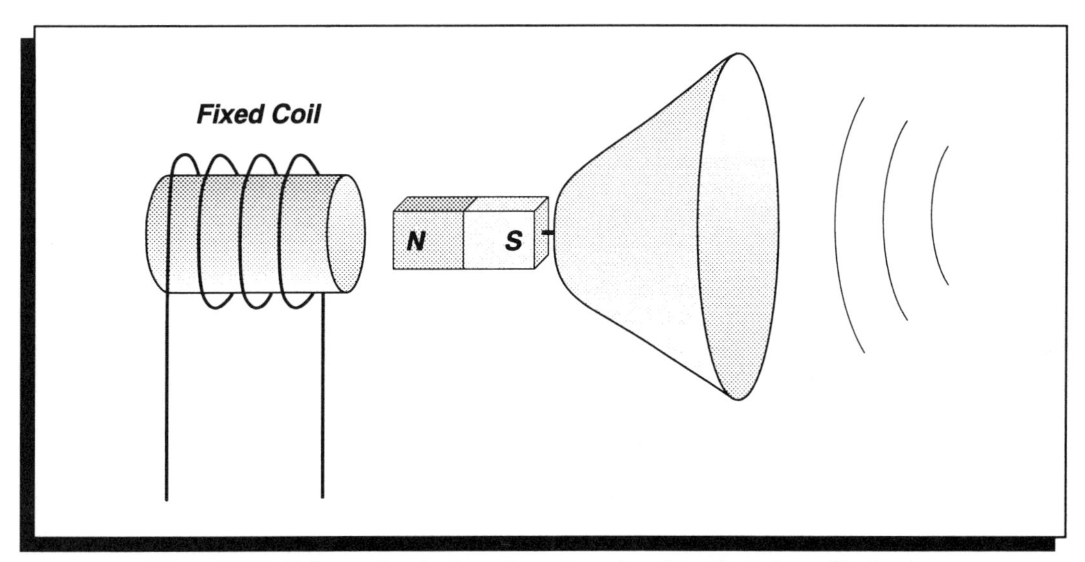

Figure 7.28 Schematic of microphone based on Faraday's law of induction.

SUMMARY OF TERMS

Electromagnet: Coil with a magnetic core which produces a magnetic field when current flows in the coil.

Faraday's Law of Induction: Change of magnetic field per unit time creates a voltage across a coil or conductor nearby.

First Left Hand Rule: Indicates direction of magnetic field produced by a coil when current flows. When fingers of left hand point in direction of current flow, the thumb of left hand indicates the north pole direction.

Heil speaker: Corrugated thin plastic sheet with current–carrying wires or strips attached to it. In the presence of a magnetic field, magnetic forces on current conductors cause an accordion–type motion and hence sound. It is based on the second left hand rule.

Magnet: Material whose magnetic domains are aligned permanently with resultant north and south poles.

Magnetic Dipole: Elementary unit in magnetism which contains a south pole and a north pole. A current loop has similar magnetic properties.

Magnetic Domain: Aligned group of atoms in a magnetic solid.

Magnetic Field: Energetic region of space around a magnet or a current–carrying conductor where a force will be felt by a test magnet.

Magnetic Planar Speaker: Uses second left hand rule where fine wires attached to a large plastic sheet in the presence of a magnetic field feel a force when a current flows.

Microphone: Device that converts sound waves into electrical signals.

Poles: Ends of a magnet which are either north or south. A basic quantity in magnetism; they always come in pairs.

Step–down Transformer: Transformer which has less turns at the secondary than the primary, thus reducing voltage of input signals.

Step–up Transformer: Has more turns at secondary than primary, thus raising the voltage level of input signals.

NAME ______________________________ DATE ____________

QUESTIONS FOR REVIEW

1 Discuss the nature of forces between magnets.

2 Why does a transformer not work with dc?

3 What is Faraday's law of induction? Where is it used in the hi–fi?

4 Explain the importance of magnetism in a cone speaker.

5 What determines the direction of the induced current in a coil by a changing magnetic field? Why?

6 How does a piece of iron get magnetized?

7 Why can a transformer step–up or step–down current or voltage, but not power?

8 What is the function of a core in a coil?

9 Discuss the origin of magnetism, starting at the atomic level.

10 How would you induce a current in a metallic ring? What happens to this current after a while?

NAME ______________________________ DATE ____________

1. A charged particle experiences a force when:
   ___A. moving along the direction of a magnetic field
   ___B. moving perpendicular to a magnetic field
   ___C. at rest in a magnetic field
   ___D. at rest near a circuit containing a current
   ___E. none of the above

2. The direction of the force experienced by a charged particle moving through a magnetic field is:
   ___A. always perpendicular to both the magnetic field and the velocity of the charged particle.
   ___B. parallel to the magnetic field lines.
   ___C. parallel to the velocity of the particle.
   ___D. parallel to both the velocity of the particle and the magnetic field lines.
   ___E. there is no force on the charge since it is electric.

3. A transformer has 5 times as many turns in the secondary coil as in the primary coil. If the electrical power going to the primary is 50 watts, what is the electrical power coming out of the secondary?
   ___A. 250 watts
   ___B. 50 watts
   ___C. 10 watts
   ___D. 0 watts
   ___E. 55 watts

4. A magnetic field can be generated by:
   ___A. an electron at rest
   ___B. a proton at rest
   ___C. an electric current
   ___D. a charged particle in motion
   ___E. both C and D are correct

5. A region within a solid where the magnetic atoms are aligned is called:
   ___A. a magnetic field
   ___B. a magnetic domain
   ___C. an electromagnet
   ___D. north pole
   ___E. induced current

6. The greater the alignment of magnetic domains in a piece of iron, the ____________ the magnetism.
   ___A. stronger
   ___B. weaker
   ___C. weaker the magnetic field causing
   ___D. faster will the electrons move and cause
   ___E. slower the electrons move and cause

7. A magnetic field:
   ___A. goes from the north pole to the south pole exterior to a magnet
   ___B. goes from the south pole to the north pole exterior to a magnet
   ___C. is generated by permanent magnets only
   ___D. is generated by electromagnets only
   ___E. has nothing to do with magnetism

8. If the direction of the current through an electromagnet is reversed:
   ___A. the magnetic field reverses direction.
   ___B. the magnetic field increases but the voltage stays the same.
   ___C. the magnetic field stays the same but the voltage changes.
   ___D. the magnetic field reverses direction and then quickly returns to its original direction.
   ___E. the magnetic field disappears.

9. If 2 amperes at 60 volts are entering the primary of a transformer, and the voltage at the secondary is 120 volts, the current in the secondary is:
___A. 1 ampere
___B. 2 amperes
___C. 0.5 ampere
___D. 4 amperes
___E. 3 amperes

10. A voltage can be induced in a coil if:
___A. a magnet moves near the coil
___B. the coil moves near the magnet
___C. the coil sees a changing magnetic field
___D. the coil is near a decreasing magnetic field
___E. all of the above are correct

11. To make an electromagnet produce a stronger magnetic field, one would:
___A. increase the current through the coil
___B. add more turns of wire to the coil
___C. add an iron core
___D. all of the above increase the magnetic field
___E. none of the above

12. To get 6 volts for the transistors in an amplifier one uses a transformer. If 120 volts from the wall outlet feed into the amplifier, a transformer can be used to change the 120 volts to the 6 volts. In that case if the primary had 600 turns, how many turns should be in the secondary?
___A. 600
___B. 30
___C. 1,200
___D. 20
___E. 720

13. The energetic region of space around a magnet is called:
___A. magnetic field
___B. electric field
___C. north pole
___D. magnetic domain
___E. electric and magnetic field

14. A wire carrying a current is placed between the poles of a horseshoe magnet. When the current is coming out of the plane of this page, the force on the wire will be:
___A. zero
___B. to the left
___C. to the right
___D. up
___E. down

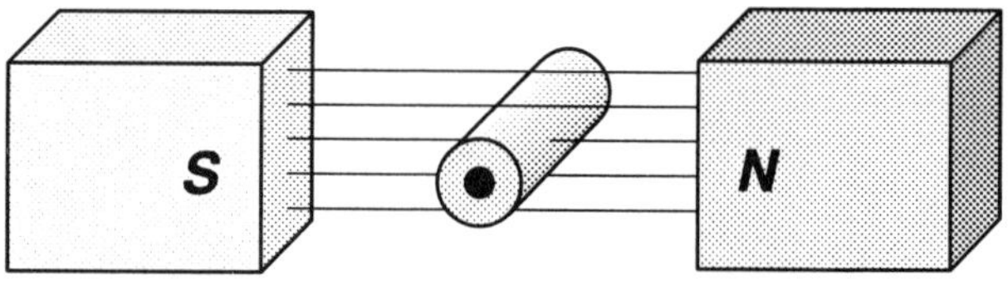

S
N

15. In which direction will the speaker diaphragm move when the current to the voice coil is as shown?
___A. to the left
___B. to the right
___C. up
___D. down
___E. to the left then to the right

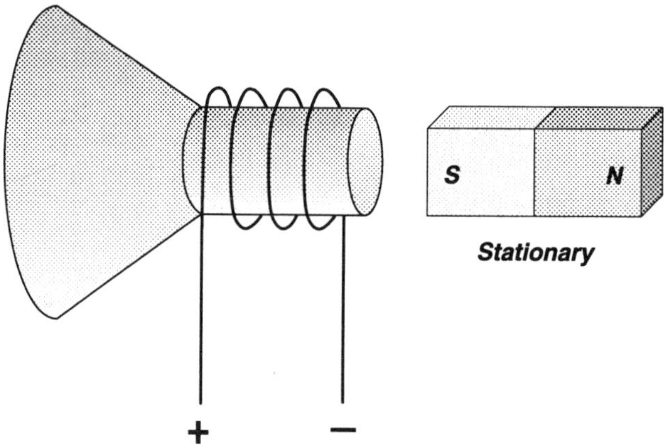

S
N
Stationary
+
−

16. If you double the number of turns in an electromagnet and triple the current through it, then the magnetic field will:
___A. be twice as strong
___B. be six times as strong
___C. be three times as strong
___D. be 2/3 as strong
___E. be 3/2 as strong

NAME ______________________ DATE ____________

17. A step–up transformer is used to:
___A. decrease the voltage
___B. increase the voltage
___C. increase ac current
___D. increase dc current
___E. decrease dc current

18. A bar magnet is dropped into an aluminum ring. As the magnet enters the ring, current will be induced that it will circulate:
___A. clockwise
___B. counterclockwise
___C. faster
___D. no current will be induced
___E. slower

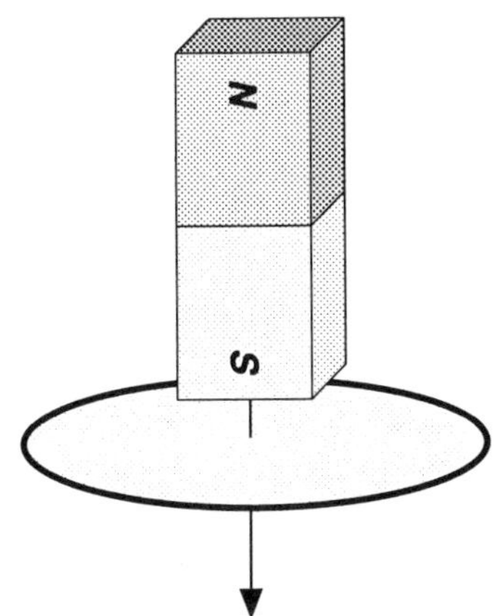

19. A bar magnet oscillating at 500 Hz near a coil induces in it 15μvolts. When it oscillates at 2 octaves higher with the same amplitude, the induced voltage will be:
___A. 30μvolts
___B. 60μvolts
___C. 7.5μvolts
___D. 2,000μvolts
___E. 120μvolts

20. What is the basic law which describes the forces between magnetic poles?
___A. like poles attract, unlike poles repel
___B. like poles repel, unlike poles attract
___C. like poles repel, unlike poles repel
___D. like poles attract, unlike poles attract
___E. there is no force between magnetic poles since they come in pairs

21. A 12–volt battery is attached to the primary of a step–up transformer. The primary has 50 turns and the secondary has 2,000 turns. The steady state output voltage at the secondary will be:
___A. 480 volts
___B. 0.3 volt
___C. 12 volts
___D. 0 volt
___E. 600 volts

CHAPTER 8 ELECTROMAGNETIC WAVES AND TUNERS

This chapter starts with an introduction to electromagnetic waves, their production and detection and some applications. The radio–frequency part of the spectrum has important applications in radio communication. Audio signals modulate radio–frequency electromagnetic waves which are then broadcast. In the reception stage, an antenna combined with a tuner converts the received signal into audio information. The various methods of broadcasting will be examined together with their relative merits.

8.1 GENERATION OF ELECTROMAGNETIC WAVES

Electromagnetic waves were predicted in 1860 by Maxwell as a result of his analysis of the existing facts on electricity and magnetism. They were confirmed experimentally by Hertz about 27 years after their prediction. This discovery opened the way for TV, radar, and radio communications detected by the tuner in the hi–fi.

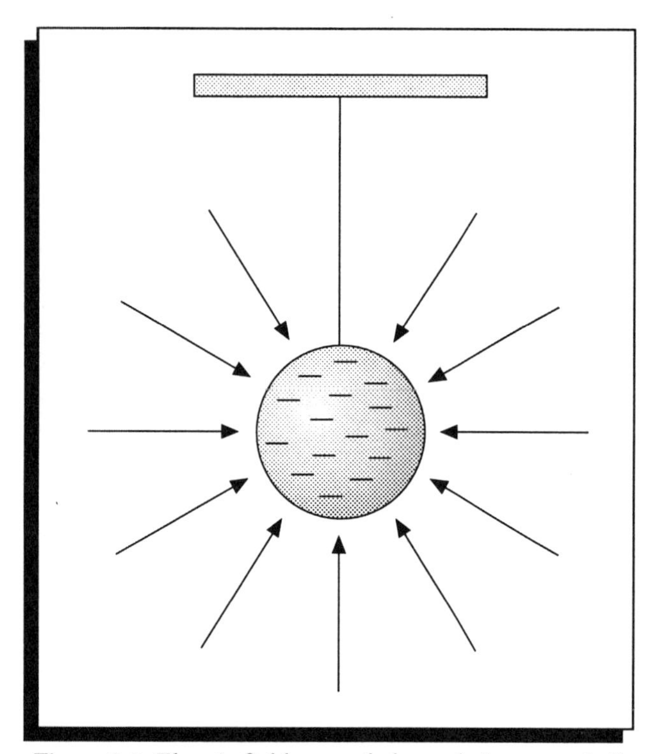

Figure 8.1 Electric field around charged ping-pong ball.

In developing this subject it is instructive to start with an explanation of an electric charge on a body, such as a ping–pong ball, as in Figure 8.1, and to explore what happens around the charge when it is at rest and when it oscillates. The results are presented as facts.

Fact 1: Around the ball there is an electric field. This can be verified by bringing in a test charge and measuring the force on it.

Fact 2: When the ball is set into oscillation, there will be an oscillating electric charge because of the displacements of the charged ball, Figure 8.2.

Fact 3: Since the charged ball is oscillating, the charge is moving and this constitutes an electric current.

Fact 4: An electric current produces a magnetic field. Hence around the oscillating ball there will be an oscillating magnetic field.

Fact 5: From Faraday's law of induction a changing magnetic field will induce an electric field. Hence, there are induced electric and magnetic fields around the oscillating ball. These fields induce each other.

Results: An oscillating charge produces electric and magnetic fields which will leave the charge, and at some critical speed this mutual induction of electric and magnetic fields will continue indefinitely, and will generate a disturbance which has oscillating electric and magnetic fields.

Maxwell calculated the critical speed of an oscillating charge to be almost 3×10^8 m/s. This calculated speed is the same as the speed of light. Hence,

Electromagnetic wave **= disturbance which has oscillating electric and magnetic fields, travelling at the speed of almost 3×10^8 m/sec. It is produced by an oscillating electric charge.**

■ ELECTROMAGNETIC SPECTRUM

All electromagnetic waves travel with the same speed in vacuum. They do not need a medium to travel in because of the nature of the wave, electric and magnetic fields. They differ from one another in frequency (and hence in wavelength). The frequency of the oscillating charge will determine the frequency of the electromagnetic wave. There is a

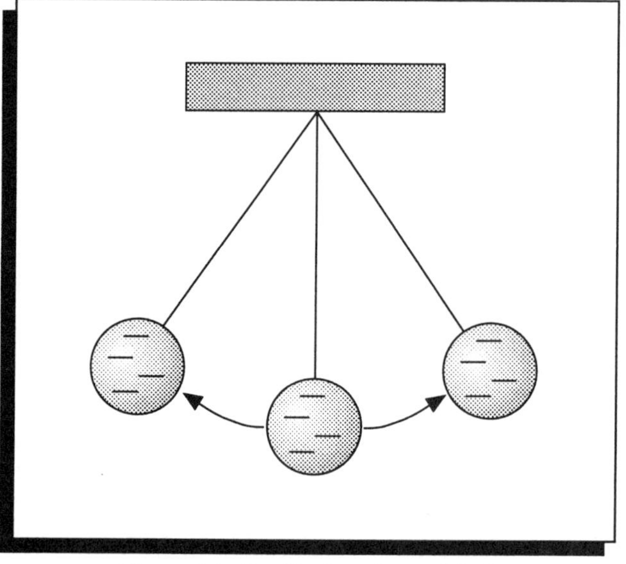

Figure 8.2 Oscillating charged ball.

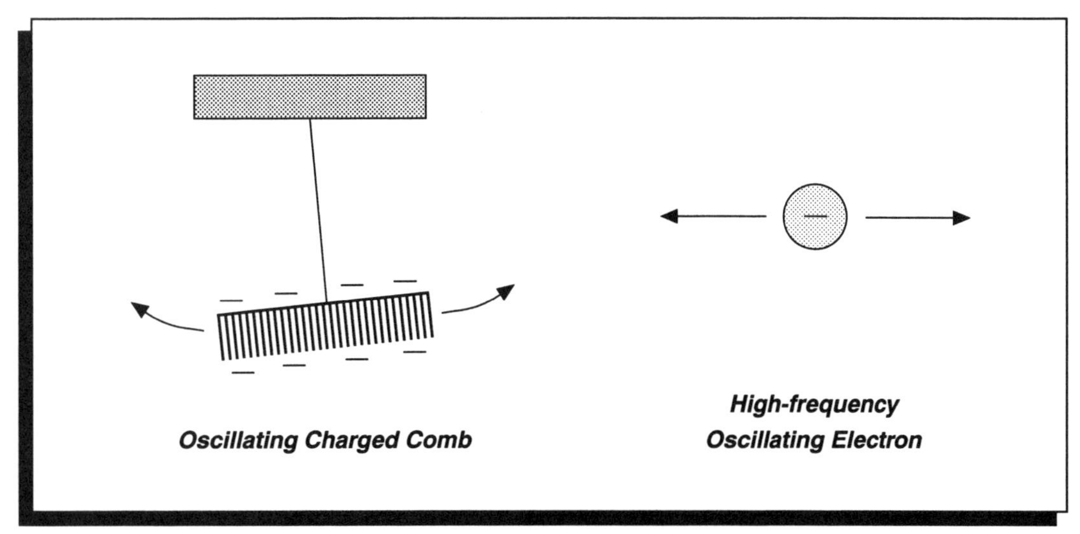

Figure 8.3 Generation of electromagnetic waves at two different frequencies.

very wide range of frequencies possible, such as moving a comb which has been charged electrically by combing hair and producing electromagnetic waves with very low frequencies, to frequencies so high that they are beyond the X–ray range. Figure 8.3 illustrates two possibilities.

Because of the wide range of frequencies possible for electromagnetic waves, the wavelengths will vary accordingly. The range of frequencies can be broken down into small groups with their particular characteristics. This is known as the spectrum of electromagnetic waves. Figure 8.4 illustrates this spectrum with the popular names associated to each frequency range.

It is interesting to look at the wavelength of some of the groups of waves in this spectrum. For each frequency the wavelength can be calculated from the wavelength times frequency formula. For example:

— wavelength of standard AM radio waves
 ~300 meters
— wavelength of standard FM radio waves
 ~3 meters

The frequency ranges are also interesting as their names are very specific to these ranges:

AM radio frequencies = 535 kHz to 1605 kHz
FM radio frequencies = 88 MHz to 108 MHz
red light = 4.7×10^{14} Hz
blue light = 6×10^{14} Hz

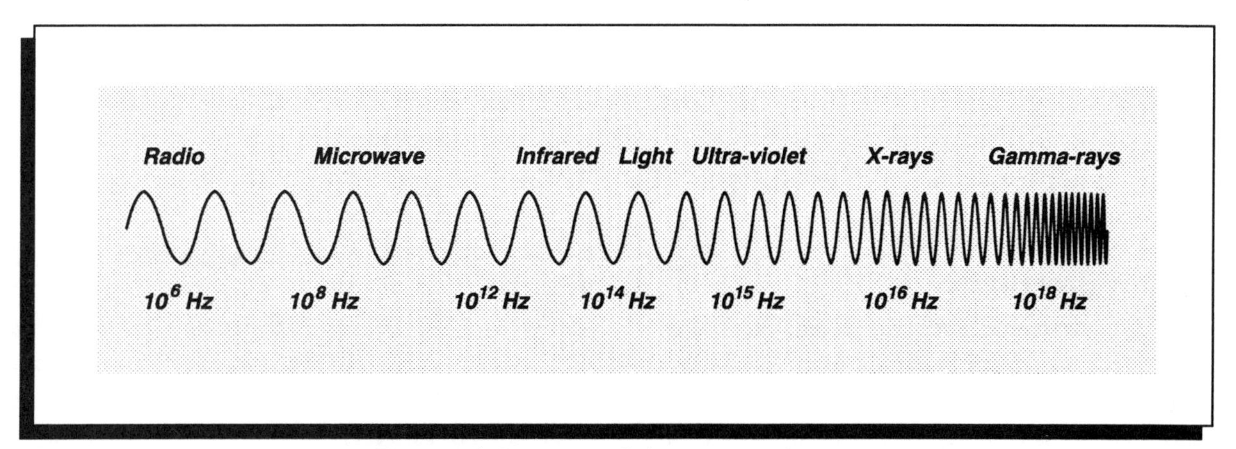

Figure 8.4 Spectrum of electromagnetic waves.

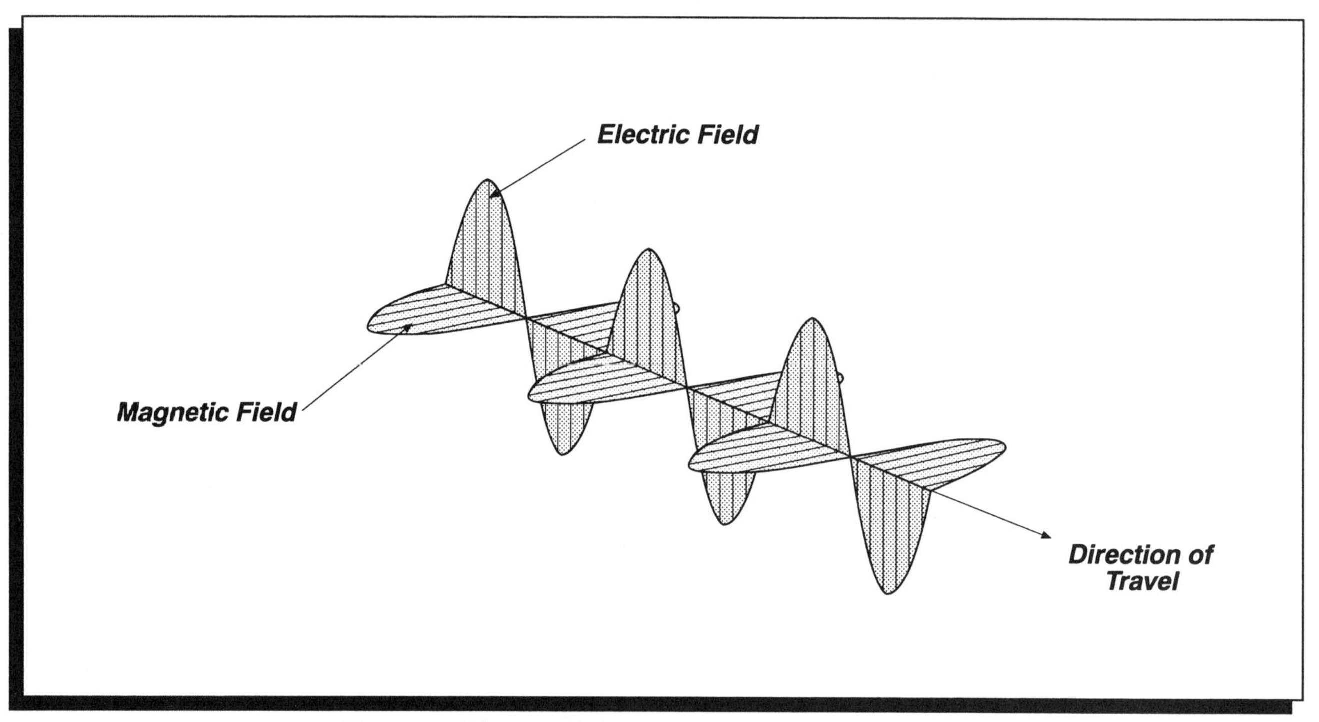

Figure 8.5 Electromagnetic waves are transverse waves with oscillating electric and magnetic fields.

CHARACTERISTICS OF ELECTROMAGNETIC WAVES

Although the range of frequencies is enormous, varying by many factors of ten, electromagnetic waves have the following common features:

— They travel at the speed of light which is almost 3×10^8 m/sec in vacuum.
— They consist of electric and magnetic fields that are always perpendicular to each other. Figure 8.5 illustrates this.
— There is always an electric and magnetic field and they are intimately connected to each other
— They are all produced by oscillating (or accelerating) electric charges
— They do not need a medium to propagate as they can travel in a vacuum

PRODUCTION

It is instructive to look into the details of how an electromagnetic wave can be generated. These ideas can then be applied to radios and communications.

Because we are interested in producing very high frequency waves it is impossible to use an oscillating ping–pong ball as its inertia will prevent it from oscillating very fast (no chance for MegaHertz frequencies). One must look for extremely light charges, like electrons, which can then be made to oscillate and radiate electromagnetic waves. A good source of electrons are the conduction electrons in a metal wire. Consider a metal wire which will have electrons oscillating when attached to an alternating voltage source. Figure 8.6 shows two of these wires attached to a voltage source. The wires are called antennas.

At some instant, one terminal of the voltage source is positive and the other one is negative. The negative part of the terminal will cause electrons to flow into that branch of the antenna, while the positive terminal will cause electrons to flow into that terminal. This causes the electrons to move down. A half cycle later, when the voltage source is reversed, the flow of electrons in the antenna will be in the other direction. If this voltage oscillation due to the voltage source is continued, the electrons will oscillate up and down the antenna at the frequency of the voltage source. Since this provides oscillating electric charges, electromagnetic waves will be produced and they will travel away from the antenna.

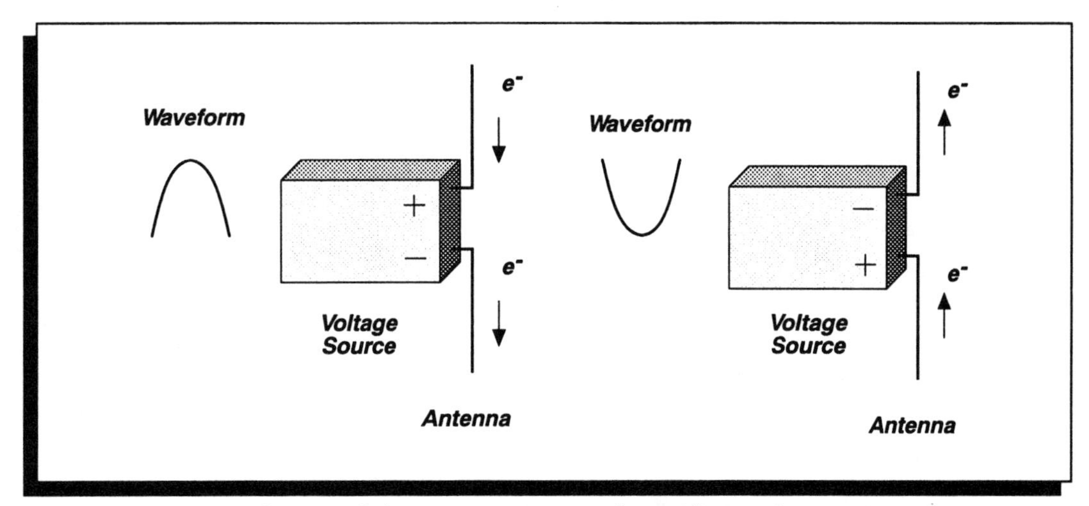

Figure 8.6 Production of electromagnetic waves by oscillating electrons in an antenna.

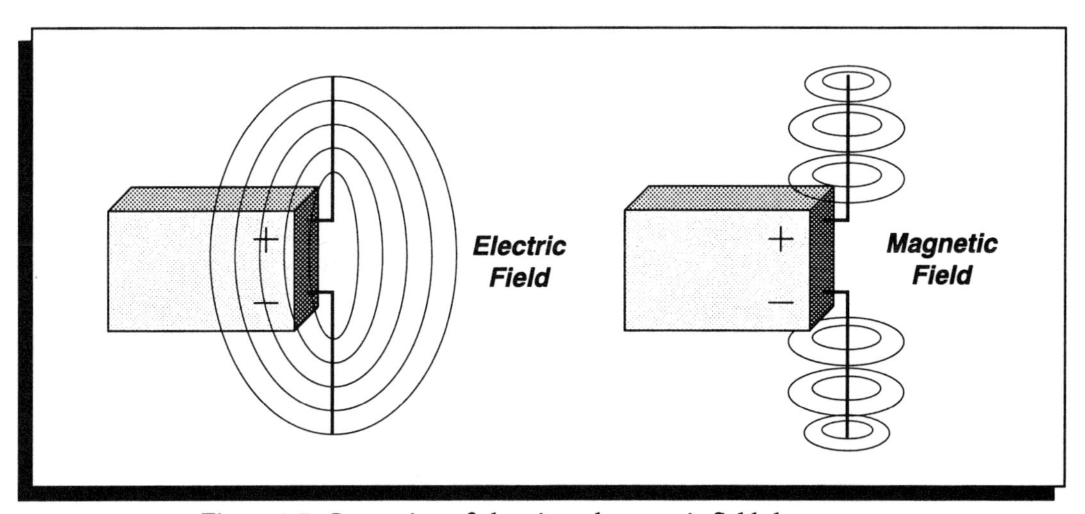

Figure 8.7 Generation of electric and magnetic fields by antenna.

Next we consider how the electric fields and magnetic fields are generated by the antenna. Figure 8.7 shows the antenna attached to an alternating voltage source. At some instant of the cycle there are charges on the antenna and they will cause an electric field around the antenna. However since the charges are not stationary, but are moving along the antenna, they also create a magnetic field. As this process keeps on going, the electric and magnetic fields detach themselves and move away from the antenna in the form of electromagnetic waves, as shown in Figure 8.8.

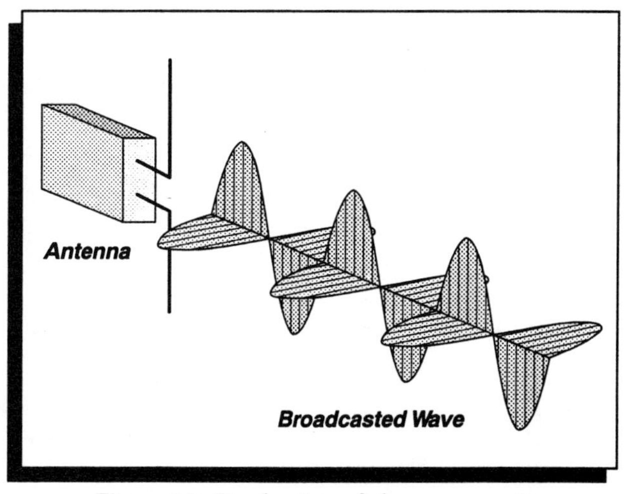

Figure 8.8 Production of electromagnetic waves by antenna.

8.2 BROADCASTING

It would be convenient to put audio information on an electromagnetic wave and then broadcast it as radio waves. However, before doing this, some technical details have to be addressed.

MODULATION

This is the method used for all information conveyed to us. Modulation means a change and this is how we communicate. A few general examples will illustrate the concept.

- ***Writing:*** a straight line produced by a pen is modulated i.e. changed, to present words.
- ***Visual perception:*** the whiteness of a canvas is modulated by a brush to present a picture.
- ***Sound:*** when it is produced, the atmospheric pressure is modulated at its frequency.

Figure 8.9 illustrates the three cases of modulation presented here. It is natural to extend this to broadcasting where an audio signal modulates an electromagnetic wave of radio frequencies

The electromagnetic wave(s) used for this application is called the carrier.

carrier **= usually a radio frequency wave of a particular frequency which carries the audio information.**

As we are in a position to modulate a carrier by audio information there are different ways of doing this.

In broadcasting the two types of modulation usually used are:

- ***amplitude modulation:*** the audio information changes the amplitude of carrier. This is known as AM.
- ***frequency modulation:*** the audio information changes the frequency of carrier. This is known as FM.

AMPLITUDE MODULATION

The objective here is to have an audio signal modulate a carrier, and then the modulated carrier is broadcast. In amplitude modulation, the amplitude of the carrier is changed according to the pattern of audio information. Consider Figure 8.10 where an audio signal (of one frequency, for simplicity) amplitude modulates a carrier creating a new waveform. This waveform is then broadcast. In the real situation, each station on your radio has a definite carrier, i.e. a specific frequency. For example there could be a carrier at 1,000 kHz, which is amplitude modulated. Another station might have a carrier at 1,400 kHz which is also amplitude modulated by the audio signal. Figure 8.11 illustrates the role played by each carrier as it carries the audio information.

Going back to Figure 8.10, a carrier at some fixed frequency f becomes amplitude modulated by the audio signal. From an analysis of this modulat-

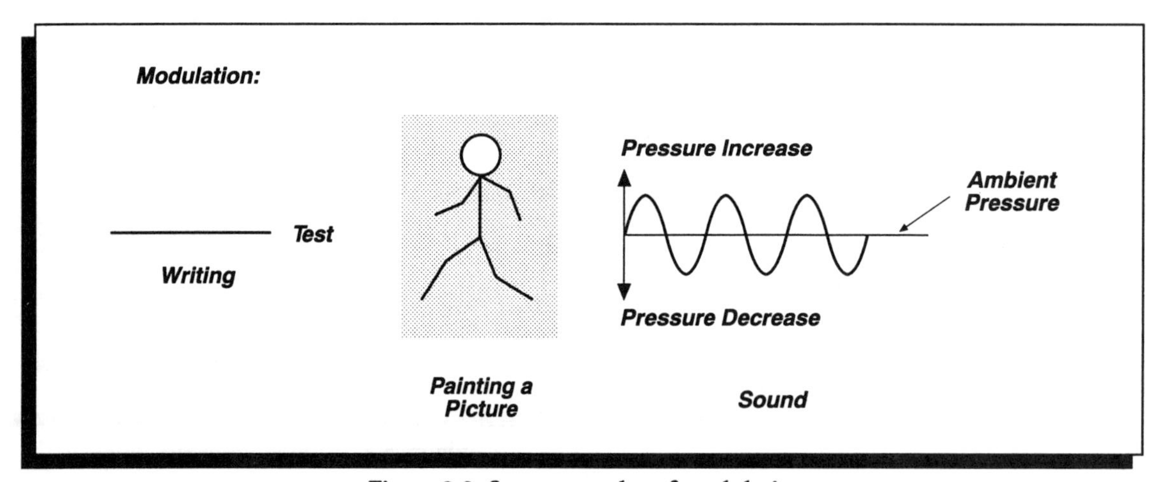

Figure 8.9 Some examples of modulation.

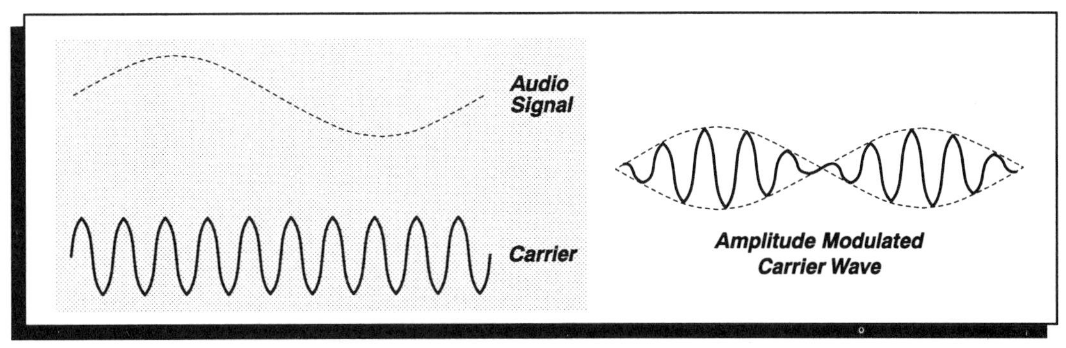

Figure 8.10 Amplitude Modulation.

ed waveform there are three frequencies present:

carrier at f
carrier $f+$ audio frequency
carrier $f-$ audio frequency

This is illustrated in Figure 8.12. This is not surprising since the carrier changes in shape by the modulation. It is almost like a distortion of the single frequency carrier. For example, when the carrier is 1,000 kHz and the audio signal is at 1 kHz, mod-

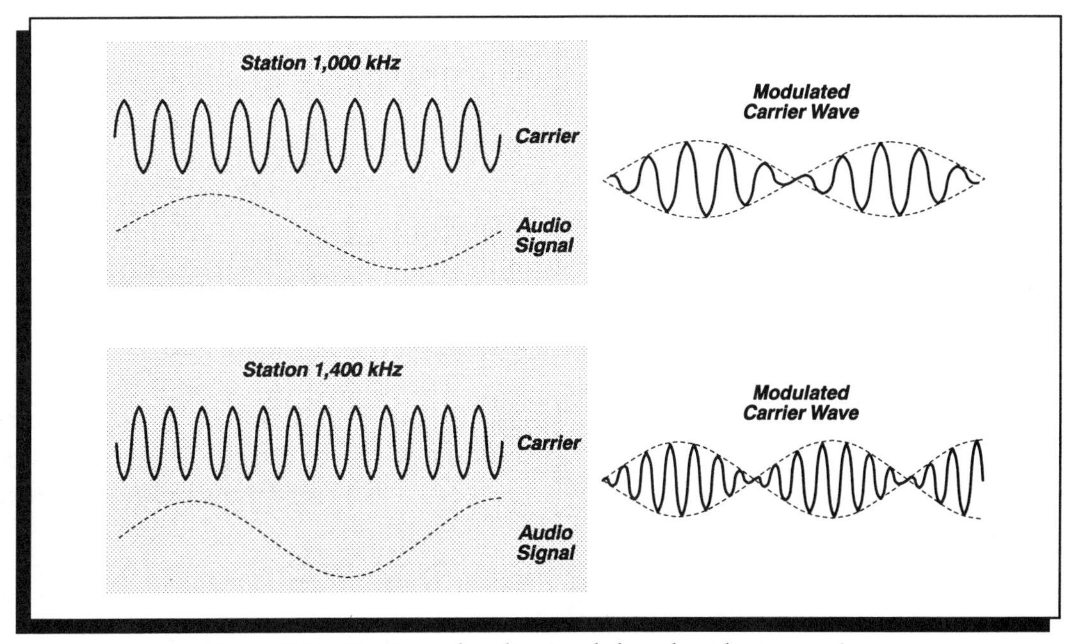

Figure 8.11 Carrier and audio signals broadcast by two stations.

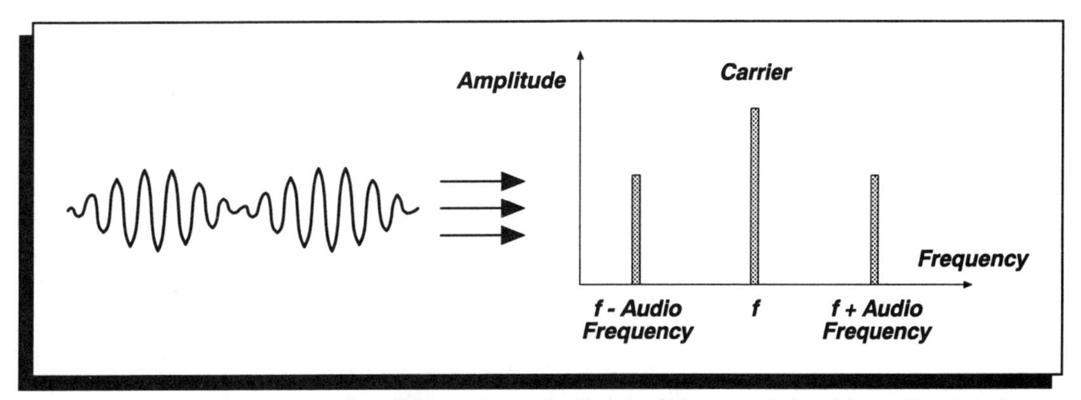

Figure 8.12 Spectrum of an AM carrier at frequency *f* when modulated by audio signal.

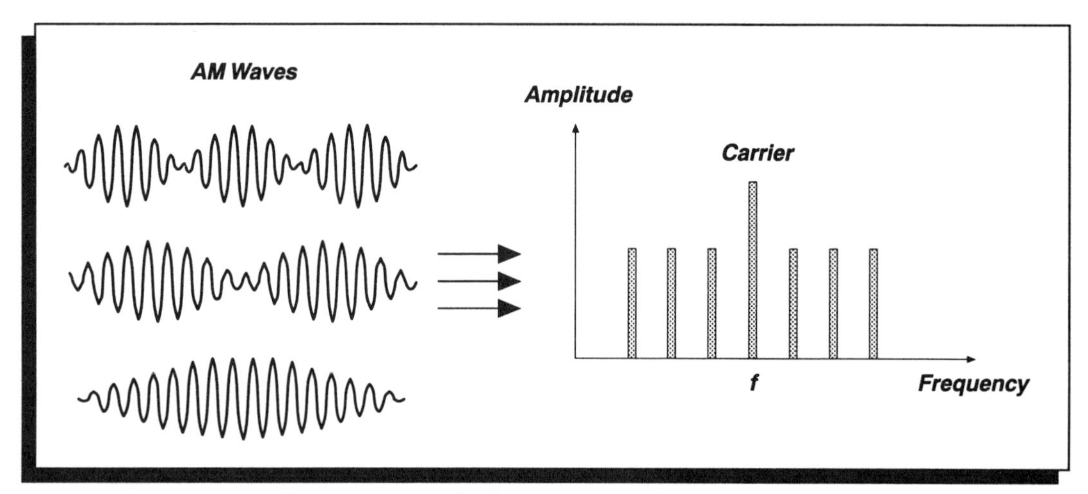

Figure 8.13 Audio frequencies modulating carrier.

ulation will yield an amplitude modulated carrier which has the frequencies 1,000 kHz, 1,001 kHz, and 999 kHz. When there is more than one frequency in the audio signal amplitude modulating the carrier, there will be a distribution of [carrier frequency + audio frequencies] and [carrier frequency – audio frequencies]. In this case the spectrum will be more complicated, as shown in Figure 8.13.

The spread of frequencies on each side of the carrier are called the upper and lower sidebands. They appear as a result of the modulation. The presence of the sideband frequencies imposes a restriction on which frequencies are available for carrier frequencies. This will become clearer by considering audio frequencies going up to only 5 kHz. When all the frequencies up to 5 kHz modulate the carrier, the sideband frequencies will appear as in Figure 8.14. At what frequency can the next carrier be? It is clear that it has to be away from the sidebands and it also should be far enough away that its own sidebands do not interfere with those of carrier *f.*

In AM the audio frequencies which are broadcast range from 200 Hz to 5 kHz only. Of course, that is not hi–fi, but it is useful for communications. Let us assume that one carrier is at 1,000 kHz; at what frequency can the next carrier be? Figure 8.15 illustrates this concept.

From the figure, it is clear that the next carrier frequency could be at 1, 010 kHz. In other words, we need a spacing of at least 10 kHz between station carriers. The Federal Communications Commission

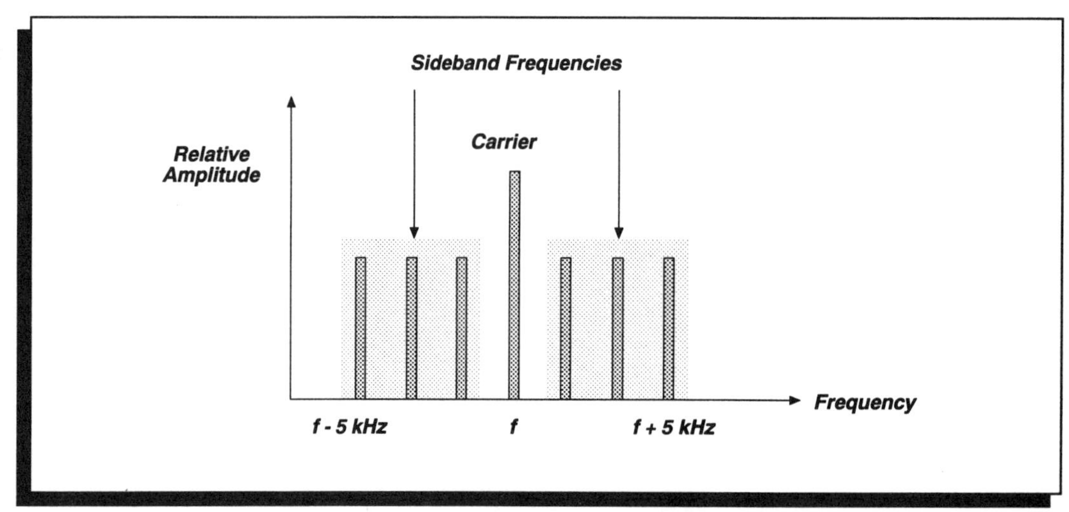

Figure 8.14 Spectrum of frequencies on carrier for audio frequencies up to 5 kHz.

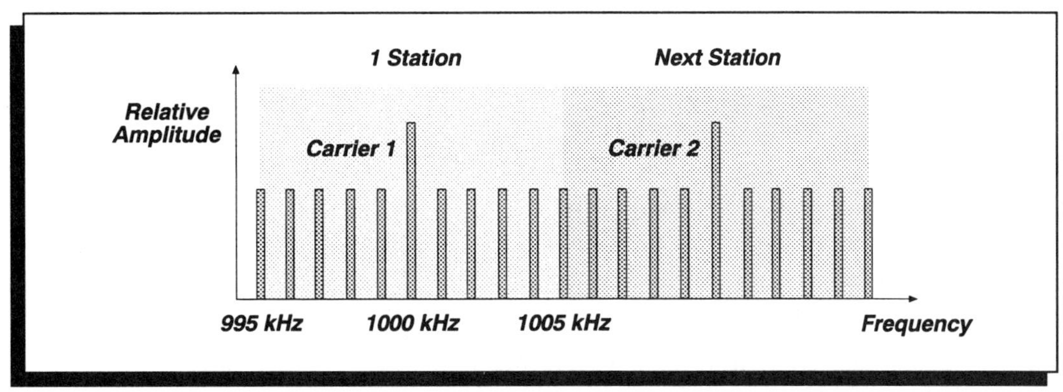

Figure 8.15 Spectrum of frequencies due to modulation of carrier.

(FCC) allows in a given area a spacing of 20 kHz (10 kHz + 10 kHz safety) so that there will be no overlap between stations. Hence, the maximum allowable number of stations in a given area is:

(1605 – 535 kHz) / 20 kHz,
which is just about 53 stations.

Since AM is not hi–fidelity, is it possible to have something better? The answer is FM.

FREQUENCY MODULATION

This was invented as an improvement in the quality of sound over AM. The range of audio signals which are broadcast is: 30 Hz to 15 kHz. This is a substantial improvement over AM. Here the modulation of the carrier will be very different from AM. The method of frequency modulation, or FM, is characterized by:

The frequency of carrier changes according to audio signal, but the amplitude of carrier remains fixed.

The larger the amplitude of the audio signal, the more the frequency of the carrier will increase and decrease. The higher the frequency of the audio, the more often the frequency of the carrier will change. Figure 8.16 illustrates this.

Both the amplitude and the frequency of an audio signal are represented on an FM wave. This is shown in Figure 8.17 where two audio signals of

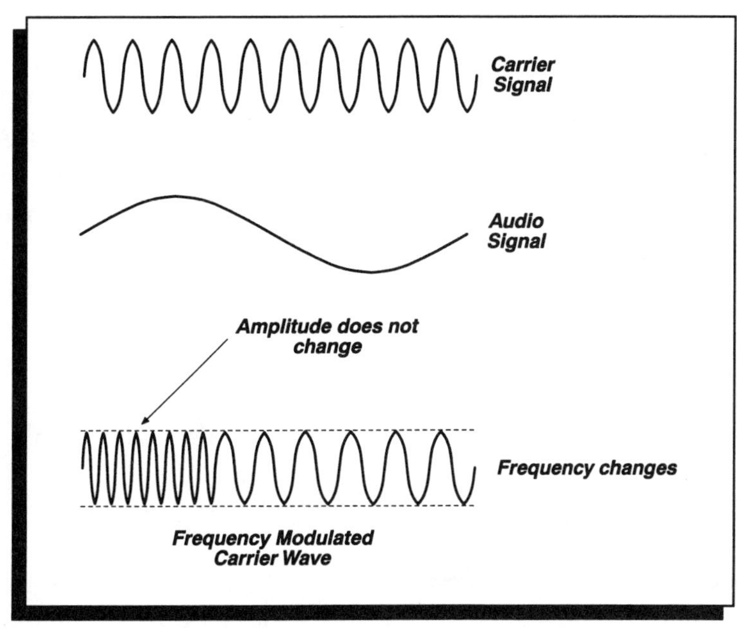

Figure 8.16 Frequency modulation (FM).

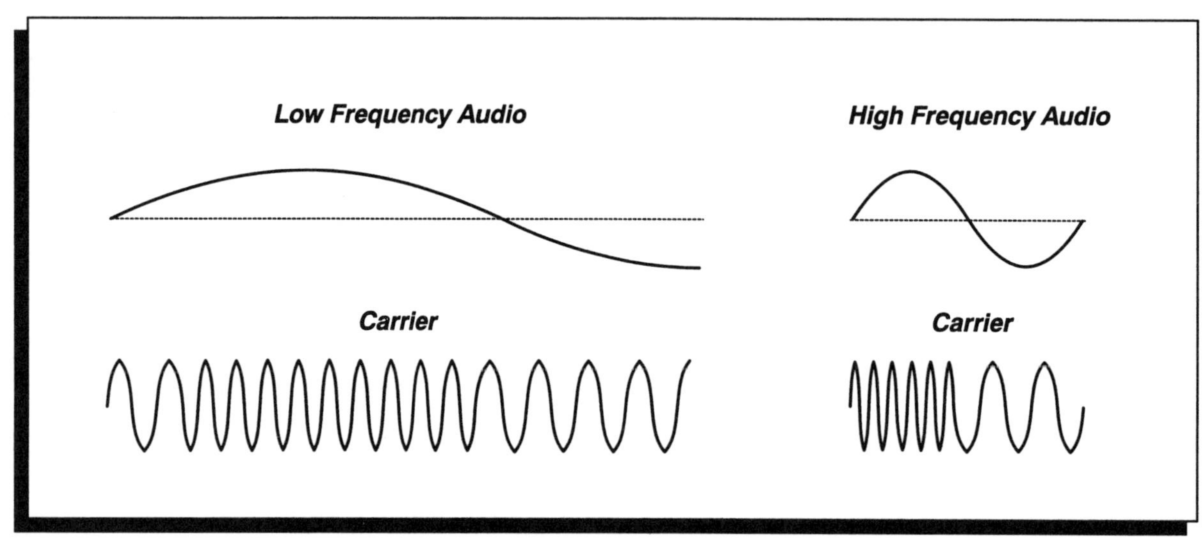

Figure 8.17 A low frequency and a high frequency audio signal frequency modulating a carrier.

different frequencies modulate the carrier. Amplitude information is shown in Figure 8.18 for a loud and a quiet audio signal. In this method of modulation, the FCC allows the modulation per station to be by ±75 kHz. In other words the carrier frequency f can go up to a maximum of $f + 75$ kHz and a minimum of $f - 75$ kHz. This will provide enough "elbow" room for the station, but the next one should be far enough away so it doesn't overlap. As a safety precaution, the FCC has an extra ± 25 kHz. Hence the spacing between stations in a given area, must be at least 200 kHz. This means that when an FM station is at 100 MHz, the next one cannot have a frequency closer than 100.2 MHz (and even that will not be allowed in a crowded area).

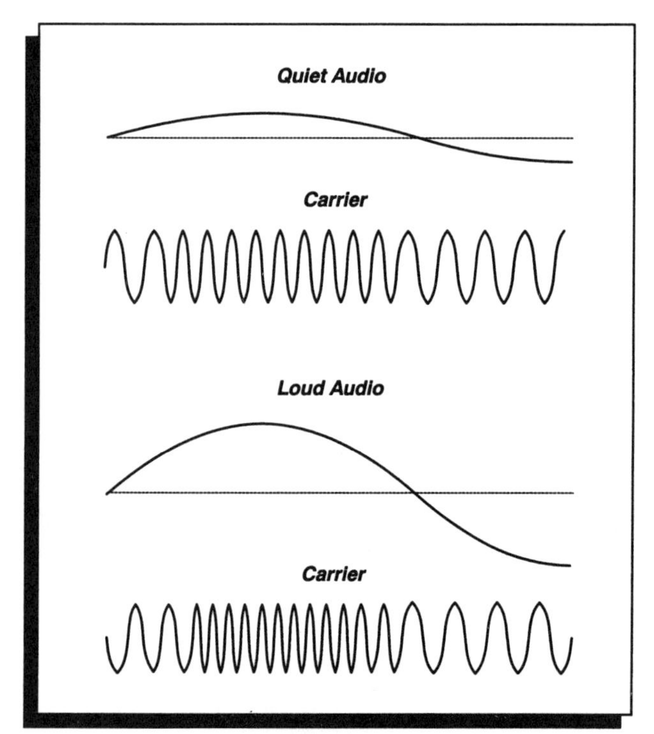

Figure 8.18 A loud and a quiet audio signal frequency modulating a carrier.

The carriers for FM can only be in a certain range of frequencies. The FCC has allocated this range to be 88 MHz to 108 MHz. The maximum number of stations which could exist in an area is: (108 – 88 MHz) / 0.2 MHz which is 100 stations.

COMPARISON OF AM AND FM

Historically AM was invented first, but it does not provide high–fidelity sound for two reasons:

1. The limited range of audio frequencies, 200 to 5,000 kHz.
2. Electrical disturbances and noise will amplitude modulate the carrier and this mixes in with the audio signal.

It is still useful because:

— News, information, and announcements do not need to be in high–fidelity.

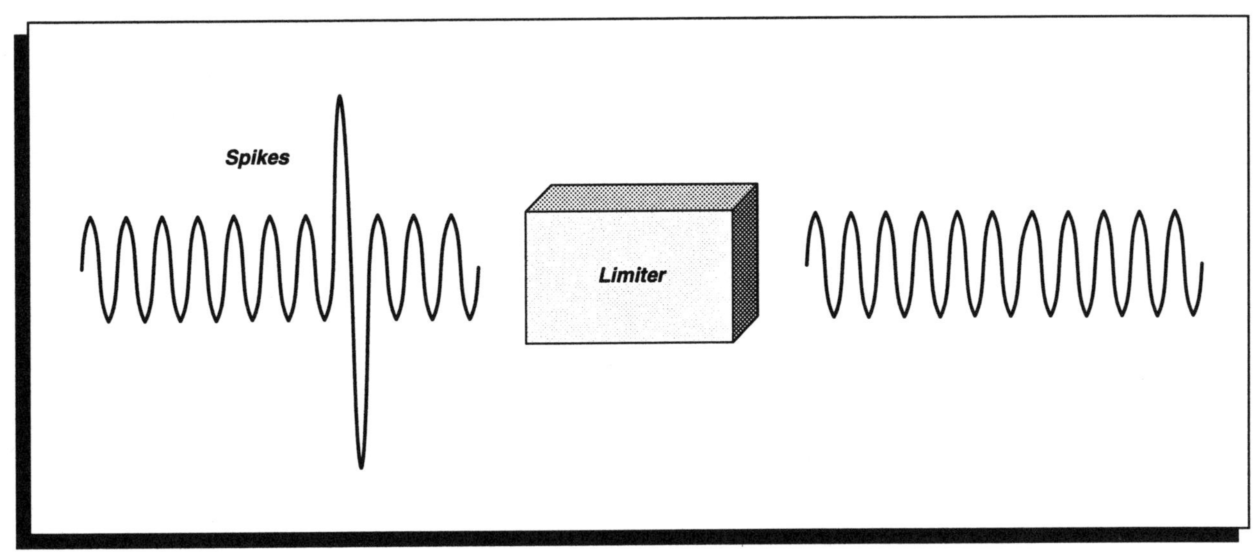

Figure 8.19 Action of limiter in FM.

— It covers a larger area than FM since low frequencies carriers do not get attenuated as easily.

FM is considered high–fidelity (even though it stops at 15,000 Hz) because it provides a fairly wide frequency range for audio information. Also, it has a limiter, Figure 8.19, where any amplitude spikes due to disturbances are suppressed. This adds to the high quality sound that one finds in FM and it does not take away any information since the amplitude has to remain constant. All the information is in the frequency modulation. This circuit is in the tuner.

PRE–EMPHASIS AND DE–EMPHASIS

In order to maintain the broadcast noise–free and static–free, FM takes special precautions during broadcasts and reception. Anticipating that once broadcast, the radio waves will pick up noise mainly at high frequencies, two steps are taken to reduce this. First, before broadcasting the audio information is modified: high frequencies above 1 kHz are boosted in amplitude according to the curve shown in Figure 8.20. This is known as pre–emphasis. The modified audio information is put on a carrier and then broadcast.

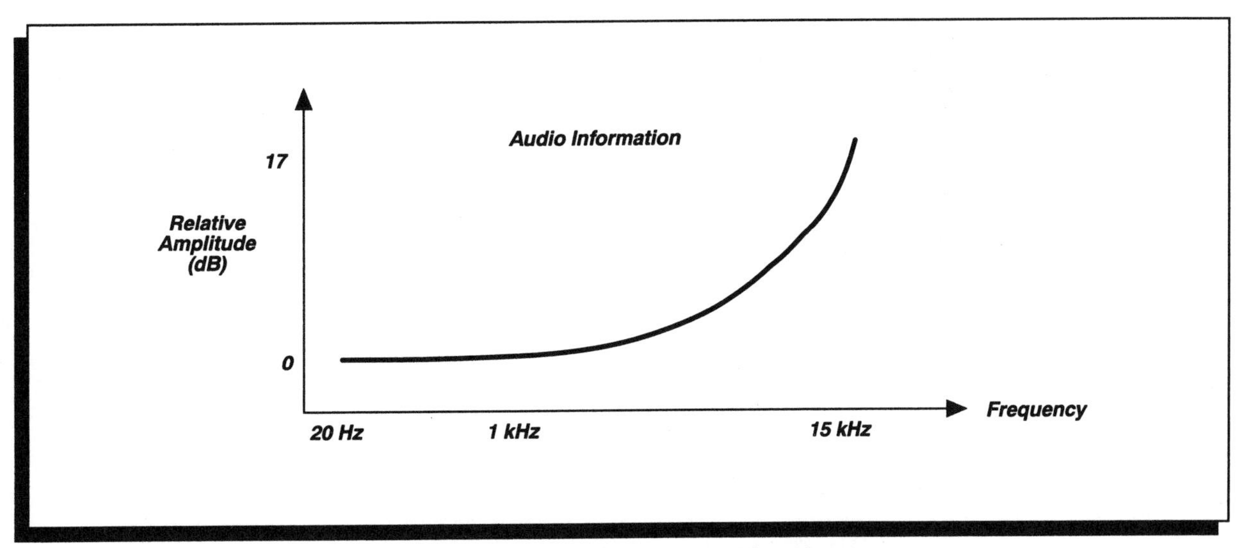

Figure 8.20 Pre-emphasis in FM broadcasting.

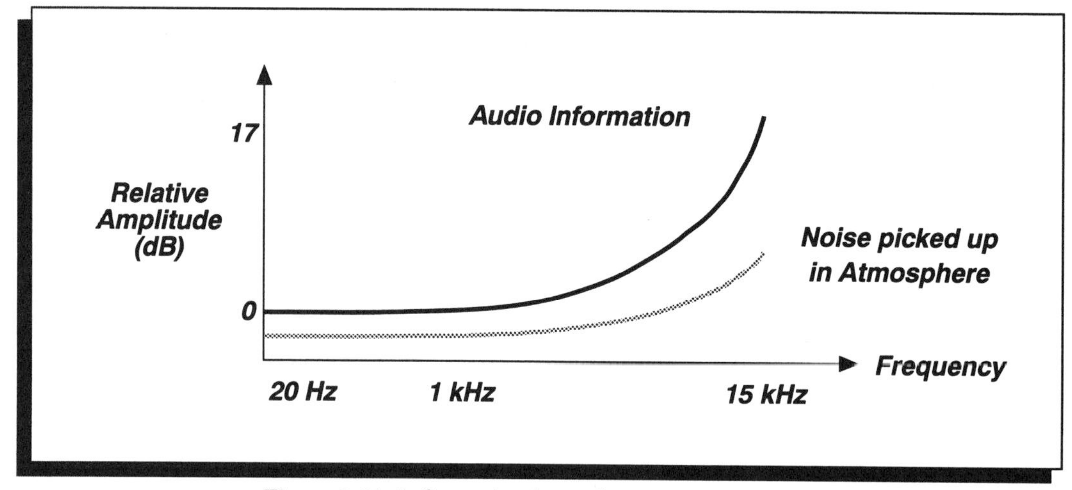

Figure 8.21 Information brought to tuner on carrier.

As an airwave, the information on the carrier picks up some of the noise present in the atmosphere. Hence the carrier wave arriving to the tuner antenna will contain the information shown in Figure 8.21.

Second, this information has to be reconstructed back so as to look like the original and have reduced noise. Figure 8.22 illustrates the step the tuner must take at its input: to go back to the original, i.e. a de–emphasis of the high frequencies. The net effect after de–emphasis will be:

— audio is restored to original spectrum
— high frequency noise picked up in atmosphere is greatly reduced

Such a procedure is performed in all tuners for FM broadcasts.

8.3 TUNERS

The basic elements needed to transmit messages, speech, or music from one location to another by radio are:

— a transmitter to generate radio frequency waves
— a transmitting antenna to radiate the waves into space

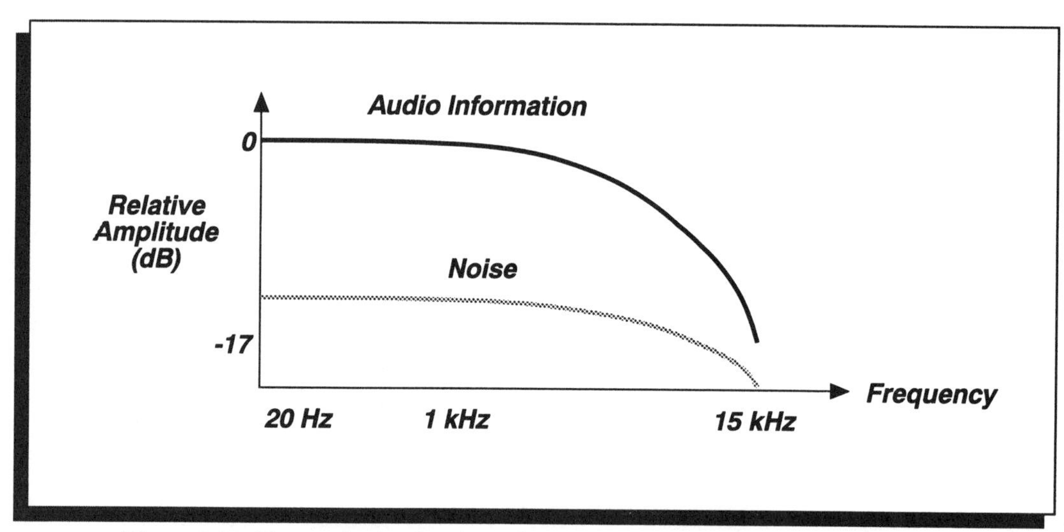

Figure 8.22 De-emphasis of audio information to reduce high frequency noise.

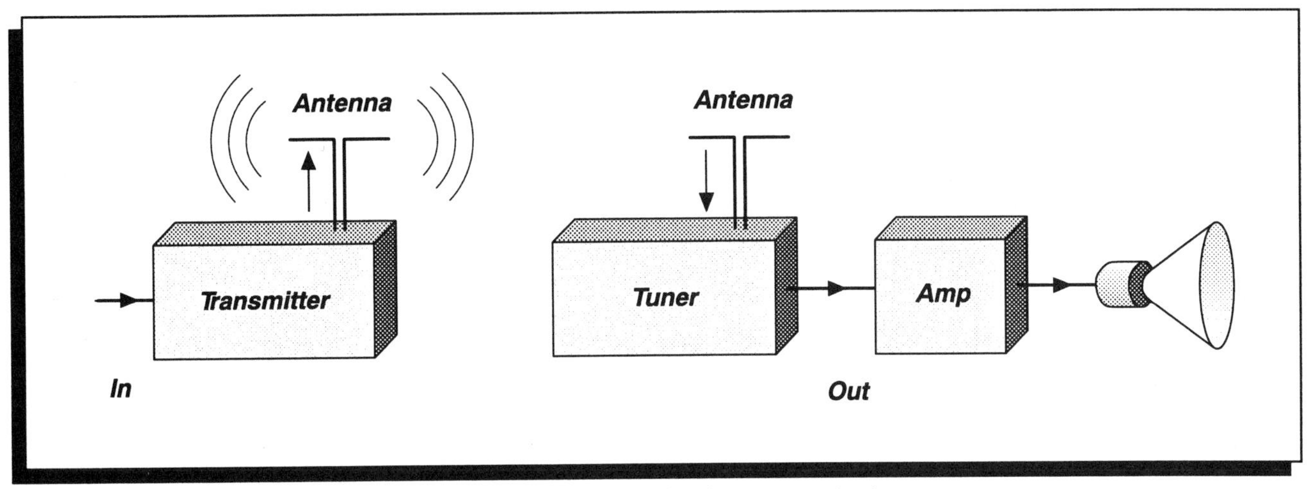

Figure 8.23 Elements of radio communication.

— a receiving antenna to intercept a portion of the radiated wave
— a tuner to select the desired transmitter signal

The whole process is shown in Figure 8.23 and it is applicable to FM and AM.

Let us now look at the tuner for AM. Its role is:

— select the desired carrier with its audio information.
— amplify the signal.
— extract the audio from the rf wave
— filter the audio, amplify it and then send it to speaker.

To effectively do this, a special type of tuner has been developed. This is used universally with some special features. This tuner is called a superheterodyne receiver, and this feature makes the tuner a simple and straightforward unit. Figure 8.24 shows the design; it is so simple that almost everyone is using it. An amplitude–modulated signal is picked up by the antenna and amplified so it raises its level. The signal then goes into the mixer, where it beats against an incoming signal from the internal local oscillator. The output of the mixer is a beat signal whose frequency is adjusted to be at 455 kHz. Since it is a beat signal, it is derived from the frequency of the internal oscillator minus that of the incoming carrier. The beat frequency signal at 455 kHz is known as the intermediate frequency (IF) signal. It is amplified by stages tuned only to the IF frequency and then it is rectified and filtered to yield the original audio information.

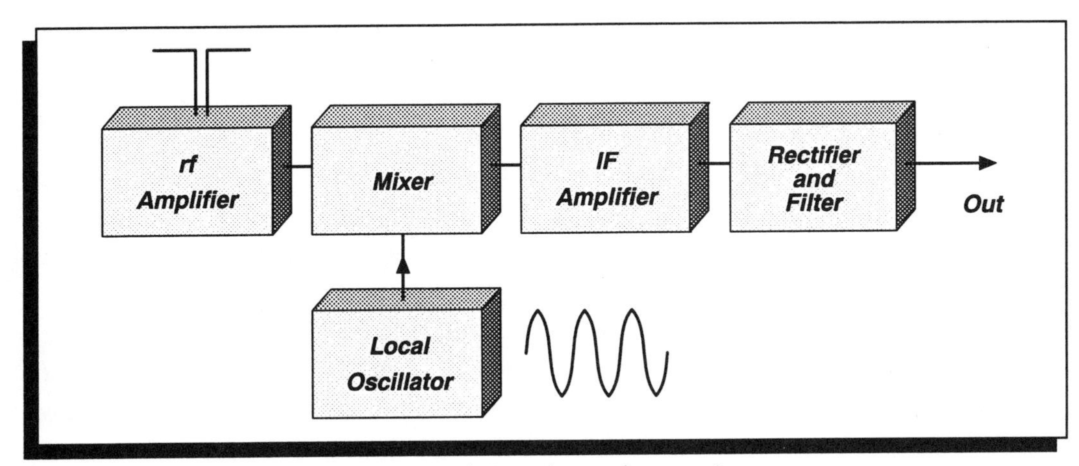

Figure 8.24 Superheterodyne receiver.

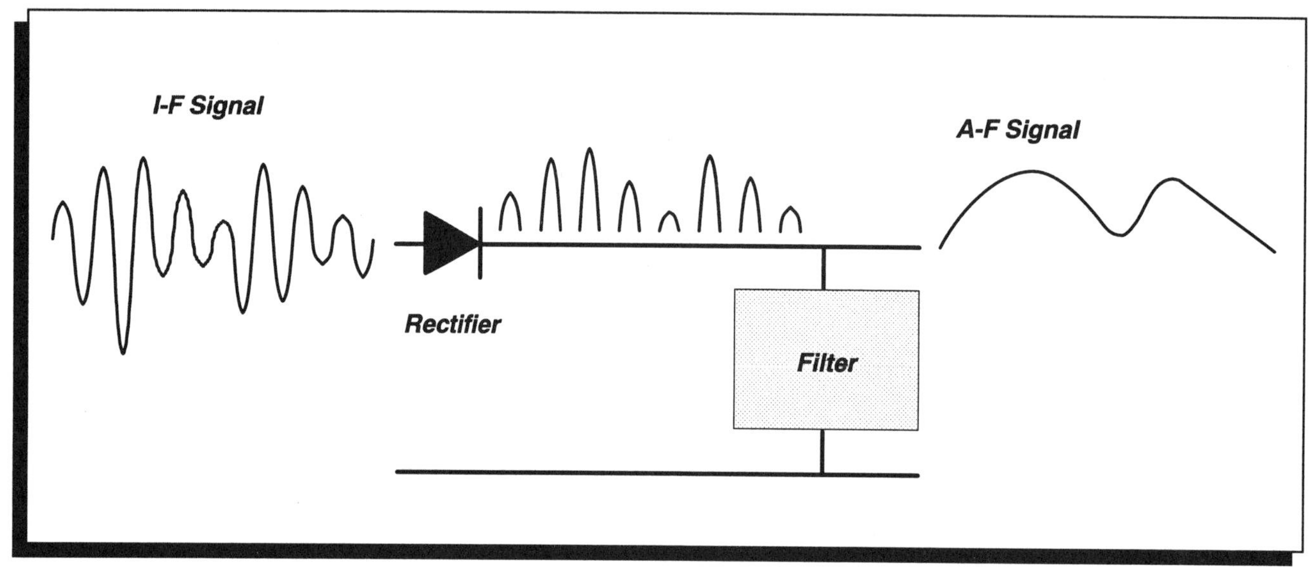

Figure 8.25 Processing part of AM signal with a simple diode and filters.

Figure 8.25 shows the signal from the output of the IF amplifier with the audio part passing through a rectifier and a filter. All commercial tuners work this way. Turning the dial on the local oscillator corresponds to selecting a frequency which will beat against the received signal to give a new carrier at the IF frequency of 455 kHz.

A similar series of steps are taken with FM reception except that then the IF frequency is 10.7 MHz. For example, when trying to catch a carrier at 100 MHz, the dial attached to the internal oscillator is rotated until the oscillator puts out 110.7 MHz. The beat frequency then becomes (110.7 MHz minus the 100.0 MHz) with the carrier being changed from 100 MHz to 10.7 MHz inside the tuner with the audio signal still present. By filtering out the IF signal the audio signal is extracted.

It is useful to remember the intermediate frequencies:

IF for AM = 455 kHz
IF for FM = 10.7 MHz

The simplicity of the superheterodyne approach is demonstrated by the fact that only one knob needs to be turned during tuning. The IF amplifiers are designed to work only at and around the intermediate frequency and hence they do not need any tuning. In effect, every tuner has its own internal oscillator, one for AM and one for FM. The superheterodyne approach is used in all commercial tuners.

STEREO BROADCASTING

Since each radio station has only one carrier, how is it possible to broadcast and receive stereo? Historically, it was decided that a stereo signal would be encoded into a single carrier signal before broadcasting, and then received like a monophonic signal. With a stereo tuner, it can be transformed back to stereo. Part of the reason for such a decision was to keep the broadcasts compatible with those only interested in monophonic information. To broadcast stereo, a radio station does the following steps:

1. adds together the LEFT channel and the RIGHT channel signals = (L+R).
2. takes the difference between the LEFT channel and the RIGHT channel signals = (L–R).

The sum (L+R) is equivalent to a monophonic signal, where both channels are added together. This part is broadcast directly on the carrier by modulating it.

The difference (L–R) signal is the information the two channels do not have in common, the difference between them. Since this information covers the same range of audio frequencies as the (L+R) information, it cannot modulate the carrier directly, or the (L+R) information and (L–R) information would get mixed up. To solve this problem, every stereo station has a 38 kHz subcarrier which is amplitude modulated by the (L–R) information. This will create sub–bands from 23 to 53 kHz (since 38 kHz – 15 kHz is 23 kHz; and 38 kHz + 15 kHz is 53 kHz). The net effect is that the audio frequencies of (L–R) information are shifted to a new range of frequencies which will not get mixed up with the frequencies of the (L+R) information. All this information now goes on the carrier together with half the frequency of the subcarrier, which is 19 kHz, and is known as the pilot frequency. Hence, on the carrier there will be:

— (L+R) information in the range 30 Hz to 15 kHz
— (L–R) information now in the range 23 kHz to 53 kHz
— a pilot frequency at 19 kHz.

This is shown in Figure 8.26. When the antenna in the tuner picks up the carrier, a sequence of operations takes place in the tuner:

— the pilot frequency is used to switch a "stereo" light on in the tuner indicating that a stereo signal has arrived.
— the pilot frequency is multiplied by two to get the original 38 kHz for restoring signals.
— add (L+R) and (L–R), giving 2L.
— subtract (L–R) from (L+R), giving 2R.

The 2L information now goes to the LEFT channel of amplifier while 2R goes to the RIGHT channel.

The steps above apply to FM where stereo is an important feature of this type of broadcasting. There have been plans to get AM in stereo, but so far the attempts have not been successful.

8.4 ANTENNAS

An antenna is basically a device which converts alternating currents at radio frequencies to electromagnetic waves of the same frequencies. It is also a device which does the inverse, the conversion of electromagnetic waves to radio frequency currents. In the first case, the antenna is used for broadcasting. The transmitting antenna forces radio waves into space. In the second case, the antenna is used for their reception. The receiving antenna receives electromagnetic energy from space.

To be effective an antenna must be of the proper size to match the space. If a resonance can be set up on the antenna, then the antenna performance will be enhanced. This is the approach that we will pursue. However, before discussing the principles

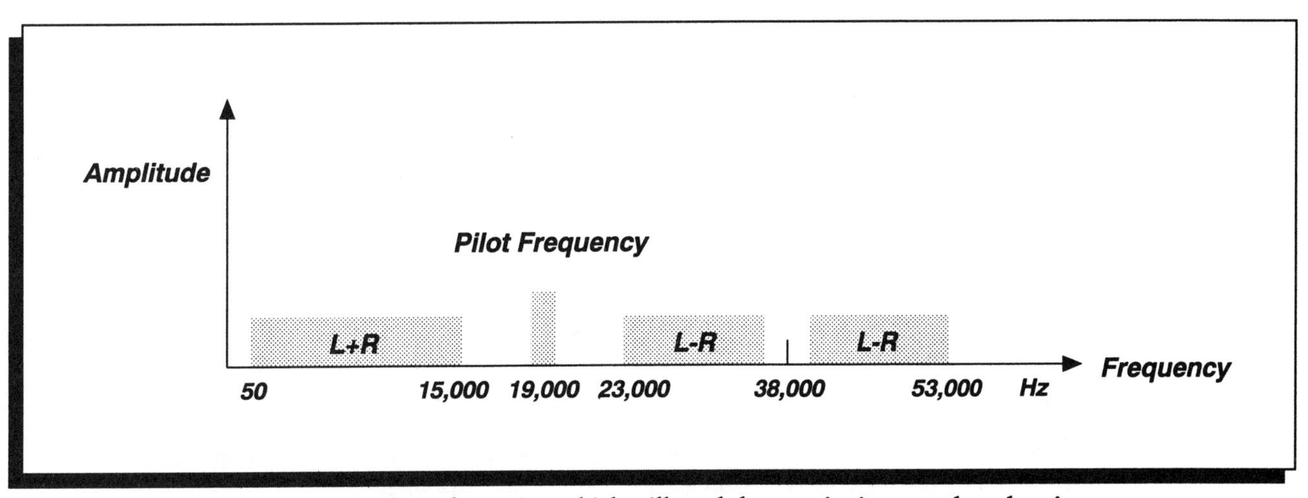

Figure 8.26 Audio information which will modulate carrier in stereo broadcasting.

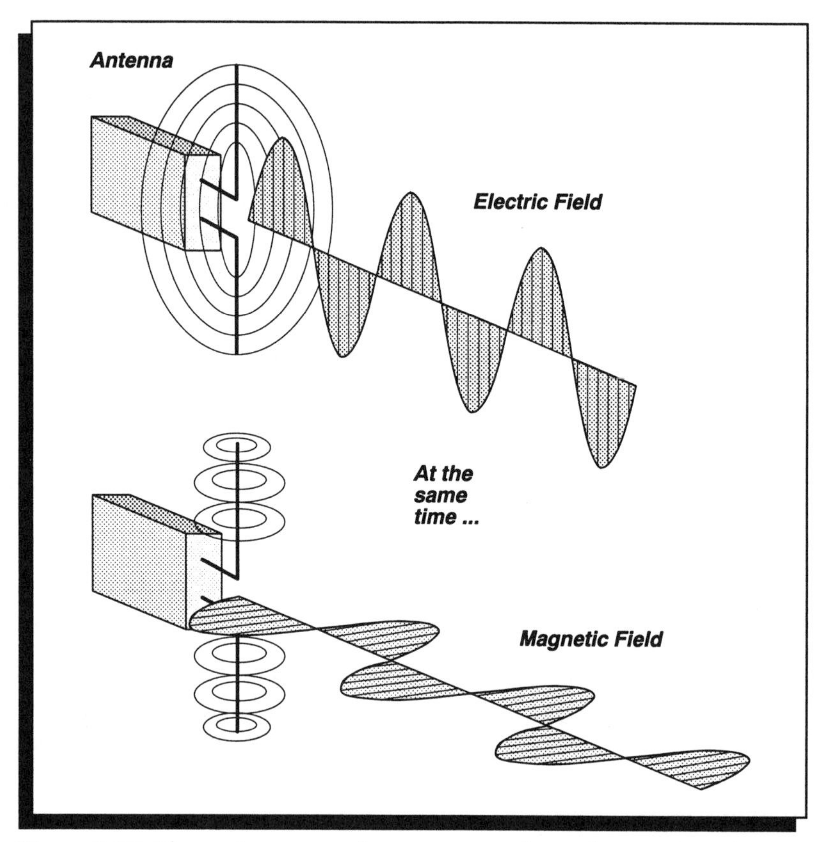

Figure 8.27 Alternating current in antenna produces electromagnetic wave.

of an antenna, let us briefly recall how it generates an electromagnetic wave. Figure 8.27 shows an oscillating current in two wires which is the source of both electric and magnetic fields around them and of the electromagnetic waves it generates. The electric and magnetic fields are always perpendicular to each other. Figure 8.27 shows the situation at one instant in time where the current in the antenna alternates in sign with the terminal reversing, also in sign. As this process is repeated over and over again, an electromagnetic wave will continue to be produced at that frequency.

It is no surprise then that there is an electric field around the antenna and that it behaves like the plates of a capacitor with charges on it. Figure 8.28 illustrates this point. Likewise, it is not sur-

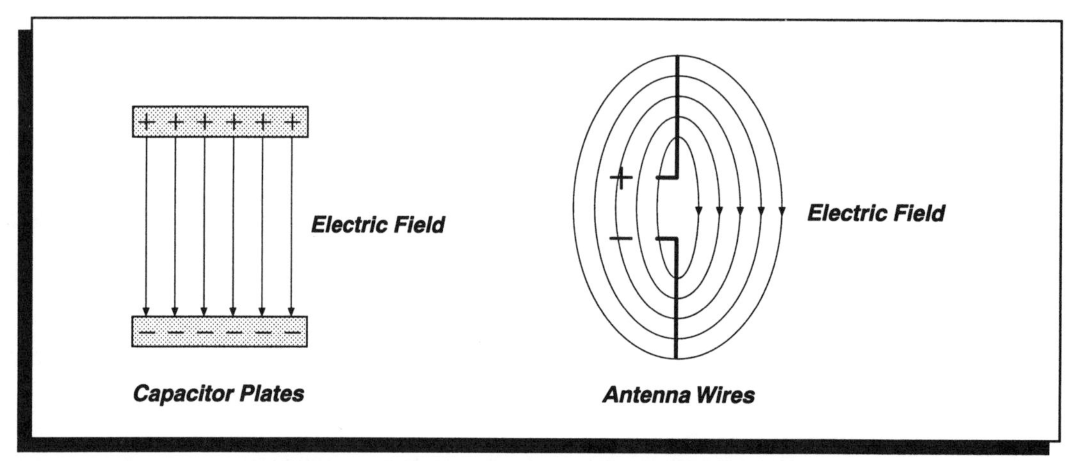

Figure 8.28 Electric field around charged antenna wires is similar to that between charged capacitor plates.

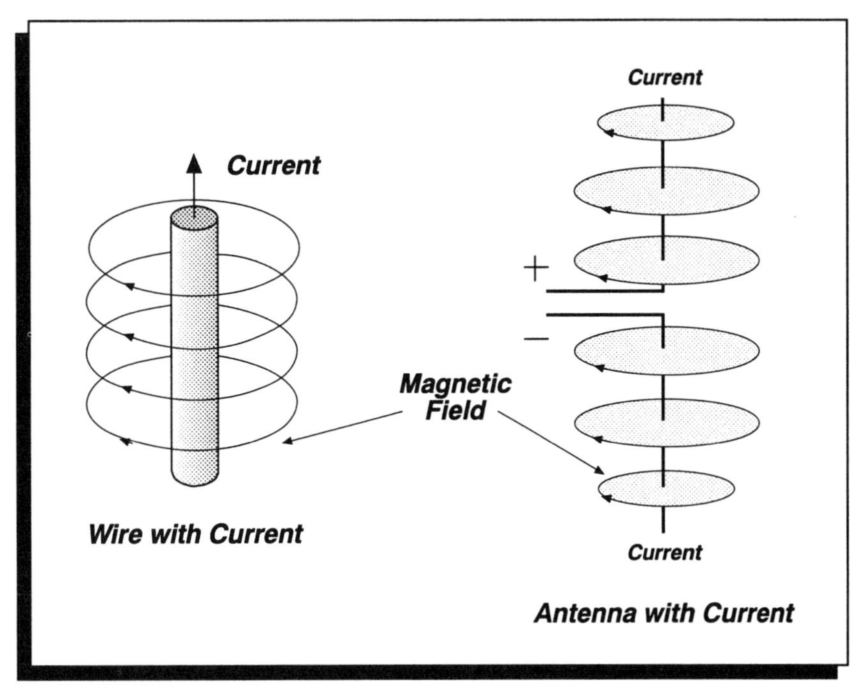

Figure 8.29 Magnetic fields around a wire and antenna with current.

prising that there is a magnetic field around the antenna when a current flows in it. Figure 8.29 compares the antenna to that of a wire carrying a current. The direction of the current determines the direction of the magnetic field.

In order to set up a resonance on the antenna and achieve greater signal levels, certain conditions have to be met. Consider the two–wire antenna in Figure 8.30, with an alternating current set–up at its center. The current is like a wave and it travels along the wire until it arrives at the end where it gets reflected. As this continues there will be waves travelling in one direction with reflected waves travelling in the opposite direction. Constructive

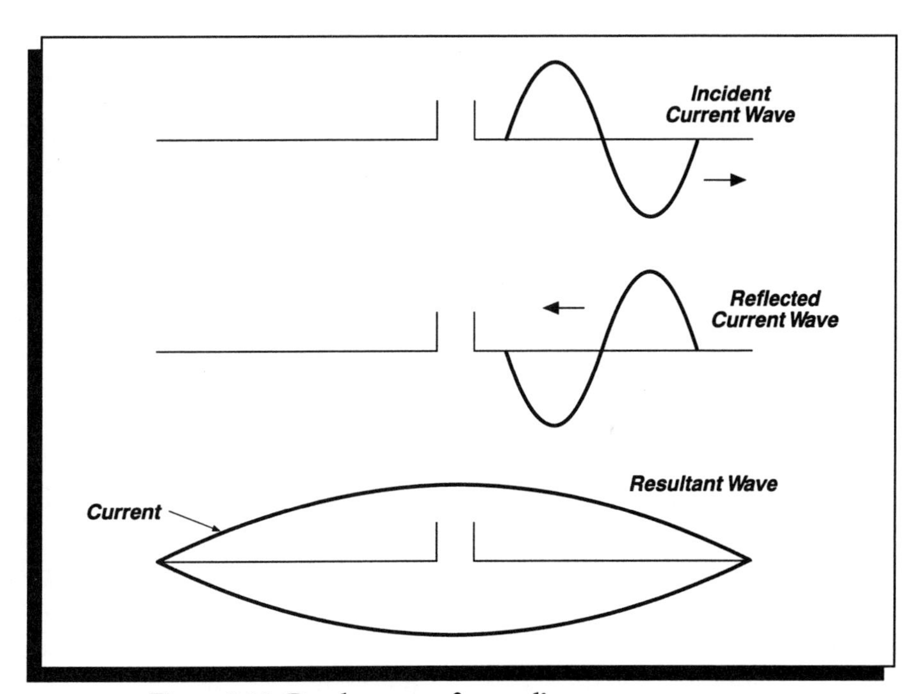

Figure 8.30 Development of a standing wave on antenna.

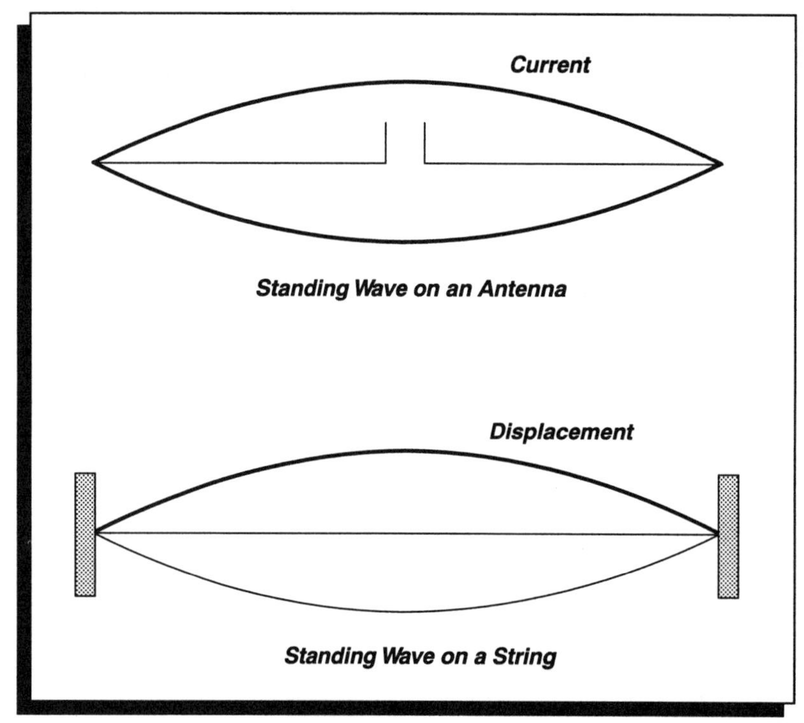

Figure 8.31 Comparison of standing wave on antenna to that of a string.

and destructive interferences will then occur setting up standing waves on the antenna.

The timing (the phasing) has to be just right for this to occur. If the time that it takes the wave to reach the end corresponds to the time for the current in the antenna to increase from its minimum value to the maximum value, a standing wave will occur for the fundamental frequency corresponding to this motion. In this case, the current distribution will look as in Figure 8.31, which is similar to the displacement on a string where a standing wave can also be set up. It is simple to calculate the fundamental resonant frequency from frequency × wavelength = speed:

— on a string,
frequency of resonance = (speed on string) / (2 × length of string)

— on an antenna,
frequency of resonance = (speed on antenna) / (2 × length of antenna)

The speed on an antenna corresponds to the speed of an electromagnetic wave in that antenna, which is a few percent less than in vacuum, because the wave interacts with the electrons in the wire of antenna.

Example: for an antenna whose length, tip to tip, is 1.5 meters, what is the resonant frequency?

Since a standing wave at the fundamental corresponds to half a wavelength, we have:

$\frac{1}{2}$ wave = 1.5 meters
1 wave = 3 meters
resonant frequency = $\sim 3 \times 10^8$ Hz / 3 meters
= $\sim$ 100 MHz

The approximate sign is shown because the speed of light in a vacuum was used for the speed along the antenna.

ONE–HALF WAVE DIPOLE ANTENNA

The type of antenna just discussed is called a half–wave dipole antenna. On resonance at its fundamental frequency the electric current sets up a standing wave whose half–wavelength is equal to the length of the antenna tip–to–tip. The distribution of the current is that of a standing wave and this corresponds to two travelling waves going in opposite directions. The speed of the wave on the wire depends on the cross–section of the wire, and it is closest to the speed in vacuum when the wire is thin and the interaction with the wire is less. This is a very important antenna for practical applications and theoretical studies of electromagnetic radiation. The resonant condition for the current distribution will determine the magnetic field distribution around the antenna.

When a voltage is impressed upon an antenna at the input leads at its center, the antenna's electrical impedance will determine how much current flows into it. For the half–wave dipole antenna, the impedance at the center wires is 73 ohms. In order to have optimal performance of the antenna with maximum power transfer, the amplifier connected to it must match its own impedance to that of the antenna's.

The pattern of the radiation around this antenna is directional and it is characteristic of this type of antenna. Figure 8.32 shows this pattern for an antenna which is horizontal. The radiation spreads out perpendicular to the antenna axis; it looks like an expanding "doughnut." A more accurate representation of the radiation pattern is the one shown in Figure 8.33 on a polar graph. There is no radiation in the direction of the antenna axis, i.e. in the 0° and 180° directions. That is why this type of antenna is called dipolar, its radiation pattern is that of a dipole. This type of antenna is useful for the high frequency range, but at low frequencies it tends to be very long and some other forms of antenna need to be developed.

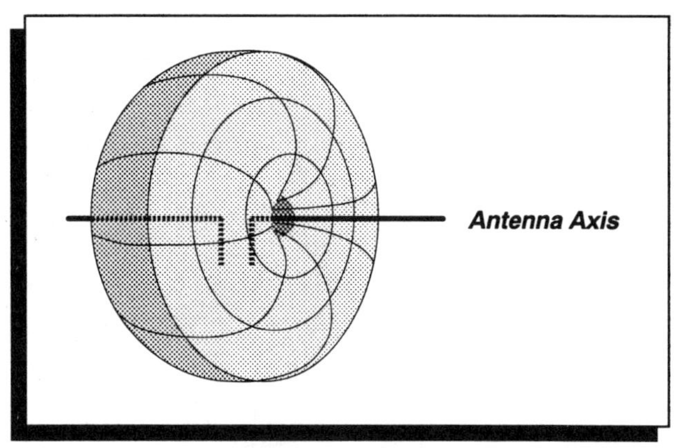

Figure 8.32 Radiation pattern of electric field around half-wave dipole antenna.

QUARTER WAVE OR GROUNDED ANTENNA

The directional properties of an antenna, half–wave dipolar type, are determined by a pattern as in Figure 8.33. One way to improve the directional property is to have the antenna directed vertically and then the radiation will propagate in all directions in the plane of the earth. For low fre-

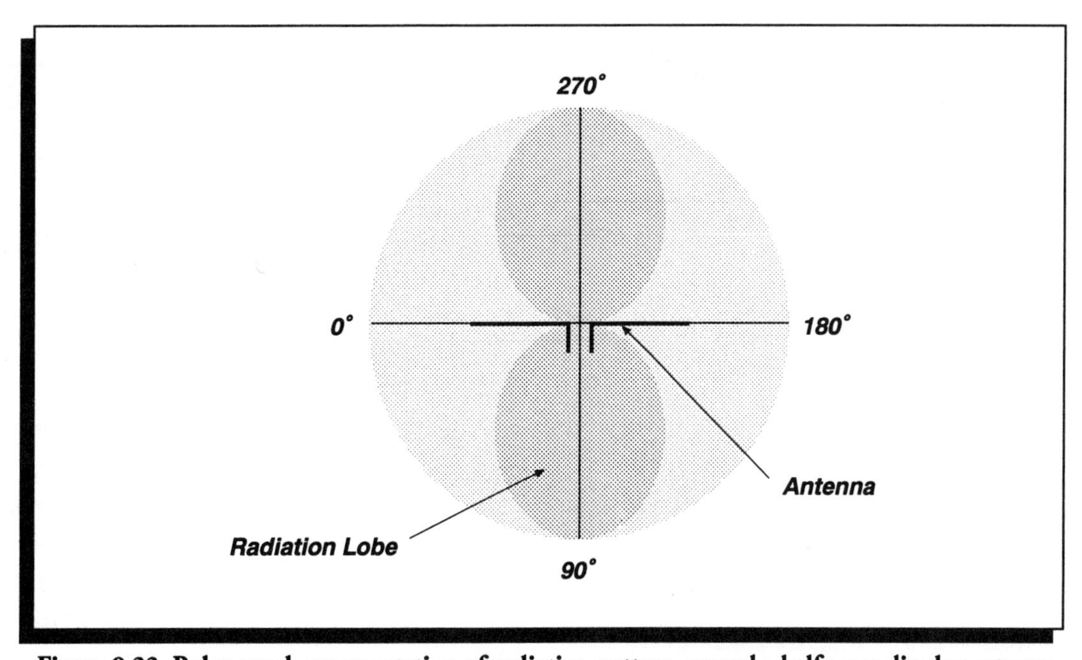

Figure 8.33 Polar graph representation of radiation pattern around a half-wave dipolar antenna.

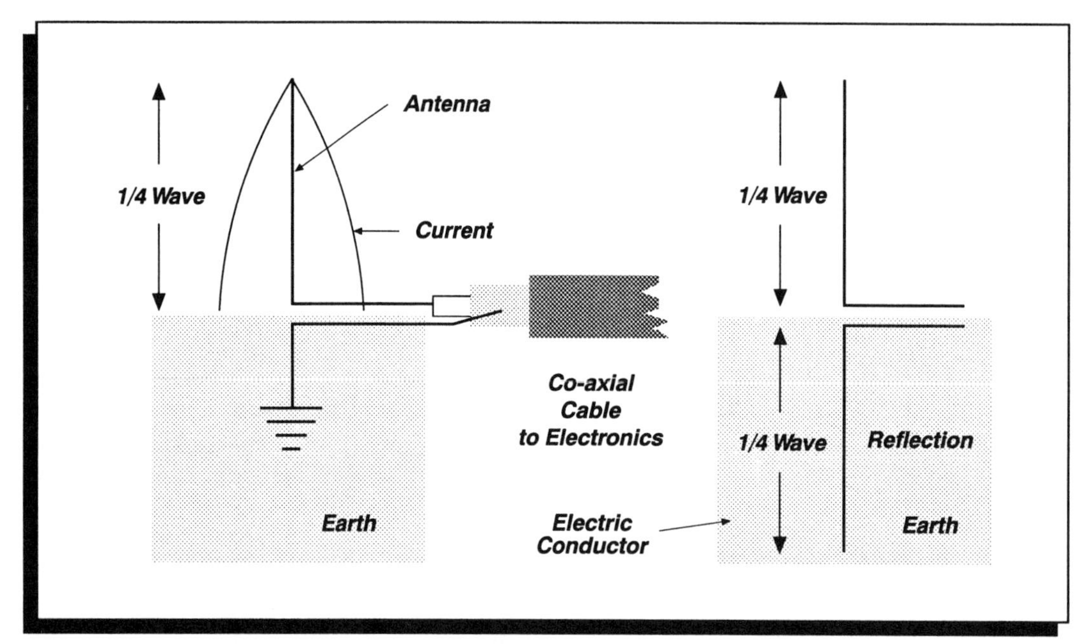

Figure 8.34 Basic elements of a grounded vertical antenna.

quencies, the half–wave dipolar antenna tends to be very long. One way to make it shorter is to use a quarter–wave antenna, mounted vertically, as in Figure 8.34. In fact, this looks like a half–wave dipole antenna whose other quarter wave part of the antenna is reflected in the earth. This is possible because the earth is a conductor of electricity and acts as a mirror to the quarter wave antenna pointing vertically. The resonant frequency can be calculated as before, because now the antenna is a quarter– wave long. Therefore:

Frequency at resonance = (speed on antenna) / (4 × length of antenna)

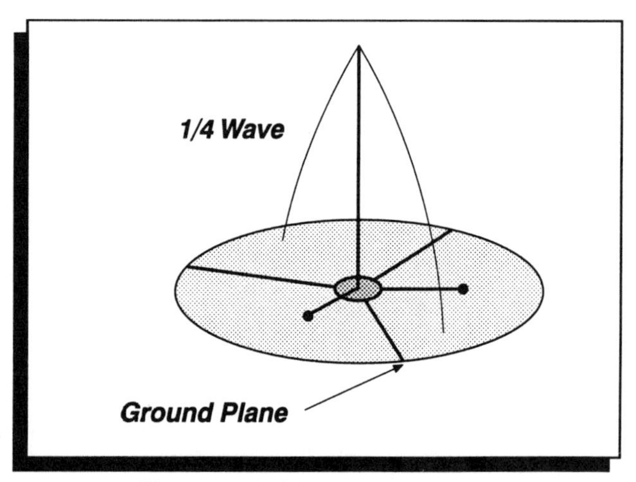

Figure 8.35 Quater-wave antenna.

Should there be no earth nearby, the antenna can be modified by placing it on its base conductors which act like the earth. This is shown in Figure 8.35. This arrangement is particularly useful for situations like in a satellite, where there is no earth in the vicinity of the antenna. A series of three or more conductors, each a quarter wavelength long, are arranged circularly around the vertical antenna, and they act electrically like the earth.

The radiation pattern is almost omnidirectional as the vertical conductor sits in the middle of the doughnut and the radiation spreads out radially away from the antenna.

The grounded vertical antenna is used in cars. To make its length shorter, as a quarter wave antenna is still a long antenna, a coil is attached in series with a short antenna and this maintains an effective quarter–wavelength, Figure 8.36.

■ LOOP ANTENNA

The two previous antennas used directly the electric–field part of the electromagnetic wave. In fact, for a vertical direction of the electric field of the electromagnetic wave, as in Figure 8.37, the receiving antenna should also point vertically because the electric field of the radio wave will force the electrons in the antenna to go up and

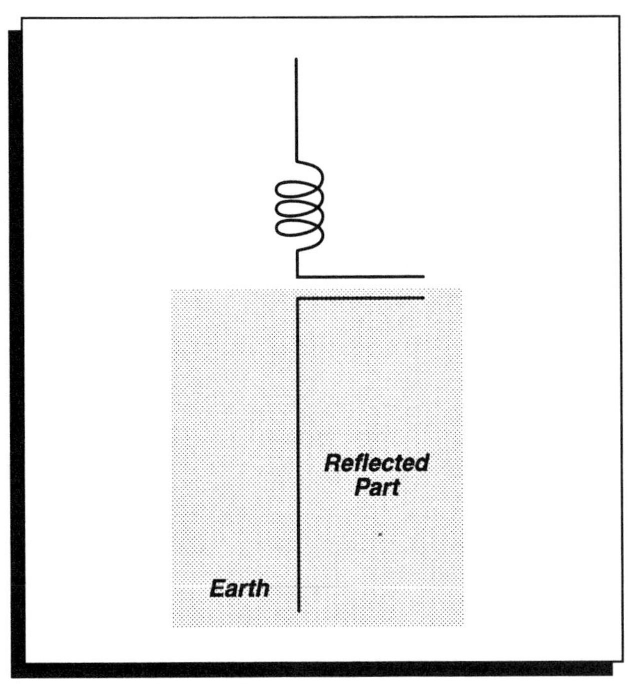

Figure 8.36 Total antenna length is made shorter by inserting a coil in series.

down and set up a current standing wave. With the radio wave there is also a magnetic field associated with the electric field. In certain situations it may be more convenient to detect the magnetic field with a loop antenna. Figure 8.38 shows an antenna with one loop. The oscillating magnetic field of a radio wave induces a voltage across the loop antenna which is then amplified by the tuner. This type of antenna is especially useful for long radio waves where a half wave dipole antenna would have to be prohibitively long. It is usually found on home tuners for the AM range. The example presented in the section on half–wave dipole antenna showed that a 1.5 m long half wave dipole antenna is suitable for the 100 MHz range, which is FM. A typical AM frequency is 1 MHz, which means that a half wave dipole antenna would have to be 150m long; this is very long and impractical. A loop antenna in this case is more convenient. Improvement in this type of antenna,

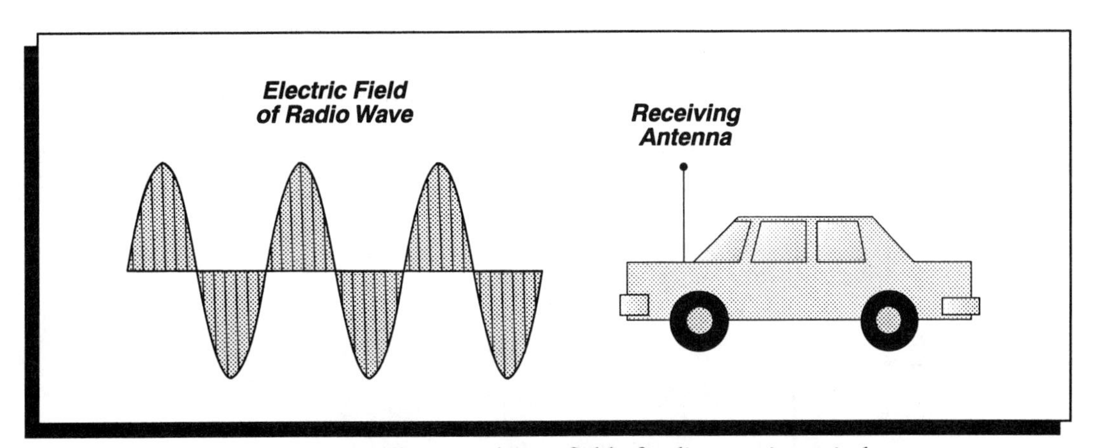

Figure 8.37 When the electric field of radio wave is vertical, the receiving antenna should also be vertical.

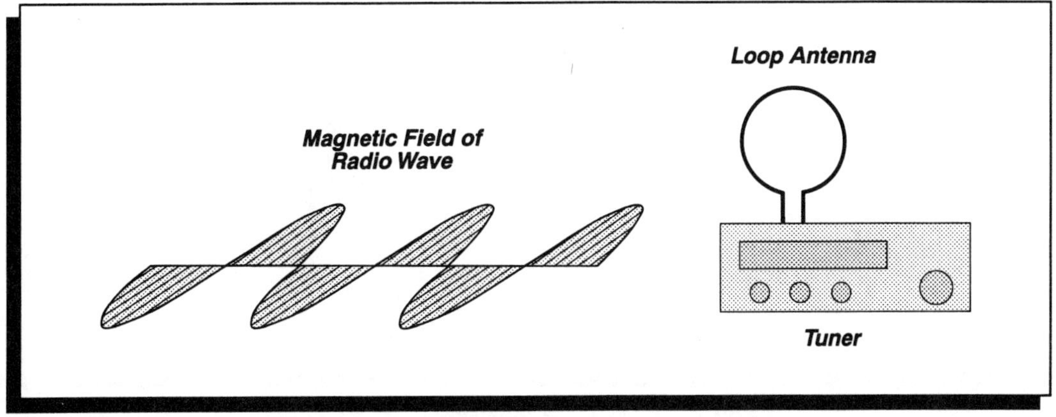

Figure 8.38 Loop antenna detects the magnetic field part of radio wave.

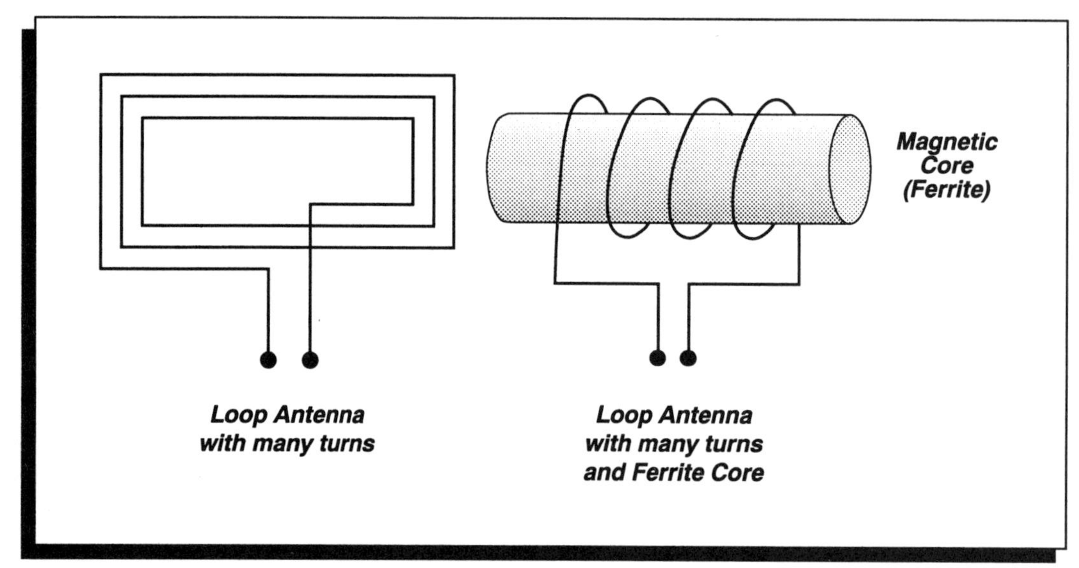

Figure 8.39 Two common loop antennas.

as to sensitivity, can be made by increasing the number of turns in the loop and even adding a magnetic core. Both types are shown in Figure 8.39. The core is ferrite, a special material which is magnetic and insulating. A conducting magnetic core would attenuate the induced signal in the coil by the induced currents in the material according to Faraday's law of induction. It is not suitable in this situation.

■ POLARIZATION

An antenna with oscillating electrons will produce an electromagnetic wave. Let us suppose that a half–wave or quarter–wave antenna points vertically relative to the earth, as in Figure 8.40. In this case the electric field of the electromagnetic wave will also be vertical. This orientation is described as vertical polarization. The receiving antenna will be effective when it points parallel to the direction of

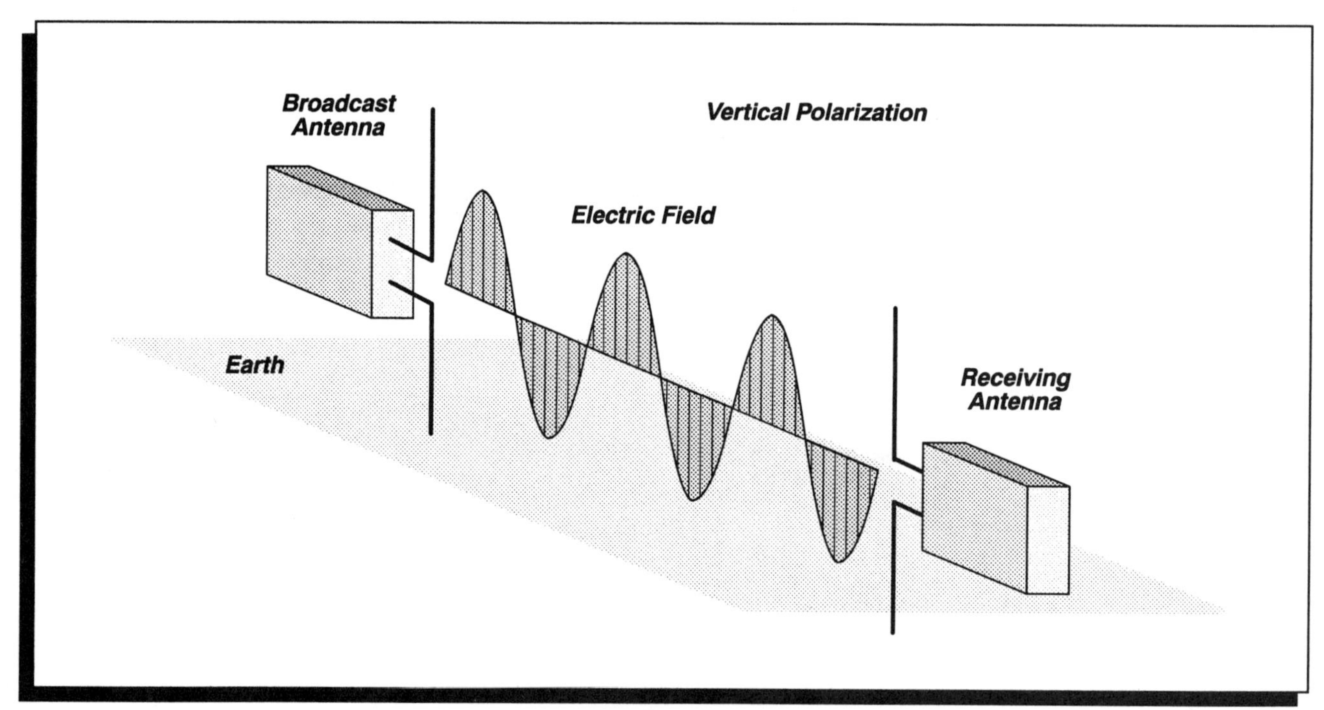

Figure 8.40 Vertically polarized radio wave.

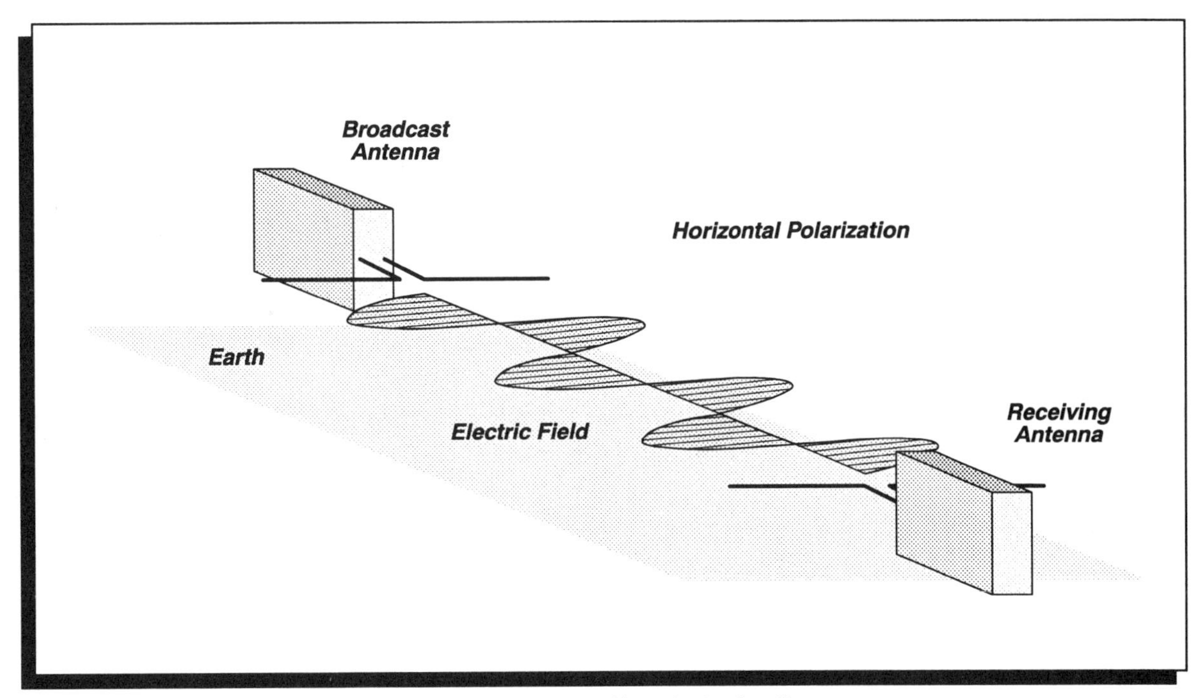

Figure 8.41 Horizontally polarized radio wave.

polarization. Of course, the magnetic field part of the wave is automatically horizontal, since it must be perpendicular to the electric field.

Suppose that a half–wave antenna is horizontal. The oscillating electrons will generate electromagnetic waves where the electric field is also horizontal, relative to the earth. Figure 8.41 shows this. The receiving antenna must then be parallel to this direction since the radio wave is horizontally polarized. The magnetic field will then be vertically oriented.

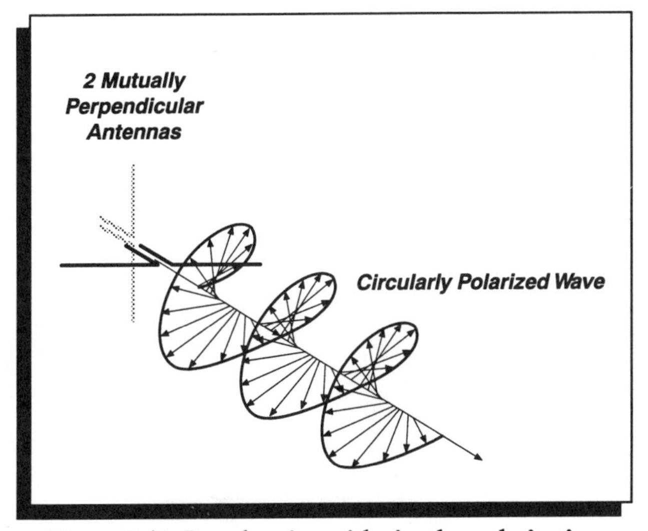

Figure 8.42 Broadcasting with circular polarization.

It is important to know the polarization of the broadcasted wave so that the receiving antenna can be aligned parallel to it for optimal reception.

Some radio stations broadcast with circular polarization, where the electric field (and the magnetic field) continuously rotate relative to the source. Such an approach reduces interference problems due to reflections. This is shown in Figure 8.42.

8.5 WAVE PROPAGATION

Radio waves broadcasted by an antenna can travel along three different routes, depending on the frequency. The three routes are:

1. along the ground; the ground wave.
2. in a straight line; the line–of–sight wave.
3. up to the ionosphere which reflects it back to earth; the sky wave.

■ GROUND WAVE

Amplitude modulated, AM, waves are always broadcast with vertical polarization. They propagate by following the curvature of the earth and

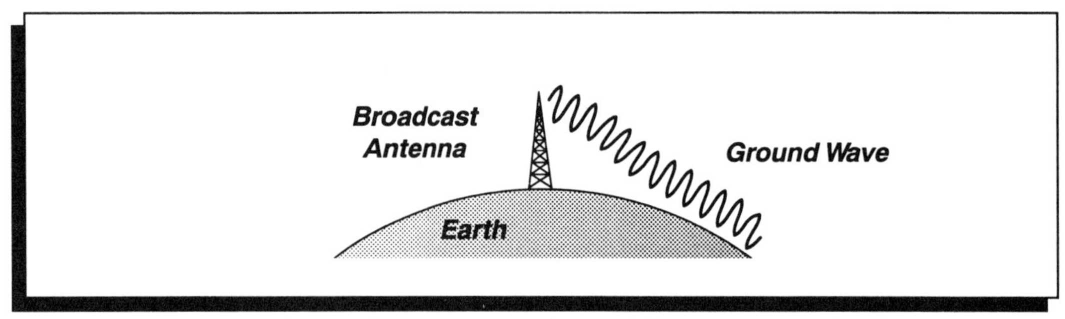

Figure 8.43 Low frequency ground wave follows curvature of earth.

they induce currents in the earth, thereby re–radiating the power (Figure 8.43), as the waves follow the earth's curvature.

Because the earth is a conductor of electricity, the waves will suffer losses and they will be attenuated. The attenuation increases with the frequency, limiting such wave propagation to low frequencies. The range for AM waves is just over 100 miles, depending on the surface conditions. Because of their higher frequencies, FM waves are totally absorbed by the earth in this mode of propagation.

LINE–OF–SIGHT WAVE

In this mode of propagation waves travel directly in an almost straight line from transmitter to the receiver. There are some departures from an exact straight line due to refraction. It is a very important mode of propagation for high frequency waves like FM and TV signals as well as radar and microwave relays. Figure 8.44 illustrates this mode of wave propagation. It is the only mode of propagation for FM and TV.

THE SKY WAVE

At one time this was the only method of communication with the other side of the earth. It is tricky, subject to fading and erratic changes with season, as well as dependent on the varying conditions between night and day. This method relies entirely on the ionosphere surrounding the earth. The ionosphere covers an area of 30 to 300 miles above the earth, where the air is easily ionized by solar and cosmic phenomena. At such distances above the earth, the density of air is very low. Electrons are removed from the outer shells of oxygen and nitrogen molecules and tend to remain for a long time as free electrons and positive ions. Since the atmosphere is not dense, there are very few collisions causing recombinations. The ionosphere affects radio wave propagation. To under-

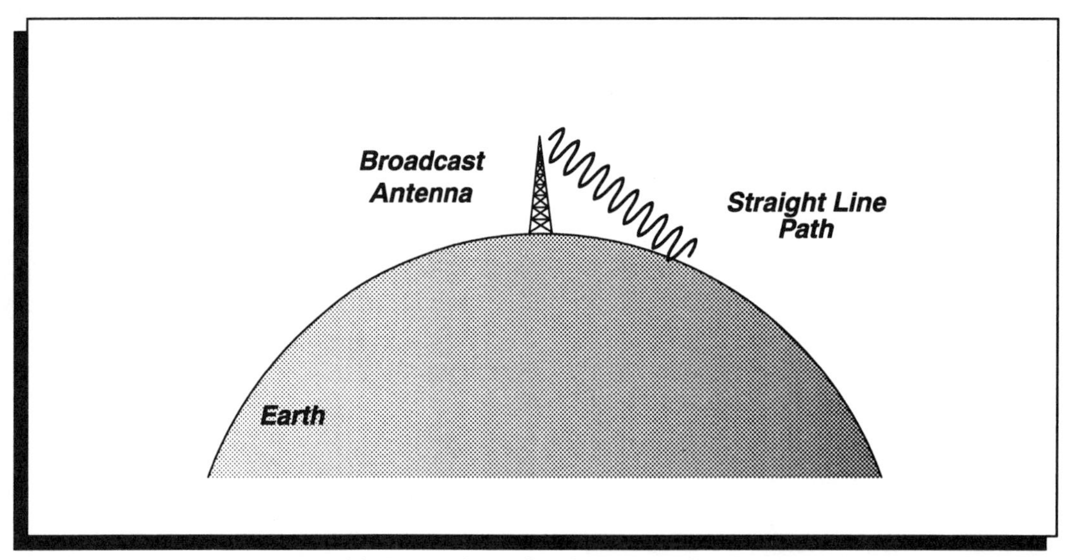

Figure 8.44 Direct (line-of-sight) mode of propagation.

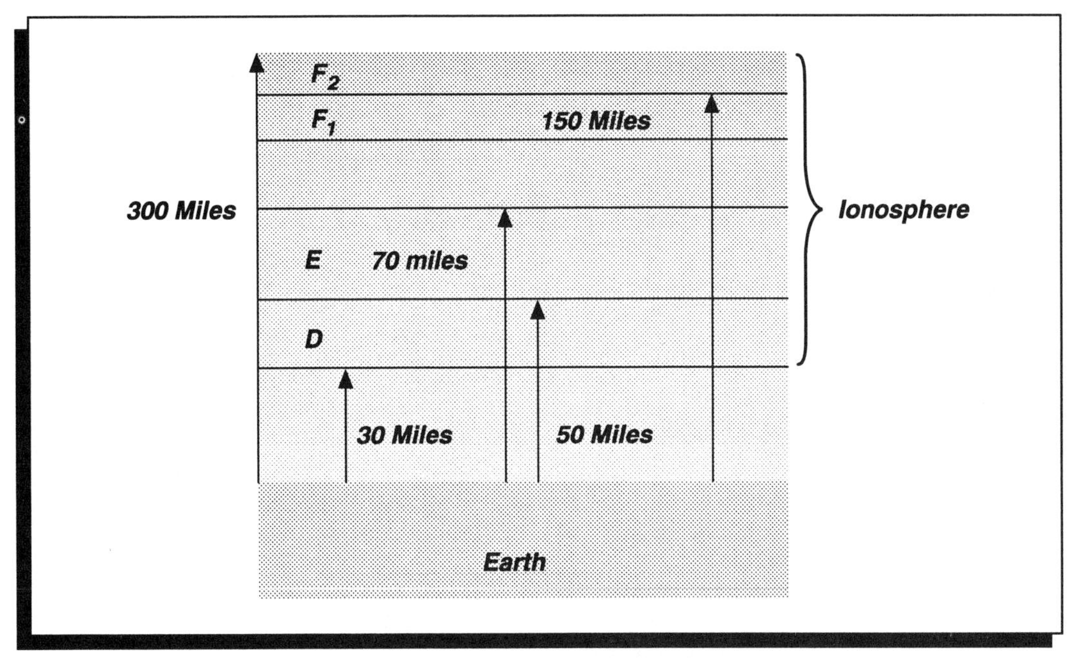

Figure 8.45 Earth's ionosphere layers.

stand how, consider Figure 8.45 which classifies the different layers of the ionosphere. Usually it is divided into four layers, called D, E, F_1 and F_2, each at a specific height above Earth. The D–layer disappears at night and its ionization achieves maximum intensity at midday. The E–layer is used for radio communications. At night most of it disappears. Reflections from the two F–layers of the ionosphere are the principal methods used for world communication, prior to satellites. This method of propagation is based on the fact that radio waves are refracted by the ionosphere and those that arrive above a certain critical angle are totally internally reflected back to Earth. The ionosphere acts like a mirror and the wave is reflected to some spot on the Earth quite far away from the source. This is shown in Figure 8.46. When the frequency is very high the waves are not reflected back, they continue straight out into space. Sometimes there can be multiple reflections

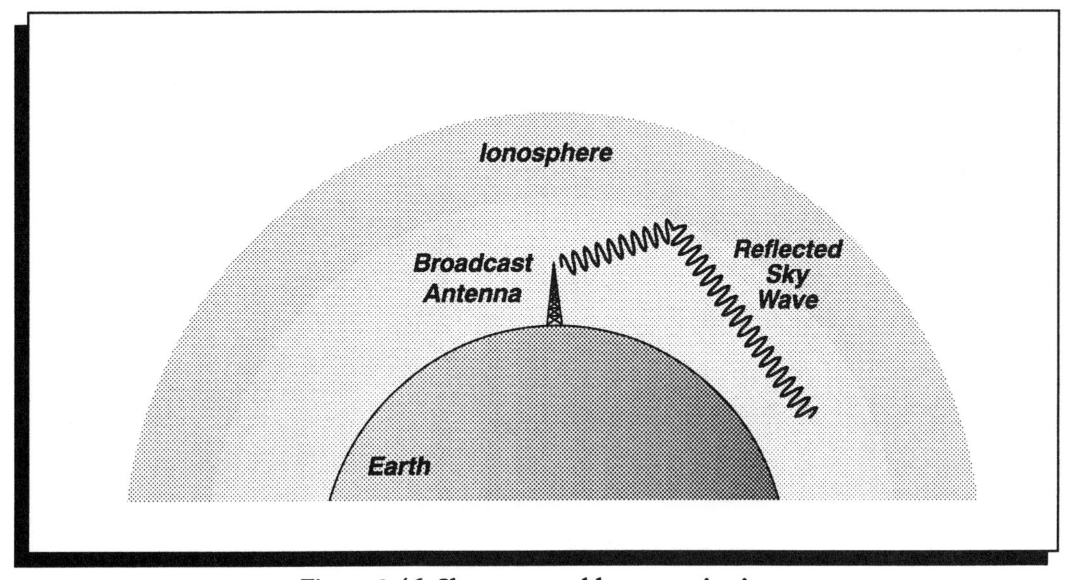

Figure 8.46 Sky wave world communications.

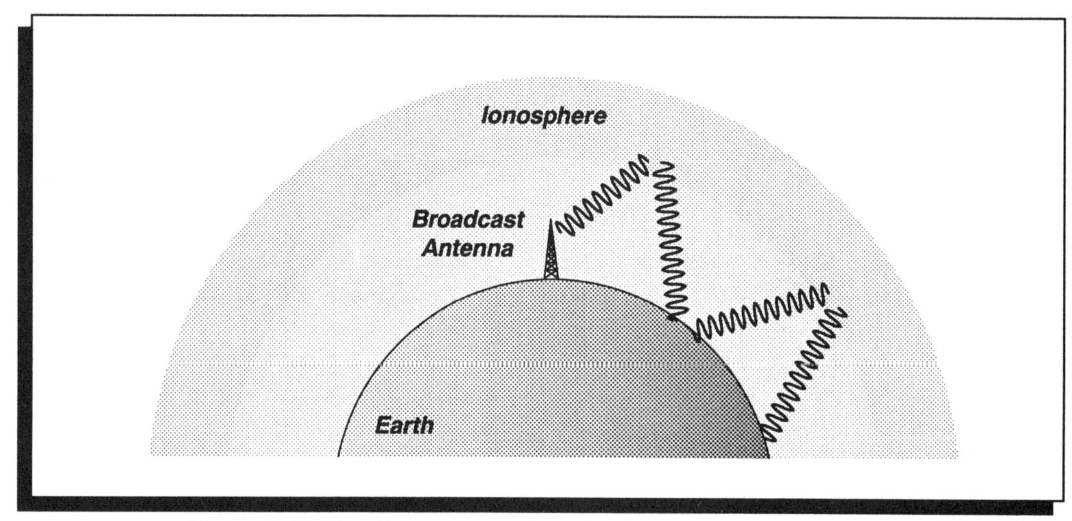

Figure 8.47 Two-hop transmission of radio wave using ionosphere.

where large distances around the Earth can be covered. The most suitable range of frequencies for such communications is the range of 3 to 30 MHz. Figure 8.47 shows a two–hop transmission using the ionosphere.

Satellite Communications

This method relies on sending the signal on a microwave carrier up to a satellite that is in a geostationary position above the equator, and rebroadcasting it to a desired spot on Earth, Figure 8.48. The satellite is placed in an orbit so that its position relative to Earth is always fixed as the Earth rotates. At a height of 35,786 km, the time it takes for the satellite to go around the Earth is 24 hours (actually it is 23 hours, 56 minutes, and 4.1 seconds). Being in a fixed position relative to the Earth, its coverage of the Earth is also fixed, providing a popular form of communication that is used extensively.

8.6 Tuner Specifications

Because AM broadcasts do not broadcast with hi–fidelity, the specifications on tuners presented here will relate only to FM. The specifications presented here for a tuner, possibly similar to those of your tuner, provide an opportunity to

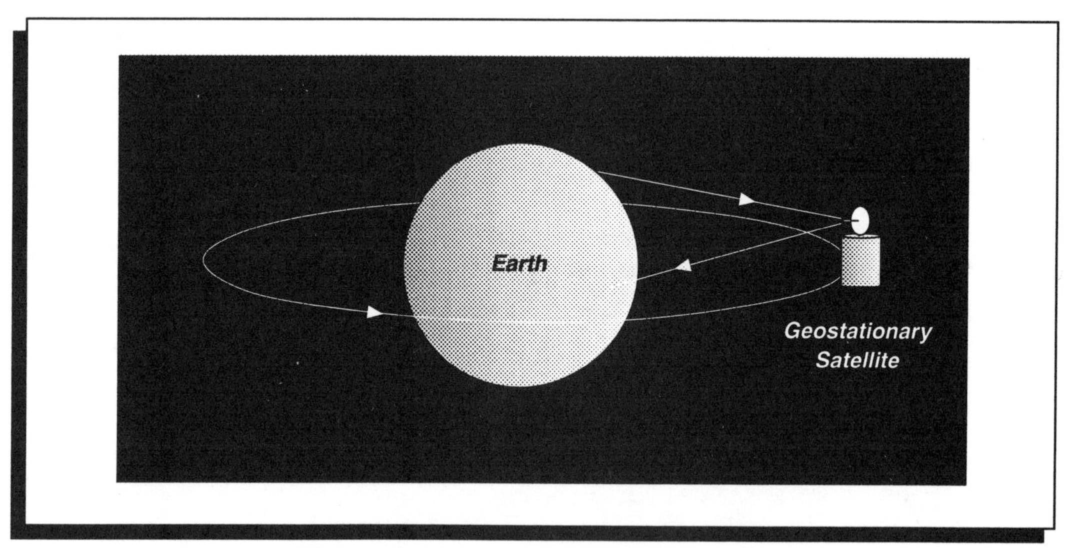

Figure 8.48 Communication using a satellite.

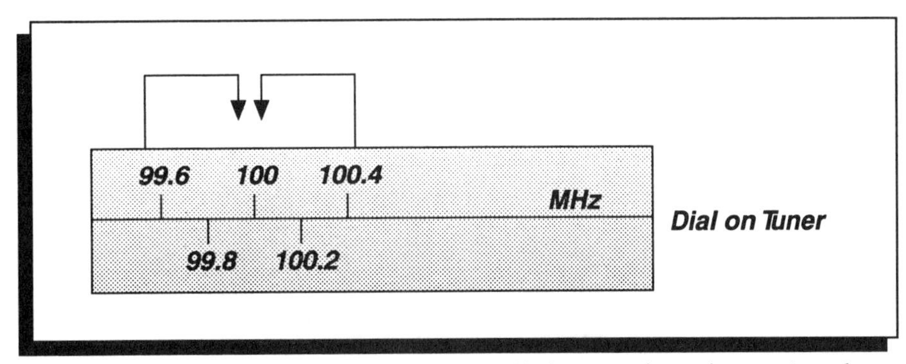

Figure 8.49 Selectivity relates to how well alternate channels are rejected.

learn about what the specifications mean and to think about possible improvements.

■ SENSITIVITY

This is probably the most important specification which relates to the quality of a tuner. Essentially, it tells how weak a radio signal can be detected and still be turned into satisfactory audio information. The specification describing this states how small a signal in microvolts (μ volts) coming out of the antenna will produce an audio signal which is 50dB above the noise and distortion. A higher sensitivity will mean that the signal coming from the antenna is even smaller, but it still can produce an acceptable audio signal. This specification is known as the sensitivity for 50dB quieting. Look for 44 μvolts or less on stereo to produce 50dB of audio above noise and distortion.

The sensitivity specification is especially important when you are located far away from radio stations.

■ SELECTIVITY

This specifies how well the tuner will pick up one station and reject all the others. Figure 8.49 shows how important it is to reject alternate channel stations. When tuned to a station at 100 MHz, there are no stations at 100.2 and 99.8 MHz, but stations at 100.4 and 99.6 MHz, known as alternate channels, could sneak into the 100 MHz information. The rejection of alternate channel stations is referred to in dB, and it means: how many more decibels of signals in the alternate channels must be produced above the present channel level so that the audio of the present channel is 30dB above that for the alternate channels. Look for 60dB or more. This specification is important in a crowded areas where there are many stations.

■ SIGNAL–TO–NOISE RATIO (S/N)

This compares the audio signal to the noise level, in decibels, when the radio signal is at a certain level, approximately 1,000 μvolts. Look for a S/N of 60dB or higher in the stereo mode.

For example, when listening to radio sounds at 100dB on a system whose S/N is 65dB, the noise level would then be 100dB – 65dB = 35dB.

■ STEREO SEPARATION

It is similar to the same specification on amplifiers. It relates as to how well one channel is kept isolated from the other one. Look for 30dB or larger. Up to now the FCC required that radio stations broadcast with a minimum separation of 30dB. Any improvement in the tuner will not improve the performance of stereo separation, at this point.

■ TOTAL HARMONIC DISTORTION

It tells how much the original signal is distorted, thus creating new unwanted harmonics. It is expressed in a percentage, and one should look for 1% or less on stereo.

■ CAPTURE RATIO

It is the ability of the tuner to reject unwanted signals coming in at the same frequency from the station one is trying to listen to. The main source of problems here is that when a station broadcasts, part of the signal will arrive directly to the tuner,

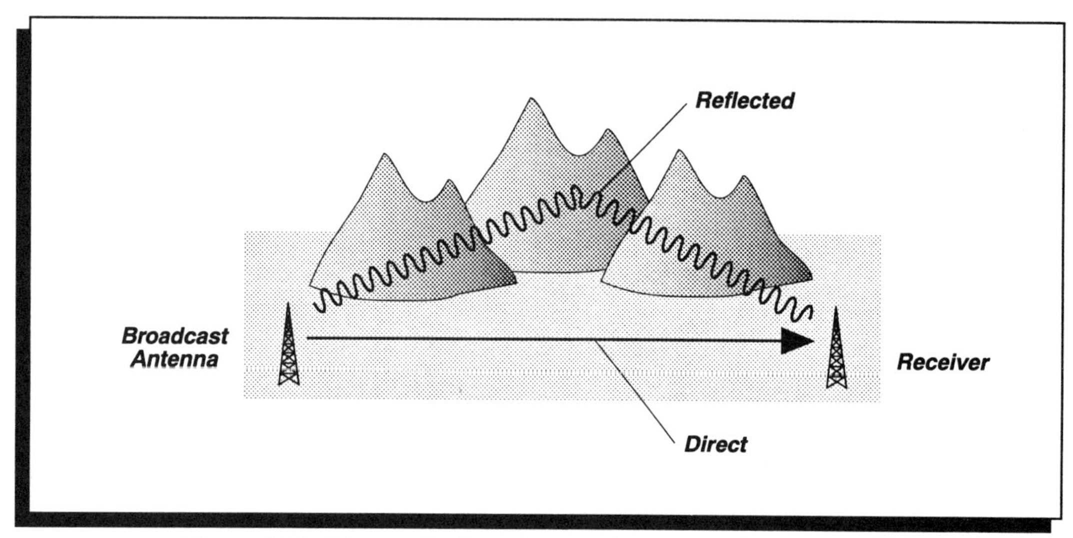

Figure 8.50 Direct and reflected waves from a broadcasting station.

while some of it is reflected by buildings, mountains, or bridges. When the reflected signal arrives to the tuner slightly after the direct one, the delay causes some fuzziness in the music. This is similar to "ghosts" on TV. Figure 8.50 illustrates this problem of multipath interference. A tuner has the ability to reject the weaker of two signals coming in at the same frequency. The smaller the difference between the two signals, expressed in dB, means that the tuner is better than one which would require a larger difference between the two signals. Look for a capture ratio of 3dB or less. Figure 8.51 shows how the weaker signal gets rejected.

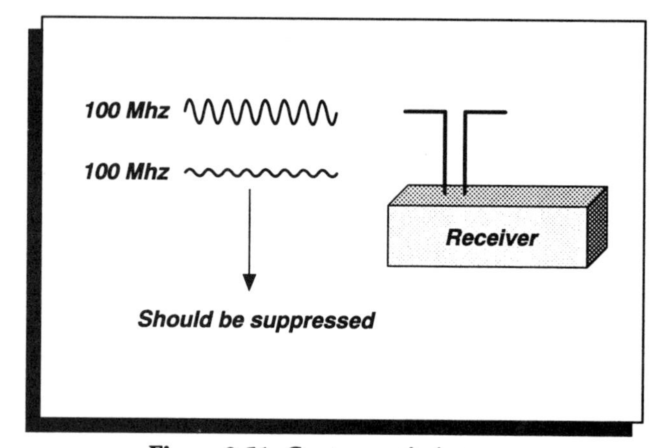

Figure 8.51 Capture ratio in tuner.

■ Frequency Response

FM stations deal with audio frequencies of 30 Hz to 15,000 Hz and this is considered high–fidelity. Look for a frequency response of 30 Hz to 15,000 Hz ± 1dB or a smaller ± dB.

Summary of Terms

Amplitude Modulation: A method of making changes on a carrier so that audio information is carried by the carrier. In this method of modulation, the amplitude of the carrier is changed according to the pattern of audio information.

Antenna: a device which converts radio frequency currents along a wire into electromagnetic waves and vice versa.

Audio signal: Representation of a sound wave as an electrical signal at frequencies ranging from 20 Hz to 20,000 Hz.

Capture Ratio: Ability to reject the weaker of two signals at the same carrier frequency arriving in a tuner. This is usually caused by reflections off buildings, bridges, and mountains.

Carrier: Electromagnetic wave which will carry audio information. It is at the frequency of the broadcasting station.

Fifty (50) dB Quieting Sensitivity: Indicates how well a tuner can convert a radio wave signal into sound. It refers to how small a signal from the antenna can produce an audio signal 50dB above noise and distortion.

Frequency Modulation: A method of making changes on a carrier so that audio information is carried by the carrier. In this method of modulation, the change in the carrier frequency corresponds to the audio information.

Horizontal Polarization: Electric field of radio wave is horizontal.

Limiter: Part of the circuit in an FM tuner which suppresses any amplitude variations of the signal.

Multipath Interference: Unwanted reflected radio waves which have been reflected by mountains, buildings, and bridges.

Radio Frequency Signal: Electromagnetic wave which is in the frequency range between 535 kHz and 1605 kHz for AM and 88 MHz and 108 MHz for FM as used in tuners.

Subcarrier: A 38 kHz signal in a studio broadcasting in stereo. It carries the LEFT minus RIGHT channel information before being put on the main carrier for broadcasting.

Tuner: A hi–fi component which converts radio frequency currents to an audio signal.

Vertical Polarization: electric field of radio wave is vertical.

NAME ______________________ DATE ____________

QUESTIONS FOR REVIEW

1 How does amplitude modulation differ from frequency modulation?

2 Explain how an AM signal is broadcast.

3 How does a light wave differ from a radio wave?

4 Explain how a superheterodyne receiver works.

5 How many octaves of audio does AM broadcast? FM?

6 What is an electromagnetic wave and how is it produced?

7 Explain why there has to be frequency spacing between radio stations. What determines this?

8 How are the amplitudes and frequencies of audio signals conveyed by frequency modulation?

9 How are the amplitudes and frequencies of audio signals conveyed by amplitude modulation?

10 Explain how an antenna works.

11 How would you produce a circularly polarized radio wave?

NAME____ **DATE** ____________

Please select one answer.

1. The ability of a tuner to pick up very weak radio signals and convert them to listenable sound is called:
   ___A. sensitivity
   ___B. selectivity
   ___C. frequency response
   ___D. S/N ratio
   ___E. THD distortion

2. Typical voltages that an antenna feeds to a tuner are between:
   ___A. 1mV and 1,000mV
   ___B. 1 and 10 volts
   ___C. 1 μvolt to 1,000 μvolts
   ___D. 50dB
   ___E. 120 volts

3. Which of the following broadcasts with the shortest wavelength carrier?
   ___A. AM
   ___B. FM
   ___C. IF
   ___D. 15 kHz
   ___E. 455 kHz

4. If you live near mountains and your tuner is picking up a lot of reflected "ghosts" at FM frequencies, you would want a tuner with a ____________ rating.
   ___A. large selectivity
   ___B. large sensitivity
   ___C. small capture ratio
   ___D. large capture ratio
   ___E. good frequency response

5. Your tuner has a S/N ratio of 70dB. This means that considering only the tuner, in your system, the audio output is __________ times louder than the noise.
   ___A. 128
   ___B. 70
   ___C. 64
   ___D. 10^7
   ___E. 10^{-7}

6. If you live far from radio stations, you should buy a tuner with a good ___________ .
   ___A. capture ratio
   ___B. sensitivity
   ___C. selectivity
   ___D. stereo separation
   ___E. frequency response

7. The specification for tuners which tells how well they reject the weaker of two signals arriving at the same frequency is called the:
   ___A. capture ratio
   ___B. IF
   ___C. selectivity
   ___D. sensitivity
   ___E. distortion

8. Which are the audio frequencies transmitted by AM?
   ___A. 30Hz – 15 kHz
   ___B. 200 Hz – 5 kHz
   ___C. 20 Hz – 20 kHz
   ___D. 535 kHz – 1605 kHz
   ___E. 20 Hz – 10 kHz

9. The range of carrier frequencies for FM are:
   A. 88 MHz to 108 MHz
   B. 535 kHz to 1.605 MHz
   C. 30 Hz to 15 kHz
   D. 455 kHz to 10.7 MHz
   E. 20 Hz to 20 kHz

10. The wavelength of a typical AM carrier is:
   ___A. 10 times longer than that of FM radio waves
   ___B. 100 times longer than that of FM radio waves
   ___C. 100 times shorter than that of FM radio waves
   ___D. 10^6 times shorter than that of light waves
   ___E. the same as that of FM radio waves

11. When a radio station sends out an FM broadcast it ___________ .
   ___A. pre–emphasizes the audio low frequencies
   ___B. pre–emphasizes the audio high frequencies
   ___C. de–emphasizes the audio high frequencies
   ___D. pre–emphasizes the carrier frequency
   ___E. neither pre–emphasizes nor de–emphasizes the audio signal

12. When a 500 Hz audio signal is broadcast on a carrier by varying the frequency of the carrier 500 times per second, this is known as:
   ___A. AM transmission
   ___B. FM transmission
   ___C. flat frequency response transmission
   ___D. 500 Hz carrier wave
   ___E. intermediate frequency modulation

13. When broadcasting audio information by FM, if the amplitude of the audio increases, the:
   ___A. carrier will have a larger amplitude change
   ___B. carrier will have a smaller amplitude change
   ___C. carrier will have a larger shift in frequency
   ___D. carrier will have a smaller shift in frequency
   ___E. carrier frequency will not change

14. When a superheterodyne tuner is picking up a radio wave at 102 MHz, the frequency of the internal oscillator must be at:
   ___A. 102 MHz
   ___B. 10.7 MHz
   ___C. 112.7 MHz
   ___D. 455 kHz
   ___E. 102.455 MHz

15. FM radio stations do not broadcast audio frequencies above:
   ___A. 15 kHz
   ___B. 5 kHz
   ___C. 1 kHz
   ___D. 18 kHz
   ___E. 10 kHz

16. To broadcast a 1,000 Hz audio signal, the amplitude of the radio wave is changed 1,000 times per second. This type of broadcasting refers to:
   ___A. FM
   ___B. AM
   ___C. IM distortion
   ___D. THD
   ___E. horizontal polarization modulation

17. For FM radio broadcasts, the higher the frequency of the audio signal, the:
   ___A. more times per second the frequency of the carrier must change
   ___B. fewer times per second the frequency of the carrier must change
   ___C. more times per second the amplitude of the carrier must change
   ___D. fewer times per second the amplitude of the carrier must change
   ___E. fewer times per second the intermediate frequency must change

18. While listening to an FM broadcast at a level of 100dB, the noise level will be at ___________ for a tuner S/N of 65dB.
   ___A. 65dB
   ___B. 165dB
   ___C. 35dB
   ___D. 100dB
   ___E. 50dB

NAME ______________________________ DATE ____________

19. The tuner specification which expresses by how many decibels larger the signal is than the noise is called:
___A. S/N ratio
___B. N/S ratio
___C. selectivity
___D. capture ratio
___E. THD

20. When trying to capture a radio station, the internal oscillator inside your tuner is putting out 1.955 MHz. The frequency of the station that you are catching is:
___A. 1,500 kHz
___B. 12.655 MHz
___C. 445 kHz
___D. 10.7 MHz
___E. 8.745 MHz

21. How long should a half–wave dipole antenna be for a radio wave at 100 MHz?
___A. 3 meters
___B. 1.5 meters
___C. 100 meters
___D. 6 meters
___E. 3×10^6 meters

22. A half–wave dipole antenna for AM radio waves should be about ___________ .
___A. 100 times shorter than for FM radio waves
___B. 100 times longer than for FM radio waves
___C. 10 times shorter than for FM radio waves
___D. 10 times longer than for FM radio waves
___E. 100 times longer than for the highest audio frequency

23. A 500 Hz audio signal amplitude modulates a carrier at 1 MHz. During one wave of the audio signal, how many waves of the carrier will there be?
___A. 2,000
___B. 1,000,000
___C. 500
___D. 0.002
___E. 4,000

24. The AM antenna on a car points vertically because:
___A. broadcasted AM waves have vertical polarization
___B. broadcasted AM waves have horizontal polarization
___C. broadcasted AM waves have magnetic fields which are vertical
___D. AM waves are always broadcast with horizontal polarization
___E. its wavelength is so long

25. What is the wavelength of the broadcasted wave by an antenna if it takes electrons 10^{-8} second to move up and down the antenna?
___A. 10^8 meters
___B. 10^{-8} meters
___C. 3 meters
___D. 0.3 meter
___E. 30 meters

CHAPTER 9

Analog Recording and Playback

An important feature of a hi–fi system is that musical information can be stored and then played back with excellent tonal qualities at some convenient time. In this chapter we will look into two very important methods of recording and playback, the phono disc and magnetic tapes. On a phono disc the audio information is stored mechanically in the form of grooves cut in a plastic disc. This form of storage and reproduction has had a long–lived success, and it was not until recently that the digital approach with CDs replaced it. In magnetic recording, very small magnetic particles on a tape are aligned according to the audio signals. Both methods have a common feature; the recording is almost an identical replica, mechanical or magnetic, of the original sound waveform. This is the analog approach.

9.1 Phono Records

In 1888, the flat phonograph record was introduced for the gramophone. For more than 100 years this system was one of the most important methods of storing and playing back music. Its success and development show how the field of hi–fi has evolved. Although this topic is a closed chapter in the history of hi–fi, it is briefly presented here as a demonstration of the physics principles involved.

The principle of the phono record is based on cutting a groove into a plastic record in such a man-

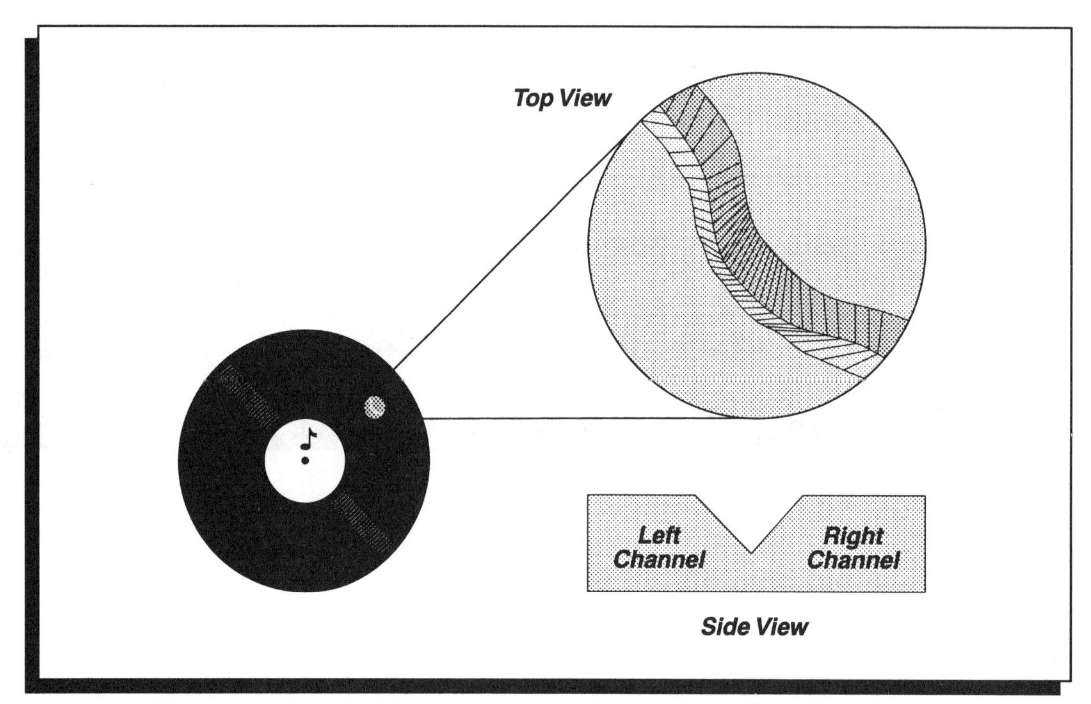

Figure 9.1 Record with grooves representing mechanically engraved waves.

ner that the groove represents the wave that it recorded. This is shown in Figure 9.1. The groove is for a mono signal. In fact, the groove should be a close replica of the complex wave representing the sound that is recorded. The grooves spiral inward toward the center of the record. The properties of a wave are well represented since the amplitude of the wave corresponds to the lateral displacement of the groove. The wavelength corresponds to the separation of the lateral displacement along the groove. For stereo recording each side of the groove corresponds to a channel, Figure 9.1.

To retrieve the signal from the groove, the record is made to rotate on a turntable, and a fine stylus follows the undulations of the grooves. The oscillations of the stylus have to be converted into electrical signals, which represent the waves in the grooves. There are two types of playback systems: a piezoelectric cartridge and a magnetic cartridge, both shown in Figure 9.2. As the stylus moves in the groove, it stresses a piezoelectric block and thus it develops an electrical signal. Remember that with piezoelectricity a stress across the material develops an electrical signal. This type of cartridge was popular in the early days of the phono records, but it was replaced by the moving magnet type of

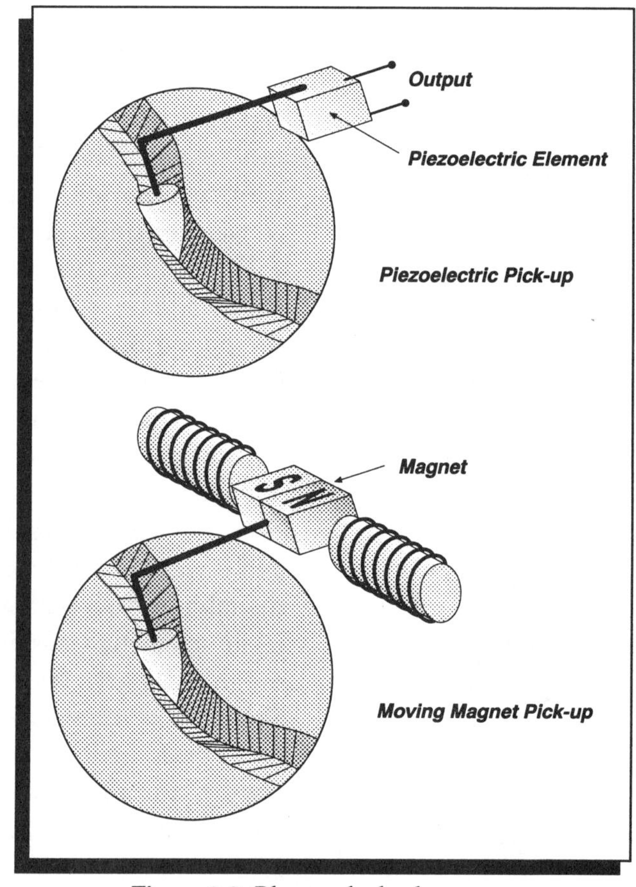

Figure 9.2 Phono playback systems.

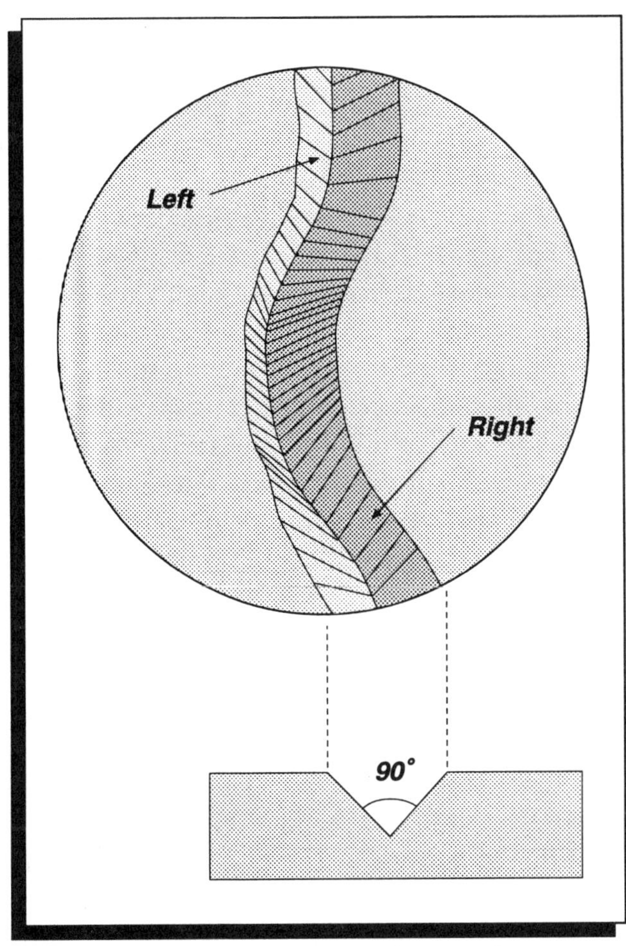

Figure 9.3 Stereo with only one stylus.

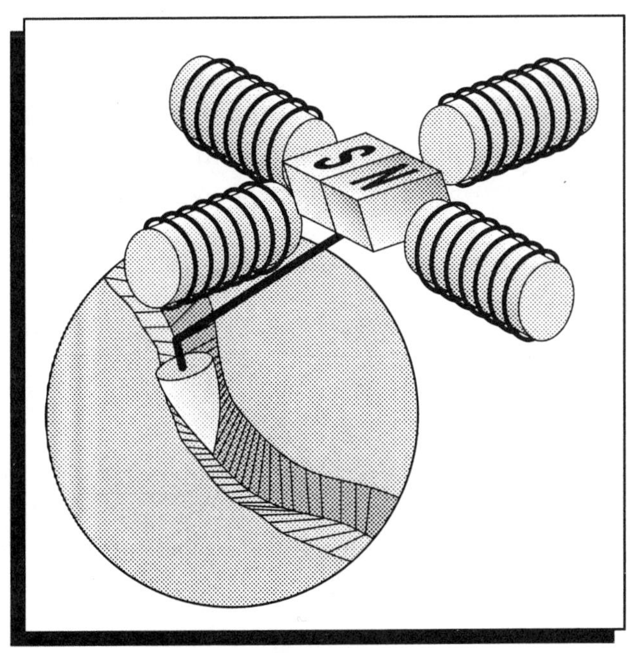

Figure 9.4 A stereo moving magnet phono cartridge.

cartridge, which gives a superior performance level. It is based on Faraday's law of induction. As the stylus moves, it causes a small bar magnet to oscillate and induces an electric voltage in a set of coils placed near the oscillating bar magnet. The frequency of oscillation of the stylus can be calculated once the stylus speed and the wavelength are known. It is given by:

$$\textit{Frequency} = \frac{\textbf{speed at which groove goes by stylus}}{\textbf{wavelength}}$$

The speed of the groove past the stylus is different for different parts of the record. For a 33⅓ revolutions/minute long–play record, the speed of the groove is typically 49 cm/sec at its edge. When the wavelength of the recorded information is 0.049 cm, the frequency will then be (49 cm/sec) / 0.049 cm = 1,000 Hz.

As technology improved it was possible to record stereo information on one groove. This information could then be retrieved in stereo with only one stylus. Figure 9.3 illustrates how this is possible. The groove is cut so that the sides are at 90° to each other. For example, information for one channel is stored in the LEFT wall of the groove, and the other channel in the RIGHT wall of the groove. The stylus is designed for this mode of operation. This is shown in Figure 9.4, a single stylus can read both the LEFT and RIGHT channels. The information on the left wall of the groove gives information for that channel by exerting a force on the stylus, likewise information on the right wall of the groove exerts a force on the stylus. In effect the groove force on the stylus make it move in all directions, inducing signals in the LEFT channel coil and in the RIGHT channel coil; this provides a stereo playback.

The main problems associated with the phono record system were:

— friction between stylus and groove caused the grooves to get worn out quickly and loose their high frequency information.
— grooves would get easily scratched. In the playback there was no way of telling the dif-

ference between the music and the scratches. That is a weakness of the analog recording, as the noise cannot be distinguished from the signal.

— dynamic range for recorded music was limited. It was impossible to playback very loud sounds since the stylus could not follow large groove variations. At very low levels, noise from scratches, dirt, and dust was the limiting factor in sound reproduction.

In 1983 the introduction of the CD brought an end to the long and popular era of the phono record.

9.2 Magnetic Recording

Magnetic recording was discovered in 1898 when Poulsen magnetized a piano wire by an electromagnet that responded to his voice. By passing the wire with the recorded information near a suitable detector, his voice could be reproduced. Developments in this area brought fine tapes with many channels of information that could be stored on them. In this chapter we will only discuss the analog aspect of magnetic recording, concentrating on the physical principles, the achievements, and the limitations. To do this, we go back to the fundamentals of magnetism.

■ Magnetism

Some atoms, like iron, are magnetic because a few electrons in their outer orbits do not cancel their magnetism which comes from their spinning behavior. In iron, four such electrons align to cause the atom to be magnetic and to behave like a small bar magnet. When a solid is formed out of magnetic material, the atoms often align parallel to each other over relatively large distances, forming magnetic domains. Usually the domains in a solid tend to be random in direction, leaving the material unmagnetized. This is shown in Figure 9.5. When the material is aligned, all the domains point in one direction and the magnetization is then non zero. The magnetization is a measure of how many domains are aligned in one direction. It is curious that once aligned, the domains (or most of them) remain aligned. We will use this property for magnetic recording, since it can be used for memory storage.

Let us look in detail at the process of magnetization of a magnetic material. Consider a coil by itself; when a current flows through it, a magnetic field is produced, Figure 9.6. The result is presented on a graph of the resultant magnetic field as a function of current in the coil. As the current increases, the magnetic field also increases. When the current is reduced to zero, the magnetic field will also become zero.

Now, a magnetic material like iron is introduced into the coil. We want to study its behavior when a current is passed through the coil. As the current increases from zero, the coil magnetic field will tend to align the magnetic domains. However, the domains are pinned to their positions and do not want to move. The current in the coil is further

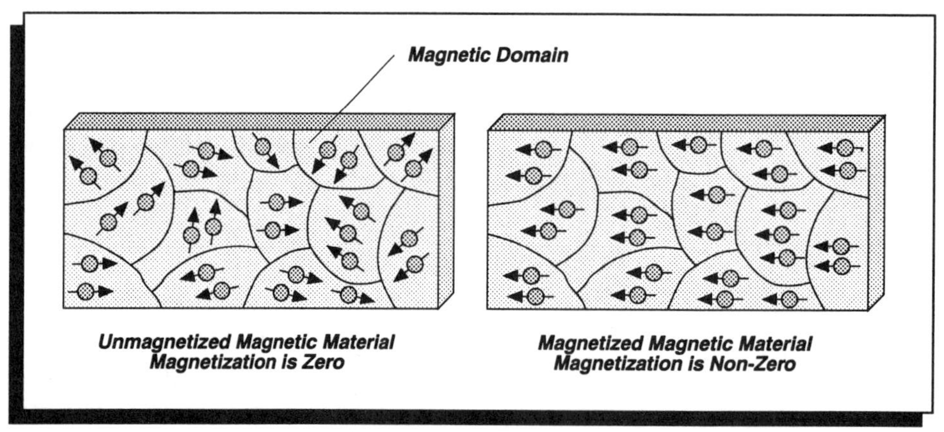

Figure 9.5 Unmagnetized and magnetized magnetic material.

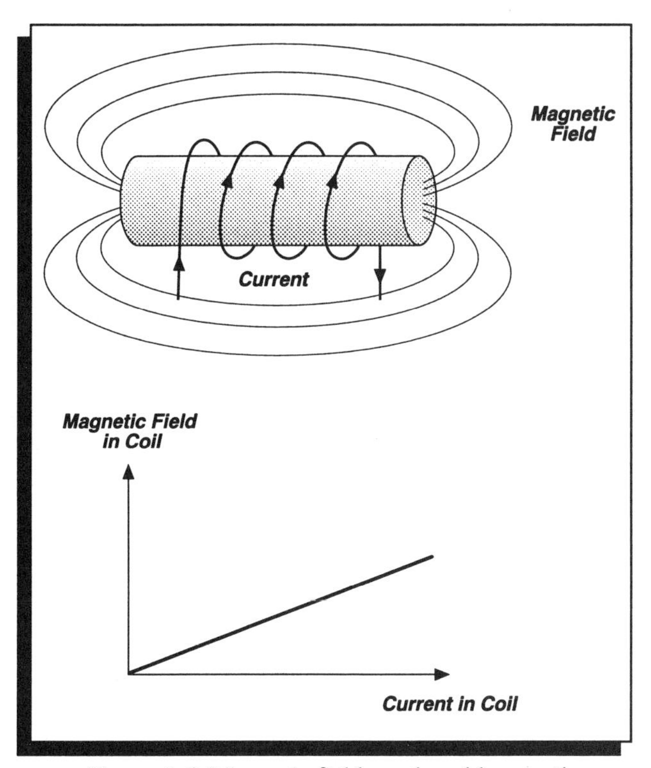

Figure 9.6 Magnetic field produced by a coil when current flows through it.

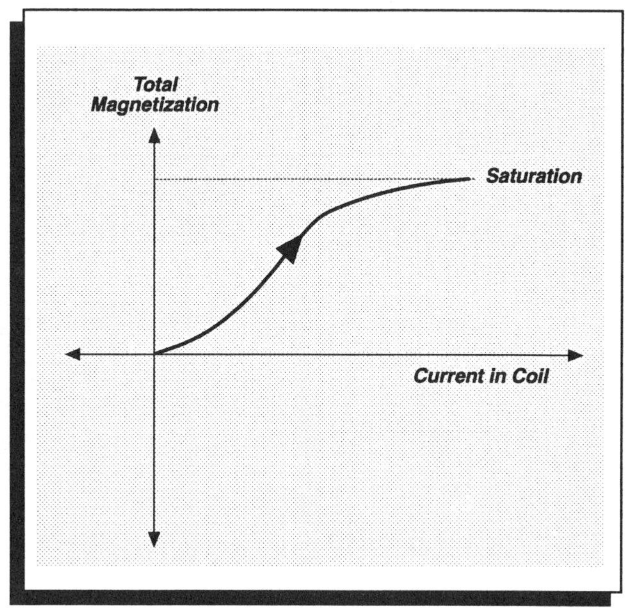

Figure 9.7 Behavior of magnetic material in a coil whose current is increased until saturation occurs.

increased, and eventually the domains will begin to align. This now causes an increase in the magnetization of the sample. As the current in the coil is further increased there will be more magnetic domains becoming aligned, until eventually they all get aligned. This state is known as saturation. Further increase in coil current will lead to almost no further increase in the magnetization – all the domains are almost aligned. This is presented in Figure 9.7.

Now slowly reduce the current in the coil. The magnetization will trace a different curve, Figure 9.8. With no current in the coil, most of the domains will remain aligned. This is not surprising, since that is how a bar magnet is made. When the magnetic field of the coil is returned to zero, there still remains a magnetization due to the domain alignment. The amount that still remains is called the retentivity.

***Retentivity* = Amount of magnetism retained by sample after magnetic field of external coil is reduced to zero.**

The retentivity acts like a memory element. That is, how much magnetism the sample or tape retains.

In order to destroy the magnetism which has been retained, one could simply apply a magnetic field in the opposite direction, Figure 9.9. Hence, pass a current in the coil in the opposite direction until that field flips all the domains in a random configuration where there is no magnetization. The amount of magnetic field in the coil necessary to destroy the memory is called coercivity.

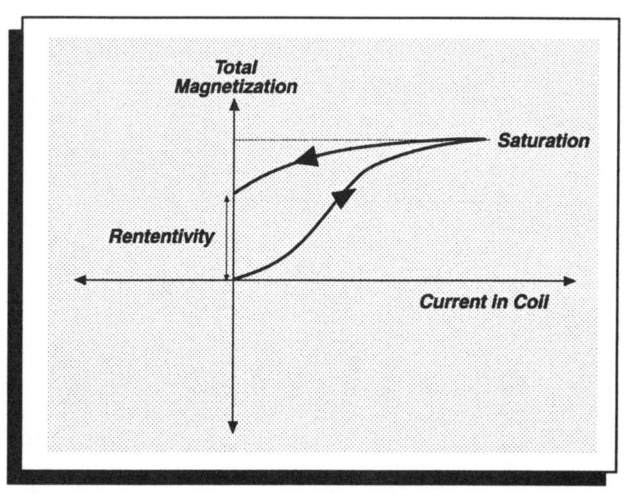

Figure 9.8 Behavior of magnetic material in a coil whose current is increased and then decreased to zero.

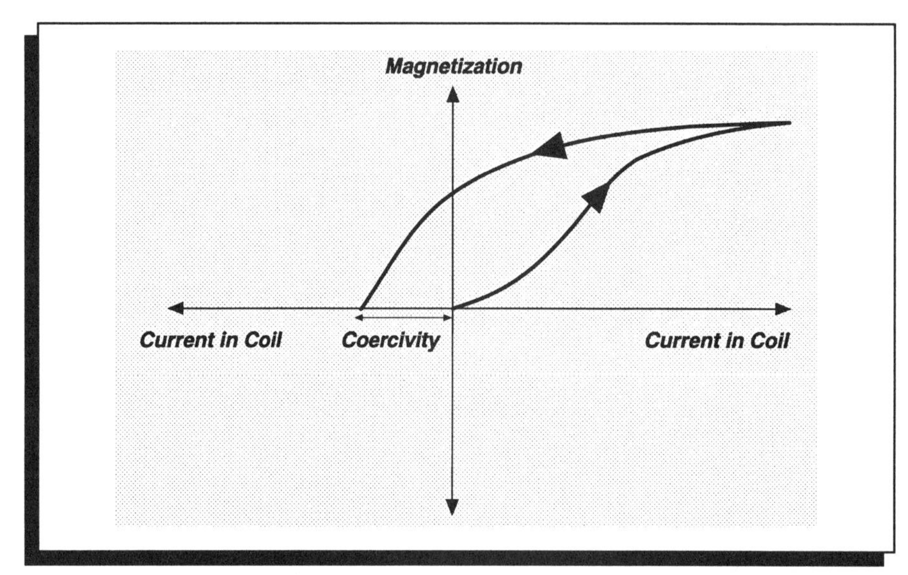

Figure 9.9 Memory is destroyed by reversing current in coil above critical value.

Coercivity **= Magnetic field needed to reduce magnetization of material to zero.**

To create a magnetization in this direction it will require an even larger current in the coil. This will again align the magnetic domains, now in a direction, opposite to what it was before, until saturation occurs in that direction. The whole cycle repeats itself again. The curve shown in Figure 9.10 describing the whole situation is called hysteresis. Therefore:

Hysteresis **= Non–linear behavior of magnetic material in a magnetic field.**

In summary, retentivity is related to the amount of memory, and coercivity is related to how easy (or difficult) it is to erase it. The sample has to be coerced to give up its magnetism.

There are two groups of magnetic materials and they are shown in Figure 9.11. There are "soft" magnetic materials and "hard" magnetic materials. Soft materials have a very narrow hysteresis curve

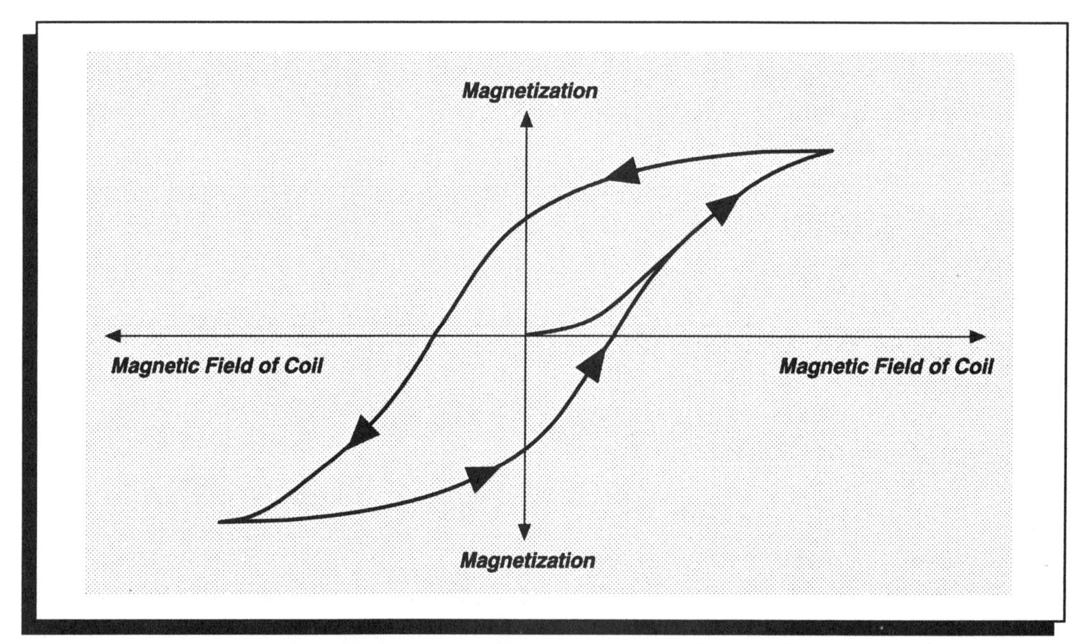

Figure 9.10 Hysteresis curve of magnetic material.

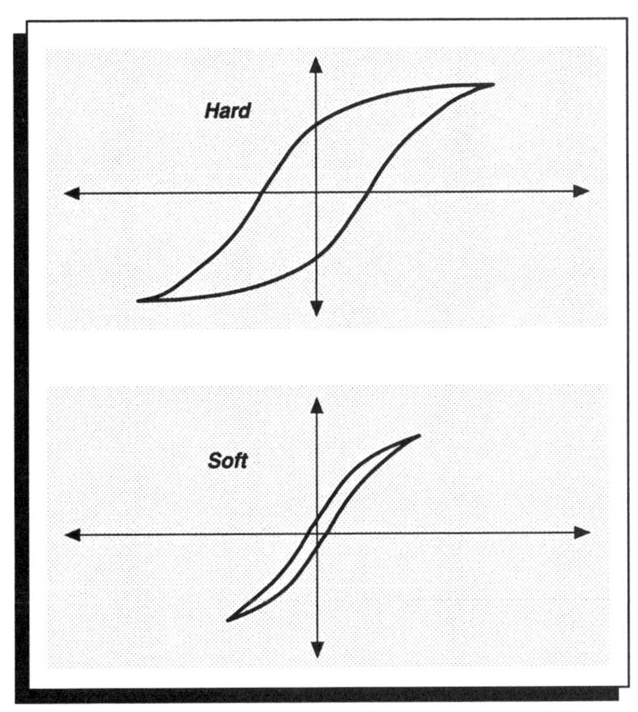

Figure 9.11 Broad groups of magnetic materials.

with essentially almost no retentivity and coercivity. The hard materials have large a retentivity and coercivity. The application will dictate which group is most suitable. The characteristics of the materials will also depend on impurity content, dislocations, and heat treatment. Some of these prevent the magnetic domains from moving easily.

■ TAPES

Tape is a polyester film that has fine magnetic particles (such as Fe_2O_3, Fe_3O_4, CrO_2, etc.) attached on one side. The principle is that a magnetic field can align the domains in the magnetic particles. Since the magnetic particles have retentivity, they will retain their alignment, and the memory element of a tape will allow the music to remain recorded. The magnetic material is prepared so that it has sufficient coercivity to avoid erasure due to stray magnetic fields. In the playback, the tape will move under a specially designed head, where the changing magnetic field of the moving magnetic domains will induce a signal, according to Faraday's law of induction. This information can then be erased by applying a variable magnetic field larger than the coercivity of the tape to randomize the magnetic domains.

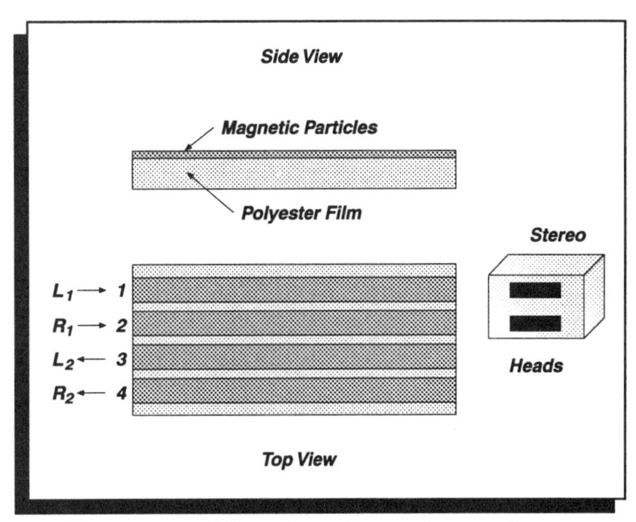

Figure 9.12 Side and top views of magnetic tape.

Figure 9.12 shows a magnetic tape, a side view as well as a top view with the arrangement of tracks on it.

The standard tape is 5/32" wide for regular audio tape decks, and comes in different lengths and consequently different thicknesses. A tape with an indication of C60 means that there is a total of 60 minutes of storage time, 30 minutes for each side. One side has tracks 1 and 2 for the recording and playback of information, and when that side is completed, the tape can be played backwards to record or playback information on tracks 3 and 4. In fact the tape is like a highway going two ways; information is never recorded on the underside of the tape.

It is important to note that the magnetic particles on the tape do not move or rotate. They are stuck with adhesive or plastic to the tape. The motion in a magnetic field refers to the alignment of magnetic domains within the particles themselves.

The magnetic particles are in the shape of short needles, as shown in Figure 9.13. Typical dimen-

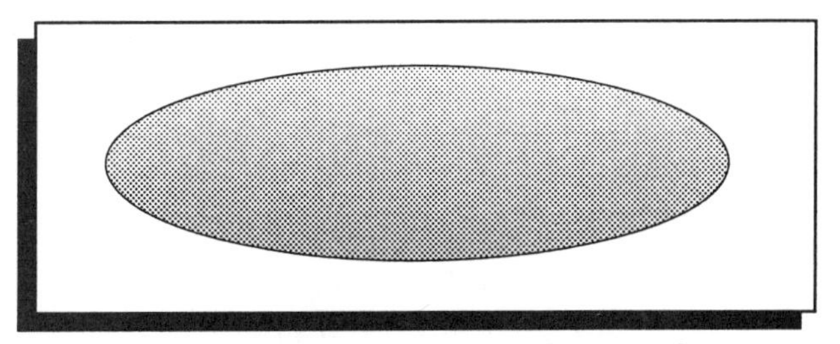

Figure 9.13 Magnetic particle of gamma-ferric oxide (g–Fe_2O_3) used on tapes.

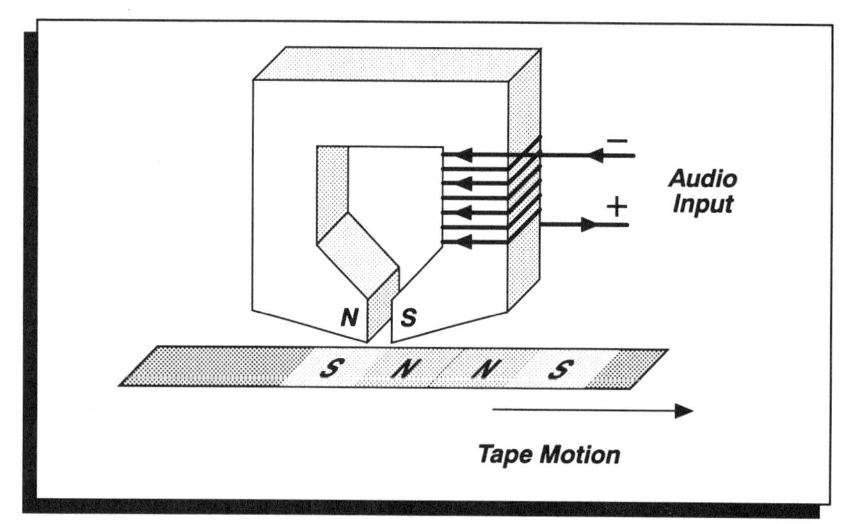

Figure 9.14 Recording head aligning magnetic domains on tape.

sions for magnetic particles that are commonly used, such as gamma–ferric oxide ($\gamma\text{–}Fe_2O_3$) are 0.25 to 0.75 μm length; 0.05 to 0.15 μm wide; and a length to width ratio of 5–10:1.

Its coercivity is 300 Oersteds (a unit of magnetic field), and it stops being ferromagnetic at temperatures above 600° C. This critical temperature, above which a material ceases to have domains is called the Curie temperature, T_C .

Tapes are also made with cobalt–modified $\gamma\text{–}Fe_2O_3$, with Chromium Dioxide and with metallic Iron Particles. Their magnetic properties are different and the specific application will dictate which is the ideal magnetic material. For instance, the coercivity of Chromium Dioxide (CrO_2) is 450 Oe, while for metallic iron particles it is 1450 Oe. This means that it will be easier to erase $\gamma\text{–}Fe_2O_3$ tape than tape with CrO_2 or with metallic iron.

Recording and Playback

In order to record on a magnetic tape, a magnetic recording head has to be designed to do the following tasks:

— Magnetize magnetic particles on a tape
— Area of tape to be magnetized has to be small so that we can put a lot of information on it
— The head has to impose to the tape the information coming to it, which is in electrical form
— The head should not retain information once it has been recorded on tape

A typical design for a recording head is illustrated in Figure 9.14. This design utilizes the first Left–Hand Rule. The recording head is an electromagnet with a coil, a core, and a very fine gap. When an electrical current flows in the coil of the head, a magnetic field is produced which is concentrated at the gap. It can magnetize the magnetic particles when the tape is close enough to the gap.

According to the first Left–Hand Rule, the left pole will be a north pole for the current as shown. Hence the right pole will be the south pole. This will induce on the tape a north pole on the right and a south pole to the left (unlike poles attract). What is the wavelength of the information recorded?

For a typical tape deck speed of 1⅞ inches per second (i.p.s.), when the frequency is 1,000 Hz, the wavelength recorded on the tape will be:

$$\begin{aligned} \text{wavelength} &= \text{speed / frequency} \\ &= 1\tfrac{7}{8}\text{ i.p.s. / 1,000 Hz} \\ &= 0.001875\text{ inch} \end{aligned}$$

How does the wave look on the tape? We want to align magnetic domains in such a way that they represent the wave; the simplest approach will be to make it analog, so that it will “look” like the original wave with a continuous alignment of domains. Figure 9.15 shows both a square wave and a sine wave recorded on a tape.

Important features are:

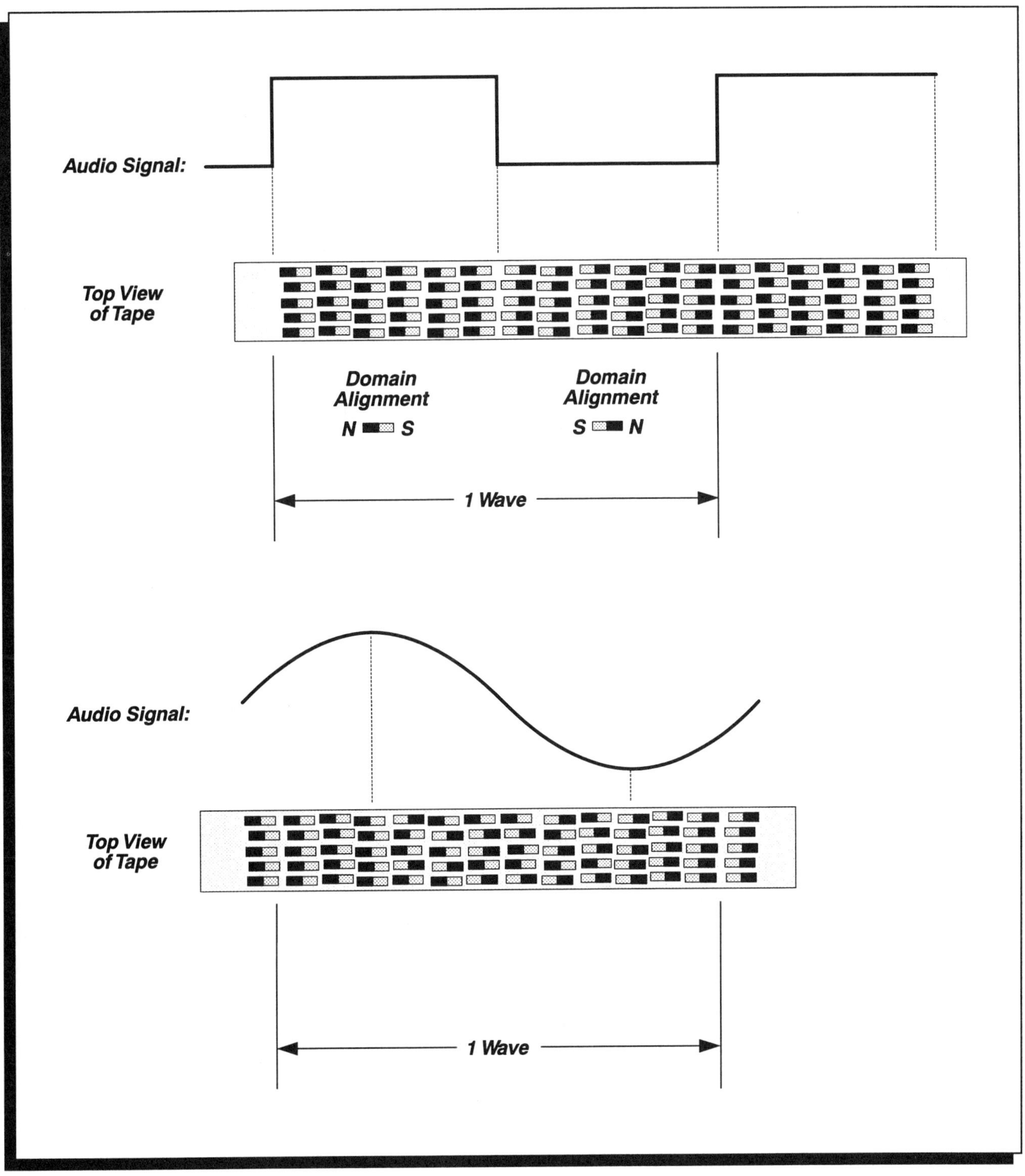

Figure 9.15 Analog recording on a magnetic tape.

— The greater the alignment of the domains, the larger is the amplitude of the recorded signal.
— The frequency is determined by how often the magnetization changes direction.
— When all the magnetic domains are aligned the tape is saturated, and hence the amplitude cannot be made larger.
— The alignment of domains is maintained because the tape has coercivity, and stray fields will not erase it easily.

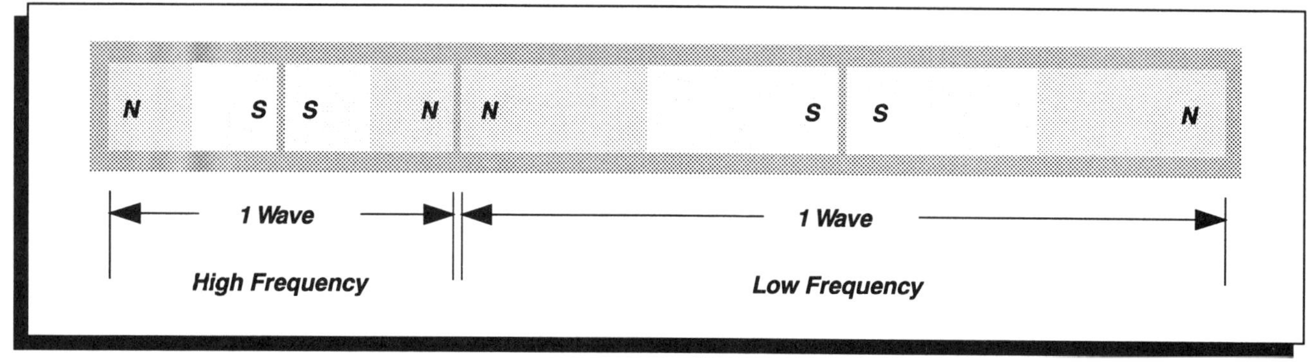

Figure 9.16 Recorded information on magnetic tape.

Taking a look at a recorded tape, it will appear as in Figure 9.16, with magnetic domains pointing to various degrees left and right.

In the playback part a magnetic signal on the tape will have to be converted to an electrical signal. Remembering Faraday's Law of Induction, a changing magnetic field caused by moving the tape past a playback head will induce a voltage in the head. Figure 9.17 shows how the head can read the information on a tape. It is of the same shape as the recording head, but it has a different function. The changing magnetic field on the tape has to be coupled to the head gap so that the coil in the head will sense a changing magnetic field and deliver an induced voltage at its output. When the tape stops, no signal is induced in the head and so there is no output. This is consistent with Faraday's law of induction.

This approach for the read–out is simple and effective. However there are a couple of problems that have to be addressed. Consider Figure 9.18 where the playback head is reading a high frequency signal. A changing magnetic field at the head gap induces a voltage. But which way is the magnetic field changing for the example in Figure 9.18? The gap in the head sees a half wave pointing one way and the other half wave pointing the other way. There is destructive interference of the two half–waves across the gap and consequently, on the average there is no induced voltage, for the example shown.

The gap in playback head can read waves down to a size whose half wave is equal to the gap, and any waves shorter than that will start cancelling out. We can formulate this situation as:

> when a half wave = gap size, signal is still well read but there is a severe limitation on higher frequencies.

Example: What is the highest frequency that a playback head can read if its gap is 0.0005 inch, and the tape speed is $1\frac{7}{8}$ i.p.s.?

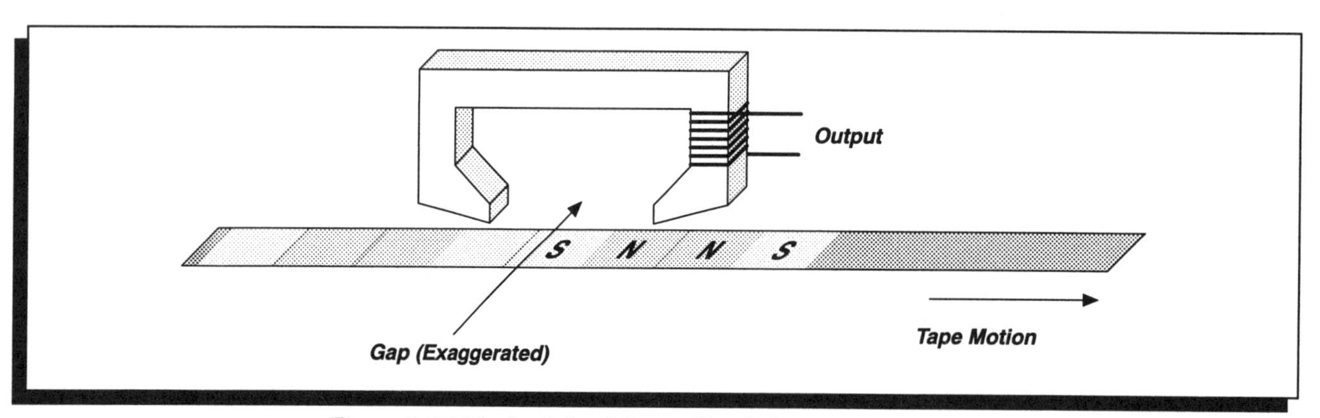

Figure 9.17 Playback head for reading information on a tape.

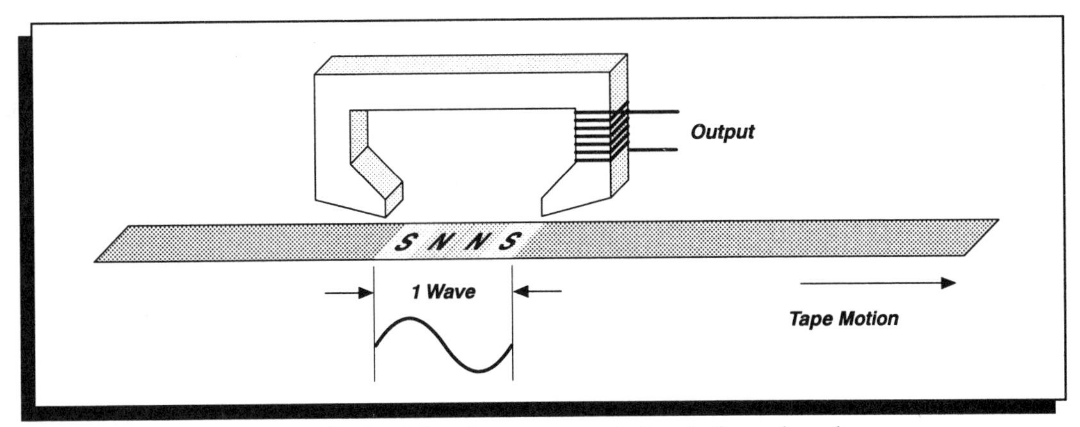

Figure 9.18 Playback head trying to read signal of wavelength = gap.

0.0005 inch = ½ wave
0.001 inch = 1 wave
frequency = speed / wavelength
= 1.875 i.p.s. / 0.001"
= 1,875 Hz

Frequencies above this limit will be severely attenuated. In order to be able to read high frequencies on a tape there are two possible solutions:

1. Use a very fine gap in the playback head.
2. Use a higher recording speed, since the waves on the tape will then be longer.

Both approaches are taken. Small gaps are used, but they can get plugged easily. Higher tape speeds are used in reel–to–reel tape decks, as in professional recording, and this leads to superior performance.

There is another problem and this is shown in Figure 9.16. According to Faraday's Law of Induction, a rapidly changing magnetic field will induce a larger voltage than a slower changing magnetic field of the same amplitude. Thus, a 1,000 Hz magnetic signal will induce a voltage which is 10 times longer than for a 100 Hz magnetic signal of the same amplitude. Special compensation will have to be taken, in the form of equalization, to take care of this problem.

The erase head is also an electromagnet. Its function is to create an oscillating magnetic field at its gap so that it can randomize the magnetic domains on the tape when it passes under its gap. This field will have to be larger than the coercivity of the magnetic material on the tape to effectively cause a random distribution of the orientations of the magnetic domains. Here as well the first Left–Hand Rule is used.

The order of heads is as shown in Figure 9.19.

An interesting question will arise about the core material used in the heads. Since it is magnetic material, it will have a hysteresis curve. What type of magnetic material should be used? Should it

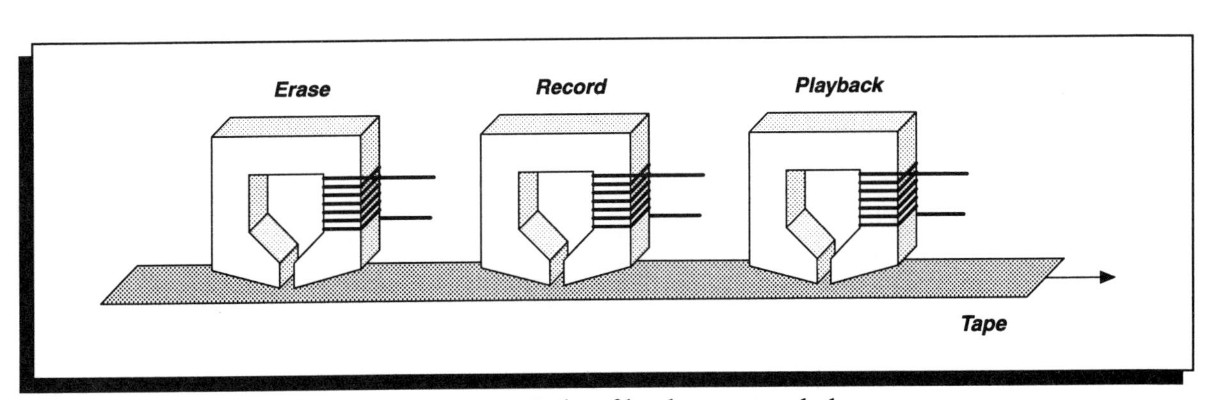

Figure 9.19 Order of heads on a tape deck.

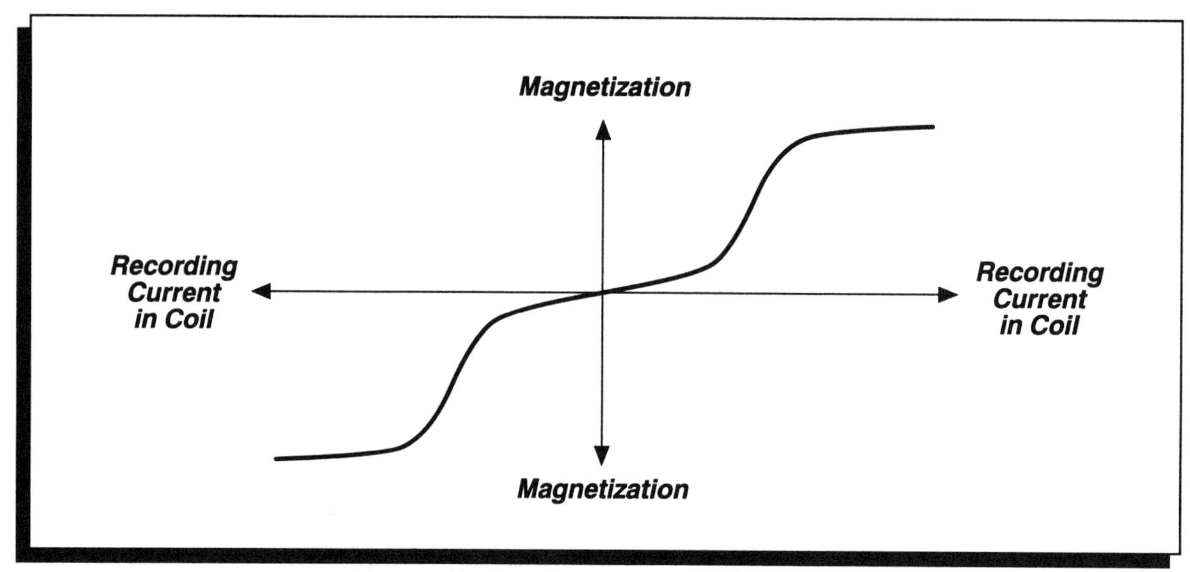

Figure 9.20 Recording on a tape with hysteresis.

have a large retentivity? Since the function of heads is either to deliver alternating magnetic fields or to read changing magnetic fields, the cores should retain no information (i.e. their retentivity and coercivity should be close to zero). Hence soft magnetic materials are used as cores in heads.

■ BIAS

The recording process on magnetic tapes relies on the magnetic hysteresis curve for the magnetic material on the tape. Figure 9.20 shows one such curve. It is a complicated–looking curve and there is a question of how well it will perform in recording audio information. Will it provide good high–fidelity quality? This is not obvious, since the hysteresis curve is not linear. Consider the beginning of such a curve as in Figure 9.20 where a recording current can be positive or negative.

Let us look at the kind of recording we will get. Consider Figure 9.20 with the hysteresis behavior of a tape; the signal to be recorded goes to the recording head. The tape characteristics in Figure 9.20 show that at low input currents to the recording head the resulting magnetization is very small. Not much is recorded. At higher input currents the magnetization suddenly starts to increase rapidly and hence a large signal is then recorded. However, because of the non–linear characteristic of the tape,

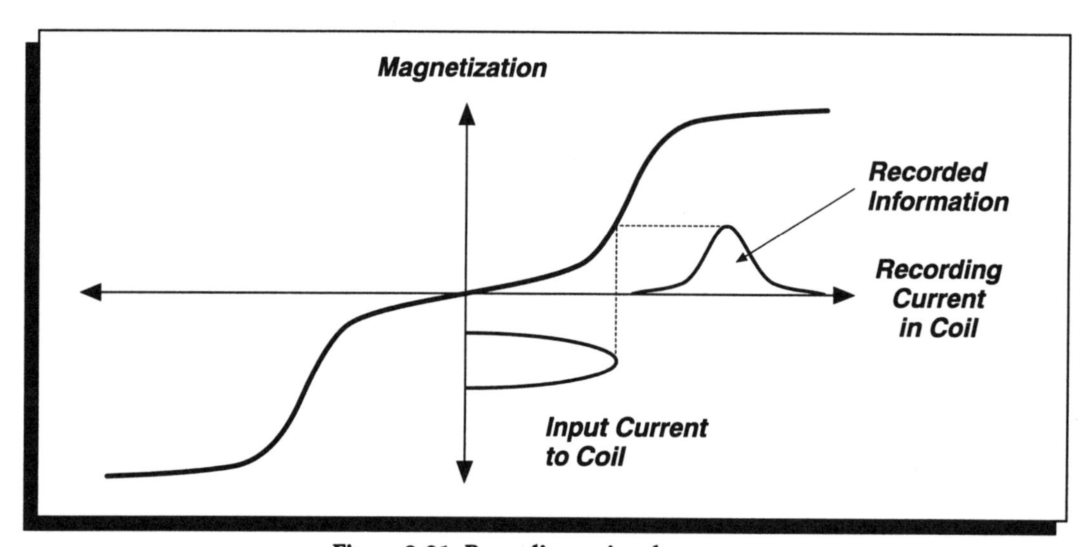

Figure 9.21 Recording a signal on a tape.

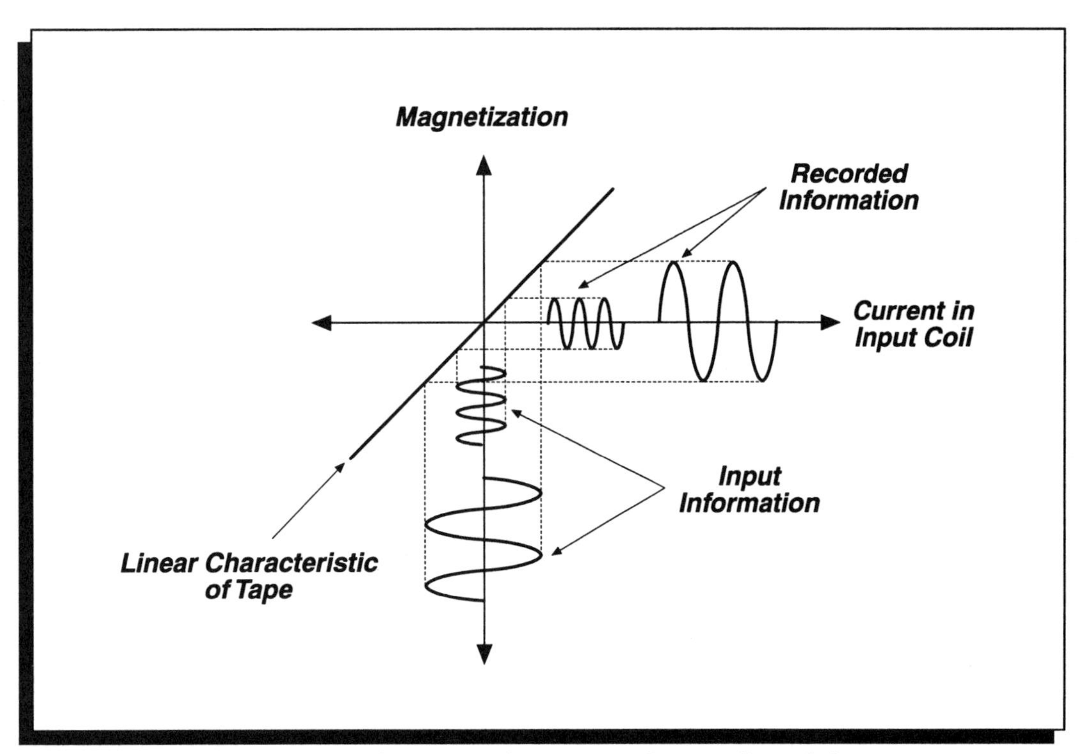

Figure 9.22 Ideal magnetic characteristic for tape, linear behavior.

the recorded signal is not a replica of the input. The recorded signal is distorted, and it certainly will not be high–fidelity. Actually the ideal curve of the produced magnetization as a function of input current would be a linear one, as in Figure 9.22, and there would be no distortion. Figure 9.21 is far from Figure 9.22. At this point one can compromise and use at least part of the hysteresis curve, the part that is linear. Let us locate on the hysteresis curve the region which is most linear in Figure 9.23. We certainly want to avoid the region near the origin where the domains are sluggish in responding to a magnetic field. The solution to this problem, which occurs only in recording, is:

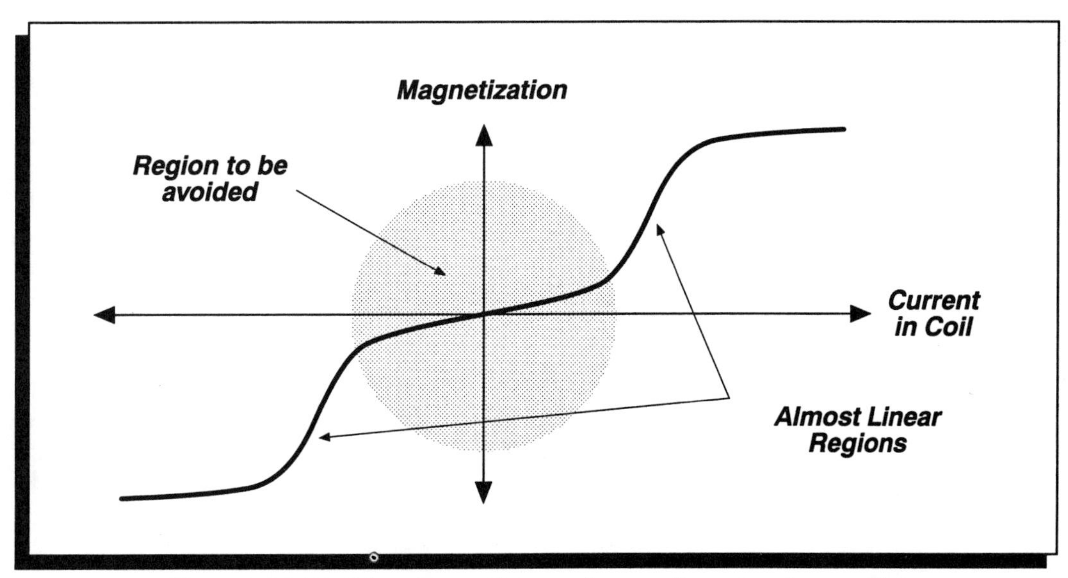

Figure 9.23 Useful regions on hysteresis curve for magnetic recording.

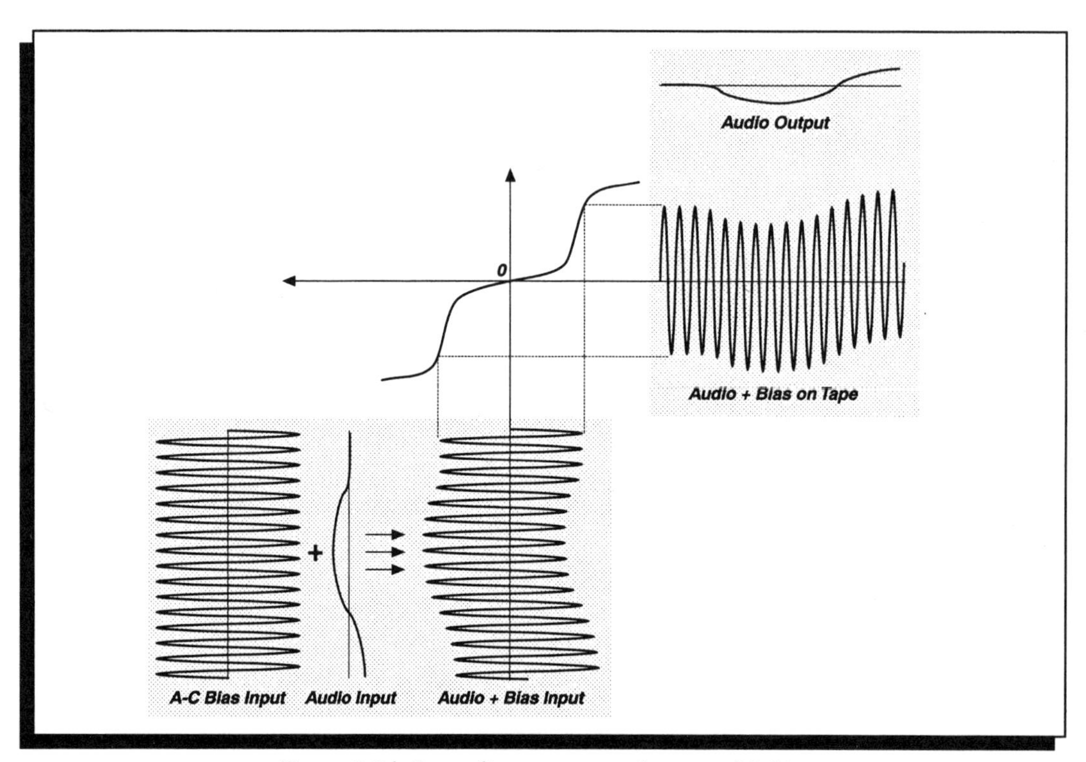

Figure 9.24 Recording on magnetic tape with bias.

— Avoid the region near origin by adding a bias signal during recording to the audio signal and then use the almost linear region.

The bias method of recording is shown in Figure 9.24, and it contains two parts: 1) add a high frequency bias signal to audio signal, and 2) record the bias signal and audio signal together. The frequency of the bias signal should be high enough so it is not heard. Typically, it is in the range 60–125 kHz. The particular frequency chosen depends on the manufacturer. The benefits of bias are:

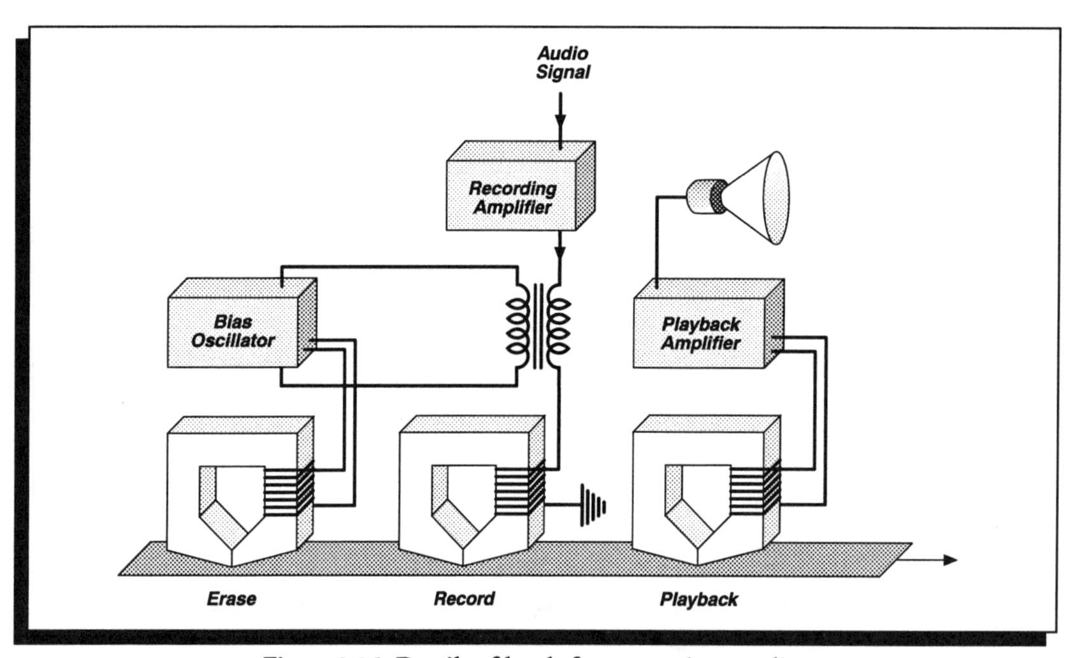

Figure 9.25 Details of heads for magnetic recording.

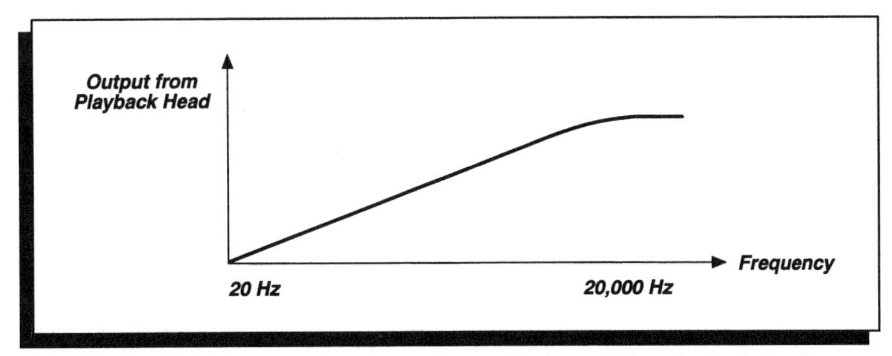

Figure 9.26 Frequency dependence of output from playback head.

— improves linearity.
— lowers distortion at low levels.
— reduces tape noise at low levels.

It is important to note that bias settings have no effect during playback since there is no bias during this process. Usually, the same frequency as the bias is used in the erase head. Figure 9.25 shows this.

EQUALIZATION

The playback head works on the principle of Faraday's law: voltage induced in the head depends on rate of change of magnetic field. The output will depend strongly on frequency, as previously discussed, shown in Figure 9.26. At high frequencies there is a leveling off due to:

— saturation effects.
— problems of finite size gap.

The finite size gap problem is made clearer by measurements of the playback head output for different gap sizes and tape speeds. This is shown in Figure 9.27.

Results like those in Figure 9.26 are unacceptable for high–fidelity (the ideal would have been an

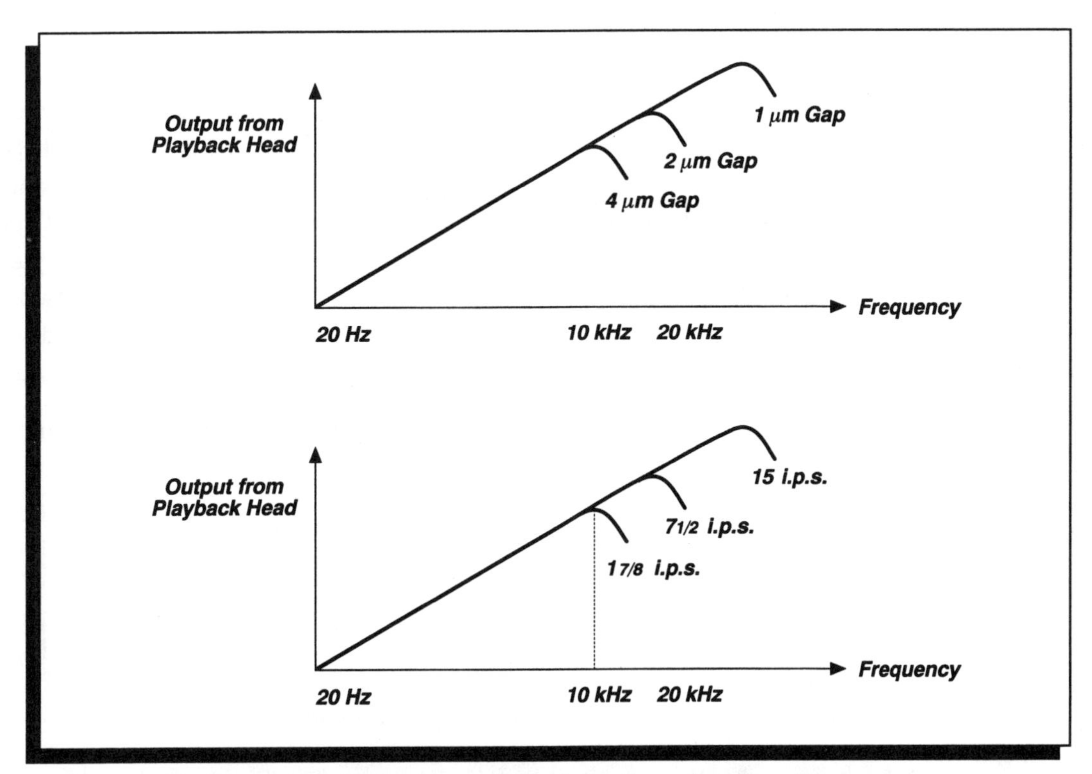

Figure 9.27 Output from playback head as a function of frequency for various gap sizes and tape speeds.

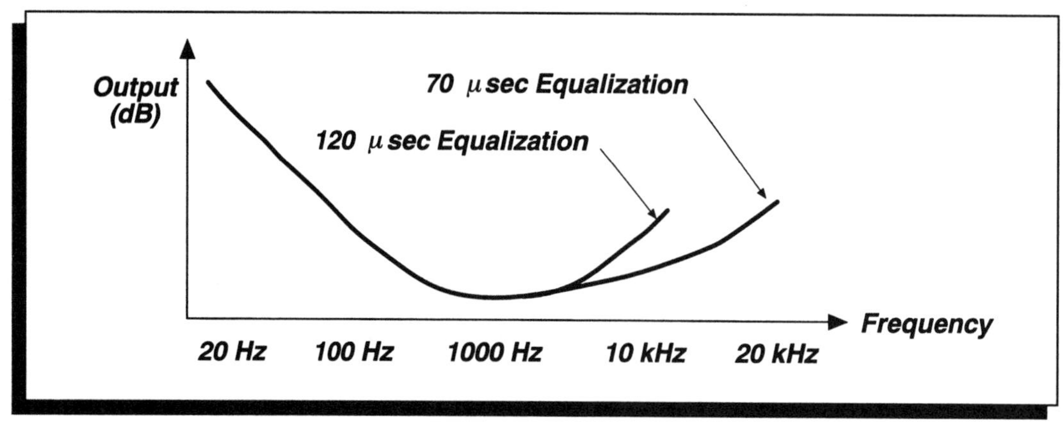

Figure 9.28 Equalization in playback.

output which is independent of frequency) recording; a compensation has to be made for Faraday's Law by using equalization in the playback.

Figure 9.28 shows an equalization curve used in playback. There is a strong bass boost with some compensation for high frequency problems.

The high frequency equalization which makes up for the high frequency problems can be pre–set depending on the type of tape used. Stated in terms of the type of filter used, 120 μ sec equalization or 70 μ sec equalization; the times refer to the turnover frequency where it becomes effective. The 120 μ sec equalization is used with ferric tapes while the 70 μ sec equalization is used with CrO_2 tapes. The high frequency equalization has to be used carefully, otherwise it can boost the high frequency tap hiss.

Since there are always problems at high frequencies with tape recording, additional equalization is used in the recording process. The effect is shown in Figure 9.29.

DOLBY NOISE REDUCTION

All tapes, even unrecorded ones, produce noise, known as hiss, when played back. Such hiss is annoying, especially during quiet portions of a recording. The Dolby noise reduction system is designed to reduce this type of noise.

First, let us see why there is always hiss in tape playback. When all the domains have a totally random distribution, the average magnetization at any point is zero. For a large sample this is true, but the playback head gap samples a very small portion of the tape. What is the chance that all the domains are completely random? The fluctuations in the random distribution of the domains then become more significant. These fluctuations lead to a small magnetization varying over the playback head gap resulting in tape hiss. The smaller the number of domains (for cases of small tracks and small gaps), the larger the hiss. If a sample has 100 domains and if due to fluctuations, two of them point the same way, the effect on the varying mag-

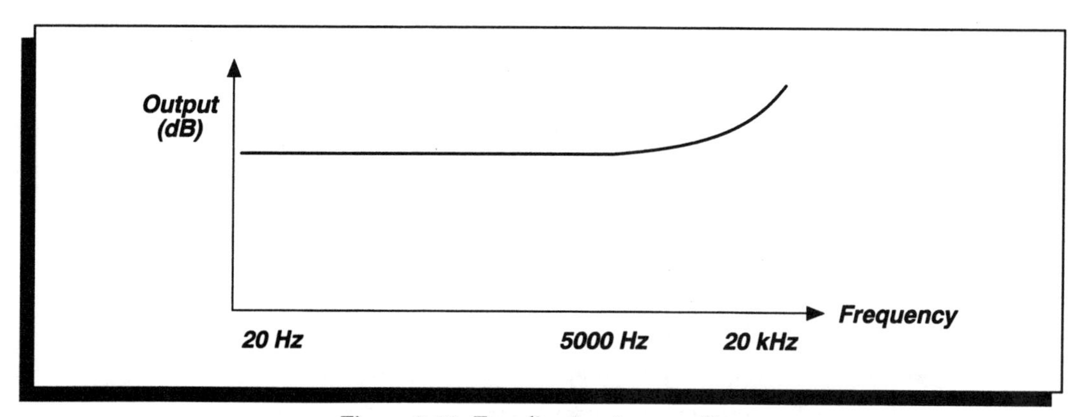

Figure 9.29 Equalization in recording.

netization and hiss will be more important than for a sample of 1,000 domains and two of them being aligned. The Dolby approach to reduce hiss is quite effective and so it is widely accepted.

The Dolby process is effective during quiet passages when tape hiss has a strong effect. Since hiss is mainly a high frequency disturbance, the Dolby process takes place at high frequencies in a two–step process.

For Dolby B:

1. quiet passages between 2,000 Hz and 20,000 Hz are boosted by 10dB (the amount depends on level of passages) and then recorded.
2. in playback, when Dolby B is in, the high frequency quiet passages are reduced by 10dB. Since the hiss is with the high frequency quiet passages, it also is reduced by 10dB.

For Dolby C:

1. quiet passages between 1,000 Hz and 10,000 Hz are boosted by 20dB and then recorded

2. in playback, with Dolby C in, the quiet passages (which were boosted by 20dB) are cut by 20dB including the hiss. This essentially eliminates all hiss (by 20dB). In fact, the noise due to hiss is made four times quieter than without the Dolby C, which is a significant improvement.

It is important to note that Dolby is not a general noise reduction system: it only eliminates the hiss from the tape onto which you are copying the music. If you are making a copy of a phono record onto a tape and the record has scratches, Dolby will not reduce the noise due to the scratches. It reduces only the hiss from you tape.

It is possible to playback a Dolby recorded tape without Dolby in the playback. In this case the music tends to sound bright, as all the quiet passages were made louder.

RECORDING METERS ON TAPE DECK

The level at which a music selection is recorded on a tape is very important. At too low a level in recording, the sound will contain much hiss. At too high a level in recording, there will be distortion, due to saturation. Let us look at a typical music spectrum, Figure 9.30.

There are two types of meters on a tape deck to help with the recording level. They are:

— ***VU meter:*** it is a volume units meter and it gives an average loudness
— ***dB meter:*** a decibel meter, which is fast acting and it give readings of transients.

Both meters have a scale ranging from negative values to positive values passing through zero.

The zero of the VU meter is such that there is a 1% distortion of the signal. The zero of the dB meter is such that at 0 dB there is a 3% of distortion of the signal.

Some tape deck units have both meters. Quite often the dB meter is a Light Emitting Diode (LED) which flashes when the level is at 0 dB.

In recording, both meters have a maximum

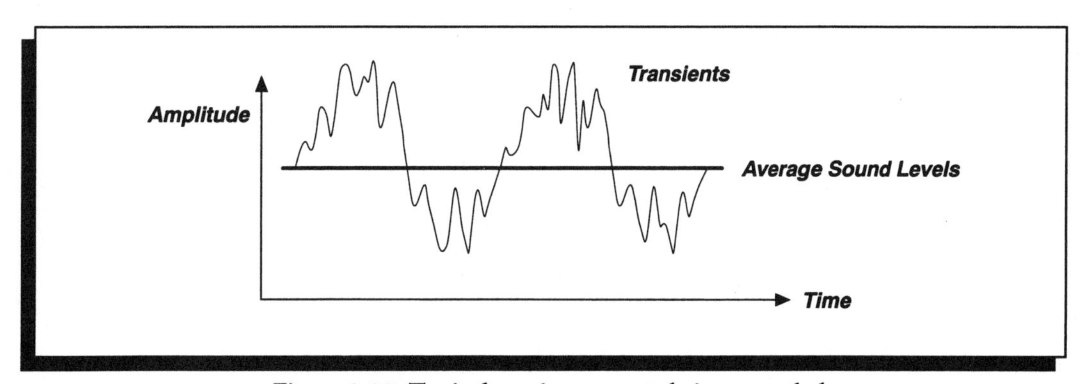

Figure 9.30 Typical music spectrum being recorded.

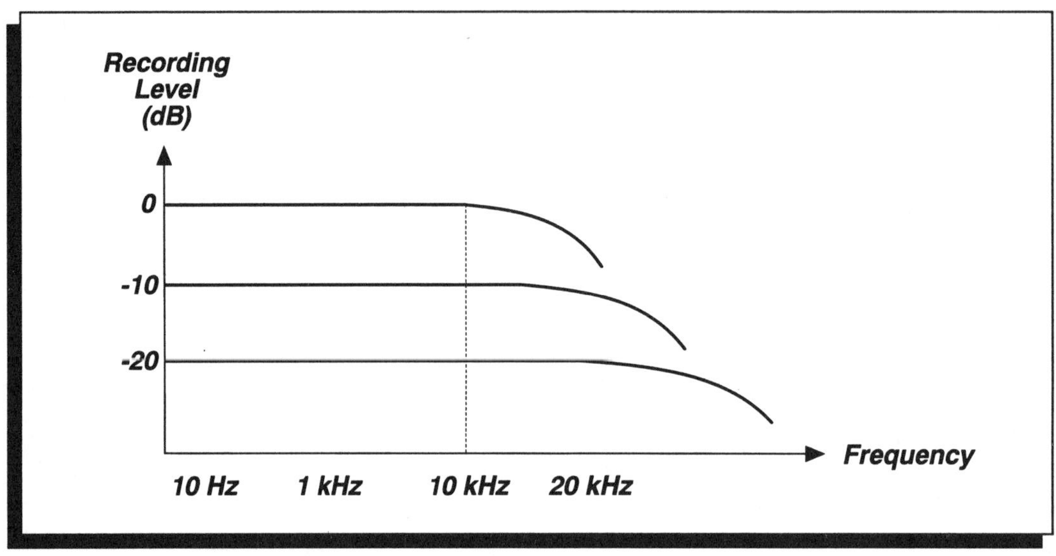

Figure 9.31 Frequency response for different recording levels.

value, 0VU and 0dB, and they should not be exceeded. The two zeros do not coincide. In fact the 0dB is approximately 8dB greater than 0VU. In recording it is important to maintain a good S/N ratio and a good frequency response. Figure 9.31 illustrates this. By reducing the recording level, the frequency response improves (because of saturation at high frequencies), but the S/N is reduced. Hence a good balance between the two is necessary.

TAPE DECK SPECIFICATIONS

— ***Signal to Noise Ratio (S/N):*** look for 50dB or higher and for 55–60dB or higher with Dolby on.

— ***Wow and Flutter:*** refers to variations in speed. It is clear from frequency × wavelength = speed that variations in speed will give variations in frequency. Hence the tape speed must be constant. Look for 0.2% or less.

— ***Stereo Separation:*** look for 40dB or better

— ***Motors:*** used to drive the system. There are two types. The synchronous one has its speed locked to the frequency of the 120 volt source, which is 60 Hz. Variations of voltage will not effect the speed, and since the frequency of 60 Hz is kept constant, the motor speed is kept constant. There is a lower quality motor called the induction motor. Its speed is locked to the voltage of the outlet. Since this varies quite a bit, the speed is not so well regulated leading to wow and flutter.

— ***Frequency Response:*** depends on the recording level.

SUMMARY OF TERMS

Bias signal: A high frequency signal added to the audio signal in recording to reduce distortion.
Coercivity: Magnetic field which will destroy recorded magnetic information and reduce the magnetization to zero.
Dolby Noise Reduction: Reduces tape hiss from the tape you are copying onto.
Magnetic Hysteresis: Behavior of a magnetic material in various magnetic fields.
Playback Head: Device on tape deck which reads information on tape. It converts a magnetic signal on tape to an electrical signal using Faraday's law of Induction. It consists of a coil around a core in the shape of a horseshoe with a small gap where the signal is read.
Recording Head: Device on tape deck which records audio information on tape by aligning magnetic domains on the tape. It is an electromagnet.
Retentivity: Ability of magnetic material to retain a magnetization when external magnetic field is removed. Serves as memory.
Tape Saturation: Alignment of most of the magnetic domains.
Wow and Flutter: Fluctuations in a magnetic tape's speed. Wow is slower changes in speed which occur between 0.5 to 6 times per second. Flutter is more rapid variations which occur at 6 to 200 times per second.

NAME ____________________ DATE __________

Questions for Review

1 Why should the core of a playback head and a recording head be made from soft magnetic material?

2 Which magnetic properties of a magnetic material are useful for recording?

3 Why does higher tape speed provide better tape performance?

4 How does tape saturation limit the performance of a tape?

5 Which principles are used in: the recording head, the playback head?

6 How does the gap of a playback head limit the performance of the head?

7 Why is it necessary to have equalization in the playback of a tape?

8 Explain the difference between soft and hard magnetic material.

9 What would be the advantage of vertical recording over that of the conventionally longitudinal recording?

10 Explain what limits the dynamic range in recording audio on a tape.

NAME ______________________________ **DATE** ______________

Please select one answer.

1. Which type of motors, used in turntables and tape decks, have their speed locked to the frequency of the electrical outlet voltage (60 Hz)?
   ___A. Induction motors
   ___B. Synchronous motors
   ___C. direct drive motors
   ___D. electromagnetic motors
   ___E. none of the above.

2. If the magnetic tape moves from left to right, what is the order of the heads (from left to right) on a 3–head tape deck?
   ___A. erase, playback, record
   ___B. erase, record, playback
   ___C. record, playback, erase
   ___D. record, erase, playback
   ___E. playback, erase, record

3. How does a 3 kHz wave recorded on a magnetic tape compare with a wave at 300 Hz?
   ___A. It is 10 times shorter.
   ___B. It is 10 times longer.
   ___C. It is 5 times shorter.
   ___D. It is 3,000 times shorter.
   ___E. Both are of the same length.

4. For the magnetic tape being magnetized by a recording head, what will be imprinted upon the tape? Refer to figure on the right.
   ___A. Point X will become a north pole and Y a south pole.
   ___B. Point X will become a south pole and Y a north pole.
   ___C. Point X will become a south pole and Y a south pole.
   ___D. Point X will become a north pole and Y a north pole.
   ___E. There will be no signal imprinted on the tape since the head has a soft core.

5. If during the recording, the level is set very low, the tape recording will have:
   ___A. a good S/N ratio
   ___B. a poor S/N ratio
   ___C. a poor frequency response
   ___D. poor cross–talk
   ___E. high THD

6. If during the recording, the level is set too high, the tape recording will have:
   ___A. a poor S/N ratio
   ___B. a good frequency response
   ___C. less than normal distortion
   ___D. a large amount of distortion
   ___E. a small amount of distortion

7. What is the highest frequency that a playback head can read on a tape, when the tape speed is $1\frac{7}{8}$ i.p.s and the gap size is 1.875×10^{-4} inch?
   ___A. 10 kHz
   ___B. 5 kHz
   ___C. 20 kHz
   ___D. 15 kHz
   ___E. 1 kHz

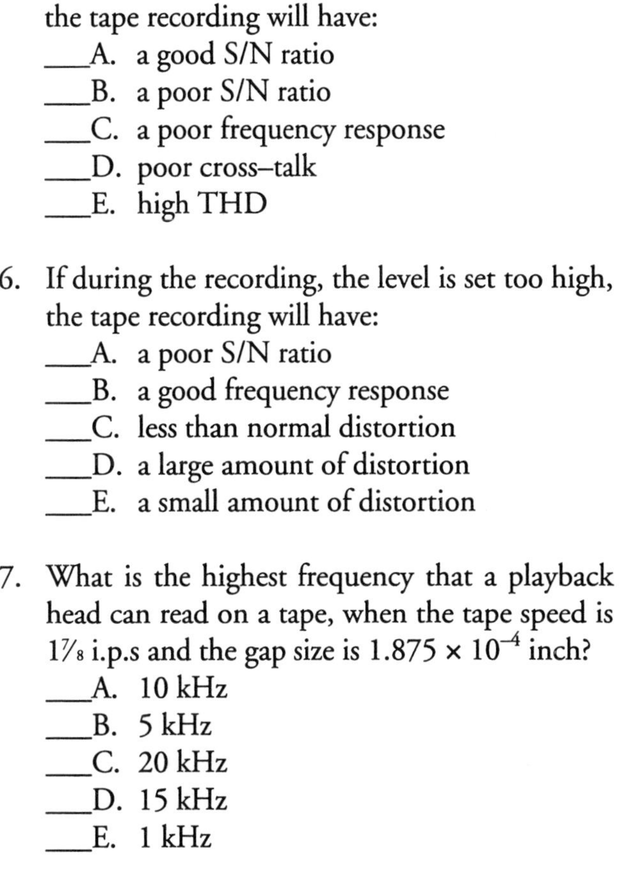

8. How does tape speed affect the performance of a tape?
   ___A. Faster speeds generally produce better performance.
   ___B. Faster speeds generally produce poorer performance.
   ___C. The performance of a tape deck does not depend on speed.
   ___D. Slower speeds provide enough time to produce a magnetization on a tape.
   ___E. Slower speeds produce longer waves which are easy to handle.

9. The amplitude of an audio signal is translated onto a tape as:
   ___A. the degree of alignment of magnetic domains.
   ___B. how long a wave is on a tape.
   ___C. how often the domains change directions.
   ___D. how well the signal is retained on tape.
   ___E. how fast the tape was recorded.

10. The purpose of a bias current in a magnetic recording head is:
   ___A. erase all previous information and start again.
   ___B. clean the tape of dust and foreign objects.
   ___C. add a bass boost on the recorded music.
   ___D. shake up the magnetic domains on the tape so that it is easier to magnetize the tape.
   ___E. add distortion.

11. The recorded information on a magnetic tape is:
   ___A. longitudinal
   ___B. transverse
   ___C. vertically polarized
   ___D. horizontally polarized
   ___E. electromagnetic

12. Before playback equalization, the signal induced in a playback head by a 12 kHz signal is ____________ one produced by a 2 kHz signal of the same amplitude
   ___A. 6 times larger than
   ___B. the same as
   ___C. 6 times smaller than
   ___D. 3 times larger than
   ___E. 3 times smaller than

13. What is the frequency, typically, of tape bias and its purpose?
   ___A. 20 kHz. Bias reduces the S/N ratio.
   ___B. 100 kHz. Bias reduces tape distortion.
   ___C. 100 kHz. Bias reduces cross–talk.
   ___D. 100 kHz. Bias improves the frequency response.
   ___E. 20 kHz. Bias improves the equalization.

14. Magnetic tapes should:
   ___A. have a small coercivity.
   ___B. have a large coercivity.
   ___C. have a small retentivity.
   ___D. have a large retentivity.
   ___E. Both B and D are correct.

15. The wavelength for a 1,000 Hz tone recorded on a tape moving at 1⅞ i.p.s. will be ____________ for a 4,000 Hz tone recorded on a tape moving at 7.5 i.p.s.
   ___A. 4 times shorter than
   ___B. 4 times longer than
   ___C. 2 times longer than
   ___D. 2 times shorter than
   ___E. the same as

16. If you played a tape without using the playback equalizer in the tape deck, the music would sound:
   ___A. as having too much bass and too little treble.
   ___B. as having too little bass and too much treble.
   ___C. as having too little bass and too little treble.
   ___D. as having too much bass and too much treble.
   ___E. perfectly natural.

NAME ______________________________ DATE ____________

17. Because of saturation on a magnetic tape used in analog recording, there will be limitation on:
   ___A. the minimum amplitude of a signal recorded on a tape.
   ___B. the noise of a tape.
   ___C. the longest wavelength of a signal recorded on a tape.
   ___D. the maximum amplitude of a signal recorded on a tape.
   ___E. the cross–talk.

18. The Dolby noise reduction system in a tape deck can be used:
   ___A. to reduce the noise from noisy tapes that you are trying to copy onto your tape.
   ___B. to reduce noise from scratchy records.
   ___C. to increase the high–frequency noise of pre–recorded tape.
   ___D. to boost all mid–frequencies at low levels.
   ___E. none of the above.

19. On a tape, a recorded wave X has a wavelength of 1.875×10^{-2} cm and it is followed by wave Y of wavelength 9.375×10^{-4} cm; both have the same amplitude. When played in a cassette deck, the voltage induced in the playback head, before equalization, will be ———————— for Y than (as) X.
   ___A. 2 times larger
   ___B. 2 times smaller
   ___C. 20 times larger
   ___D. 20 times smaller
   ___E. the same

20. Two magnetic tapes have different hysteresis curves. Recording on them:
   ___A. the bias should be kept the same.
   ___B. the amount (amplitude) of bias should be different for each.
   ___C. the bias frequency should be different for each.
   ___D. the recording speed should be different.
   ___E. the erasing frequency should be different for each.

21. When recording music on a tape deck what advantage is there to setting the recording level at –20dB rather than 0dB?
   ___A. better S/N ratio
   ___B. less wow and flutter results
   ___C. there is no advantage
   ___D. better frequency response
   ___E. less cross–talk

22. Recording heads are:
   ___A. permanent bar magnets.
   ___B. made of magnetically "hard" material.
   ___C. electromagnets.
   ___D. high speed transformers.
   ___E. permanent horseshoe magnets

23. Approximately what is the erase frequency used in most tape decks?
   ___A. 20 Hz
   ___B. 20 kHz
   ___C. 88 MHz
   ___D. 535 kHz
   ___E. 60–125 kHz

24. Which type of meter responds well to the sudden peaks or transients in the music?
   ___A. a VU meter
   ___B. a dB meter
   ___C. a LED meter
   ___D. both answers B and C are correct.
   ___E. both answers A and B are correct.

25. Approximately how much does the Dolby C noise reduction reduce the tape hiss?
   ___A. 1 or 2 dB
   ___B. 10dB
   ___C. 20dB
   ___D. 4dB
   ___E. Dolby noise reduction boosts only the high frequencies.

CHAPTER

10 Digital Recording and Optical Playback

In this chapter, limitations of analog methods will be presented as an introduction to the digital approach, which will be developed and applied to optical playback. The digital process is a new way of storing, handling, and playing back audio information. It sprung from many technological advances and it has become one of the most successful developments in the field of audio information storage and reproduction. Two digital approaches are presently available: optical and magnetic. Digital developments in radio wave communications are expected to be implemented soon.

Examples of commercially available audio units based on optical methods and magnetic ones are:

Optical: compact disc (CD)
mini–disc (MD)

Magnetic: digital audio tape (DAT)
digital compact cassette (DCC)
mini–disc (MD)

The reason for the superior sound of these recordings is that they overcome many of the limitations of analog recording. By their own nature analog methods had reached the limit of their potential and could not provide further advancement in the field.

10.1 Digital Recording

Limitations of Analog Method

The approach presented in previous chapters was based on the ana-

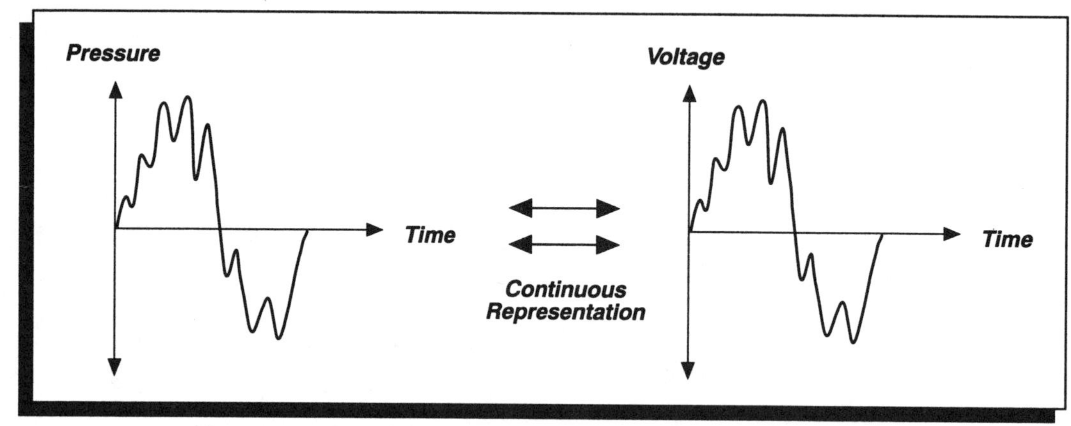

Figure 10.1 Sound wave and its analog representation as a voltage.

log representation of an audio signal and its recording, storage, and reproduction. Analog means that the wave, the sound wave here, shown in Figure 10.1, is ***continuously*** represented or modeled by some continuous parameter. Evidently this parameter is chosen so that it is easy to control. It can be the alignment of magnetic domains in a magnetic medium like a tape, or the pressure variations about the equilibrium pressure of air around us, or even an engraving of the wave in a record which is a mechanical duplication or "photograph" of the wave. Of course, we all know what happens to photos when they are not handled carefully. They get marked by fingers and sometimes they even get crumbled by mistake. At this point the photo certainly looks bad, and will not be an accurate replica of the original image. Had a supermarket receipt been crumbled it would still have been possible to read without too much loss the numbers on it that correspond to the prices. This is an important point:

> **A numerical representation of a quantity is less susceptible to damage than an analog one.**

The grooves on a record provide an illustration of some of the problems with analog recording. Consider Figure 10.2, which shows three interesting situations in dealing with an analog signal such as on the record groove:

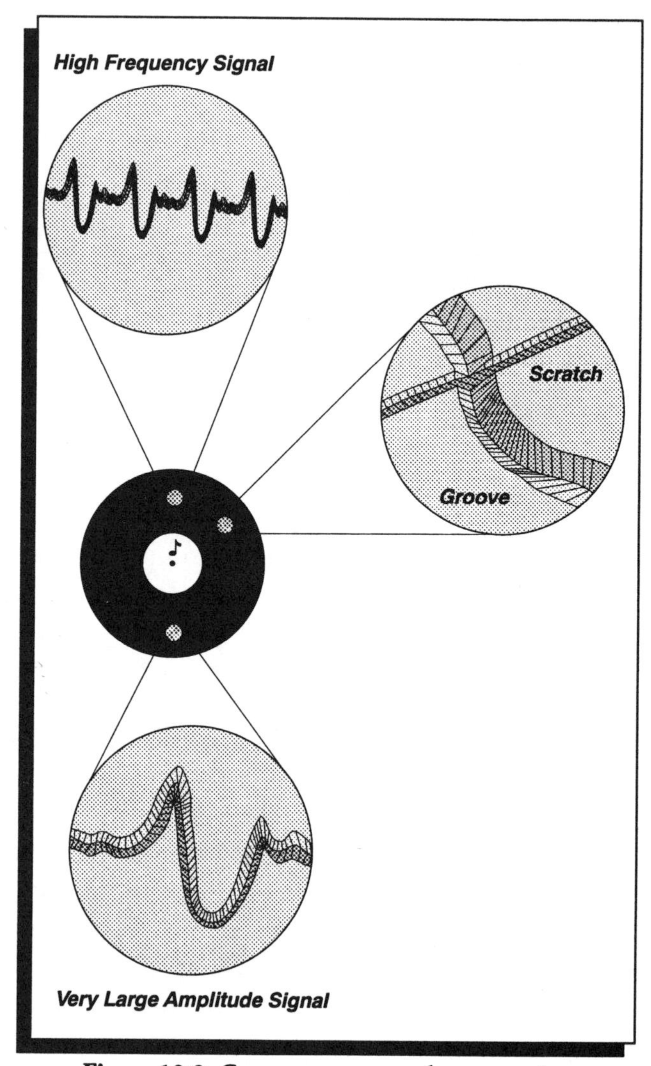

Figure 10.2 Grooves on a record representing analog signals.

— a high frequency signal
— a scratch
— a very large amplitude signal (due perhaps to the canon going off in the 1812 Overture).

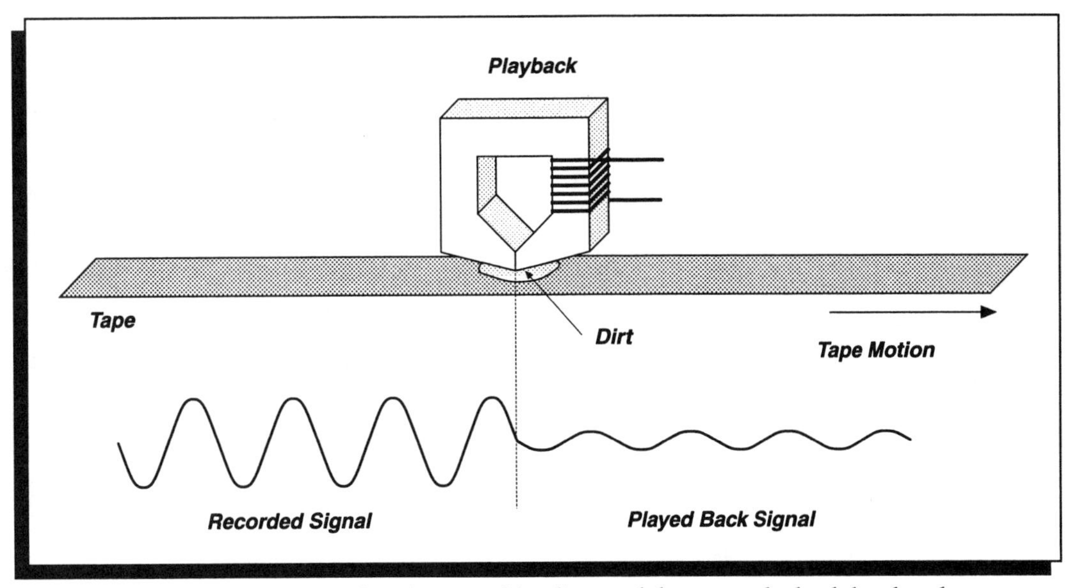

Figure 10.3 Distortion of analog signal by dirt stuck between playback head and tape.

The high frequency signal represented by sharp spikes in the groove will most likely be modified when played, or even damaged when the stylus goes through them and some of the sharp spikes representing part of the wave will be knocked out. In fact, the act of reading the analog information on a record with a stylus wears out the information. The scratch of course is undesirable; it cannot label itself as being different from the analog signal and the stylus will reproduce it along with the signal. Consequently, this will cause noise on top of the signal, an undesirable effect. As to the very large amplitude signal, here as well there will be problems, as the stylus will most likely not be able to follow such a sudden change in amplitude; it will probably jump over this passage. In fact, because very large signals will not be tracked properly, the dynamic range of the audio signals will be limited. Typically, the dynamic range for a record is 50 to 55 dB. Similar problems arise in magnetic recording when analog signals of large amplitude cause tape saturation; this will limit the dynamic range and it will certainly lead to distortion. Also, should a bit of dirt get stuck between the tape and the gap of the playback head, as shown in Figure 10.3, the signal will be reduced in amplitude and it will not sound like it was originally recorded.

In summary, some of the problems with analog recording and playback are:

— limited dynamic range
— poor S/N
— continuous degradation of signal by reading process
— distortion
— cross–talk between channels
— limited frequency response
— degradation of sound quality during signal processing for recording.

THE DIGITAL WAY

In this method, a wave as in Figure 10.1 is expressed by numbers and this information can then be processed, stored, and played back in the digital domain. Looking at the wave in Figure 10.1, the quantity that will have to be expressed by numbers is the amplitude. Hence the thing to do is to measure the amplitude, express it as a number, and then record it. It appears simple enough, but when it comes to do it, all sorts of questions appear. How often should the amplitude be measured and expressed by a number? What kind of numbers should be used?

It would seem that the more measurements of the amplitude are taken in recording, the better the recording will be. But it is impossible to get an infinite number of amplitudes expressed numerically to represent the wave. Even if this were possible, where would such a vast quantity of numbers be stored? A compromise has to be made. In fact it is the same type of problem as trying to make a movie of an action–filled scene. The film shooting takes many frames and the scenes with the most action will require the largest number of frames per second to follow all the details of the action. A familiar example is the rotation of wheels of a stagecoach for a western movie. At times, they seem to rotate backward. If the wheels rotate at the rate of 8 spokes per second, then more than 8 frames per second are required to cover this action. A minimum of 16 frames per second is needed. Should the wheel rotate at 9 spokes per second, then only 7 spokes per second would be seen. The same reasoning applies to digital audio recording where the number of measured amplitudes will be critical. Actually instead of taking measurements of the amplitude the equivalent phrase, used in this field, is sampling. One of the next sections will deal with the rate at which a signal should be sampled for digital recording. We define sampling as:

***Sampling* = Measuring the amplitude of a wave at some specific time**

The numerical representation of amplitudes of a wave should be in a form so that it is easy to store and read by machines. Although the decimal number system is the one we are most familiar with and it could be used for representing amplitudes of a wave, a more convenient one for digital recording is the binary system. It is based on two numbers which could relate to "on" or "off" in a switch.

The advantages of expressing amplitudes digitally are:

- When reading the recorded numbers they will not be changed to other numbers. Abuse could destroy them but regular usage will not affect the values, especially if the binary system is used. Figure 10.4 shows how a worn out number 2 still looks like a 2.
- It is just as easy to read a large number as a small one and hence there will be no limitation on achieving the expected dynamic range in sound recording.

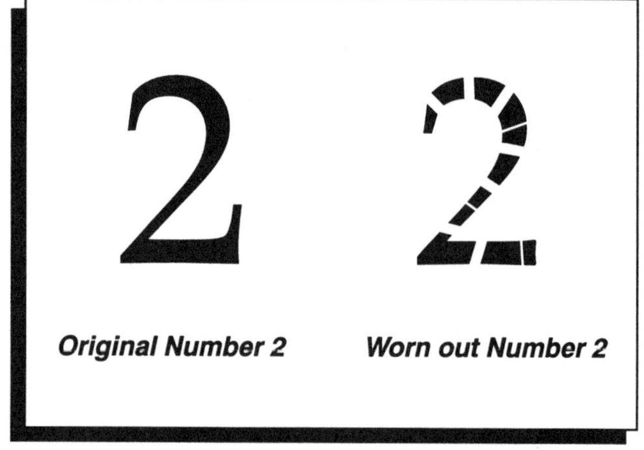

Figure 10.4 Original number 2 and worn out number 2. The basic information is not lost when number is worn out.

- The presence of one number does not affect the ones next to it. Cross–talk between channels will be small.
- No limitation on the wavelength which exists on an analog tape when being read by a playback head with too large a gap compared to the wavelength. All numbers will be of the same length on the tape.
- By having all the information in digital form it will be easy to do signal processing on it, and even make error corrections.

■ NUMBER SYSTEMS

We are used to dealing with numbers in the decimal system which is based on 10 numbers ranging from 0 to 9. It would seem natural that such a system could be used to describe the amplitudes of a signal to be recorded. The decimal system is not practical for this, and a different number system should be used. Even in our every day experience we use different number systems.

For example, we buy eggs by the dozen; there are 60 seconds to the minute with 60 minutes in an hour; we have 12 months in a year. These are a few examples that show we do use the most convenient system of numbers.

For the amplitudes of a wave which are to be recorded, the number system that will be used will be the one which is easy to read by a machine. In principle it would correspond to an ON or OFF signal, i.e. a number system based on two numbers, a "1" and a "0". However, it would seem that such a number system, based on 2 and called binary, would not give enough numbers to cover the wide range of amplitudes that will occur in audio recording. The binary number system was invented by Leibnitz in 1689, and its importance was only recently appreciated.

The binary number system can be developed by comparing it to the decimal system. Consider the number 1994 in the decimal system. It means the following:

$$1\times10^3 + 9\times10^2 + 9\times10^1 + 4\times10^0$$

where 10^0 is chosen to be one. Of course, this operation is not performed explicitly in every day usage but that is what it means. It has four decimal places.

To understand how to construct a number based on only "1" and "0", an example will be given. What does 1101 in the binary system mean? Keeping in mind how the decimal system works, the binary number is developed by replacing 10 by 2 and using 1 or 0 instead of numbers from 0 to 9. Here it means:

$$1\times2^3 + 1\times2^2 + 0\times2^1 + 1\times2^0$$

Its value in the decimal system would correspond to:

$1\times8 + 1\times4 + 0\times2 + 1\times1$,
which is 13.

Here 2^0 is defined to be 1. The binary number in this example has four ***bi**nary digi**ts*** and hence it has four ***bits***. We define:

***bit* = binary place in binary system, a binary digit.**

Now, what is the maximum value of a 4–bit number, 1111? It corresponds to:

$1\times2^3 + 1\times2^2 + 1\times2^1 + 1\times2^0$,
which has the value of 15 in the decimal system.

There is a quicker way of getting the maximum value. By examining this, it is clear that since the next bit will correspond to 2^4 (equal to 16), then the maximum value of a 4–bit number is 2^4-1. For an n–bit number, its maximum value is 2^n-1.

If the amplitudes are going to run from small to very large, enough bits will be needed to represent such amplitudes. For CD recording, the presently accepted and used maximum value of bits for an amplitude is 16 bits. In the future it may be extended to 18 or even 20 bits.

To implement such a system, one way of doing this is shown in Figure 10.5. Part (a) shows the analog signal as a function of time with amplitudes represented by decimal numbers; part (b) is its presentation in the binary system. The "1" and "0" values can be represented by some quantity which can easily be recorded or detected, optically or magnetically.

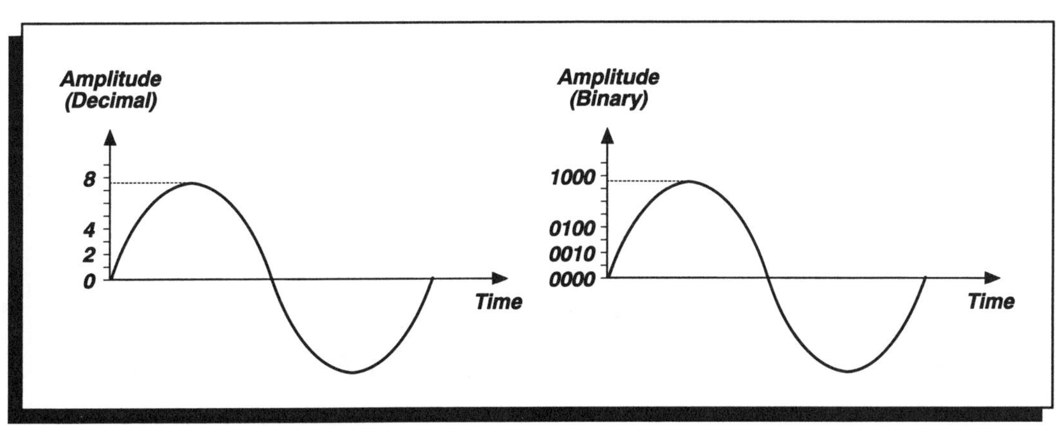

Figure 10.5(a) Analog signal, decimal scale

Figure 10.5(b) Analog signal, binary scale

SAMPLING RATE

As mentioned earlier, it is important to select a suitable sampling rate (the number of times per second that the amplitude will be measured):

— which will preserve the waveform but
— which will not present an overwhelming amount of data as this requires much storage space.

One fixed sampling rate will be chosen for all the audio frequencies of interest, otherwise it would be too complicated to have a sampling rate which changes with frequency.

The audio frequencies which will be sampled range from 20 Hz to 20 kHz, the audible range. A low frequency wave will get more samples per wave than a high frequency one. For example consider Figure 10.6, showing a 20 Hz wave and a 200 Hz wave. The 20 Hz wave will have more samples per wave than a 200 Hz wave, in fact ten times more. From this example, it is clear that the tricky part will be for the highest frequency wave, as it will get the least number of samples per wave. For the highest frequency of interest, 20 kHz, would one sample per wave be sufficient to describe the wave? The answer is no, since the wave goes up and down but two samples per wave might just do it. Nyquist had analyzed this problem, together with others, and as a result, the minimum sampling rate for preserving all the information in a signal can be specified. This very important result is known as the Sampling Theorem and it usually is stated as:

The highest frequency of interest must be sampled at least twice in order to preserve all the information, i.e. the sampling frequency should be twice the highest frequency of interest.

Two samples per wave would be sufficient for the 20 kHz audio signal, since that is the highest frequency of interest. This means that a sampling rate or sampling frequency of 40 kHz would cover all the audio range, including the 20 kHz signal. This result may be puzzling at first because what if the 20 kHz signal were a complex wave with many harmonics. It does not matter since people, on the average, do not hear above 20 kHz, so the harmonics of 20 kHz do not have to be recorded. As an example of an acceptable sampling rate for CD, recording is made at a sampling frequency of 44.1 kHz, slightly higher than the needed 40 kHz. This value was chosen so as not to work at the limit of the system and for technical reasons. The technical reason for choosing this rate is that it is the video scan rate for TV.

Example: When an audio signal at 1764 Hz is recorded for a CD, one wave of that signal will be sampled how many times? Therefore:

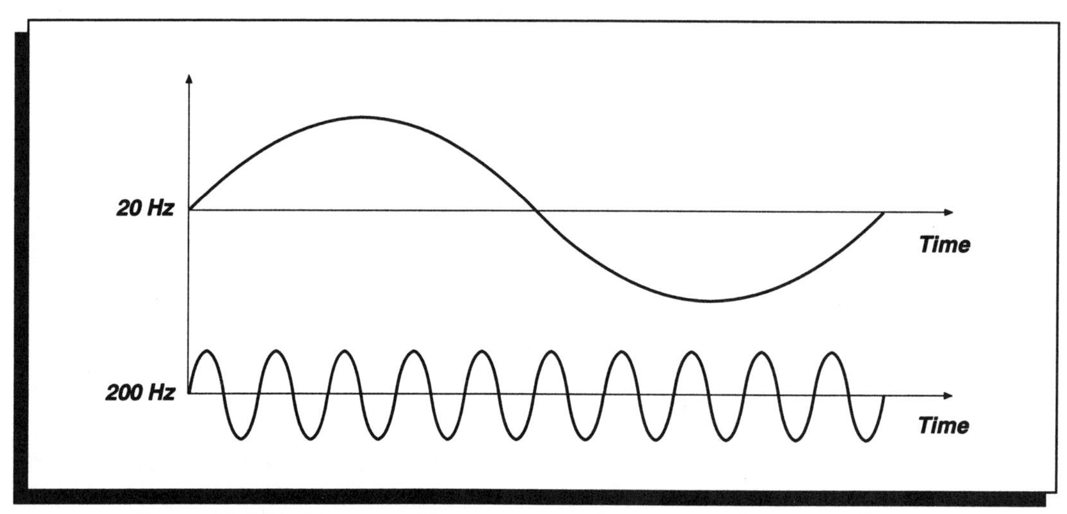

Figure 10.6 20 Hz wave will get more samples per wave than a 200 Hz wave.

44,100 times/second × 1/1764 second = 25 times since one wave lasts 1/1764 second,

while one wave at 17.64 kHz will be sampled:

44,100 times/second × 1/17,640 second = 2.5 times since here one wave lasts 1/17,640 second.

Later on we will see that for other types of recording different sampling rates will be used; however, they all meet the conditions specified by the Sampling Theorem.

ALIASING

An interesting question can arise about what will happen if the Sampling Theorem is not obeyed and in particular, what if a frequency appears that is higher than half the sampling frequency? The answer to this goes back to the example of filming a stage coach and trying to capture the rotation of the wheels when the number of frames/second is not adequate. Unrealistic effects occur and the wheels appear to be turning backward. The same thing happens in audio and this is shown in Figure 10.7. A wave is shown where its frequency is higher than half the sampling frequency. Once sampled, another waveform can be reconstructed by passing through the sampling points and in effect a new wave of lower frequency than the original one is created. This wave did not exist before, and it has sneaked into the scene as an alias frequency. This is a severe form of distortion. A new wave not present originally has been introduced into the audio information because of the aliasing of this signal. We define:

Aliasing **= Introduction of new frequencies when sampling rate is inadequate or when input frequencies are higher than half the sampling rate.**

To understand this, it is useful to remember what was learned about modulation in the chapter dealing with electromagnetic waves. When signals modulate a carrier, as in amplitude modulation AM, sideband frequencies are created in the process which are the sum and difference of frequencies of the carrier and of each frequency modulating it. Here this situation is similar in that the audio signals modulate the sampling pulses. Sideband frequencies are created as shown in Figure 10.8. When the maximum signal frequency F is equal to half the sampling frequency (F_s) there will be no aliasing, however, when the lower sideband frequencies overlap into the audio frequencies, aliasing occurs. In that case F is greater than $F_s/2$. This is to be avoided by making certain that:

No frequencies above half the sampling frequency be allowed to get into the system. This is

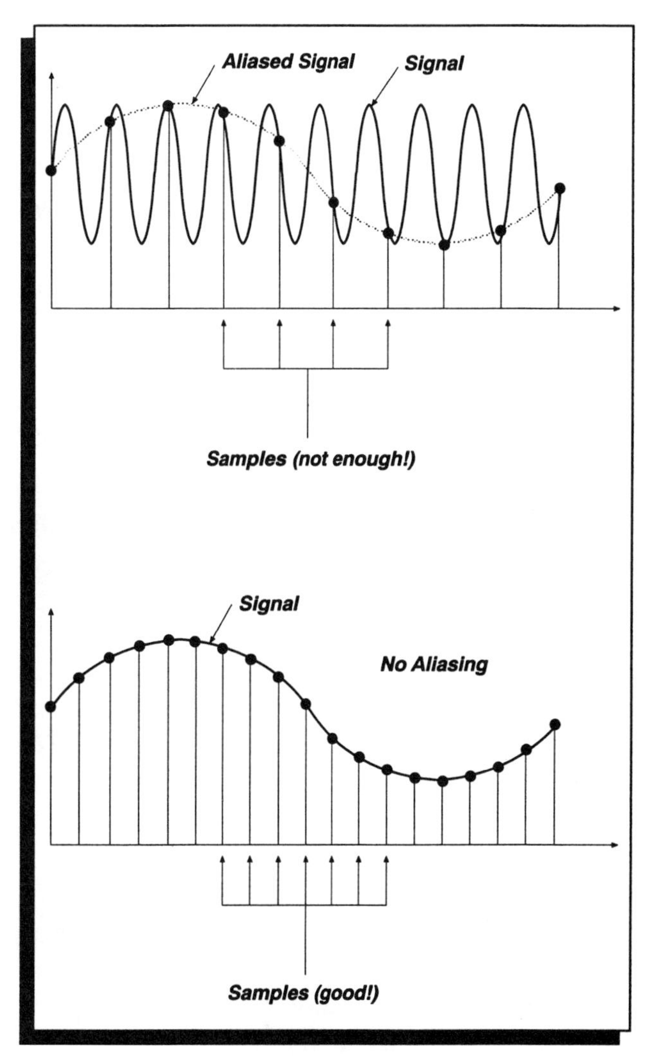

Figure 10.7 Aliasing due to inadequate sampling rate.

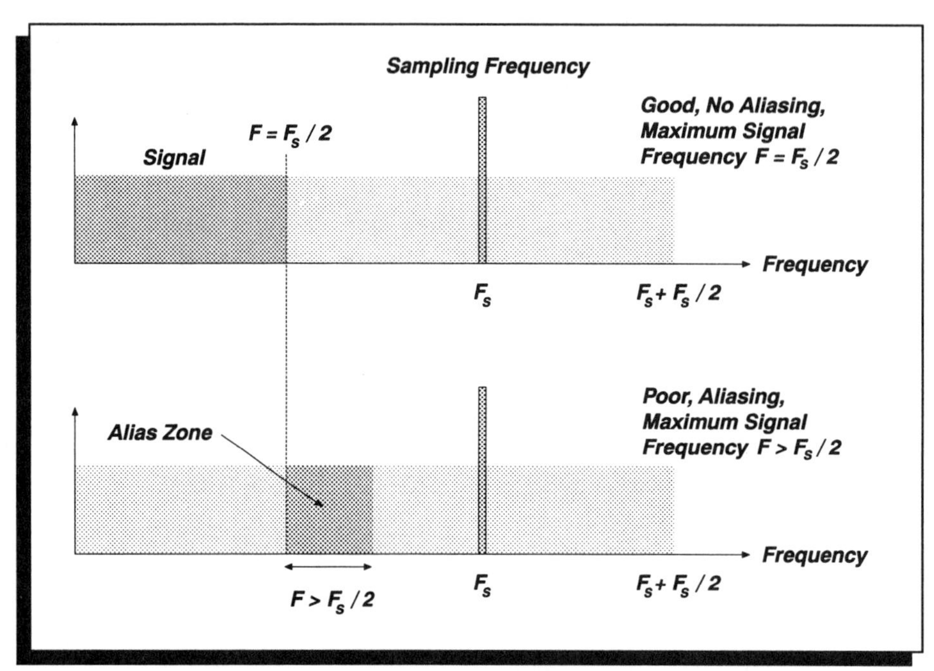

Figure 10.8 Audio spectrum and sideband frequencies due to sampling.

achieved by using a low pass filter at the input to the system.

Example: While recording digitally for a CD, an unwanted frequency of 37.1 kHz gets into the system. This will cause an alias frequency of 44.1 kHz–37.1 kHz = 7 kHz, which will then be introduced into the audio information. A low pass filter is needed here to remove the 37.1 kHz signal and cut away signals above a certain value in frequency.

DIGITIZING FOR RECORDING

The goal in digitizing is to take audio signals, which are analog, and to convert them to digital form so that they can be recorded digitally. To achieve this, the audio signal first goes through a low pass filter to make sure that no frequencies higher than half the sampling frequency get through. This prevents aliasing problems. Then the amplitudes of the signal are sampled so that they can be converted to digital form. This process is not instantaneous and after the signal is sampled it must be held at this amplitude long enough to digitize it. The signal goes through a sample and hold circuit and during the hold time it is converted to digital form by an analog–to–digital converter, A–D converter. The time for this operation is short because when the sampling rate is 44,100 Hz, the time allowed to take one sample is:

$$1/44{,}100 \text{ samples/second} = 0.0000227 \text{ sec.}$$
$$\text{which is } 22.7\ \mu \text{ second.}$$

During 22.7 μ second the signal must be sampled and digitized with 16–bit accuracy (for a CD) and only then can the next sample come along. This is shown in Figure 10.9. The A–D converter is the main element in the digitizing process. It must be

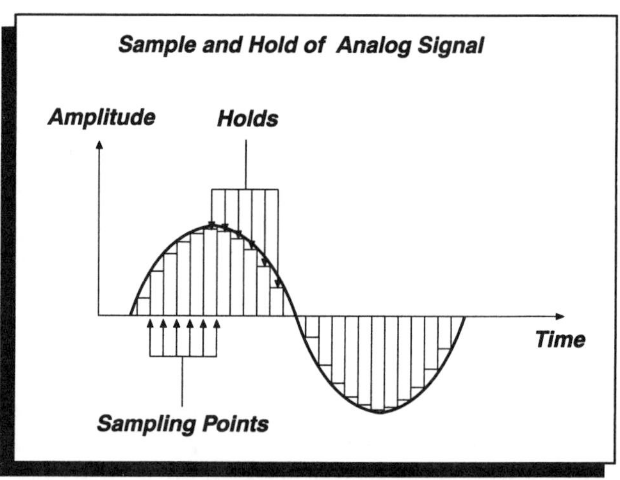

Figure 10.9 Sample and hold of a signal for digitizing.

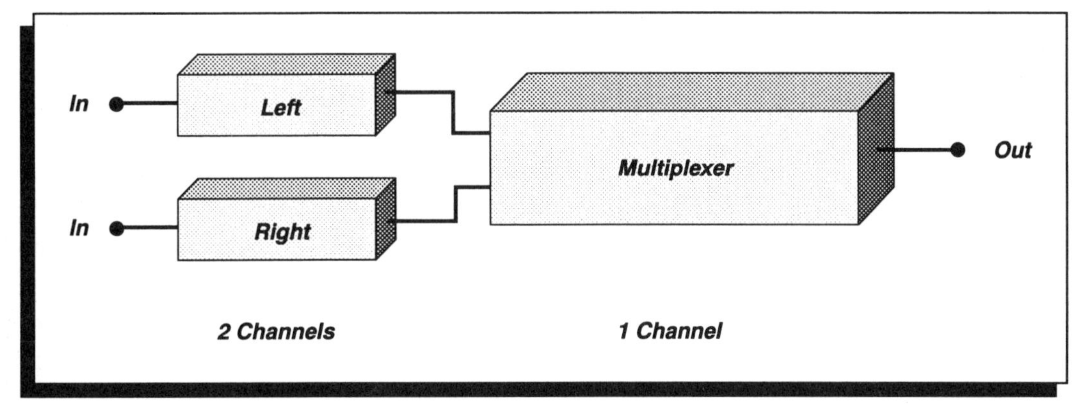

Figure 10.10 Multiplexing of left and right channels.

fast and it must provide the necessary accuracy for expressing amplitudes with 16–bit accuracy. At this point an interesting problem arises: if the signal is stereo, with two channels, how is this information stored? Because the information is in digital form, it is easy to store in any way that is convenient. Here the two channels, the Left and Right, are merged together in an orderly way, one after the other, by a multiplexing circuit. This is shown in Figure 10.10. When these steps are finished, the signal is ready to be digitally stored on whatever medium is convenient: an optical system or a magnetic one. Figure 10.11 summarizes the basic steps taken to get to this point.

PLAYBACK OF DIGITAL INFORMATION

The playback of stored information, whether optical or magnetic, will follow steps which are the reverse of those discussed in the previous section. After the signal is read, it is de–multiplexed to get it back to the original two channels, Right and Left, and then it must be converted back to an analog signal that will eventually go to a loudspeaker. In this sequence of steps, conversion is performed by a Digital–to–Analog converter, a D–A converter. This electronic device has been developed for all sorts of applications; it must be fast and it must maintain the accuracy expected from 16–bit values. Figure 10.12 shows the output of a D–A converter

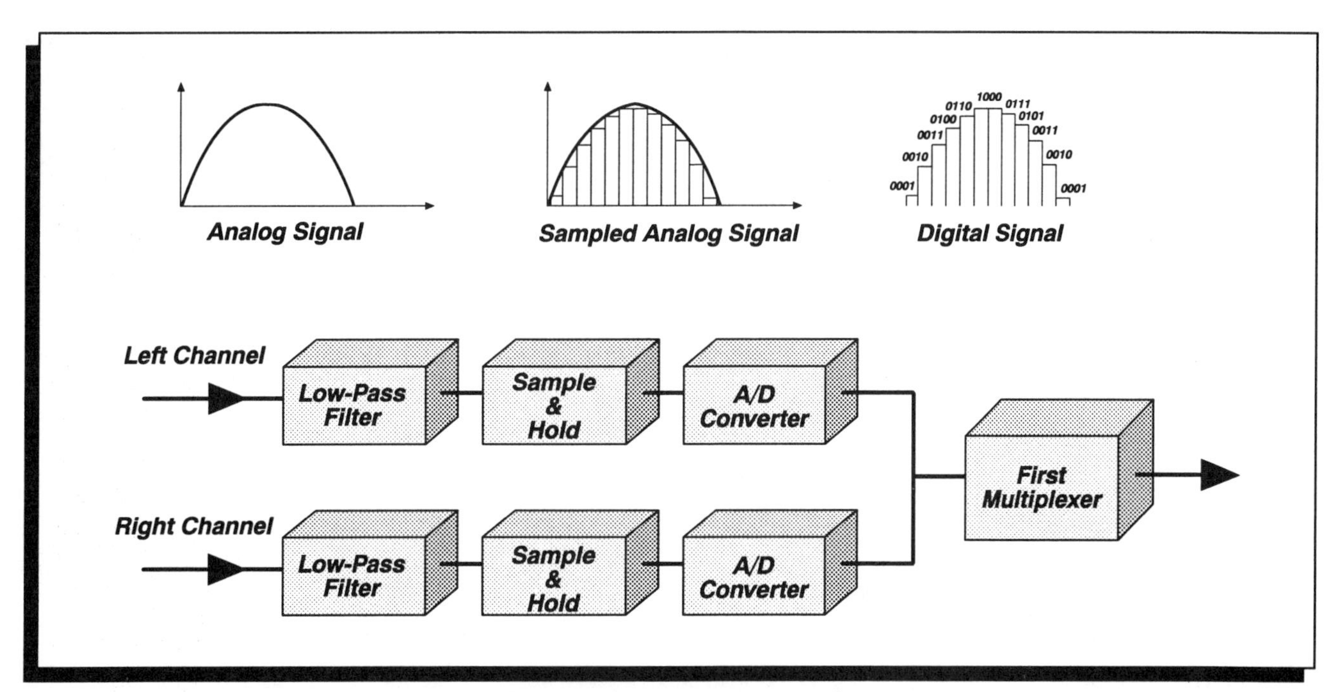

Figure 10.11 Digitizing a signal.

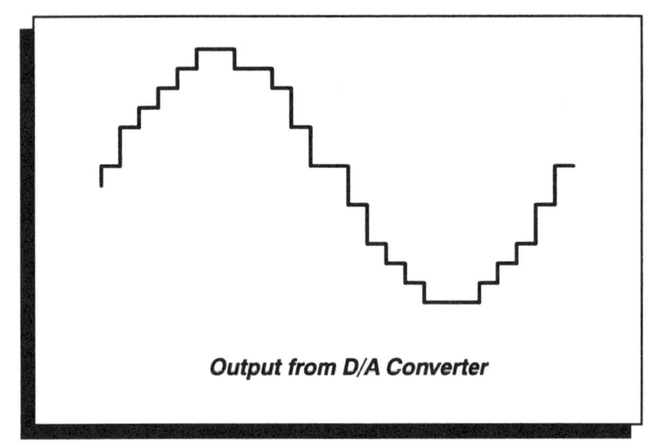

Figure 10.12 Output of D–A converter.

that has been through a sample and hold circuit (to give a cleaner signal). At this point, the amplitudes are in analog form but they still consist of a series of steps, containing high frequency harmonics. A smooth curve without the sharp steps can be achieved by passing the signal through a low pass filter. Now the signal is in analog form and it can be sent directly (via amplifiers) to loudspeakers to reproduce the original sound. Figure 10.13 shows this process.

In summary, the digital signal, regardless of how it was stored or retrieved, goes through a series of steps which convert it back to an analog signal. This is shown in Figure 10.14. The basic steps of the digital process were presented. Now we will consider specific systems.

10.2 OPTICAL RECORDING AND PLAYBACK: THE CD

Presently there are two ways that digital information can be stored and played back using optical methods. The first method, which has been around for 10 years, is the compact disc, the CD system, where the information is stored digitally on a disc by the manufacturer and the consumer can play it back. The second method, which just came out recently is known as the mini–disc. It operates in two modes. The information stored by the manufacturer can be played back. In the second mode new information can be written in and played back by the consumer. This information can be erased and new information can be written in. The readout is performed by a focused laser beam, which is reflected from a series of digital patterns stored within the transparent material of a plastic disc.

■ COMPACT DISC

The information is stored as a series of aluminized pits and lands on a very fine spiral starting near the center hole and winding out to the outer edges of the disc. This is shown in Figure 10.15. Read out is performed by a laser beam which is reflected off the pits and lands. In order to achieve a high density of information on the CD, the pits are very small (0.5 μm wide) and they are separated by 1.6 μm. Such fine details are protected by

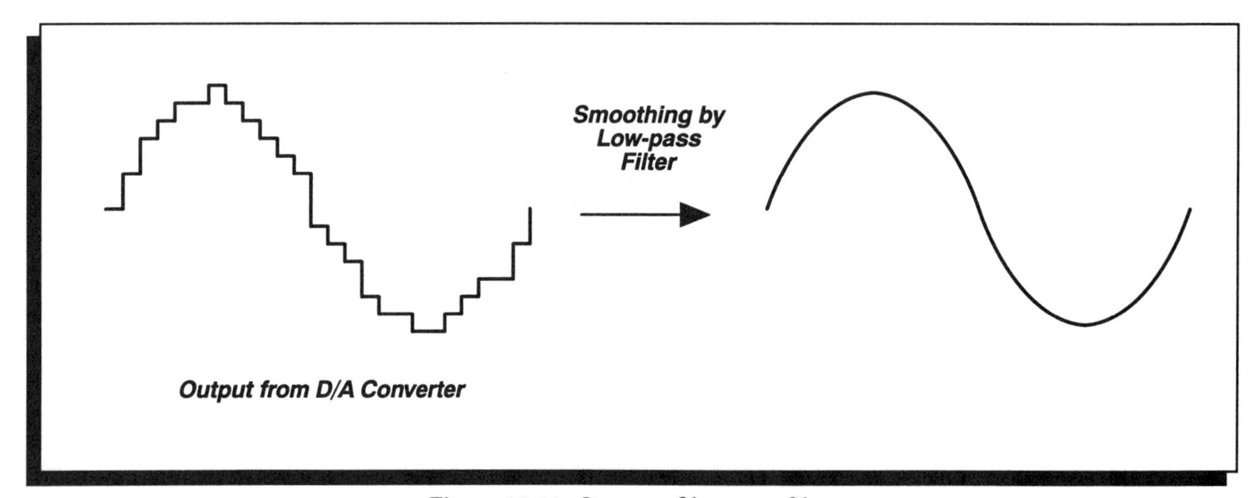

Figure 10.13 Output of low pass filter.

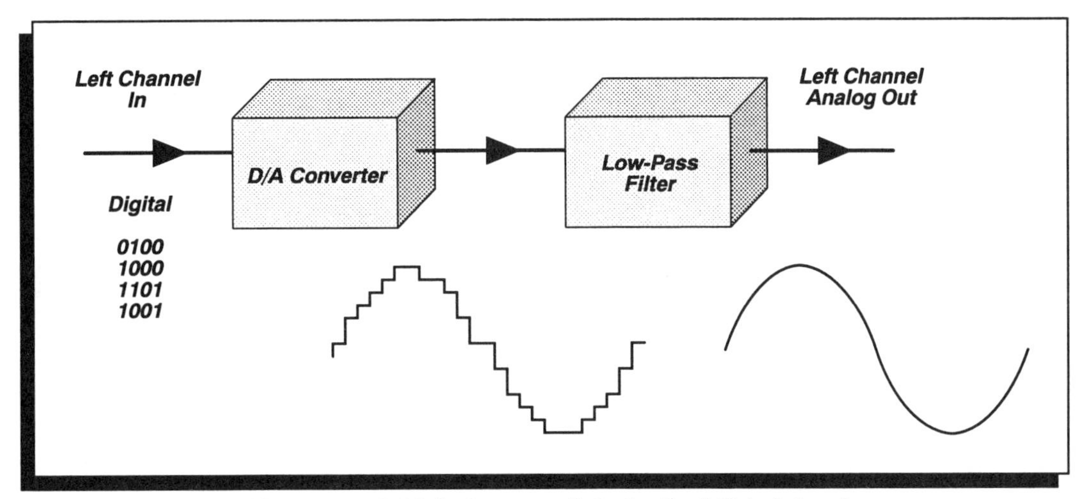

Figure 10.14 Main features of playback of digital signal.

encapsulating the pits and lands in a plastic disc. The digital information is specially encoded so that a change between a pit edge and land (leading or trailing) corresponds to a "1" and all pits or lands in between are "0".

Table 10.1 shows some of the physical characteristics of a CD.

Table 10.1 Physical Characteristics of a CD

Outer diameter:	120 mm
Thickness:	1.2 mm
Diameters for information storage:	117 mm o.d. and 46 mm i.d.
Index of refraction of disc substrate:	1.55
material:	polycarbonate (usually)

Since each track is spaced by 1.6 μm from the next, the total track length can be calculated using the information in Table 10.1. It is very long!

The average length of a pit is the same on the inner diameter of the disc as on the outer one. This means that the rotational speed of the disc must change as it moves toward the outer edge, leading to a constant linear velocity between the laser pickup and the disc. In fact, the rotational speed changes from 500 rpm to 200 rpm as the pickup moves toward the outer edge track. Electronic servo devices control the disc speed in synchronization with words on the disc. Consequently the pits are about the same length on any track, not needing to be stretched near the outer edge. This provides a higher data density on the disc.

It is interesting to note that on a phono gramophone record the waves are longer on the outside tracks and shorter on the inside ones, but the rotational speed is fixed. According to the much used relation, in this book,

$$\text{frequency} \times \text{wavelength} = \text{speed}$$

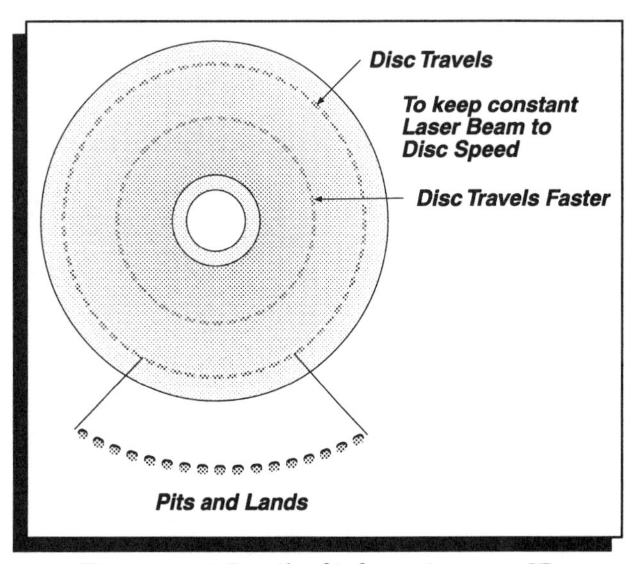

Figure 10.15 Details of information on a CD.

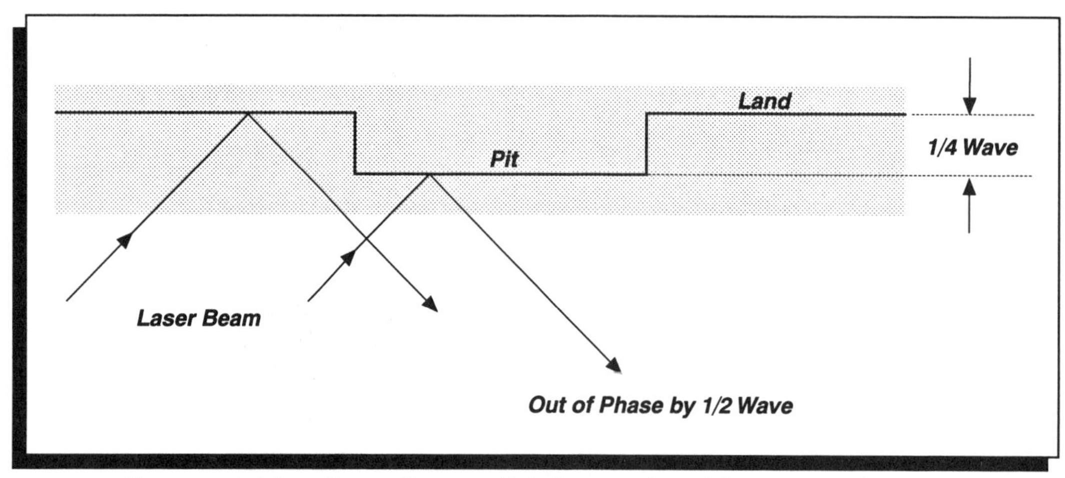

Figure 10.16 Interference between light beam reflected from pit and from flat.

The wavelength is longer, for a fixed frequency, when the speed is higher. This occurs at the outer tracks where the distance that has to be travelled is much larger than for the inner tracks.

Because the speed of light varies for different media, being slower in plastic than in air, a useful quantity is introduced, the index of refraction, expressed as n. It is defined as:

$$n = \frac{\textbf{speed of light in vacuum}}{\textbf{speed of light in medium}}$$

For example, an index of refraction for plastic of 1.55 means that the speed of light in the plastic is the speed of light in vacuum/1.55. The speed of light is slower in plastic because light interacts with the plastic.

Since the pits and lands are covered by a thin layer of aluminum (some discs have gold), the read–out relies on reflection from the pits and lands. The change from pit to land or land to pit relies on interference between various parts of the wave travelling different distances. Figure 10.16 shows that the pits are closer to the surface of the disc than the flats by 1/4 wavelength. The round trip of the laser beam going in ¼ wavelength and then retracing itself through the same distance causes a ½ wavelength path difference between the

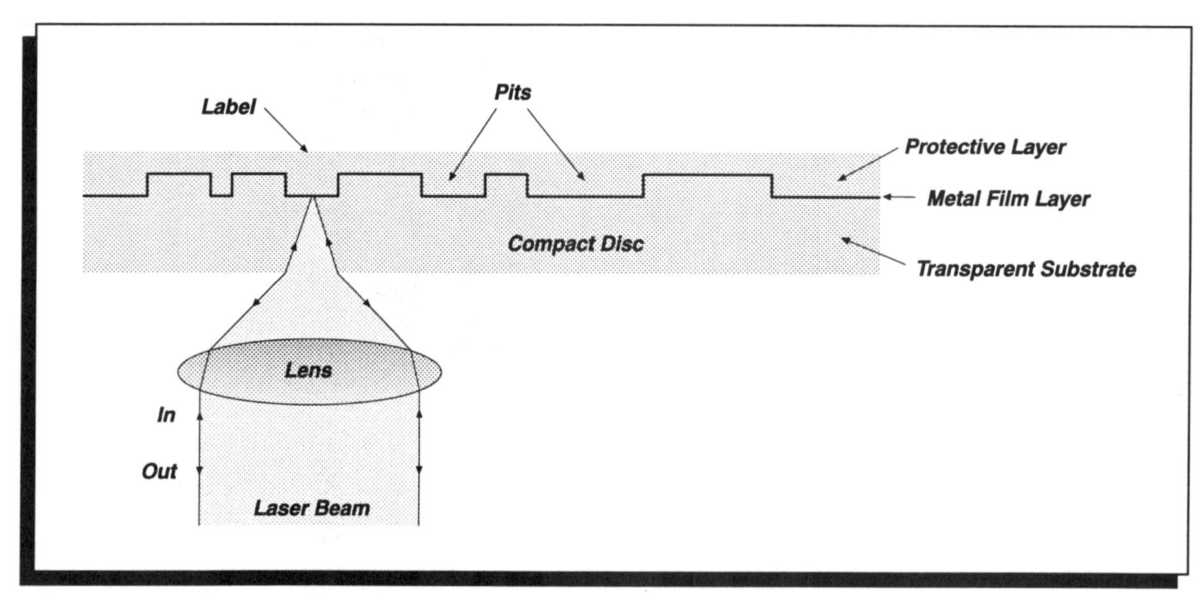

Figure 10.17 Focusing action of laser beam by disc.

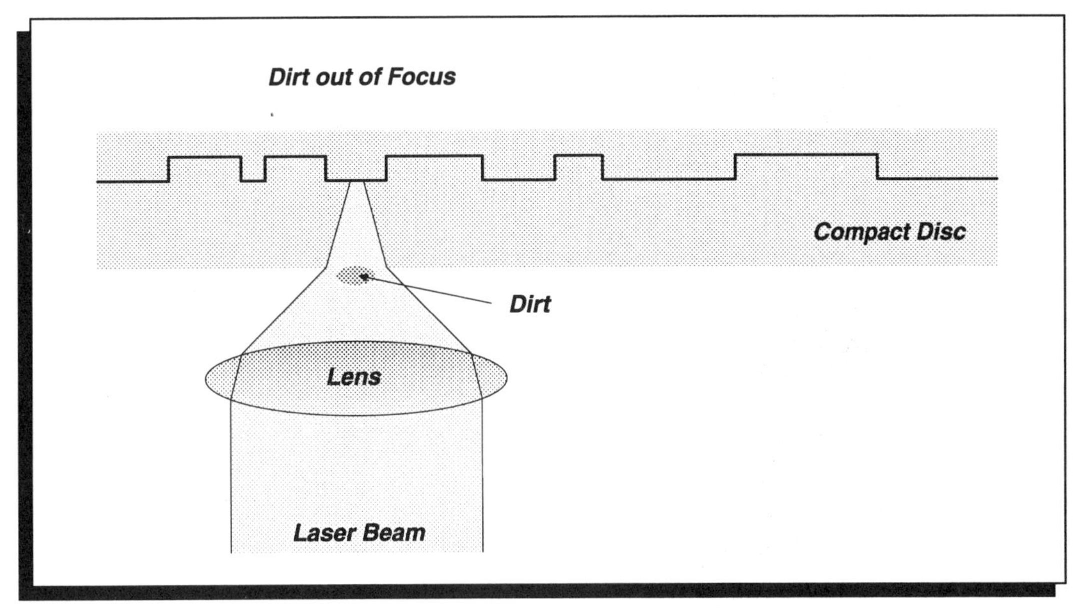

Figure 10.18 Reduced effect of surface defect on CD.

part of the wave hitting the flat and the part hitting the pits. This path difference is just enough to cause destructive interference between the two parts of the light beam.

In order to use destructive interference of the beam between pits and flats, the height variation between the pits and flats must be ¼ wavelength. The speed of light in the plastic differs from that in vacuum, since the index of refraction of the plastic is 1.55. This means that the wavelength of the laser beam in the plastic will be 1.55 times shorter than in air. The wavelength of the laser is 780 nanometers and it will become 780/1.55 nanometers in the plastic, which is 503.2 nanometers. The pit must then be displaced from the flat by 125.81 nanometers. In this case, as the laser beam is scanned from the pit to a flat, there will be a decrease in the reflected light due to destructive interference. Since the waves here are very short we use 1 nanometer = 10^{-9}m.

The large change of light speed as the beam enters from air to plastic will lead to a change of direction of the beam (when it comes in at an angle); this is refraction. Figure 10.17 shows refraction and how it can be used to reduce the effects of scratches and dirt on the surface of the disc.

Since dirt and scratches will be mainly at the surface of the CD, they will be out of focus when scanned by the laser beam which is focused on the pits and lands inside the disc.

The speed of light in the plastic of the CD can be calculated from the index of refraction n. For n = 1.55 the speed of light is:

$$3 \times 10^8 \text{ m/sec}/1.55 = 1.93 \times 10^8 \text{ m/sec}.$$

It will determine where the laser beam will be focused. This is shown in Figure 10.18 where a scratch or dirt at the surface is out of focus and its effect is greatly reduced, while the laser beam focuses on the pits.

The ability to ignore such defects is limited by their size, since above a certain limit they will affect the reading of the information. Large scratches will prevent the information from being read well.

It is interesting to look into the size of the beam when it is focused at the pits.

A fine beam will detect very small details and it will be possible to store a high density of information on the disc. The situation is very similar to the sound radiation pattern around a speaker where, for a given speaker diameter, high frequencies beam mainly forward while the low frequencies are spread around. Here a laser beam emerging from a lens and focused onto a spot will have an image which will be strongly affected by diffrac-

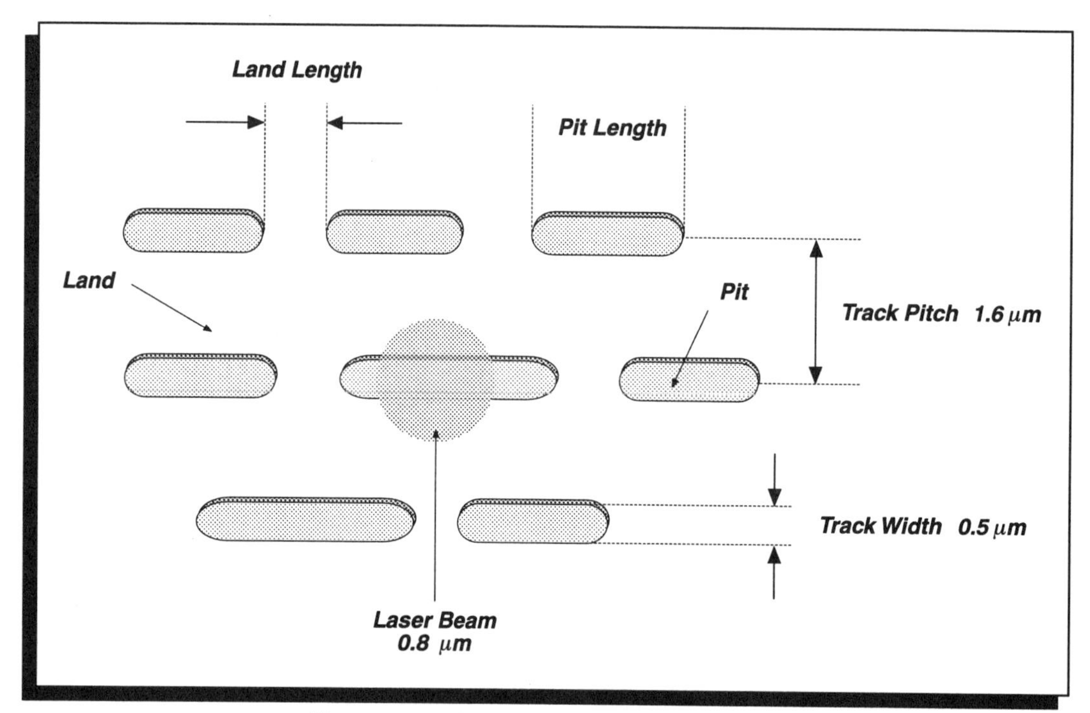

Figure 10.19 Laser spot focused on disc data.

tion. Figure 10.19 shows what the intensity distribution of this beam should be. It must be strongly peaked at the center, with a decreasing distribution that oscillates as it moves away from the center.

The intensity oscillations are due to interference from various parts of the beam emerging from the lens. In fact, we cannot get a fine point. It is instructive to consider where on Figure10.19 the tracks should be placed on the disc. One track should be covered by the high intensity part of beam. However, the adjacent tracks would be ideally situated at the first minima of the laser beam diffraction pattern. This leads to a spot size of 1.7 µm in diameter and hence the spacing between tracks should be ~1.7 µm for a laser beam working in the infra–red at a wavelength of 780 nm. A finer beam can be formed by operating with a shorter wavelength laser beam; this will then lead to a higher density of information. At the present, all CDs are limited to the infra–red range because of the convenience and low cost in using IR semiconductor lasers. The argument for the limiting diameter of the laser beam is based on a well–known study of this problem and it is usually referred to as the Rayleigh criterion for observing optically small objects. It shows that the wavelength of the light limits how small an object can be seen.

Many CD players feature three–beam laser detection. In fact, they have only one semiconductor laser whose beam is split. The three–beam is produced by a diffraction grating which generates the central beam for reading the information, and the first order beams are used for tracking, between two adjacent tracks. This arrangement is shown in Figure 10.20. The three–beam system has excellent tracking features at the expense of having a reduced intensity main beam.

The CD information is usually read from below with the laser beam directed up, and then reflected from the information inside the CD. Such a read–out is not simple because the incoming beam must not get mixed up with the reflected one. The solution to this problem is elegantly found in the application of the polarization of the light beam. When the incoming light beam has a different polarization from the reflected one then it is easy to separate the two. This is the approach taken in all CD's. A light wave is an electromagnetic wave and has transverse electric and magnetic fields. When the disturbance is confined to a plane, the wave is plane polarized. Figure 10.21 shows an unpolar-

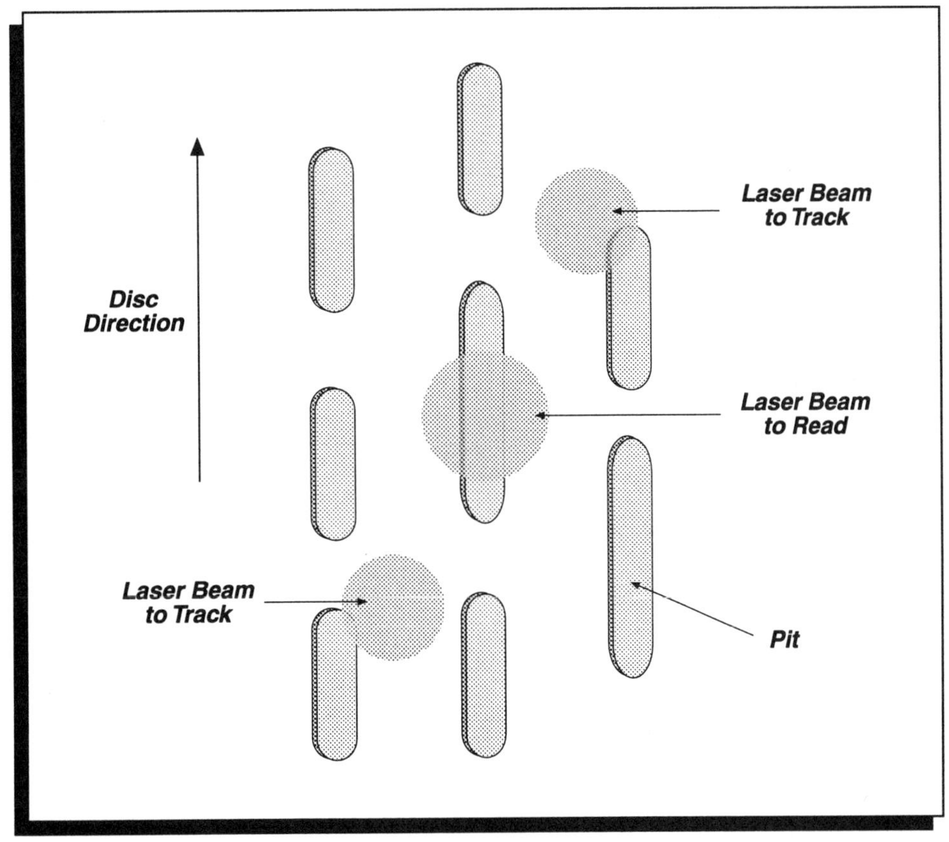

Figure 10.20 Three-beam detection, one for read-out and two beams for tracking.

ized beam and a plane polarized beam.

When the light beam emerges from the laser it is plane–polarized horizontally. It passes through a device, known as a quarter–wave plate, which changes the polarization to circular polarization. The disc information reflects the light like a mirror changing on reflection its circular polarization by 180° (i.e. it changes from a clockwise circular polarization to a counter–clockwise circular polarization). On the return path it passes through the

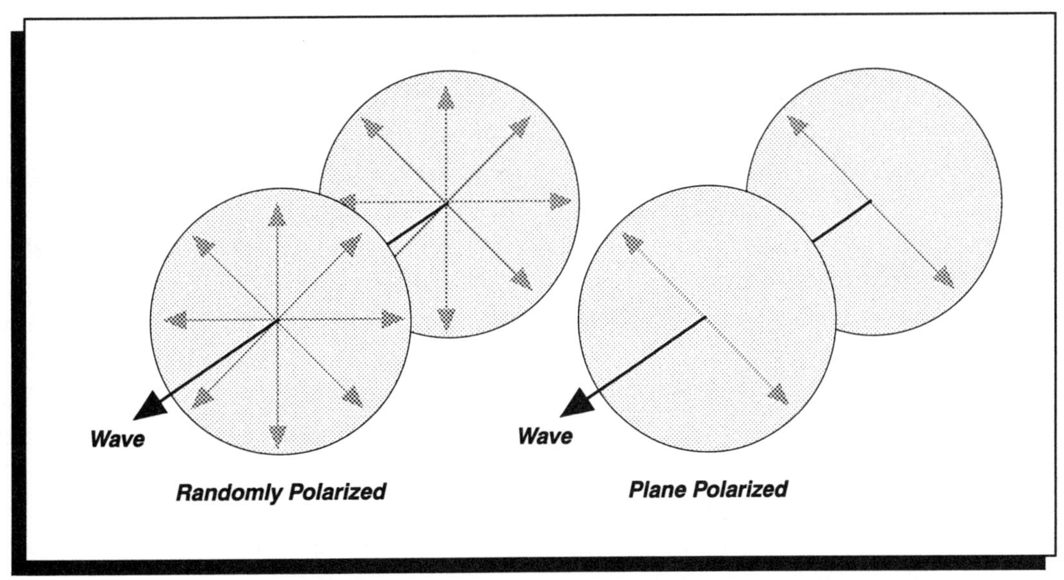

Figure 10.21 Randomly polarized beam and plane polarized beam.

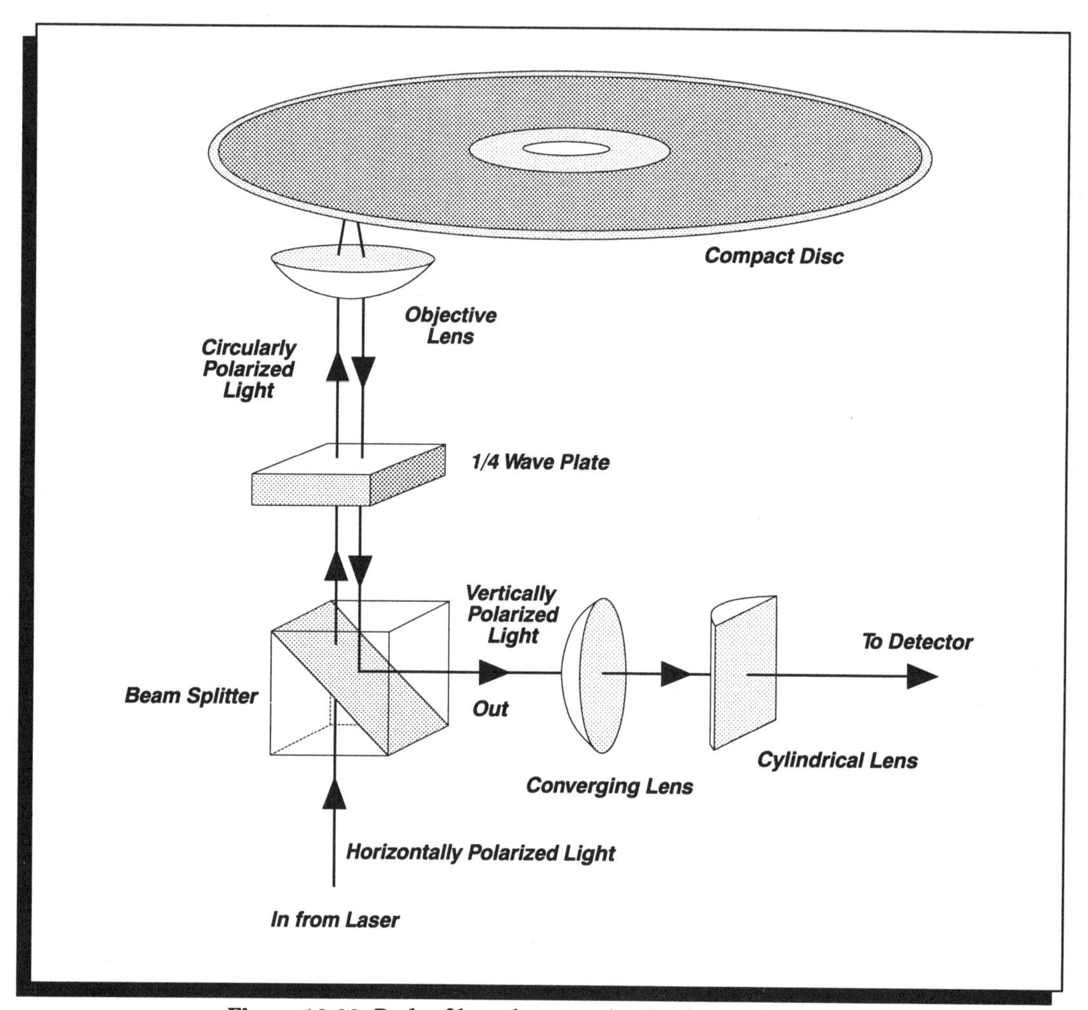

Figure 10.22 Path of laser beam and role of its polarization.

same quarter wave plate which now changes the circular polarization to vertically polarized light. On its way back this polarization is deflected by a beam splitter at 90° from its original direction, sending it to a detector. Figure 10.22 shows a simplified version of this process and the crucial role played by the polarization.

The role of the beam splitter is to transmit horizontally polarized light toward the CD and to reflect in the return trip vertically polarized light toward a photodetector. It is important that the laser beam be locked onto the data track. This is achieved by a servo motor and a system of lens which continuously maintain the beam in focus on the disc track.

The tracks are densely packed and yet they can be detected because the wavelength of light is short. This gives the CD an immense storage capability. Each pit is about 0.6 μm wide and if 5,000 of them were stacked side by side, they would only be about as wide as a letter of this text.

■ Laser

In order to accurately read the information contained in the pits and flats of the CD, a light beam is used in such a way as to have constructive and destructive interference effects from the information. This requires careful control of the phase of the light beam which can be achieved by a laser. It is a device that puts out light at essentially one frequency and which maintains phase coherence.

This is in contrast with a light beam from a light bulb where all sorts of frequencies are emitted, and each wave has a different phase relative to the others. Figure 10.23 shows the difference between a coherent beam and an incoherent one. The coherent beam is produced by a semiconductor laser which produces light waves with a wavelength of 790 ×

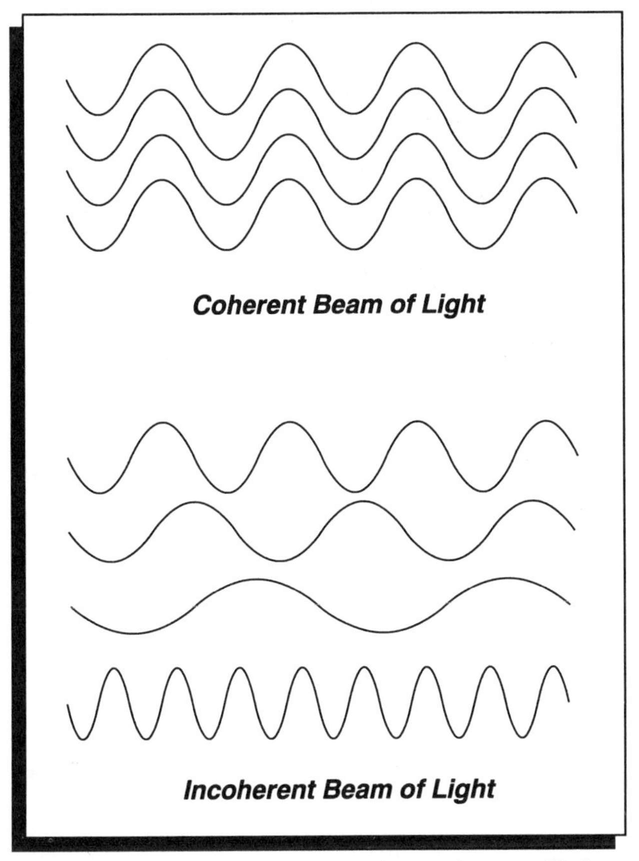

Figure 10.23 Coherent and incoherent beams of light.

10^{-9} meter. The word laser comes from **L**ight **A**mplification by **S**timulated **E**mission of **R**adiation. The basic principle of operation of this device is that electrons radiate electromagnetic waves when they are forced to change their energy; a resonance is used to enhance the effect. Figure 10.24 shows a semiconductor laser based on GaAlAs.

Atoms of the laser material are excited to higher energy states. When they drop back to the lower energy state they emit light. That light stimulates other excited atoms to drop back to the lower energy state, emitting light as well. Such multiplicative effects continue as long as energy is pumped into the system, producing laser light. Such effect is enhanced by having the beam bounce between 2 mirrors in a resonant cavity before coming out. This light is special in several ways:

— It essentially consists of a single frequency
— It comes out parallel in a straight narrow beam
— The beam is coherent meaning that each part of the wave maintains a constant phase relative to other parts of the wave.

The coherence property of a laser beam makes it attractive for interference and diffraction effects. It is for this reason that it is used with a CD where the read–out is based on interference effects between the light reflected off the pits and flats.

10.3 COMPACT DISC SPECIFICATIONS

The performance of CD players has reached such high levels of excellence that it is unlikely that there are vast differences between the

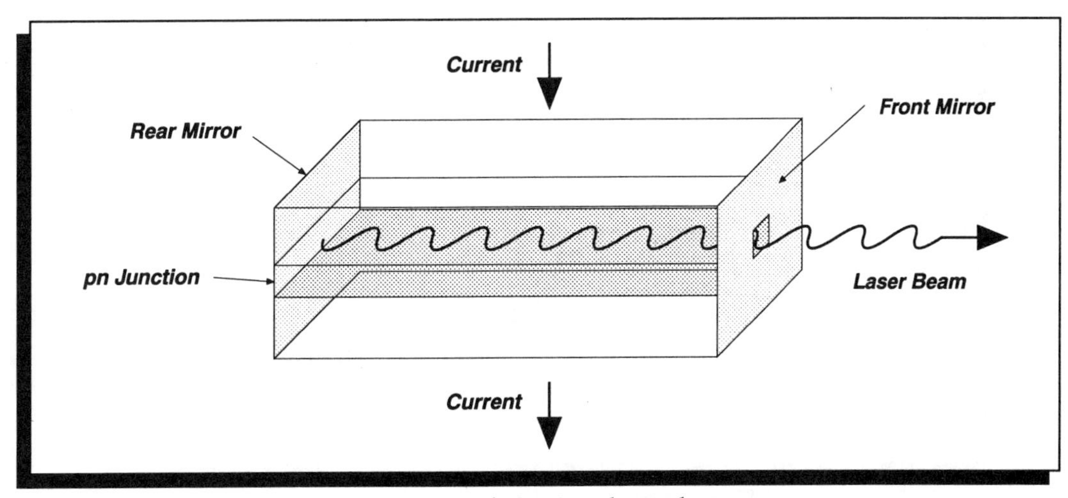

Figure 10.24 Semiconductor laser.

various CD players. The differences that do exist are very small and they refer to sophistication of controls, features, looks, and not so much with sonic performance.

The specifications surpass all the analog ones (such as for tape decks or phono records) and they have achieved very high levels of standards. Here are typical examples of what is presently available:

— ***Frequency response:*** 2Hz to 20, 000 Hz ± 0.3dB.
— ***Signal–to–noise ratio:*** greater than 100dB.
— ***Total Harmonic Distortion:*** less than 0.002% at 1 kHz.
— ***Dynamic range:*** loudest signal/softest signal in decibels. Theoretical maximum is 98dB for 16 bits accuracy.
— ***Stereo separation:*** better than 94dB.
— ***Filter:*** digital with 4–times or 8–times oversampling. This filter in the CD player essentially resamples the original digital data at a higher rate so as to make it easier then to filter. Figure 10.25 shows the effect of 2–times and 4–times oversampling of data.

In this example, the 4–times oversampling makes it easier to filter and to get back to the original waveform, which is a smooth curve without sharp edges. The purpose of a digital or analog output filter is to suppress the high–frequency components in the sharp steps of the sampled audio signal. The high–frequency components are all above 20 kHz and they cannot be heard. They should be suppressed so they do not overload the amplifier and produce distortion. That is the role of the output filter, and a digital one is the most economical approach. The quest for higher sampling rates is not really justified and it is not necessarily true that the higher oversampling rate will lead to better performance — it is all inaudible anyway. A large value for the oversampling rate should not be a reason for choosing one player over another one.

— ***Digital–to–analog converter (DAC):*** there are two types, the multi–bit converter and the 1–bit converter. The multi–bit DAC (16 or 18 and even 10–bit in some units) changes the 16–bit number on a CD to a stair–stepped waveform which is then filtered to

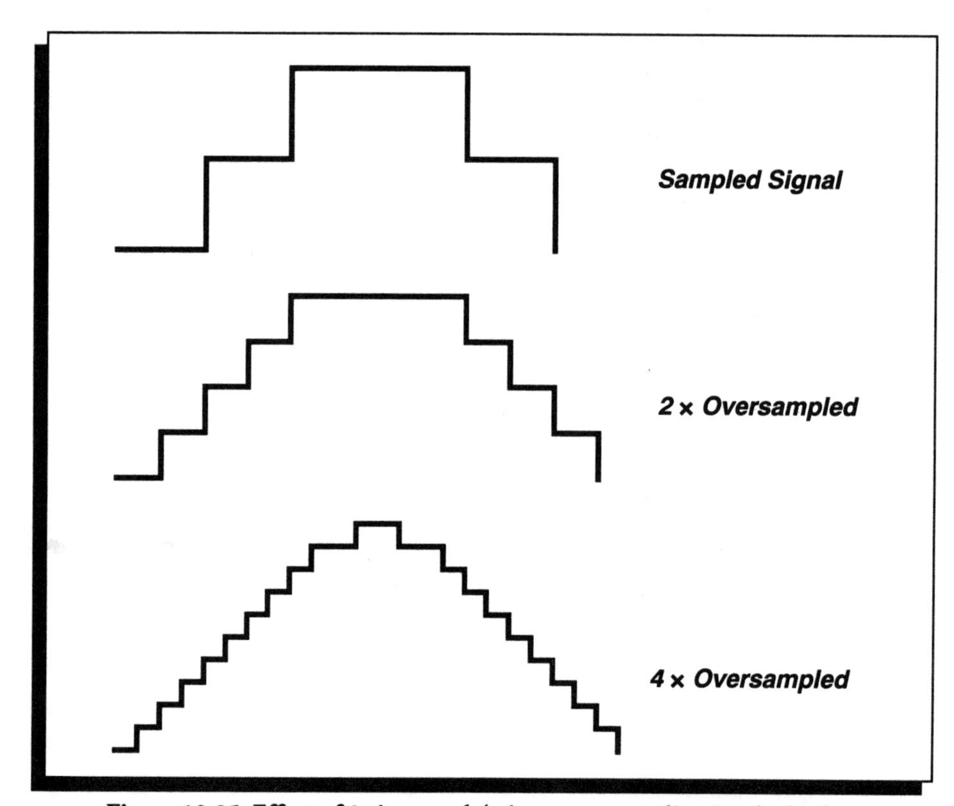

Figure 10.25 Effect of 2-times and 4-times oversampling in playback.

recreate the original smooth waveform. In a 1–bit DAC the 16– bit numbers are converted to a high frequency output of constant amplitude pulses, the number of them depending on the signal. When filtered it replicates the original waveform. The performance of both DAC can be excellent. The 1–bit DAC is simpler and it has long–term stability, which is important in low price players.

— ***Digital outputs:*** useful for use with a separate DAC, or for
direct recording on a digital tape deck or disc.

Single–disc CD players no longer dominate the market as multi–disc CD changers have become popular. Here there are two choices, the magazine–type player, which uses a cartridge of CDs, or the rotary carousel type on a rotating platter with five to six discs.

The CD player has recently been very successful in the portable format, which comes in two basic configurations: the hand–held player and the boom–box. The hand–held players usually require head phones, but they can also be plugged into a regular hi–fi system. For the boom–box configuration, no extra attachments are necessary as it has an amplifier with speakers. The specifications tend to be inferior to the regular CD players, but they still are high quality. The frequency response is typically 20 – 20,00 Hz ± 0.01 to 3 dB, depending on the model.

10.4 PRERECORDED MINI–DISC

The Mini–Disc is a new audio format which uses optical read–out. It comes in two types: prerecorded mini–disc and a recordable mini–disc. In this section we will discuss the prerecorded type (the recordable one will be left for the next chapter). It is specifically designed for portable applications but it also comes in the model for home use.

The disc is much smaller than a CD, measuring only about 2½ inches in diameter and having a surface area ¼ that of a CD; it is in a special case like a floppy disk. The disc is read by a laser beam and is capable of producing 60 or 74 minutes of music, depending on the disc. Information is recorded as pits in the disc's substrate much the same way as in a CD. It has a special encoding scheme called **A**daptive **T**ransform **A**coustic **C**oding (ATRAC), and in this way it is very different from a CD. The encoding scheme is based on:

— Psychoacoustic principles dealing with the threshold of hearing and masking.
— Audible sounds are recorded with the resolution necessary for hi–fi reproduction with a variable number of bits per sample adaptively based on the sound's audibility compared to the noise (note that audio information for a CD is recorded with 16–bit resolution for every sample).

Such a scheme allows an efficient method of encoding by ignoring sounds that fall below the threshold of hearing. Actually, only about 20% of the available data is recorded. The ATRAC system compares the input data to a psychoacoustic model and discards all the information that the listener probably cannot hear, recording only the essential information. The system in this respect is more efficient than a CD. Presently tests are being conducted to determine the validity of such an approach. Also, loud sounds are given only the necessary number of bits because they do not need more bits to accurately record their large amplitude. Because of such a scheme the audio data rate is just below 0.3 megabits per second (on a CD it is 1.4 megabits per second).

An important feature of this unit is that it is not susceptible to shocks and vibrations, which is usually a problem with older portable CD units. Here the data is read into a memory at a rate faster than it delivers. Should the playback head mistrack due to a vibrations, the unit switches automatically to the memory. Figure 10.26 shows the approach taken to achieve this shock–proof memory.

The main differences between a CD and a prerecorded Mini–Disc are:

CD	Prerecorded Mini–Disc
— every sample has 16–bit resolution.	— every sample has variable resolution using many bits only when necessary
— all sampled information is recorded	— only data which compares with psychoacoustic model is recorded
— maximum playtime: 74 minutes	— maximum playtime: 74 minutes
— thickness: 1.2 mm	— thickness: 1/2 mm
— diameter: 120 mm	— diameter: 64 mm
— linear track speeds: 1.2 to 1.4 m/sec	— linear track speeds: 1.2 to 1.4 m/sec

Figure 10.26 Shock-proof memory in mini-disc.

Typical specifications feature:

— Frequency response: 20 to 20,000 Hz ± 0.3 dB
— Dynamic range: 85.5 dB
— Distortion: 0.04%
— Signal–to–noise ratio: 88.4 dB
— Channel separation: 78 dB at 1 kHz

SUMMARY OF TERMS

Aliasing: Introduction of erroneous signals due to frequencies too high for sampling rate.
Analog: Continuous representation of signal so that it looks similar to the sound it represents.
Analog–to–Digital Converter (ADC): An electrical circuit which converts an analog signal into a digital one.
ATRAC: Adaptive Transform Acoustic Coding system used for encoding audio information on a Mini–Disc. It is based on psychoacoustic principles dealing with the threshold of hearing and masking so as to reduce the amount of acoustic data recorded.
Binary System: System of numbers based on 2 numbers, a "1" and a "0". It is much simpler for electronic devices and machines than a decimal system.
Bit: Stands for ***bi***nary digi***ts***. It represents the significant places in a binary number.
CD player: A digital source component that plays compact discs and uses optical readout.
Digital: Signal is expressed as a series of numbers each corresponding to an amplitude of the signal.
Digital–to–Analog Converter (DAC): An electrical circuit which converts a digital signal into an analog one.

Filter: A circuit that attenuates, boosts, or removes selected frequencies.
Index of Refraction: Speed of light in vacuum divided by speed of light in medium.
Laser: a device which emits a beam of coherent light. The name stands for **L**ight **A**mplification by **S**timulated **E**mission of **R**adiation.
Mini–Disc: New compact format of digital audio information with readout by laser beam. It uses ATRAC system for encoding.
Multiplexing: Merging of two channels together.
Oversampling: Digital filtering technique used in CD player to remove spurious signals from audio range.
Polarization of laser beam: Specific direction of electric field in electromagnetic wave.
Sample and Hold: Circuit which samples a signal and holds amplitude at sampled value for a time long enough for it to be digitized.
Sampling Frequency: The number of times per second the amplitude of signal is measured. For a CD it is at 44,100 times per second. It must be at least two times the highest frequency of interest.
Sampling Theorem: To digitize an analog signal, it must be sampled at a frequency which is at least twice the highest frequency of interest.

NAME______________________ DATE __________

QUESTIONS FOR REVIEW

1 What are the advantages of digital recording over analog recording?

2 What determines the dynamic range on a CD?

3 What is the maximum relative range of amplitudes available on a CD?

4 Explain the function of an ADC and a DAC.

5 What is the Sampling Theorem? Explain its importance.

6 Why is it necessary to read a CD with a laser beam rather than with ordinary light?

7 How does a prerecorded Mini–Disc differ from a CD?

8 Comment on the effects of a laser which emits blue light rather than infra–red in a CD system .

9 What role does the polarization of the laser beam play in the playback of a CD?

10 What limits the amount of information stored on a CD?

NAME ______________________________ DATE ____________

Please select one answer.

1. What is the maximum value of a 16–bit number?
   ___A. 65,535
   ___B. 16
   ___C. 32,767
   ___D. 131,071
   ___E. 32

2. If you wanted to record digitally audio information from 30 Hz to 16 kHz, what would be an acceptable sampling rate?
   ___A. 16 kHz
   ___B. 32 kHz
   ___C. 8 kHz
   ___D. 20 kHz
   ___E. 30 Hz

3. On a CD when dealing with audio information how many bits per second of data are handled?
   ___A. 705, 600
   ___B. 1,411,200
   ___C. 44,100
   ___D. 16
   ___E. 32

4. The digital information on a CD:
   ___A. starts at the outer edge and spirals in toward center.
   ___B. starts near the center and spirals outward toward the edge.
   ___C. starts at the outer edge as 2 separate channels (one for left and one for right) and spirals in toward center.
   ___D. starts near center as 2 separate channels (L and R) and spirals outward.
   ___E. is randomly located on CD.

5. Consider a compact disc which has a dynamic range of 100dB. If one compares it to a phono player whose dynamic range is 50dB, the sound produced by the CD will be __________ than by a record.
   ___A. 50 times quieter
   ___B. 32 times quieter
   ___C. 50 times louder
   ___D. 32 times louder
   ___E. 1024 times louder

6. Consider wave X digitally recorded with an amplitude 000111 at one instance while wave Y has amplitude 110001. The power of Y must then be __________ that of X.
   ___A. 7 times
   ___B. 1/7 times
   ___C. 49 times
   ___D. 1/49 times
   ___E. the same as

7. The left channel of an audio signal at 1764 Hz is to be recorded on a CD. During ½ wave of that signal, it will be sampled __________ times.
   ___A. 1764
   ___B. 44,100
   ___C. 50
   ___D. 25
   ___E. 12.5

8. Express 201 as an 8–bit binary number.
   ___A. 11001001
   ___B. 11100011
   ___C. 11001000
   ___D. 11111111
   ___E. 11001011

\9. When playing back a digitally recorded disc, a CD, the information from the disc must:
___A. remain in analog form since it really was recorded that way on disc.
___B. be sent in digital form to the speakers.
___C. go through an ADC before being sent along to the speakers.
___D. go through a DAC before being sent along to the speakers.
___E. be sampled at 22,050 Hz.

10. The digital information on a CD is:
___A. comparable in size to the wavelength of the laser beam.
___B. much larger than the wavelength of the laser beam.
___C. much smaller than the wavelength of the laser beam.
___D. 16 times larger than the wavelength of the laser beam.
___E. 16 times smaller than the wavelength of the laser beam.

11. If a CD were recorded with 20–bit resolution rather than 16 bits, which sound factor would be affected?
___A. its frequency
___B. its wavelength
___C. its speed
___D. its amplitude
___E. its harmonic content

12. If instead of using an infra–red laser, a blue laser were used, the pit depth would have to:
___A. increase.
___B. remain the same.
___C. decrease.
___D. increase ten times.
___E. decrease ten times.

13. The wavelength of the infra–red laser in a CD is 7.9×10^{-7} meter. What is its frequency?
___A. 3.8×10^{15} Hz
___B. 3.8×10^{14} Hz
___C. 1.9×10^{14} Hz
___D. 7.6×10^{14} Hz
___E. 24×10^{15} Hz

14. When recording audio on a CD, how many more samples per wave will there be for a 20 Hz sound compared to one at 20,000 Hz?
___A. 44,100
___B. 1,000
___C. 1/1,000
___D. 10,000
___E. 400,000

15. A Mini–Disc saves space by:
___A. recording not all the sound.
___B. recording certain tones with less than 16–bit accuracy.
___C. not recording some of the sounds that tend to be masked by loud ones.
___D. selecting carefully the sound that will be recorded.
___E. All of the above.

16. When a laser beam enters the plastic of the CD, its speed:
___A. decreases.
___B. increases.
___C. remains the same.
___D. decreases as well as its frequency.
___E. increases as well as its frequency.

17. The wavelength of the laser beam in the plastic of the CD is:
___A. longer than in air.
___B. shorter than in air.
___C. the same as in air.
___D. 1.55 times longer than in air.
___E. $(1.55)^2$ times shorter than in air.

18. As the laser beam moves away from the outer edge, the speed of the CD:
___A. increases.
___B. decreases.
___C. remains the same.
___D. is maintained such that there are 44,100 rotations per second.
___E. is maintained such that there are 16 rotations per second.

NAME ______________________________ **DATE** ____________

19. A laser beam emits coherent light such that each part of the beam has:
    ___A. the same frequency
    ___B. the same phase
    ___C. the same direction
    ___D. the same wavelength
    ___E. all of the above.

20. At least how many bits total does a CD contain, if it can play 74 minutes? ________ bits.
    ___A. 1,411, 200
    ___B. 6.26×10^9
    ___C. 84,672,000
    ___D. 65,535
    ___E. 3.13×10^9

CHAPTER 11

Digital Magnetic Recording and Playback

Magnetism has had extensive applications to recording of information and to its playback. Presently, a new generation of recording and playback instruments has appeared which is based on a variety of novel applications of magnetism. This chapter deals with three such units: the recordable mini disc, digital audio tape recording, and the digital compact cassette. The various sophisticated techniques used in the recording and the playback processes offer high–density data storage media and excellent high–fidelity reproduction of sound in the playback.

11.1 Recordable Mini-Disc

This unit was developed so that sound could be recorded digitally, played back as many times as necessary, and then erased for new information to be recorded. The number of times that the information can be erased and recorded is essentially unlimited.

The principles which are used in this new format are:

— digital signals are recorded magnetically using a laser beam on a thin film of magnetic material encased in a disc.
— signals are read from the disc using magneto–optical technology.
— efficiency of data storage is achieved by means of data compression technique, ATRAC.

Recordable mini discs, unlike the prerecorded type, employ magneto–optical technology on a magnetic thin film of Terbium Iron Cobalt (Tb Fe Co) which has been deposited on a plastic substrate. This film is mounted inside the plastic of the disc. Magnetic digital signals are recorded perpendicularly (vertically) on the disc as shown in Fig. 11.1 by

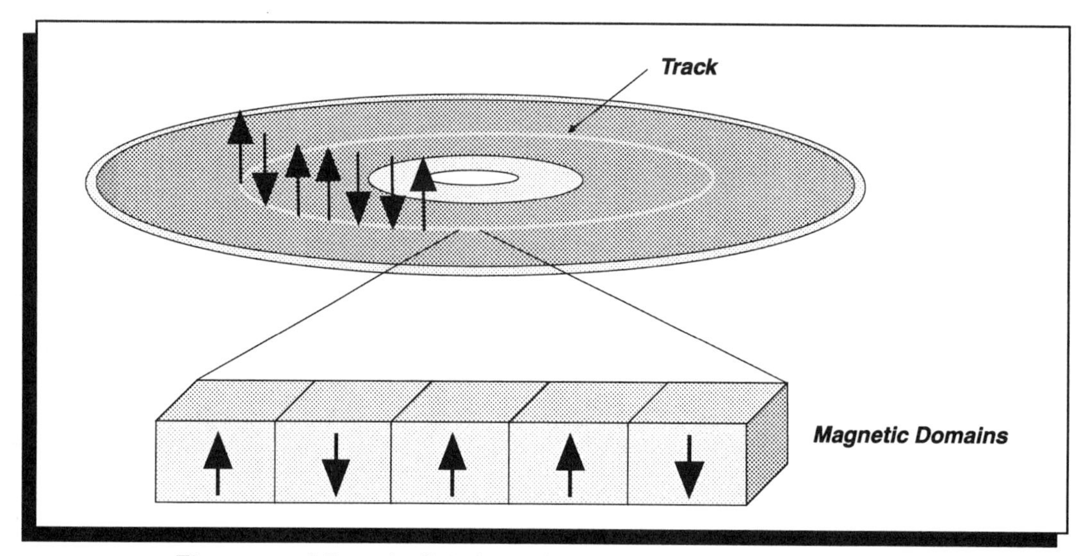

Figure 11.1 Magnetic digital signals recorded vertically on a mini disc.

a recording head on one side of the disc. Since the magnetic field of a recording head is so extended spatially, it would be difficult to achieve a high density of information. To overcome this, a laser beam is used to define a tiny spot on the disc where the signal is to be recorded. This is achieved by shining a high intensity laser beam from below the disc onto the spot where the magnetic signal is to be recorded, from above the disc. The laser beam locally heats the tiny spot on the disc above its Curie temperature (about 180°C for the Tb Fe Co film) and this reduces the coercivity to zero, making this spot lose any magnetism it could have had. At the same time the recording head produces a vertical magnetic field oriented north–south or south–north, depending on whether the signal to be recorded is a 1 or 0; the magnetic field then goes through the spot, magnetizing it. Since the disc is rotating, the heated spot moves away from the laser beam and it rapidly cools below its Curie temperature while trapping the acquired magnetization from the recording head above the laser beam. Thus a digital signal is recorded on the disc. Figure 11.2 shows the process where:

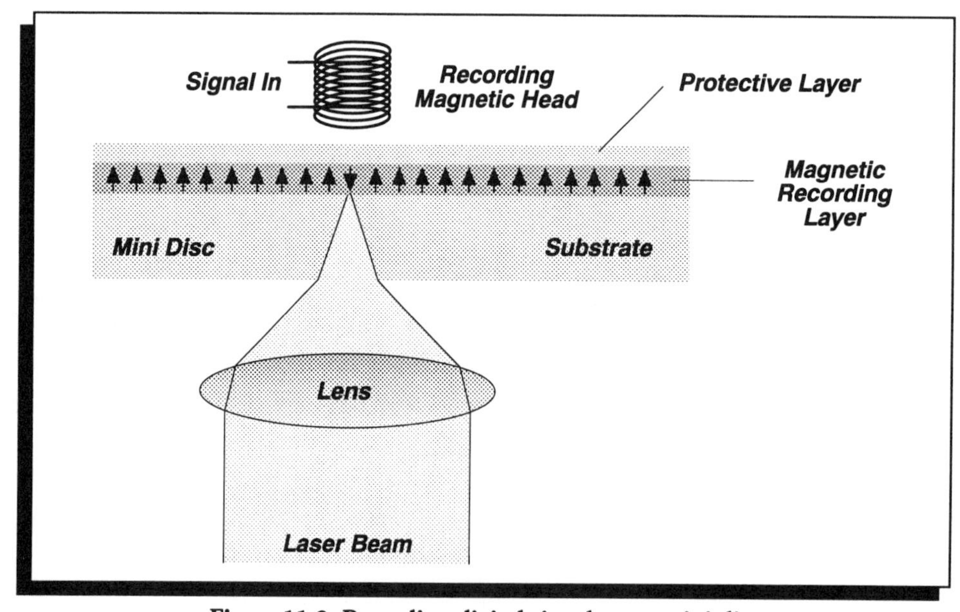

Figure 11.2 Recording digital signals on a mini disc.

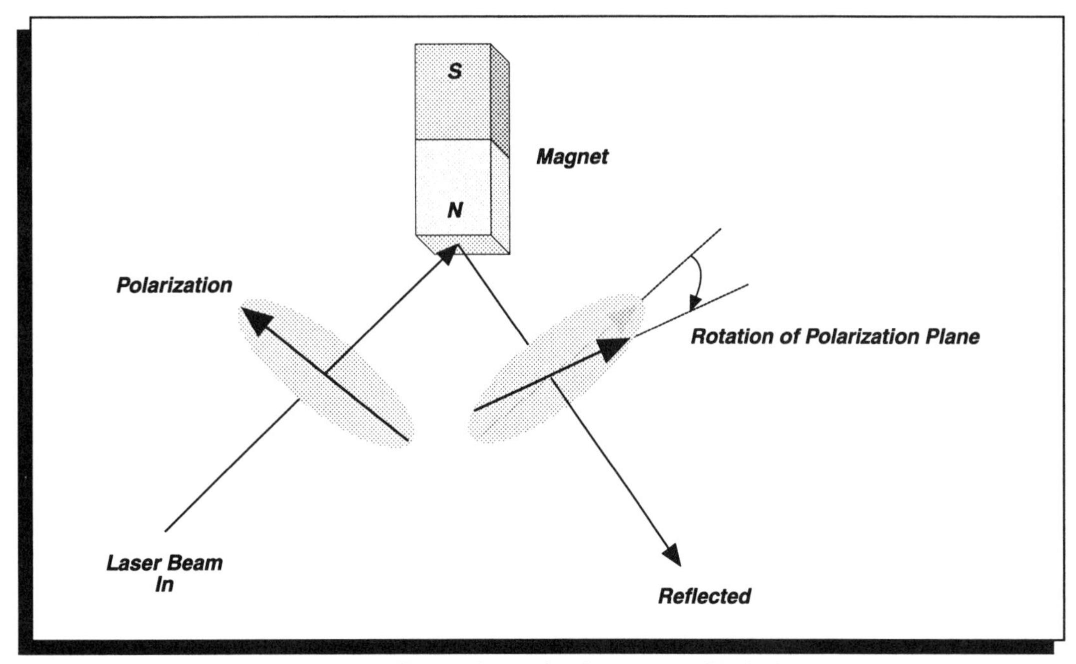

Figure 11.3 Kerr effect: plane of polarization of light beam rotates upon reflection from a magnetized surface.

— The laser erases thermally any previous signal.
— The magnetic field of recording head determines the magnetic state on disc.
— The laser ensures that the magnetized area is very small. The information space is practically the same as on a CD since a similar laser in the infra–red is used here.

The playback of the recorded information is also clever because it utilizes the effect of a magnetic field on the polarization of the laser beam, known as the Kerr effect. Figure 11.3 illustrates this principle: when a beam of plane polarized light is reflected from a magnetized surface its angle of polarization gets rotated. Hence, to read the recordable disc, which has equal amplitude magnetic fields in the directions north–south and south–north, rotations of the plane of polarization of the reflected laser beams are detected (they are about $\pm \frac{1}{2}°$). Figure 11.4 shows a read–out scheme. Although the same laser beam is used for

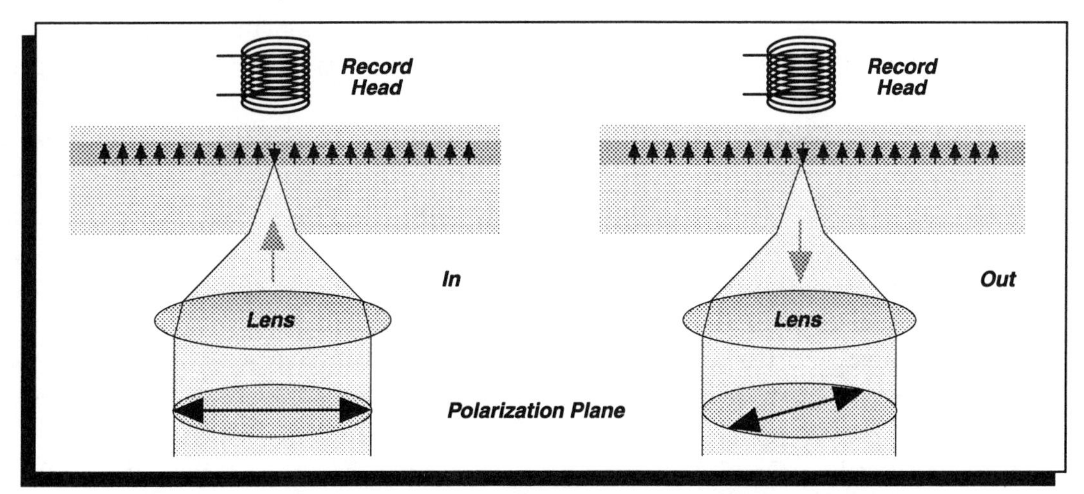

Figure 11.4 Read-out of digital information using Kerr effect. Magnetic field direction affects plane of polarization of reflected laser beam.

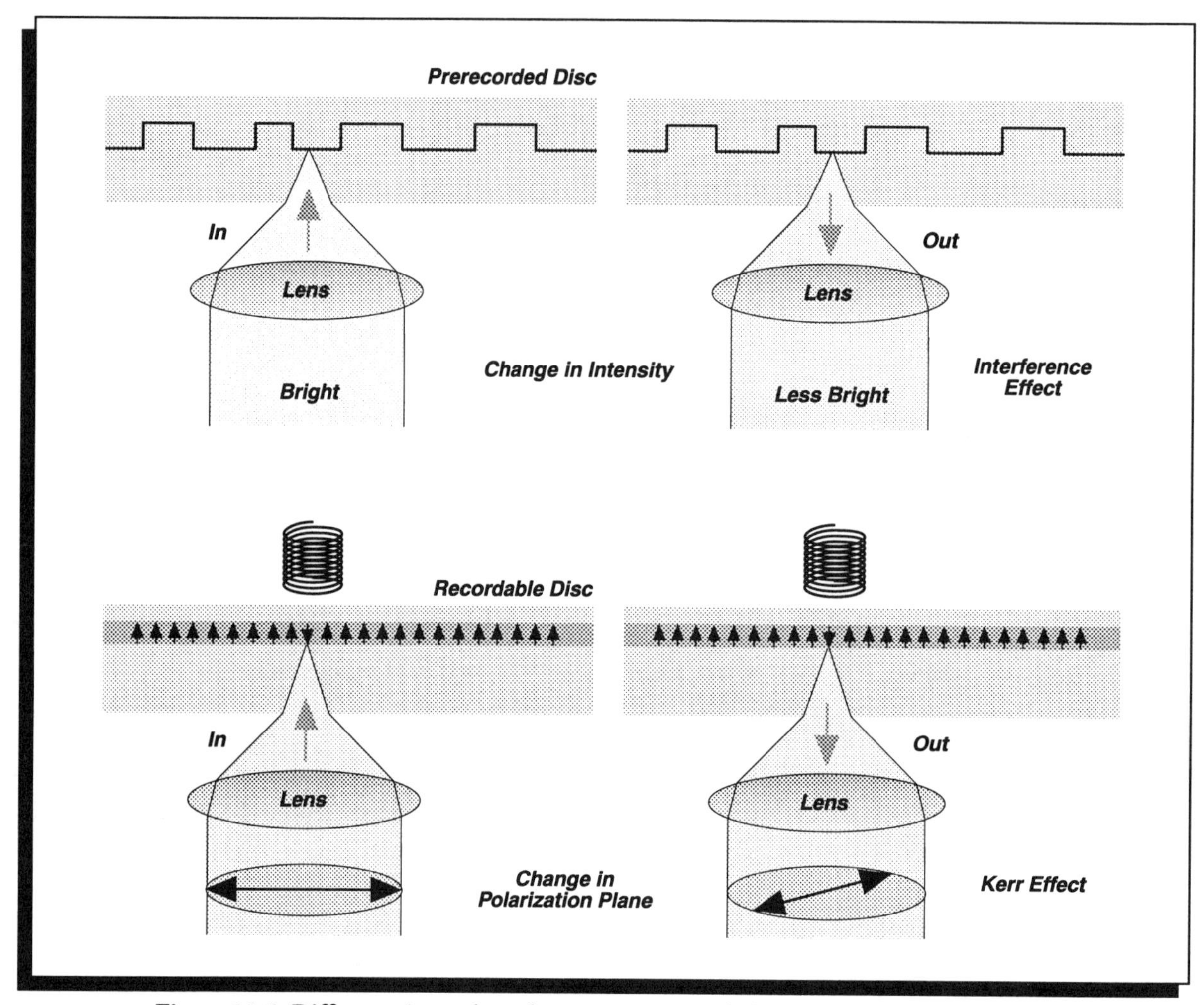

Figure 11.5 Difference in read-out between pre-recorded and recordable mini-discs.

read–out as for recording, its intensity is turned down by a large factor for the read–out, so that it will not heat the information above the Curie temperature.

It is important to note that this type of read–out is used only in the recordable mini–disc. The prerecorded disc has a read–out similar to that of a CD, where intensity variations in the reflected laser beam provide the signal read–out. This is illustrated in Figure 11.5. The Tb Fe Co film, approximately 1000 Å thick, is sandwiched between silicon nitride (SiN) layers to prevent oxidation and to enhance the Kerr rotation angle. There is also an aluminum film to help reflect the laser beam. All this is encapsulated in a plastic matrix which is pre–grooved. A prerecorded MiniDisc has the feature of random–access, where there is quick and easy access to any selection. A recordable disc does not have this capability as the final recorded sequence can be changed anytime. Hence, the recordable disc is pre–grooved for tracking and control in recording and playback. One technique, used by Sony, is to put a wobble on the grooves which create address marks. This is shown in Figure 11.6. The disc has a lead–in area recorded optically (like a CD) which contains a user table of contents. After this table there is the magnetic recordable area that ends with a lead–out area. Figure 11.7 shows the different sections on a recordable mini disc. Tests have shown that this type of disc can be recorded and erased many times as the magnetic film is very well protected and shielded from any oxygen contamination.

As in the prerecorded discs, the ATRAC system is used to achieve a high density of information based on psychoacoustic phenomena.

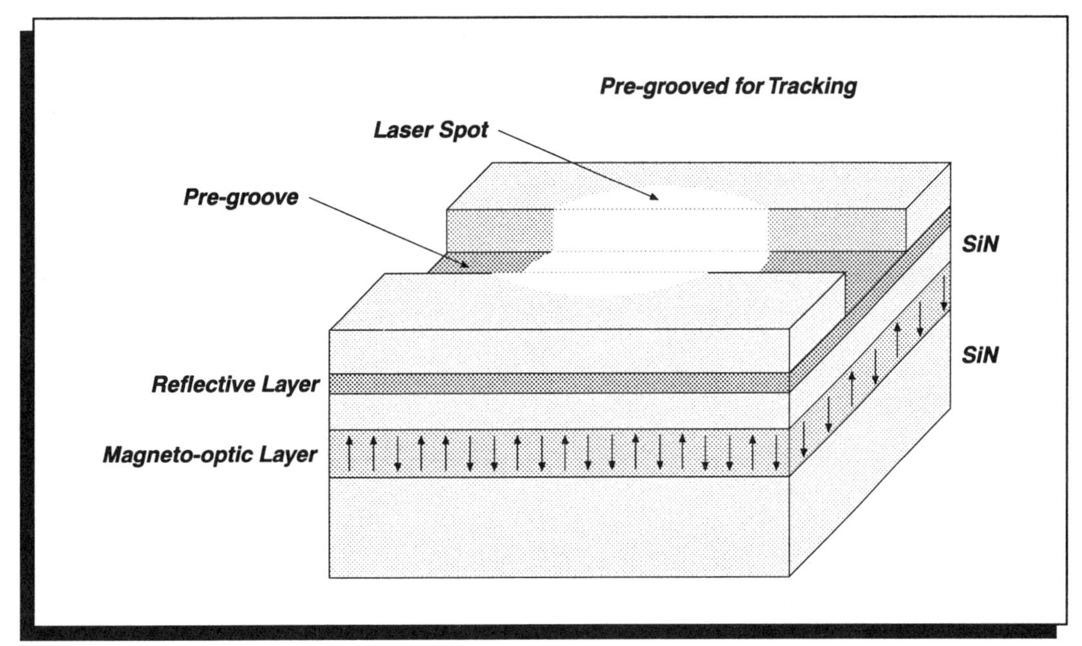

Figure 11.6 Layered structure of recordable mini disc.

The specifications of a mini disc, although slightly inferior to those of a CD, are still very impressive. Some specifications are:

— frequency response: 20 – 20,000 Hz ± 0.3 dB
— dynamic range: 84 dB
— distortion: 0.06%
— signal–to–noise ratio: 82 dB
— channel separation: 74 dB

The most important benefits are that it is recordable and very compact.

11.2 DIGITAL COMPACT CASSETTE

This unit, also known as DCC, was developed as a high quality upgrade and replacement for the analog cassette. It can record and playback magnetic tapes digitally, yet it is compatible with existing analog cassette tapes in that it can also play them back. The DCC tape is the same width as the analog cassette tape and it runs at the same speed (1⁷⁄₈ inches per second, or 4³⁄₄ cm per

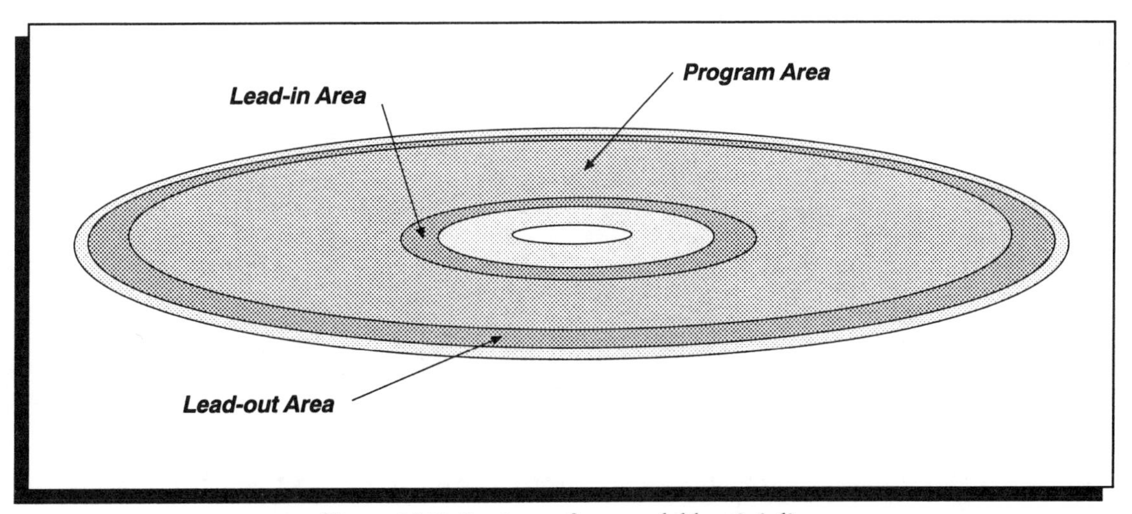

Figure 11.7 Sections of a recordable mini disc.

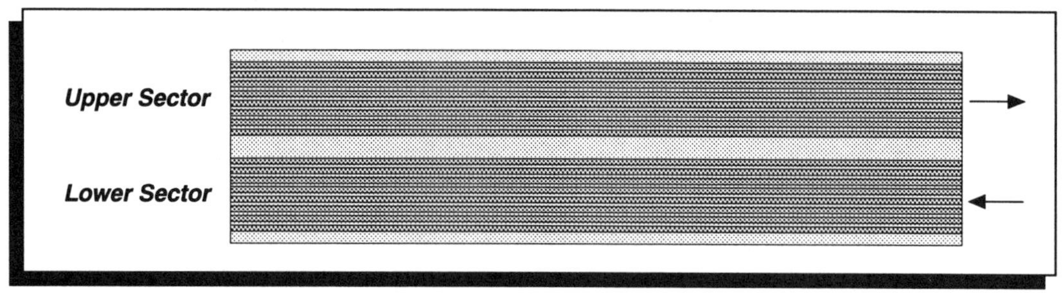

Figure 11.8 Track pattern on DCC tape.

second). This speed is too slow for recording at the same rate that a CD is read (1, 411, 200 bits per second). To increase the data transfer rate, the digital compact cassette (DCC) systems puts data on eight parallel tracks in each direction of travel. The data is recorded or played back by a stationary "thin–film" head. Even with eight tracks, the data rate is still low, about 384,000 bits per second, which is about 1/4 that of a CD. This problem is resolved by not recording all the information. A special system is used to decide how much of the sound information and which part will be recorded. It is called the Precision Adaptive Sub–band Coding (PASC) system. In this way, the data rate becomes manageable at 384,000 bits per second (plus the same rate for the digital error– correction data). All this information is divided between the 8 tracks, each carrying data at 96,000 bits per second. Figure 11.8 shows the track pattern on a DCC tape. The upper half has eight tracks for the audio and one auxiliary track for control information and non– audio data. Details of the track pattern are shown in Figure 11.9. The playback head reads only a portion of the recorded track.

The PASC is the digital coding which makes the DCC system work. It is based on two psychoacoustic principles:

- — we hear sounds only above the threshold of hearing.
- — loud sounds will tend to mask quiet ones by raising the threshold of hearing.

The PASC system encodes only sounds that can be heard, rejecting most of the rest. Figure 11.10 shows the threshold of hearing curve and its relevance for determining which sounds will be heard. Figure 11.11 shows which sounds will be recorded, based on the PASC processor, and which ones will be left out.

The DCC has three sampling rates: 32 kHz, 44.1 kHz, and 48 kHz. The 32 kHz rate will be useful for digital radio broadcasts, the 44.1 kHz rate is convenient for taping a CD or for playing prere-

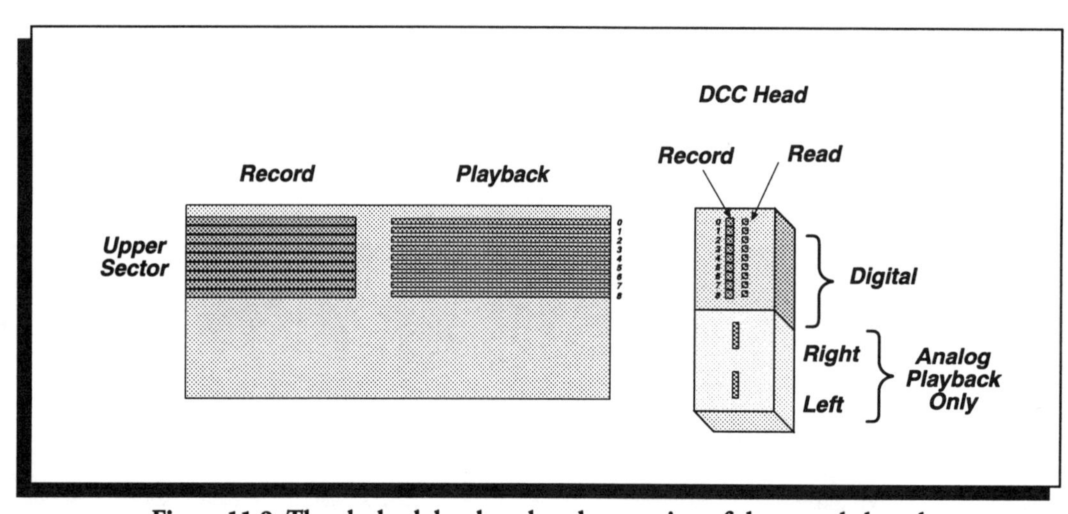

Figure 11.9 The playback head reads only a portion of the recorded track.

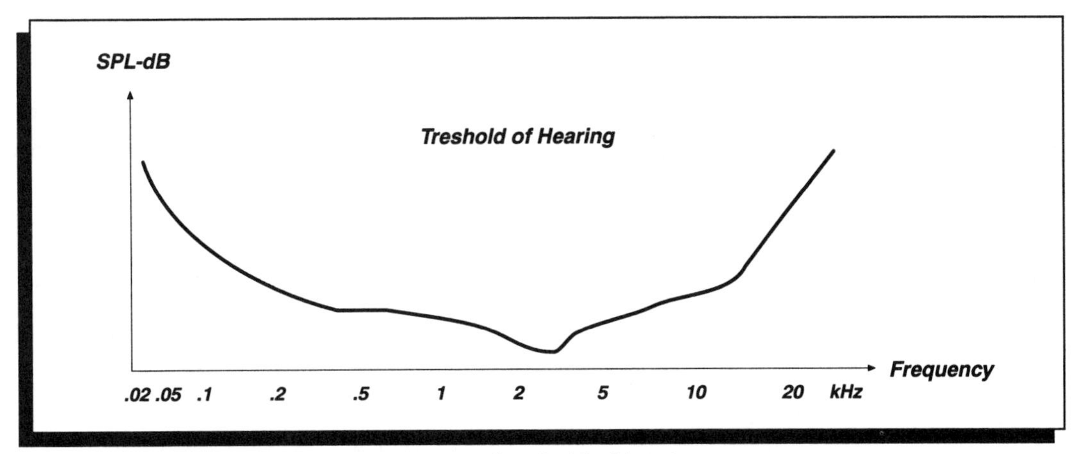

Figure 11.10 Threshold of hearing curve.

corded tapes. The 48 kHz is the standard sampling rate for any analog signal coming in to be recorded.

The recorded digital information on the tape will be as shown in Figure 11.12. It is longitudinal.

The specifications are very impressive, some of them that stand out are:

— frequency response: 5 Hz to 20,000 Hz ± 0.3 dB.
— dynamic range: 108 dB.
— channel separation: 90 dB.
— distortion: 0.003%.
— playing time: 90 minutes or 120 minutes.

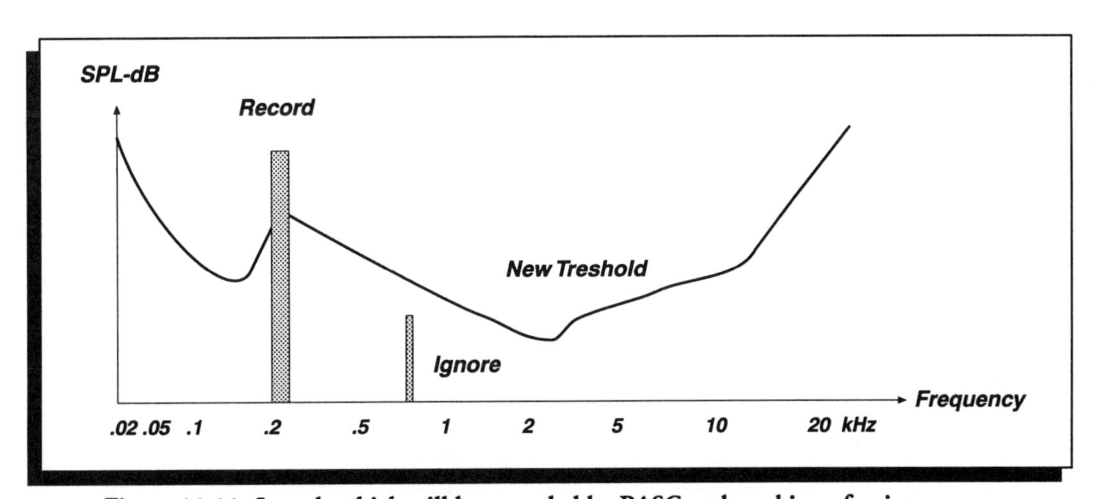

Figure 11.11 Sounds which will be recorded by PASC and masking of quiet passages.

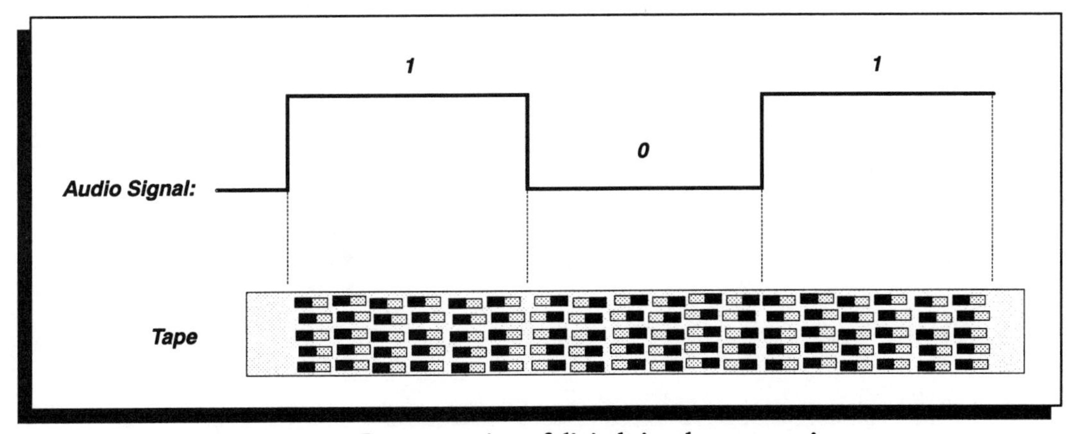

Figure 11.12 Representation of digital signal on magnetic tape.

11.3 DIGITAL AUDIO TAPE

This system was developed to provide all digital magnetic recording and playback capability and to complement the CD format. Although introduced a few years after the CD, its acceptance has been slow. It basically provides the same quality of sound as the CD with the extra benefit of being capable of recording in a variety of modes. The digital audio tape, DAT, system uses some of the video tape and digital technology.

In order to record digitally with the same sampling rate as for a CD (44.1 kHz, at 16–bit resolution), the system would have to handle 1,411,200 bits per second. For a tape moving at $1\frac{7}{8}$ i.p.s., that frequency would correspond to a wavelength of:

$$\begin{aligned} \text{wavelength} &= \text{speed/frequency} \\ &= 1.875 \text{ i.p.s}/1.4112 \times 10^{6} \text{ Hz} \\ &= 1.33 \times 10^{-6} \text{ inch} \end{aligned}$$

Hence, the gap in the playback head would, at the most, have to be half of that; such an extremely small gap is not practical. Therefore, a different approach has to be taken. A possible solution is to increase the tape speed, thus making the wavelengths longer. It is an interesting solution, but this is not practical since the process would use up a large amount of tape. A practical solution is:

— Maintain low tape speed which will not consume large quantities of tape.
— Make recording head rotate at high speed relative to tape to achieve long wavelength.

These are the approaches taken with the popular video tape systems, VCR, where helical recording is used. This is shown in Figure 11.13 with information recorded on diagonal tracks at an angle of ~ 6° to the tape edge. There are two heads 180° apart on a rotating drum. The tape is wrapped around the drum in such a way that it enters and leaves in different planes. This causes the rotating heads to record information on long diagonal tracks. Because the heads are 180° apart, neither head is in contact with the tape for half of the time. To simplify the threading mechanism for the tape and to reduce the friction between tape and drum, the tape is usually wrapped over 90° of the drum. Figure 11.14 shows how the rotating heads A and B are mounted on the drum; it rotates at 2,000

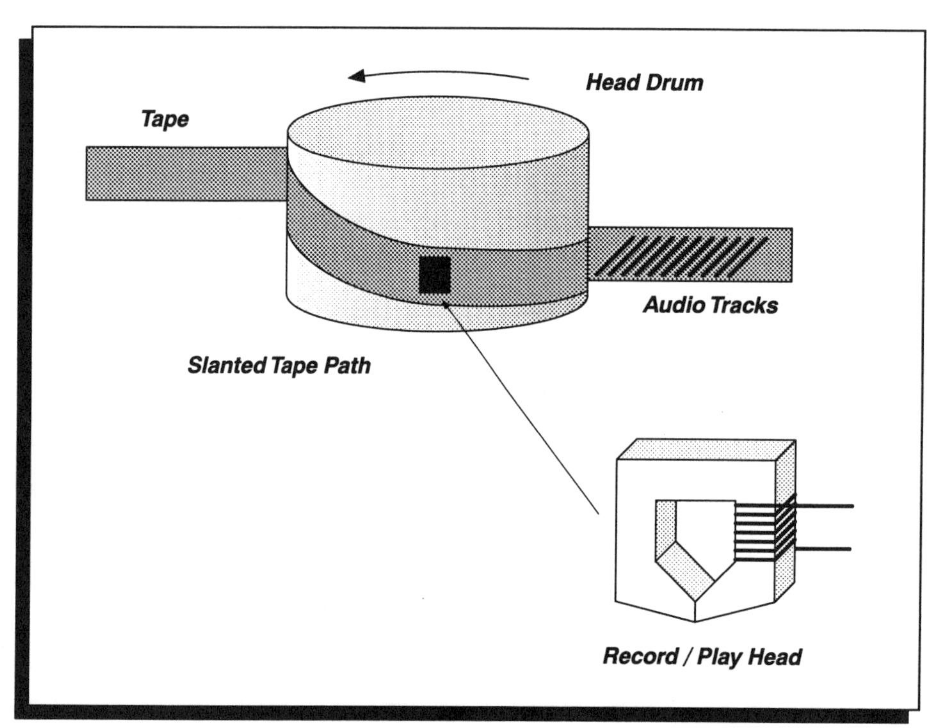

Figure 11.13 Helical recording with rotating heads.

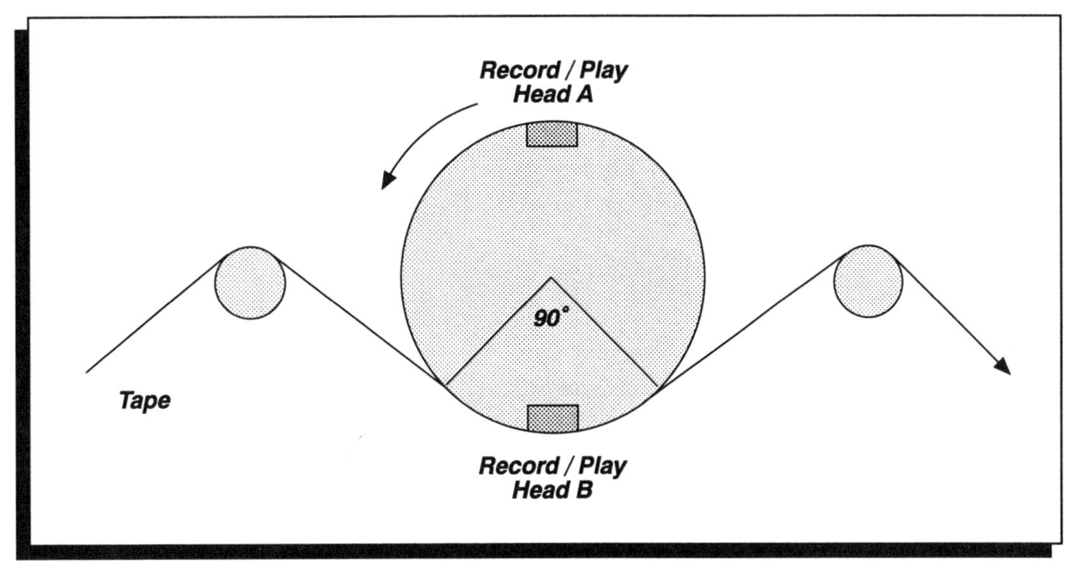

Figure 11.14 Tape contact to rotating head.

rounds per minute (r.p.m.) and its diameter is 30 millimeters. The 90° tape wrap helps reduce tape wear and head damage. Other drum diameters can be used. Some smaller units have drum diameters of 15 mm and the tape makes contact then over 180° of the drum. For the 90° wrap, time compression is used. Samples to be recorded are stored in a memory at the sampling rate but recorded at a faster rate (i.e. the information is compressed in time). This is shown in Figure 11.15.

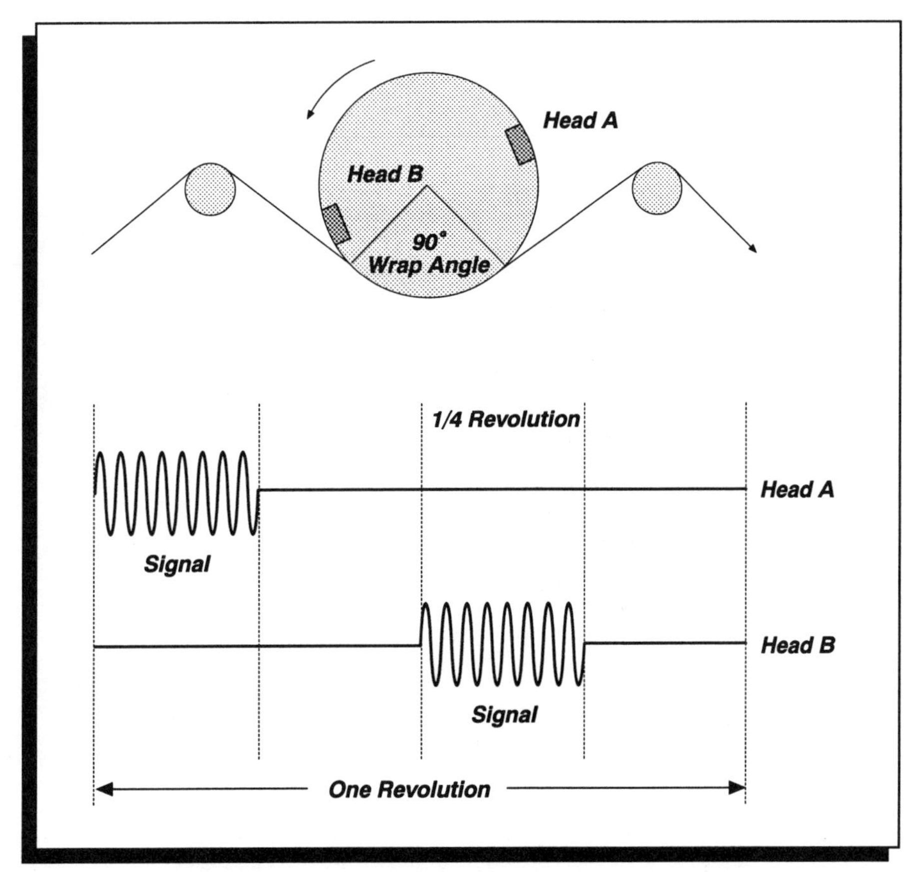

Figure 11.15 Time compression to reduce wrap angle.

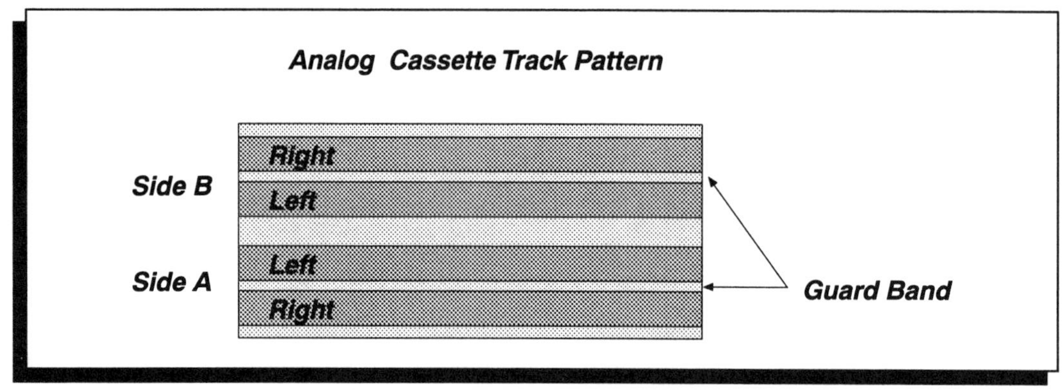

Figure 11.16 Guard band between tracks on analog tape reduces cross-talk.

Since the goal is to have a high density of information on the tape, it is important to have the tracks close to each other, without interaction between them. In regular analog tapes, there is a guard band (a space) between tracks, Figure 11.16, to reduce cross–talk; of course this is unused and wasted tape area. On a digital tape, this problem is elegantly solved by using heads which are at an angle to each other. This is known as azimuthal recording. Figure 11.17 illustrates this. The heads (record–playback) in the drum are angled to each other at ± 20° azimuth angle. This has two advantages:

1. Track B can be very close to track A. This saves tape space.
2. Track A does not interfere with head B in playback (i.e. low cross talk). The same is true for track B not interfering with head A.

Since the information on the tape is digital, 1 or 0, the recorded signal direction with saturated longitudinal signals, will be one way or the reverse, as in Figure 11.18. The gap length must be less than the bit length on the tape. Should the gap length = 2 bit lengths, the playback signal averages to zero.

Three types of signal are recorded. They are:

1. The audio signal encoded with special digital coding.
2. The subcode signal with information for playback such as number of musical selection and control signals.
3. automatic track finding (ATF) signal for tracking purposes.

Figure 11.19 shows how the different signals are located.

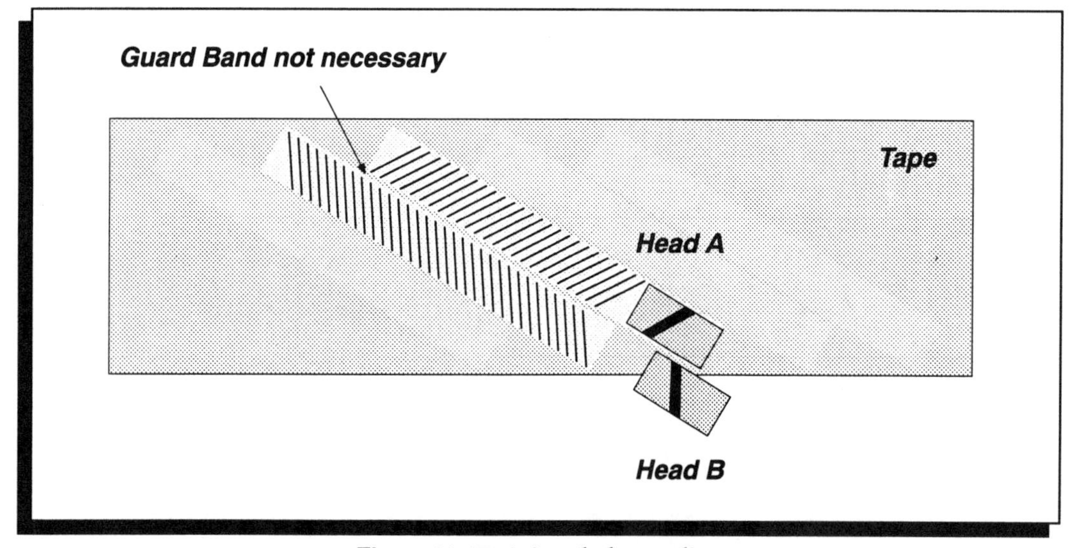

Figure 11.17 Azimuthal recording.

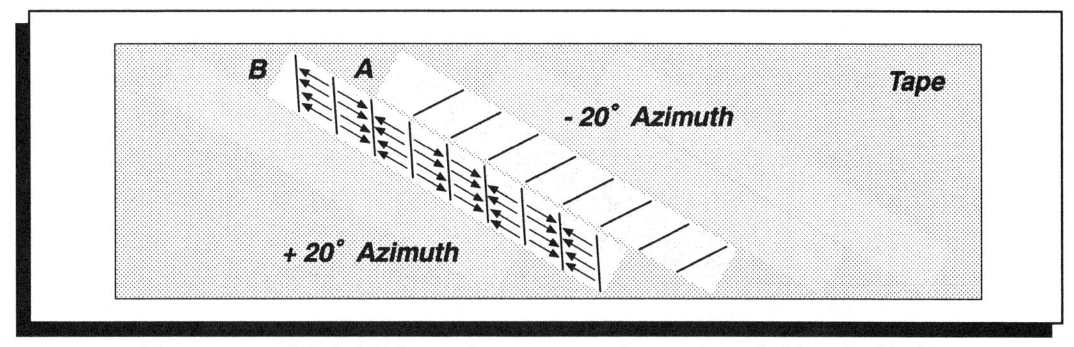

Figure 11.18 Digital information on magnetic tape recorded longitudinally.

The erasing is done by overwriting the signal. Since the information is either a 1 or a 0, one simply writes over the previously recorded signal, eliminating the need for an erase head.

The tape is kept aligned with respect to the heads, by the specially encoded signal ATF.

Each track is 13.6 microns wide (this corresponds to about 1/5 the thickness of a human hair). The space available for each bit is 0.67 microns.

For a typical sampling rate of 48 kHz and 16–bit accuracy, there are ~ 1.5 megabits/sec of data; when the extra information is taken into account the overall data rate is ~ 2.77 megabits/second. This makes it the highest density magnetic recording medium.

Usually there are four options on a DAT:

— Mandatory option: 48 kHz sampling rate, 16 bits, 120 minutes, record and/or playback
— Option 1: 32 kHz sampling rate, 16 bits, 120 minutes. This will be useful with digital radio which will use the 32 kHz sampling rate.
— Option 2: extra long play, 32 kHz sampling rate, 12 bits, linear and drum speed reduced, 240 minutes.
— Option 3: 32 kHz sampling, 4 audio channels, 120 minutes.
— Option 4: for playback of prerecorded material on a DAT tape, 44.1 kHz sampling rate, 80 minutes.

The DAT does not replace the CD; it complements it, making it possible to record audio with excellent tonal quality. The tape will eventually get worn out since it makes contact with the heads and it is fast–forwarded when searching for a selection. However, it does provide the flexibility of recording selected features for car or home use. Another convenient feature is that the cassette case is smaller than the regular analog cassette. It is 73 × 54 × 10.5 mm; the analog is 102.4 × 63 × 12 mm. The dynamic range at 96dB or more is much higher than that of an analog tape, which usually provides 50–60dB only. This unit is an excellent candidate as a replacement for the popular analog cassette deck.

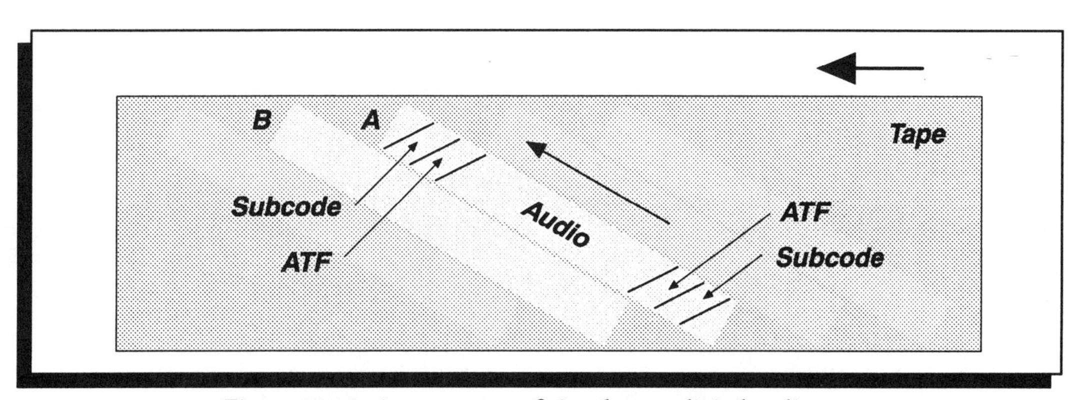

Figure 11.19 Arrangement of signals on a digital audio tape.

SUMMARY OF TERMS

Azimuth recording: Recording by heads canted to each other to reduce cross–talk.
Curie Temperature: Temperature above which a magnetic material becomes almost non–magnetic, and the coercivity goes to zero.
Kerr Effect: Plane of polarization of light beam is rotated when it is reflected off a magnetized magnetic material.
Magneto–optical reading: Data readout using the Kerr Effect, as on mini–discs.
Magneto–optical recording: Method of recording, as on mini–disc, where magnetic information is recorded with the help of optical techniques.
Overwriting: Method of erasing digital information by writing directly over previous signal. It works because the signal is digital.
Precision Adaptive Sub-band Coding (PASC): System based on psychoacoustic principles which determines what audio information will be recorded on a DCC.
Rotary head: Drum with two (or four) heads for record–playback, in a mode where drum rotates relative to tape.

NAME__________________________________ **DATE** ________________

QUESTIONS FOR REVIEW

1. Discuss how DCC recording differs from DAT recording from the point of view of audio content.

2. Why does an analog tape deck need an erase head, but the DAT does not?

3. Explain how cross–talk is reduced on a DAT.

4. Is the amplitude of each recorded digital signal on a DAT the same?

5. Why is a bias signal not used in digital magnetic recording?

6. Comment on the effect of tape saturation in digital magnetic recording.

7 What determines the dynamic range of a signal on a digital tape?

8 In one complete revolution of the drum head, how many tracks are recorded on a DAT?

9 Explain how a high density of information can be achieved with digital magnetic recording.

10 Discuss vertical magnetic recording and longitudinal magnetic recording.

NAME ______________________________ DATE ______________

Please select one answer.

1. The signal on a recordable mini–disc is ______ while on a DCC it is _______.
   ___A. longitudinal, vertical
   ___B. vertical, longitudinal
   ___C. vertical, vertical
   ___D. longitudinal, longitudinal
   ___E. vertical, longitudinal

2. Which digital magnetic recording system has the highest data rate?
   ___A. the DAT
   ___B. the recordable mini–disc
   ___C. the DCC
   ___D. the recordable mini–disc and the DCC.
   ___E. the DCC and the DAT

3. Which one of the following digital magnetic recording systems has a fixed 16–bit accuracy for all samples?
   ___A. the DCC
   ___B. the recordable mini–disc
   ___C. the DAT
   ___D. the DCC and the DAT
   ___E. the DCC and the recordable mini–disc

4. When heated above its Curie temperature, a magnetic material:
   ___A. becomes difficult to magnetize
   ___B. has a larger retentivity
   ___C. cannot be erased
   ___D. becomes very easy to magnetize
   ___E. has a larger coercivity

5. The read–out for a recordable mini–disc consists of _____________ and for a pre–recorded mini–disc of __________.
   ___A. differences in laser beam intensity, differences in beam intensity
   ___B. differences in polarization of laser beam, differences in polarization of laser beam
   ___C. differences in beam intensity, differences in the polarization of the reflected light
   ___D. differences in the polarization of the reflected light, differences in the intensity of the light
   ___E. changes in frequency of reflected beam, changes in amplitude of reflected beam

6. Mini–disc recording and playback are based on two different phenomena:
   ___A. magneto–optical effect for playback
   ___B. frequency modulation effect for recording
   ___C. Kerr effect for recording
   ___D. Faraday's law of induction for playback
   ___E. reflection of laser beam for recording

7. What will be the information recorded on a recordable mini–disc for the case shown?
   ___A. X will be a north pole and Y a south pole
   ___B. X will be a south pole and Y a north pole
   ___C. X will be a south pole and Y a south pole
   ___D. X will be a north pole and Y a north pole
   ___E. this coil cannot produce a perpendicular magnetization.

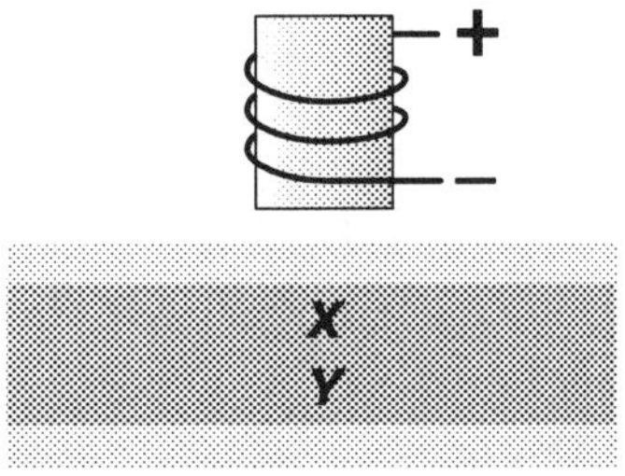

+
–
X
Y

8. Why is it necessary to use a laser beam in recording on a recordable mini–disc?
   ___A. to define a very small spot where the magnetic signal will be recorded
   ___B. to act as a bias and shake the magnetic domains so that they can overcome friction and get magnetized
   ___C. to change the magnetization on the disc from perpendicular to longitudinal
   ___D. to change the magnetization on the disc from longitudinal to perpendicular
   ___E. to prevent the domains from aligning too quickly

9. At what rate is the raw data recorded or played back on a DAT? Assume normal recording or playback sampling rates.
   ___A. 1.4112 million bits/second
   ___B. 1.536 million bits/second
   ___C. 48,000 bits/second
   ___D. 96,000 bits/second
   ___E. 44,100 bits/second

10. The reason for two heads on the rotating drum in DAT is:
    ___A. one is a spare
    ___B. one is for the LEFT channel and the other one is for the RIGHT channel
    ___C. one is for recording and the other one is for reading
    ___D. so that each head writes its own track, adjacent to each other, but canted to each other; this provides a high density of data on the tape
    ___E. one is for erasing and the other one is for record/playback

11. For the basic sampling rate on a DAT, the highest audio frequency that could be recorded is:
    ___A. 48 kHz
    ___B. 24 kHz
    ___C. 44.1 kHz
    ___D. 22.05 kHz
    ___E. 32 kHz

12. A DAT unit does not have Dolby because:
    ___A. hiss from tape is not played back digitally
    ___B. playback head does not use Faraday's law of induction
    ___C. digital tapes are of superior quality and do not produce surface noise
    ___D. it was not developed yet
    ___E. tape hiss averages out to zero over gap of playback head.

13. The two heads on a rotating head drum for DAT are __________ apart.
    ___A. 180°
    ___B. 90°
    ___C. 45°
    ___D. 360°
    ___E. not

NAME ______________________________ **DATE** ____________

14. The major difference between magneto–optic recording and standard magnetic recording is:
   ___A. standard magnetic heads orient domain at room temperature while in the magneto–optical case they are oriented at high temperatures.
   ___B. magneto–optical recording is longitudinal and standard magnetic recording is perpendicular.
   ___C. magneto–optical recording uses a laser to record while standard magnetic recording uses a magnetic head
   ___D. in the polarization of each. One has circular polarization and the other has plane polarization.
   ___E. in the medium. Magneto–optical method uses reflective medium which is not magnetic, while the standard one uses a magnetic tape.

15. The audio quality will usually be the best on:
   ___A. DCC
   ___B. recordable mini–disc
   ___C. analog tape deck
   ___D. DAT
   ___E. prerecorded mini–disc

16. What is the frequency of the laser beam used in a mini–disc?
   ___A. 38 x 10^{14} Hz
   ___B. 1.92 x 10^{14} Hz
   ___C. 3.8 x 10^{14} Hz
   ___D. 2.56 X 10^{14} Hz
   ___E. 7.6 x 10^{14} Hz

17. If a "1" stored magnetically on a mini–disc rotates the plane of polarization of the laser beam ____________ then a "0" will make it rotate __________.
   ___A. clockwise, counterclockwise
   ___B. clockwise, clockwise
   ___C. counterclockwise, counterclockwise
   ___D. clockwise, twice as much clockwise
   ___E. only the intensity of the laser beam is changed

18. On a prerecorded mini–disc, the rotational speed of the disc is:
   ___A. largest at the outer edge
   ___B. largest at the inner edge
   ___C. the same everywhere on the disc
   ___D. smallest at the inner edge
   ___E. determined by the audio selection on disc

19. On a recordable mini–disc, tracking control is achieved from:
   ___A. the pre–grooves on mini–disc
   ___B. information stored digitally on disc
   ___C. size of digital signals on disc
   ___D. a secondary laser beam
   ___E. polarization of laser beam

20. Which recording format eliminates 3/4 of the audio data and still produces sound of excellent quality?
   ___A. a CD
   ___B. an analog tape deck
   ___C. a DAT
   ___D. a DCC

CHAPTER 12 HEAT

Heat is a topic usually not associated directly with the hi–fi. A closer look reveals that heat has some very important effects on the performance of amplifiers, magnetic tapes, loudspeakers, and most components. Unfortunately, it is also quite often the cause of their failure. It is essential that heat be controlled and dissipated. However, in some applications in the hi–fi heat is essential for the proper performance of certain components. Examples are the thermo–magnetic effect in digital magnetic recording on the miniDisc and the well–known infra–red remote controls in the hi–fi system. In this chapter we will address these two extremes of heat and its role in the performance of a hi–fi. Basic concepts of heat will be introduced first, then heat will be discussed as to its origin, its measurement, and its dissipation in the various components of a hi–fi. Its role in magnetic recording and communications will also be presented.

12.1 BASICS

Heat is a form of energy that increases the vibrations of molecules in any object. The molecules of an object constantly vibrate back and forth and the greater the vibrational energy, the hotter the material is. In fact, when the vibrations are large enough the molecules leave the substance and this is known as evaporation. We observe this in water, where by raising its temperature the water molecules increase their vibrations and at some critical temperature they leave the liquid. There are three questions which have to be addressed here. What causes heating in the hi–fi system, how can it be measured, and how can it be controlled before severe damage is done to the hi–fi system? In some hi–fi components controlled heat input is essential for their operation.

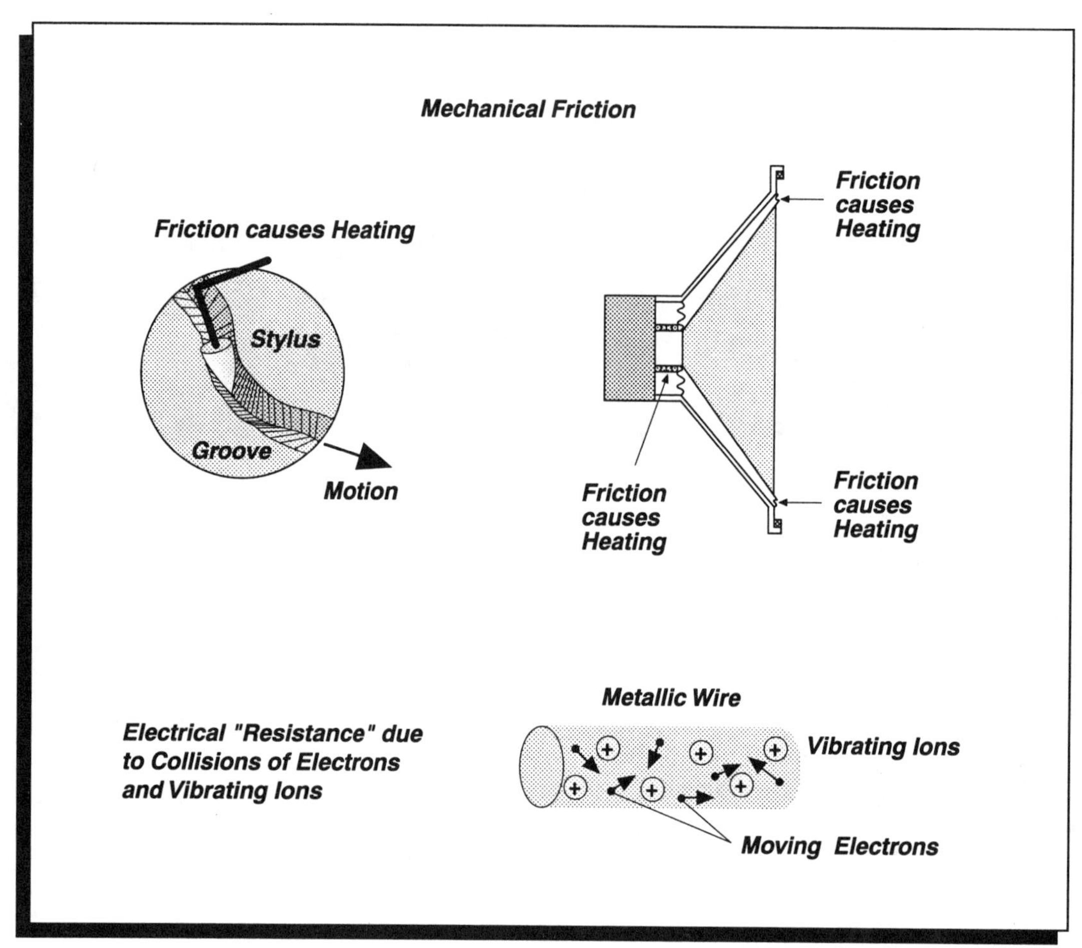

Figure 12.1 Sources of heating in a hi-fi due to mechanical friction and electrical resistance.

■ CAUSES OF HEATING IN THE HI–FI

The main cause of heating in the hi–fi is friction, which occurs mechanically and electrically. The phono stylus rubs against the walls of the groove in a record and causes its temperature to go up dramatically (at times over 1,000 degrees Fahrenheit!!). Similar mechanical friction occurs in a speaker, where the moving diaphragm and voice coil are rubbing against other parts of the speaker, thus generating heat. In electricity there is some "friction" exerted against the flow of electrons causing the wire to have resistance; when a current flows, heating occurs due to the collision of electrons with vibrating ions. The rate at which heating occurs has been calculated in an earlier chapter. It is the power dissipated in a resistor when a current flows through it:

$$\textbf{Power} = \textbf{current}^2 \times \textbf{resistance}$$

Every component that has resistance will have heating whenever current flows through it. This situation is summarized in Figure 12.1, where the mechanical friction and electrical resistance are the main causes of heat in the hi–fi. In the electrical case there are three components where heating will be important when current flows:

1. a resistor (and sometimes even in a wire).
2. a diode.
3. a transistor.

This is shown in Figure 12.2.

Is it possible to measure this heat and to dissipate it to the outside? A thermometer is a device which will measure the temperature of the amplifier or component and indicate if it is too hot.

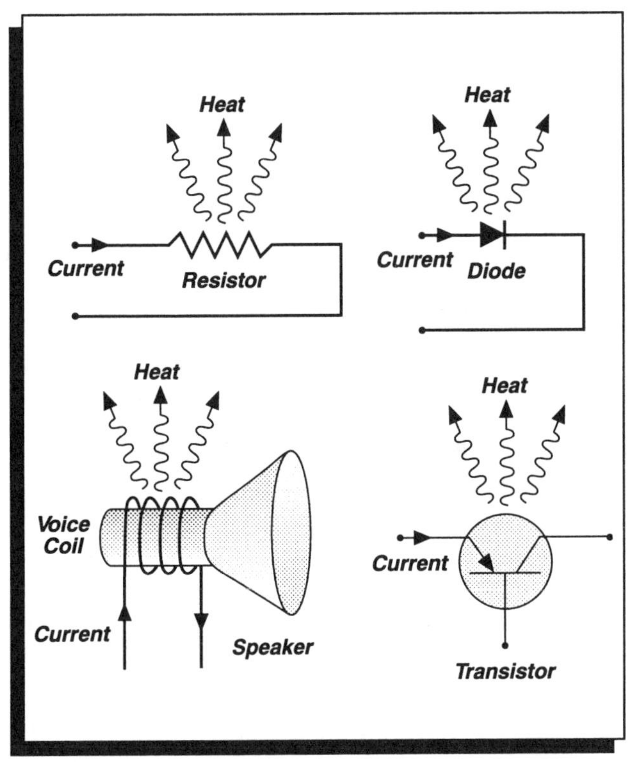

Figure 12.2 Electrical resistance causes heating in amplifier components and voice coil when current is flowing.

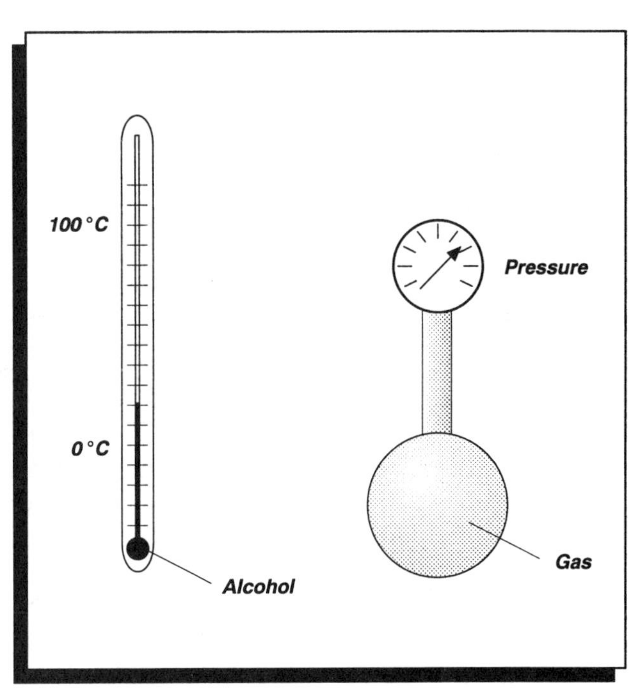

Figure 12.3 Two types of thermometers: alcohol expansion thermometer and gas thermometer.

■ TEMPERATURE

Temperature is a measure of how cold or how hot an object is; it is measured by a thermometer. Generally, when an object changes temperature, some of its properties will change as well. These changes can be in size, electrical properties, magnetic properties, or optical properties. Such changes in properties can be used to detect how hot or how cold an object is and thus to measure changes in temperature. A common thermometer is particularly convenient to measure temperature changes around room temperature. The common thermometer is based on the expansion or contraction of a column of alcohol or mercury. The Centigrade temperature scale is set by assigning 0 degree at the point at which water freezes and 100 degrees when it boils. On the Centigrade scale, there can be temperature excursions beyond the 100° C range and below 0°C. This thermometer, shown in Figure 12.3, is not practical for the hi–fi since it is too bulky and too slow in its response. We could use another type of thermometer, where the pressure of a gas in a bulb is an indication of the temperature. This is also shown in Figure 12.3 and it is known as a gas thermometer. Because large variations in temperature can occur (hopefully not inside your amplifier!), it is convenient to define a temperature scale that never goes negative, beginning at absolute zero and going up. Such a scale is known as the Kelvin scale and it is used in scientific work. The absolute zero in temperature 0° K corresponds then to the maximum order which can exist in a system and it occurs at –273.15° C.

Since the rather bulky thermometers in Figure 12.3 are not practical for application to the hi–fi, more compact thermometers have been developed. A practical one for the hi–fi is a resistance thermometer based on the change of electrical resistance of a metal or semiconductor with temperature. Figure 12.4 shows the temperature variation of the resistance of a semiconductor. For a metal, the temperature variation of the resistance (as shown in an earlier chapter in Figure 5.20) is opposite to that of a semiconductor. Because of ease in fabrication and high sensitivity, it is more convenient to use a semiconductor element as a resistance thermometer. According to Figure 12.4, as the temperature goes up, the resistance decreases; if the resistance goes up,

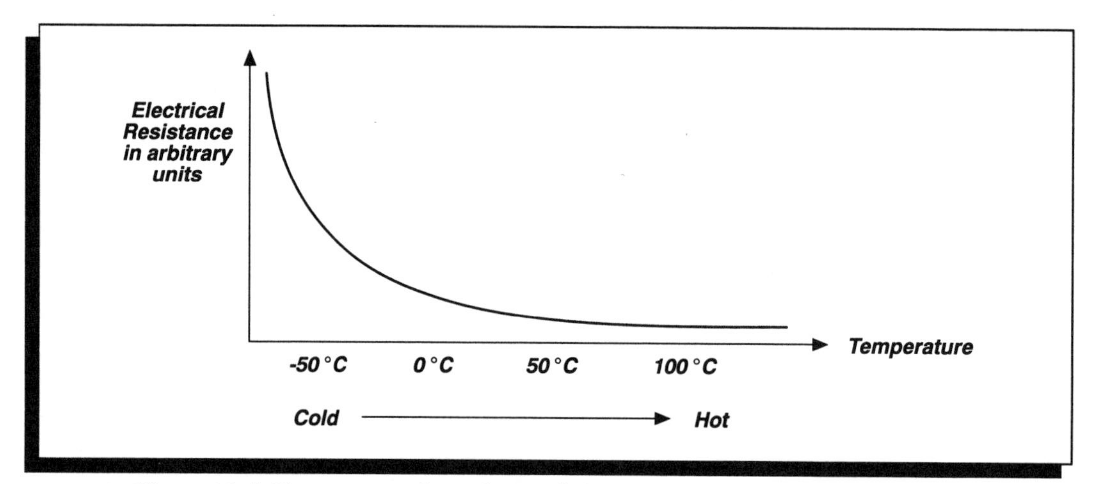

Figure 12.4 Temperature dependence of electric resistance of a semiconductor.

it means that the temperature has decreased. These concepts can be incorporated into the circuit shown in Figure 12.5 for a resistance thermometer with a simple readout.

The current in this circuit is given by Ohm's law:

current = voltage / resistance

When the resistance changes, the current will also change. Hence, according to Figure 12.4, cooling will cause a decrease in current since the resistance has increased, while heating will cause an increase in current. After calibration, which indicates how much the current changes for a certain temperature change, we have a simple, practical thermometer. When the element is thermally attached to the amplifier it will indicate the amplifier temperature and temperature change. Quite often the semiconductor resistance thermometer is referred to as a thermistor or a heat sensing resistor.

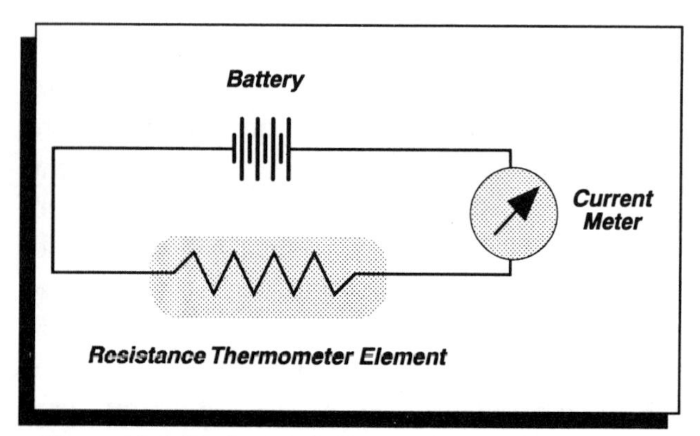

Figure 12.5 Basic circuit for a resistance thermometer.

■ CONSERVATION OF ENERGY

Energy can come in different forms. For example, in the speaker, acoustic energy is produced from electricity delivered by the amplifier. There is also energy in the form of heat which is developed in the voice coil by the current flowing through it due to electric resistance. A laser beam also has energy and it can easily heat up an object in its path. There is a very fundamental law in physics that deals with energy, whatever its form and its conversion to other forms. This law is known as the Law of the Conservation of Energy. It is a very important law as it tells us how to keep track of energy.

Law of Conservation of Energy **= energy cannot be created nor destroyed, it can only be converted from one form to another.**

In the case of the speaker,

electrical energy from amplifier	→	SOUND	+	HEAT ENERGY
		↓		↓
		for our pleasure (useful quantity)		heats up speaker (a waste)

For a 1% efficient speaker, 99% of the energy received becomes converted to heat. This means that if the amplifier is putting out 200 watts of electrical power, there will only be 2 watts of sound and 198 watts will go into heat. Note that the powers add up according to the conservation of energy (power and energy are related and they are used interchangeably here).

With so much heat produced in the voice coil, what will its temperature be? This depends on the material being heated, and on how much material there is. There is a simple relation that tells us how to calculate the temperature change in a material when it is heated. It can be written as:

> ***amount of energy absorbed by material*** **= mass of material × its specific heat × temperature difference**

Specific heat of a material is a measure of its capacity for storing internal energy. For example, it would take 7 times longer to heat water up to the same temperature as an equal mass of iron since the specific heat of water is 7 times that of iron. We define:

> ***specific heat*** **= quantity of heat required to raise the temperature per unit mass of a substance by one degree centigrade temperature change.**

The quantity of heat is specified in units of calorie (for historical reasons) which is defined as:

> ***calorie*** **= amount of heat necessary to raise 1 gram of water up 1 degree Centigrade.**

The specific heat is given in units of calories per gram per degree Centigrade. It is a very useful quantity as it tells us how much heat is required to raise the temperature of a substance. For example a voice coil with twice the mass of copper coil, as opposed to a regular coil, would take two times longer to heat up to the same temperature for the same heat input. Such a fact comes from a general equation which tells us how to keep track of thermal energy. When the temperature of an object is changed we have:

> **total amount of thermal energy used = specific heat × mass × change in temperature**

Since each substance has its own specific heat, Table 12.1 lists a few examples.

Table 12.1 Specific heat of various substances

Substance	Specific Heat (cal/gm/ °C)
water	1.00
aluminum	0.21
copper	0.092
glass	0.15
steel	0.11
TbFe	0.11
plastic (PMMA)	0.34

As an example, consider the heating by a laser beam of the magnetic film in a recordable mini–Disc. How long does it take to heat the spot illuminated by the laser to just above the Curie temperature of the magnetic film? Consider a section of the recordable disc in Figure 12.6. The TbFeCo film is initially at room temperature. To record on

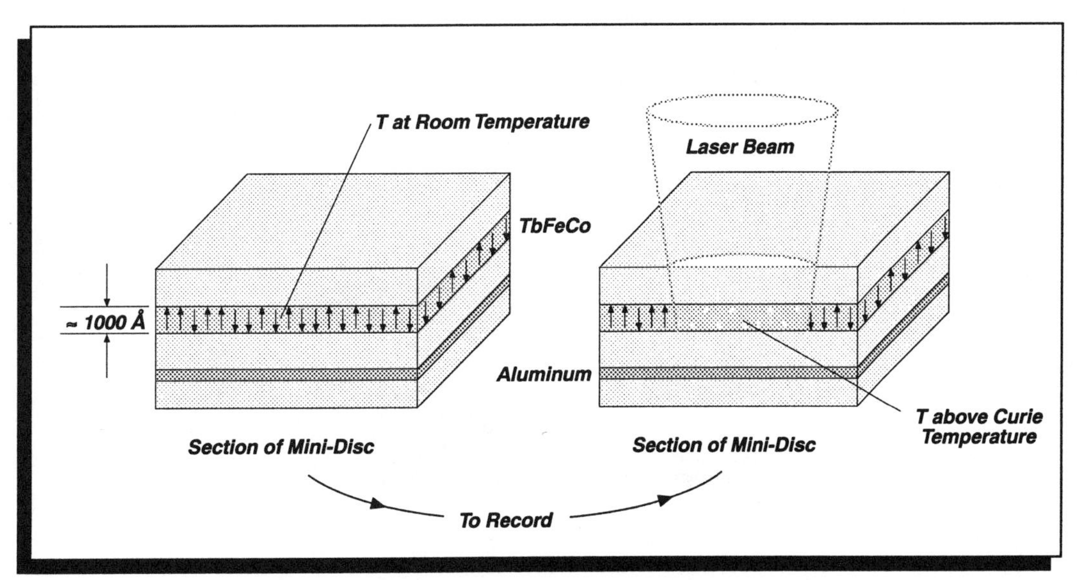

Figure 12.6 Heating of a spot on mini-disc for recording.

it the laser beam, 2μm in width, heats it. The small unit of dimension 1μm is 10^{-6}m. The heated volume of the magnetic film 0.1μm thick is:

$$\begin{aligned} \text{Volume} &= \text{area} \times \text{height} \\ &= \pi r^2 \times h \\ &\approx 0.3\ \mu m^3 \end{aligned}$$

Since the density is ≈ 7 gm/cm^3, we have a mass of (7 gm/cm^3) × 3 × 10^{-13} cm^3 (i.e. $\approx 2 \times 10^{-12}$ gm). The amount of heat necessary to raise the temperature of this mass of material from room temperature to the Curie point is:

$$\begin{aligned} \text{heat required} &= \text{specific heat} \times \text{mass} \times \text{temperature difference} \\ &= 0.11 \times 2 \times 10^{-12} \times 160^\circ C \\ &= 3.5 \times 10^{-11}\ \text{calories} \end{aligned}$$

But the heat from the laser is produced at a rate of approximately 2 millicalories per second. Hence, the time needed to warm up that spot to its Curie temperature is:

$$\begin{aligned} &3.5 \times 10^{-11}\ \text{calories} / 2 \times 10^{-3}\ \text{calories/sec} \\ \approx\ &1.7 \times 10^{-8}\ \text{second} \end{aligned}$$

which is very fast. This justifies the high rotational speed of disc.

■ HEAT TRANSFER

Having discussed how heat is produced in the hi–fi system, and how it is detected, we will now present the various methods by which heat is transferred away from the hi–fi system.

There are three ways in which heat can be transferred and it consists of:

i. conduction
ii. convection
iii. radiation

i. Conduction

In conduction, heat is transferred from the hot region to a cold region by means of the energetic vibrations of atoms and electrons which interact with their neighbors and thus send their energy to the less energetic atoms and electrons. For heat to flow in this manner there has to be a temperature difference. Heat will flow from the hot to the cold. Figure 12.7 illustrates this type of heat transfer. It is most important in a solid; it does exist, to a much smaller degree, in fluids and gases. At high temperatures atoms and electrons vibrate with large amplitude. They collide with less energetic atoms and electrons, transferring some energy to them and thus the heat is transferred down to lower temperatures along the rod. The amount of heat flow per unit time along the rod in Figure 12.7 can be calculated quite simply when a comparison is made of this situation to that of an electric circuit. We had electric current, which is:

amount of charge / time = (voltage difference) /electrical resistance

and by analogy, we have:

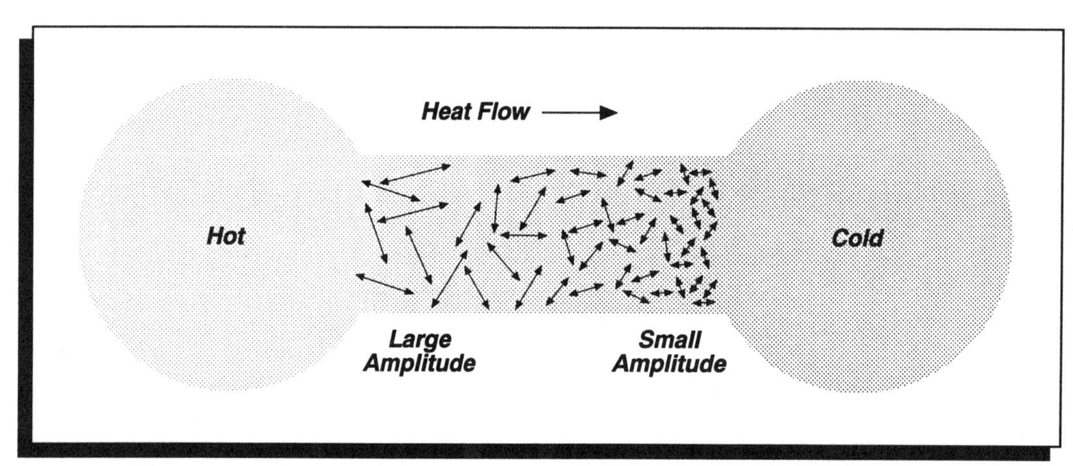

Figure 12.7 Heat conduction along a bar between a hot body and a cold one.

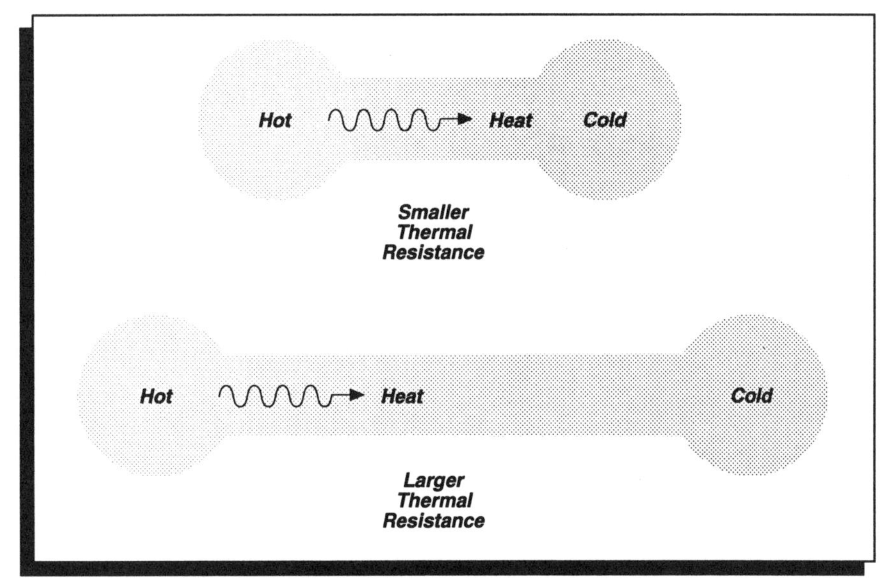

Figure 12.8 Thermal resistance depends on length of heat conductor.

amount of heat / time
= (temperature difference) / thermal resistance

The thermal resistance depends:

— on the material.
— on the length of material.
— inversely on the material cross–sectional area.

The materials vary from good conductors of heat, which have a small thermal resistance, to poor conductors of heat, which do not conduct much heat between hot and cold. Table 12.2 provides a list of the thermal conductivity of a few materials.

Table 12.2
Thermal conductivities of some materials

Material	Thermal conductivity (calories/second × cm × °C)
air (at 20°C)	5.6×10^{-4}
aluminum	0.57
copper	0.95
TbFe	9.5×10^{-2}
Plastic (PMMA)	4.8×10^{-4}
glass	2.5×10^{-3}

The geometrical factors which determine the thermal resistance are similar to electrical factors. The thermal resistance is proportional to the length of the heat conductor. Figure 12.8 shows how the thermal resistance increases as the length of heat–carrying material increases. If the length of the heat conductor is doubled, the thermal resistance doubles and there will be two times less heat flowing, for the same temperature difference.

The cross–sectional area will also determine the thermal resistance to heat flow. Consider the two cases shown in Figure 12.9. As the cross–sectional area of the heat conductor decreases there will be more thermal resistance and hence less heat flow. In fact, when the cross–sectional area doubles, the

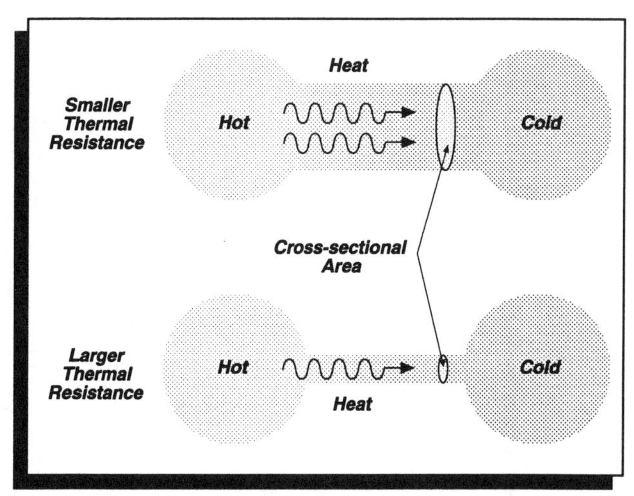

Figure 12.9 Thermal resistance depends inversely on cross-sectional area of heat conductor.

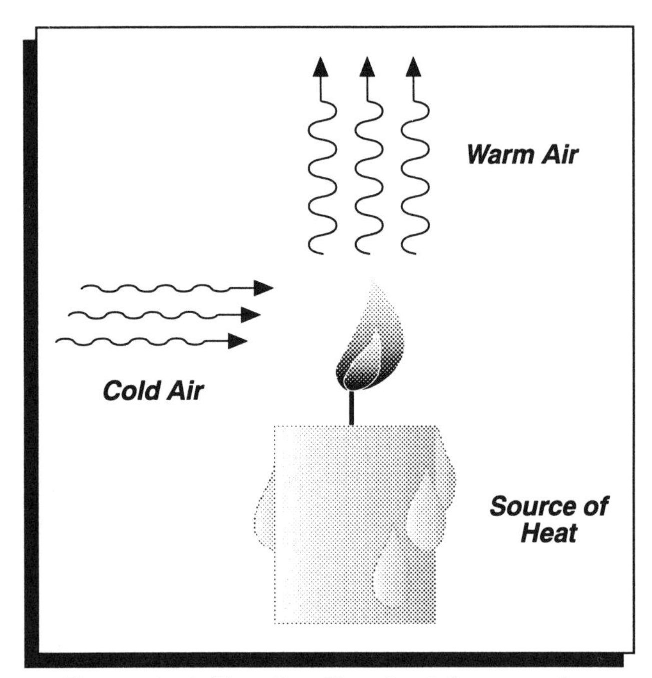

Figure 12.10 Transfer of heat in air by convection.

thermal resistance decreases by a factor of two, for a fixed temperature difference.

In the above presentation, there was heat flow because there was a temperature difference. The amount of heat flowing per unit time is directly proportional to the temperature difference.

ii. Convection

Convection is a very powerful method of heat transfer in fluids and gases. It consists of transfer of material from the hot part to the cold part. In natural convection the motion is caused by a difference in density between the hot and the cold region. Density is defined as:

Density **= mass / volume**

For example, hot air has a smaller density than cold air since its volume expands to larger dimensions and hot air will rise, while cold air goes down. Heat will warm the air near it, causing it to rise and then cold air will replace it. Such flow of air causes heat to be transferred. Figure 12.10 illustrates the principle of convection using air and a source of heat.

iii. Radiation

This method of heat transfer does not at all involve contact between the bodies transferring heat as did the previous methods. Every object at some temperature T will emit radiation which is electromagnetic in nature. Since every object has atoms (and hence electrical charges), the thermal energy of the body causes the charges to vibrate and, thus they will be emitting electromagnetic waves. The rate at which a body emits such radiation is proportional to the fourth power of its absolute temperature. Hence:

$$\textbf{power radiated by a body} \ \alpha \ T^4$$

and this is shown in Figure 12.11. The radiated electromagnetic waves are usually in the infra–red range.

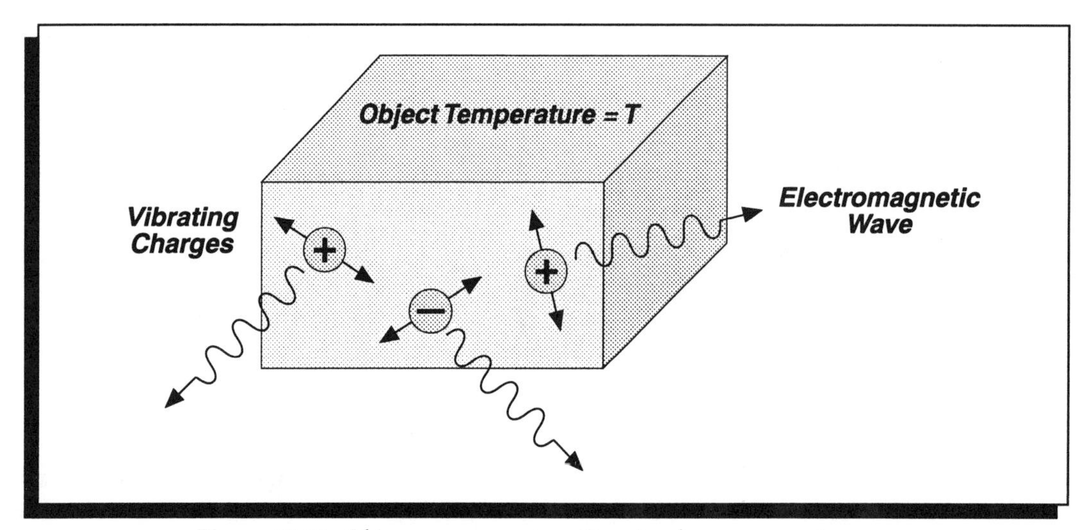

Figure 12.11 Object at temperature T emits electromagnetic waves.

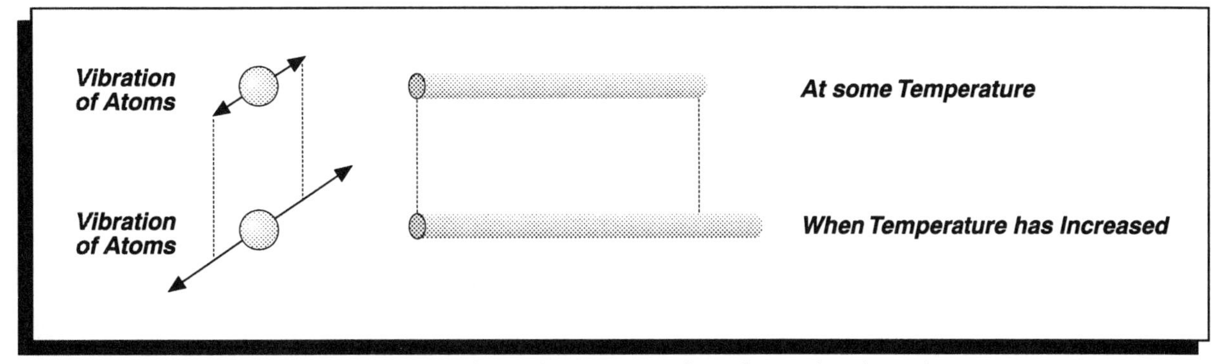

Figure 12.12 Thermal expansion of an object when heated.

An object which radiates energy also absorbs electromagnetic radiation. At equilibrium the object radiates and absorbs at the same rate, and so its temperature does not change. However, when an object is hotter than its surroundings, it will radiate more power, and so it will cool. Since electromagnetic waves do not require a medium for propagation, radiation from an object does not require any contact to it.

THERMAL EXPANSION

When the temperature of an object is increased, the atoms vibrate more vigorously with a larger amplitude. When the amplitude of vibrations increases, the average position of the atoms changes, and the object expands. Figure 12.12 shows expansion due to heating. Conversely, as the object is cooled, it contracts. The amount an object changes its dimensions as its temperature is changed is called thermal expansion. Each substance has a different expansion value, and it expands at a different rate. Thermal expansion occurs in solids, liquids, and gases.

There are many practical devices which use thermal expansion as a method for displaying effects of heat increase or decrease of an object. A popular example is the thermometer with a column of alcohol or mercury, which expands when heated and contracts when cooled. Another important class of devices is based on a bimetallic strip which is formed with materials which expand at different rates. Consider the bimetallic strip in Figure 12.13 formed with two different materials, steel and brass, and clamped strongly together. Because brass expands more than steel when heated, it will become longer than the steel, and the strip will bend as shown in Figure 12.13. Likewise, upon cooling, the reverse action will take place.

12.2 HEAT IN AMPLIFIER

The basic heat phenomena discussed above are applied to the problem of heat removal from an amplifier. Consider transistors and diodes, which produce heat when current flows through

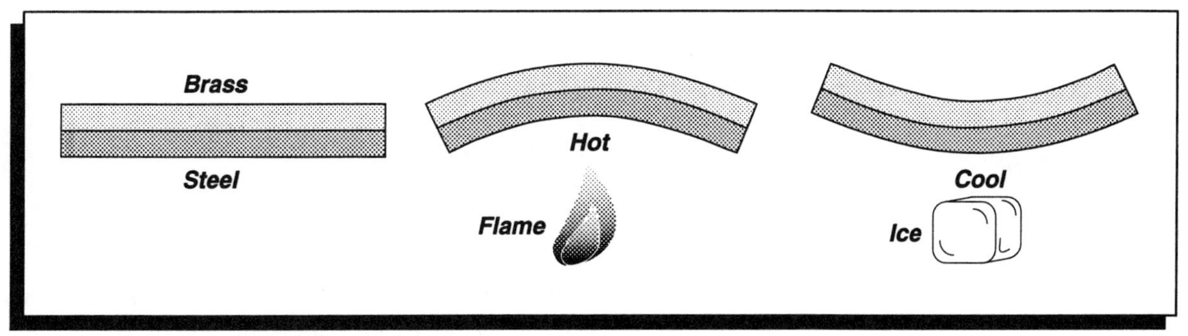

Figure 12.13 Bimetallic strip and its behavior when heated or cooled.

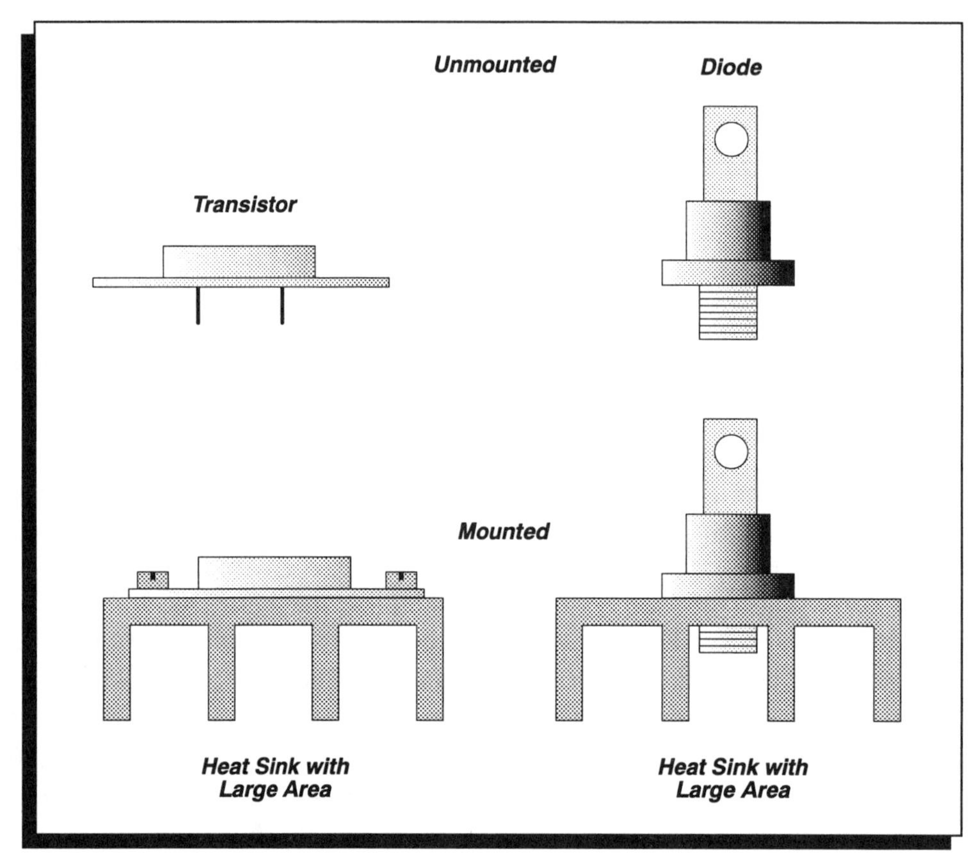

Figure 12.14 Mounting of a transistor and a diode on heat sink to transfer heat away from devices by heat conduction.

them. The resulting heat is removed by mounting these devices on a large metallic structure made out of material which can effectively conduct heat away, a heat sink. Figure 12.14 shows a transistor and a diode mounted on a metallic heat sink to conduct heat away from them. Because the heat sink has a large cross–sectional area, its thermal resistance is low and heat will be conducted away from the devices efficiently.

Heat from the heat sinks will be removed by convection and by radiation. Figure 12.15 shows how both methods of heat transfer are used in the amplifier.

Should the holes in the chassis of the amplifier get blocked (by mistake) heat will not be removed effectively and the inside temperature will rise. To prevent damage from operation at too high a temperature or when too much current is drawn from the power supply, there is a safety feature that uses a bimetallic strip; it is the circuit–breaker. This is shown in Figure 12.16, where a bimetallic strip is part of the circuit carrying the current.

When the bottom part of the bimetallic strip has a material of larger thermal expansion than the top, the device will bend upward when overheated, and this will cause an open–circuit which interrupts all the current. The circuit breaker can be reset manually by a push–button.

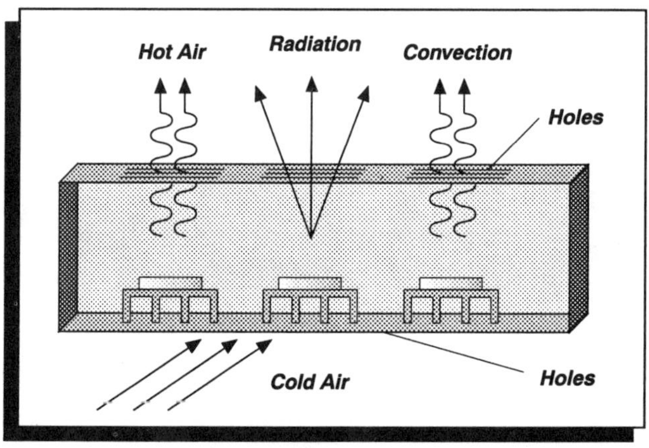

Figure 12.15 Heat removal by convection and radiation.

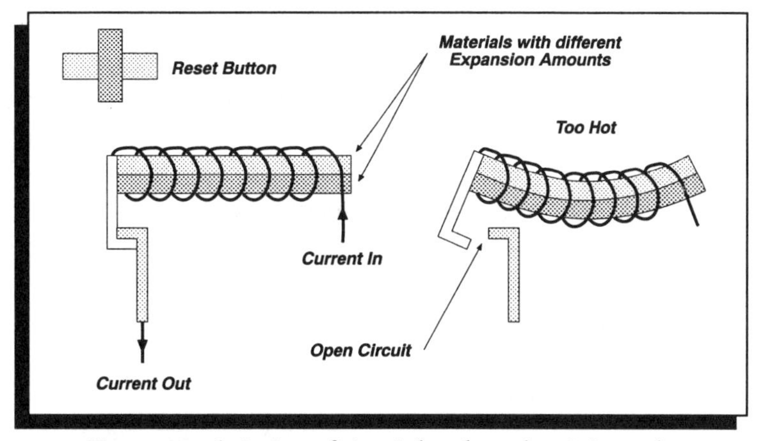

Figure 12.16 Action of circuit breaker when it is too hot.

12.3 HEAT IN MINI–DISC

The operation of the recordable mini–Disc is based on thermo–magnetic recording. The recording process requires local heating of a tiny spot on the magnetic film of the disc to a temperature above the Curie temperature. The laser beam passes through the plastic (PMMA) of the disc and part of it energy is absorbed by the magnetic film, causing heating. Because the power of the laser is limited, between 5 to 10 milliwatts, it is important that the temperature of the selected spot on the magnetic thin film will rise rapidly. This is facilitated by:

— very small mass of magnetic film region defined by laser.
— large thermal resistance between plastic substrate and air so as not to conduct heat away too rapidly to ambient surrounding.
— large thermal resistance between magnetic film and disc plastic so that heat is not conducted away too rapidly from the region defined by spot.

A previous example has shown that the spot on the magnetic film responds to the laser in much less than 1μsecond. This sets a limit on the speed of the disc. If it rotated too fast, there would be not enough time to heat the spot. On the other hand, the spot has to cool as soon as possible after the magnetic field of the write head has entered the film. This can be achieved by the very small mass of the heated spot; it cools rapidly when it is away from the laser beam, but it must cool in the presence of the write magnetic field. Figure 12.17 shows the steps that lead to recording digital information on the disc.

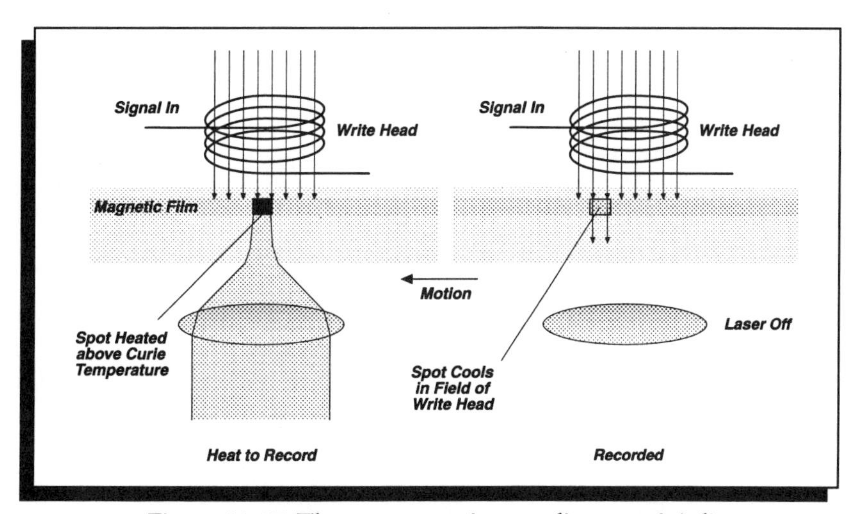

Figure 12.17 Thermo-magnetic recording on mini-disc.

SUMMARY OF TERMS

Bimetallic strip: Strip made of two materials which have different thermal expansions. When heated, the strip bends in or out.
Circuit breaker: A device which protects circuits from excessive electric currents by interrupting the current. Often a bimetallic strip is used to switch the current off; after it is triggered it can be reset.
Conduction: Natural flow of heat along an object when a temperature difference exists across it. Flow of energy occurs by vibrations of atoms and electrons that are passed on to neighbors.
Convection: Flow of heat by displacements of material due to temperature differences or other forces acting on it. Usually occurs in fluids and gases when temperature differences cause density changes.
Density: Characteristics of material which is its mass per unit volume.
Radiation: Transfer of energy from an object by emission of electromagnetic waves, usually at infra–red frequencies.
Specific heat: Quantity of heat per unit mass needed to raise the temperature of a substance by one degree centigrade.
Temperature: Measure of how much thermal activity is present in an object. It is measured in Centigrade or Fahrenheit for everyday experiences, and in Kelvin for scientific research.
Thermal expansion: Change in volume of an object when heated or cooled.
Thermal resistance: Difficulty offered to the flow of heat along an object. It depends on material and geometric factors.

NAME ______________________________ DATE ____________

QUESTIONS FOR REVIEW

1 Explain what is meant by conservation of energy.

2 How is the temperature measured inside an amplifier?

3 What factors determine how much heat flows along an object?

4 List three good conductors of heat. List three poor ones.

5 Explain the Law of Conservation of Energy and give two examples of its application in the hi–fi.

6 What effect does adding heat have on a substance?

7 Explain why a spot on the magnetic film in a mini–Disc has to be heated before recording.

8 How is heat removed from a transistor in an amplifier?

9 Why do materials expand when heated?

10 Why does a spot on the magnetic film of a mini–Disc heat up so fast when illuminated by a laser beam?

NAME________________________________ DATE ______________

Please select one answer for each question.

1. A bimetallic strip bends when heated because:
   ___A. the two metals have different masses
   ___B. the two metals are at different temperatures
   ___C. the two metals expand by the same amount
   ___D. the two metals expand by different amounts
   ___E. the two metals have different lengths

2. Convection is a process:
   ___A. where air is uniformly warm
   ___B. where heat travels in a vacuum as an electromagnetic wave
   ___C. where heat flows through a metal heat sink
   ___D. where the natural flow of air carries heat away
   ___E. which causes transistors to get hot

3. Heat energy is transferred by radiation as:
   ___A. electromagnetic waves
   ___B. sound waves
   ___C. air flow
   ___D. pressure fluctuations
   ___E. density fluctuations

4. A change in an object's heat energy is detected by:
   ___A. its temperature
   ___B. a change in its temperature
   ___C. a change in its specific heat
   ___D. a change in its mass
   ___E. an increase in its mass

5. The law of conservation of energy states that:
   ___A. energy cannot change form
   ___B. energy of a system cannot increase
   ___C. energy of a system cannot decrease
   ___D. energy cannot be destroyed nor created, but can only be converted from one form to another
   ___E. energy can be created and destroyed, but can never be converted from one form to another

6. The action of a cone loudspeaker is an example of:
   ___A. conversion of electrical energy to mechanical energy and heat
   ___B. conservation of energy
   ___C. 100% efficiency
   ___D. all of the above are true
   ___E. A and B are true

7. Conduction is a process where:
   ___A. heat flows through a material when there is a temperature difference across it
   ___B. heat flows through a material where there is no temperature difference across it
   ___C. heat flows toward the hotter end of a bar
   ___D. heat radiates as electromagnetic waves
   ___E. heat causes density fluctuations

8. When a copper bar is made 4 times shorter and its cross–sectional area is decreased 2 times, its thermal resistance will be:
___A. 4 times larger
___B. 4 times smaller
___C. 2 times smaller
___D. 2 times larger
___E. 8 times smaller

9. A transistor in an amplifier is kept cool by dissipating heat by:
___A. conduction
___B. convection
___C. radiation
___D. all of the above
___E. only A

10. When equal masses of water and aluminum are heated from 20°C to 50°C, one can say that:
___A. both had the same amount of heat energy added to them
___B. water had almost 5 times more heat energy added to it than the aluminum
___C. water had almost 5 times less heat energy added to it than the aluminum
___D. water had almost 12.5 times more heat energy added to it than the aluminum
___E. they will take the same amount of time to cool to 20°C

11. Transistors produce heat when current flows through them because:
___A. they amplify input signals
___B. they have electric resistance
___C. a heat sink is attached to them
___D. they get the heat from the heat sink
___E. they have a small specific heat

12. How much heat is needed to raise the temperature of the plastic (PMMA) in a mini– Disc by 1°C if its mass is 1 gm?
___A. 1 calorie
___B. 0.34 calorie
___C. 3.4 calorie
___D. 0.1 calorie
___E. 0.034 calorie

13. It takes a very short time to heat a spot on the magnetic film of a mini–Disc above its Curie temperature because:
___A. magnetic film is very thin
___B. disc rotation speed is extremely high
___C. magnetic film is very thick
___D. specific heat of film is high
___E. power of laser is low

14. The spot defined by the laser beam on a magnetic film of a mini–Disc radiates more power when:
___A. it is heated
___B. it is at room temperature
___C. it is being heated
___D. it is cooling
___E. there is no radiation as the laser beam is reflected back

15. A solid expands when heated because:
___A. atoms vibrate with smaller amplitude
___B. atoms vibrate with larger amplitude
___C. its specific heat increases
___D. its heat energy increases
___E. its density increases

CHAPTER 13 MECHANICS

The impressive achievements in the field of high–fidelity are due in part to the important advancements in electronics and magnetic devices, but also to the precise control of the mechanical aspects of sound reproduction. The mechanics deal with tape transport in tape decks, rotations of a CD past the laser beam, the focus of the optical head, rotating head in a DAT, and of course the sound production by a loudspeaker. This chapter covers these aspects, starting with basic definitions and their role in the hi–fi, and ending with Newton's laws of motion as applied to some of the components in the hi–fi.

13.1 MOTION IN ONE DIMENSION

All sources of audio signals in the hi–fi use mechanical principles in the production of high–fidelity signals. Here we will present some of these principles and show how they are used in the hi–fi system. First, a few important definitions will be presented within the context of the material dealing with hi–fi. Words like speed, pressure, force, friction have been used in the previous chapters without an explanation of their meaning; this chapter presents now their definition and explores their role in the hi–fi.

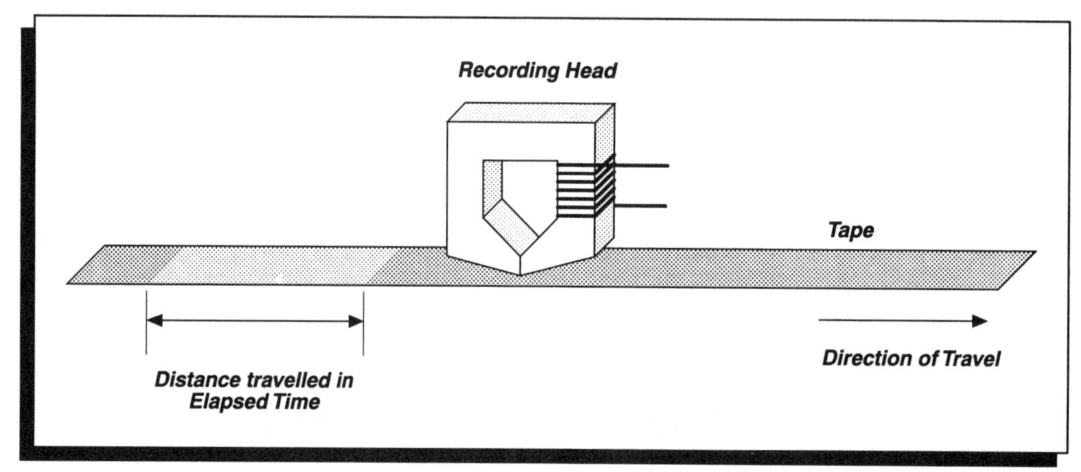

Figure 13.1 Speed of tape past recording head.

SPEED

The speed at which an object travels relates to the distance that it covers in a period of time. Consider, for example, the tape in a tape deck. It travels at some speed relative to the recording head or playback head. In Figure 13.1, a tape is recorded in analog format while it passes over the gap of the recording head.

The average speed of the tape past the recording head is:

$$\textbf{average speed} = \frac{\textbf{distance travelled}}{\textbf{elapsed time}}$$

Here, the distance travelled is actually the length of tape that went past the recording head in an elapsed time. The speed just calculated is called average because there could be some fluctuations in the rate at which the tape is pulled past the head. At times the tape could be going slightly faster, and at other times slightly slower.

We can use the definition of average speed to calculate quantities relevant to the operation of a tape deck.

Example: How long is an analog magnetic tape as used in a tape deck, when the cassette can store 60 minutes of audio information and the tape speed is $1\,\frac{7}{8}$ i.p.s. ? The cassette has a tape which can play one side for 30 minutes.

Using the definition of average speed, the total distance covered is:

total displacement
= (average speed) × total time
= ($1\frac{7}{8}$ i.p.s.) × 30 minutes × 60 seconds/minute
= 3375 inches

As there are 12 inches in a foot, this is approximately 281 feet. A cassette programmed for 120 minutes of recording will have twice as much tape.

Example: How long will it take a radio wave to go around the surface of the Earth at the equator? The diameter of the Earth at the equator is 12.76×10^3 km. Figure 13.2 presents the problem.

Using again the definition of average speed, the time that it takes to go around the earth is:

$$\textbf{total time of travel} = \frac{\textbf{distance covered}}{\textbf{speed}}$$

$= \pi \times 12.76 \times 10^6$ meters ÷ 3×10^8 m/sec
$= 1.33 \times 10^{-1}$ second

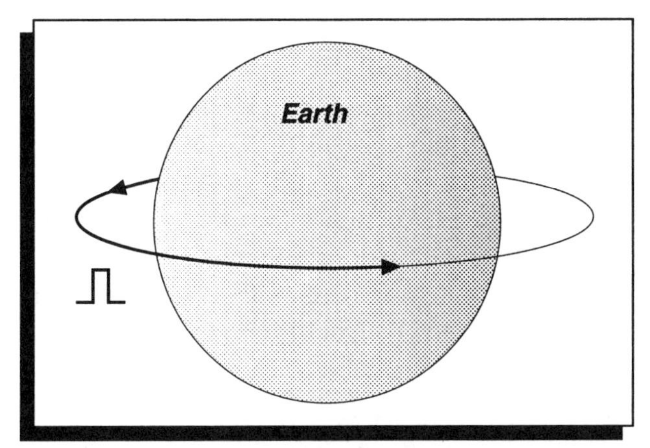

Figure 13.2 Time for a radio wave to go around the Earth at the equator.

Here, we have used the fact that the distance around a circle, its circumference is π × diameter.

In the example of the tape being pulled past the recording head, we have used the average speed. However, it is possible that at some instant of time the speed of the tape was different from the average speed. This could have been caused by the capstan whose speed of rotation had some variation or its shape was not round. In that case the instantaneous speed is a displacement divided by an extremely small elapsed time; it is the speed at that instant of time. When the tape moves across the recording head or the playback head the speed is expected to be constant. Variation of the speed in the tape deck will lead (according to frequency × wavelength = speed) to:

Speed variations in recording → wavelength variations

and

Speed variations in playback → frequency variations

Such problems in a tape deck occur when the instantaneous speed is not equal to the average speed. This causes wow and flutter; wow corresponds to slower variations in speed at about a few Hertz, while flutter corresponds to more rapid variation occurring from about 5Hz to 200Hz. Acceptable values for wow and flutter are at 0.15% or less. This means that 0.15% speed variations for a 1,000 Hz tone would cause frequency variations of 1.5 Hz.

VELOCITY

This term describes the motion of an object; its definition contains two quantities:

1. Speed.
2. Direction of travel.

It is important to note that speed and velocity have different meanings. Velocity gives the speed of an object and tells us the direction of travel. As an illustration of this consider an old phono record, as shown in Figure 13.3. It rotates at 33 ⅓ rounds per minute. A wave recorded in the groove will

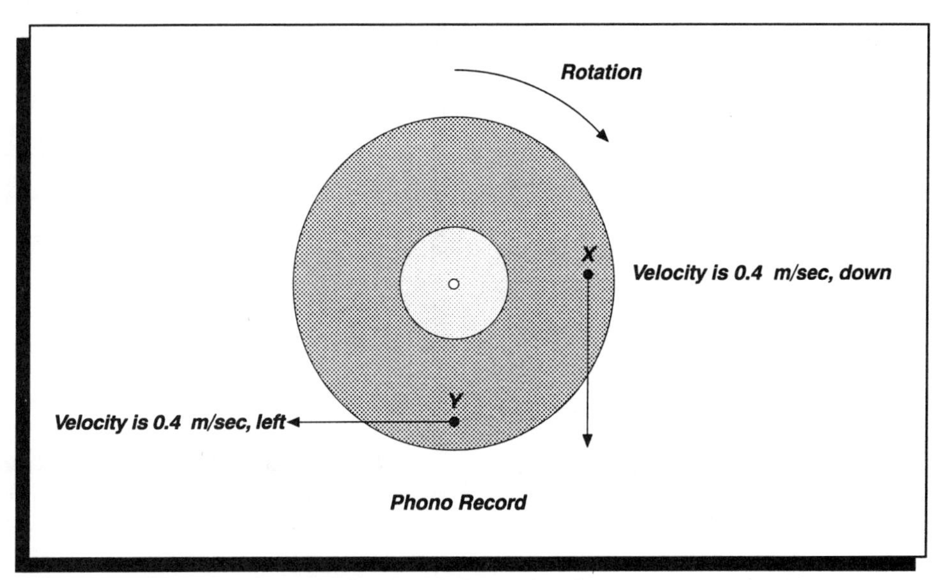

Figure 13.3 Speed of a recorded signal is the same at X and at Y their velocities are different.

have a certain speed, typically at 0.4 m/sec, when the record rotates. Its speed at X and Y is the same. However, its velocity at X is 0.4 m/sec down, while at Y it is 0.4 m/sec to the left. The velocities are different because the directions are not the same. The average velocity is defined as:

$$\textbf{average velocity} = \frac{\textbf{displacement}}{\textbf{elapsed time}}$$

Quantities which specify a direction and a magnitude (a certain amount) are called vectors; and velocity is one such example. A displacement is another example of a vector because it specifies how much an object has been moved and in which direction. There will be other examples of vectors in this chapter.

The average velocity does not necessarily tell us the exact velocity at a specific time. The instantaneous velocity would indicate how fast an object is moving and in which direction, at each instant of time. It refers to the velocity obtained by measuring the displacement per extremely short elapsed time.

■ ACCELERATION

This quantity refers to a change of velocity. From everyday experience acceleration means a change of the velocity of an object. We define average acceleration as:

$$\textbf{average acceleration} = \frac{\textbf{change in velocity}}{\textbf{elapsed time}}$$

Because velocity is a vector, acceleration is also a vector. Since velocity consists of speed and direction, changing one of them or both requires an acceleration. In the example of the signal in a record groove, it is being accelerated as it goes around and its direction changes continuously. The change in velocity is the final velocity minus the initial velocity, so the acceleration can be positive or negative, depending on whether the final velocity is larger or smaller than the initial velocity.

■ FORCE

It is a push or a pull. Because there is a direction associated with it, it is also a vector. An example where both can occur is the action of the magnet on the voice coil of a speaker when it carries a current. Figure 13.4 shows the effect of the force for two different situations due to current going in the voice coil one way or the other. The force on the voice coil then changes directions and this causes the diaphragm to vibrate creating sound. The units of force are pounds in the British system and newtons or dynes in the metric system. A pound force corresponds to 4.45 newtons. The concept of force was used earlier in the definition of compliance, which is a displacement per unit force and its units are cm/dyne.

In a tape deck, the capstan and pinch roller pull the tape at a constant speed. Figure 13.5 shows the action of the capstan and pinch roller, as they exert a force on the tape. In fact, it is the force of friction

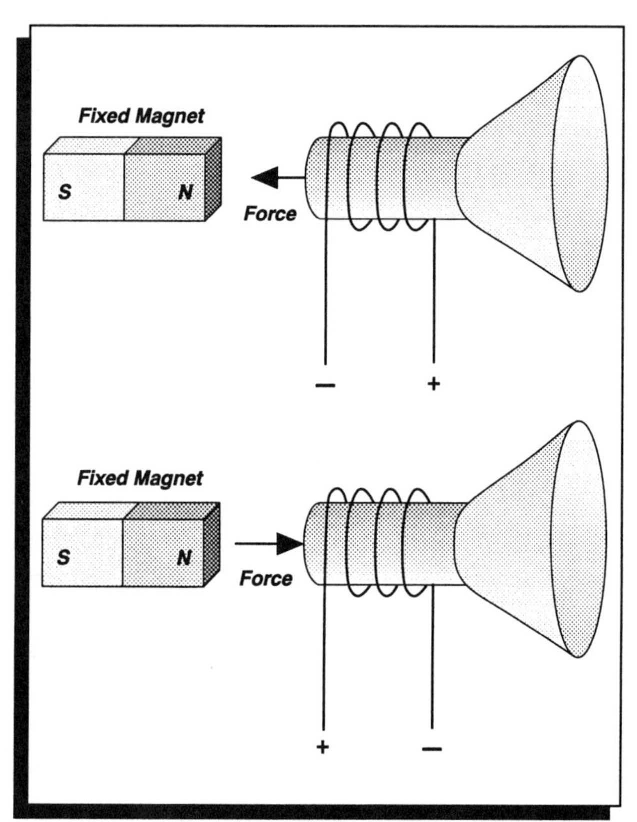

Figure 13.4 Force on voice coil giving it a push or a pull depending on direction of current in voice coil.

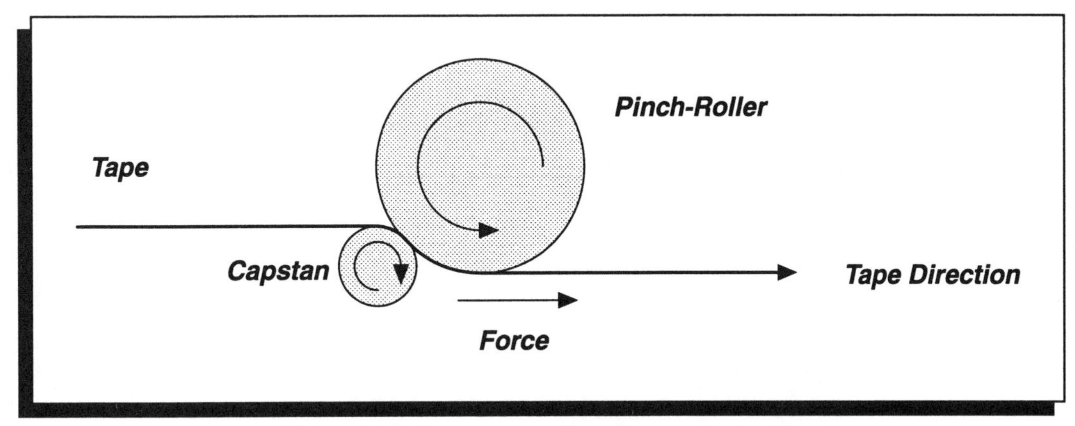

Figure 13.5 Force on tape by capstan-pinch roller.

due to the pinch–roller pressing the tape against the capstan that provides the necessary pull on the tape. Figure 13.6 shows how this frictional force originates. It is called a static force of friction because there is no relative motion between the tape and the capstan–pinch roller assembly. Because the force of friction between the tape and the capstan–pinch roller is larger than just the pull of the capstan, the tape remains stationary, with respect to the capstan–pinch roller, and hence it is pulled in the forward direction.

There is another type of frictional force and that is the kinetic force of friction which deals with relative motion between two objects. An example is the tape passing over the recording head (or playback head). This force of friction opposes the motion of the tape and it is in the opposite direction to that of the tape pull. The tension in the tape is larger than this force of friction and hence the tape moves in the direction of the pull.

■ PRESSURE

The main subject of hi–fi is that of sound. It was defined as pressure variations at various frequencies. We now define pressure as:

pressure = force / area

It means that a force is exerted over a certain area. Consider the act of removing a CD from its plastic container. The central clips holding the CD in the container have to be released by applying some

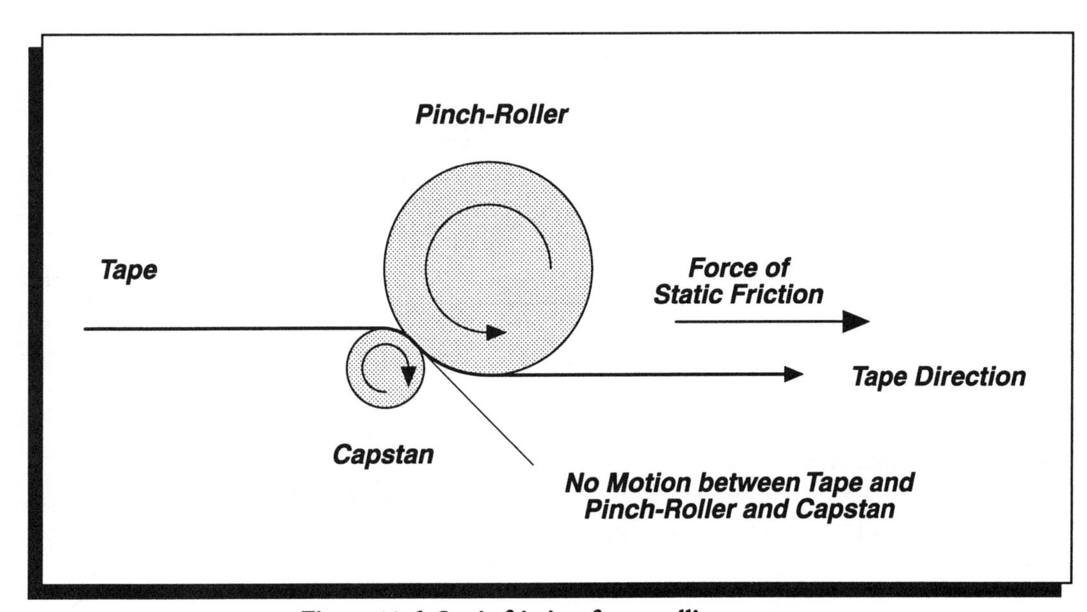

Figure 13.6 Static friction-force pulling on tape.

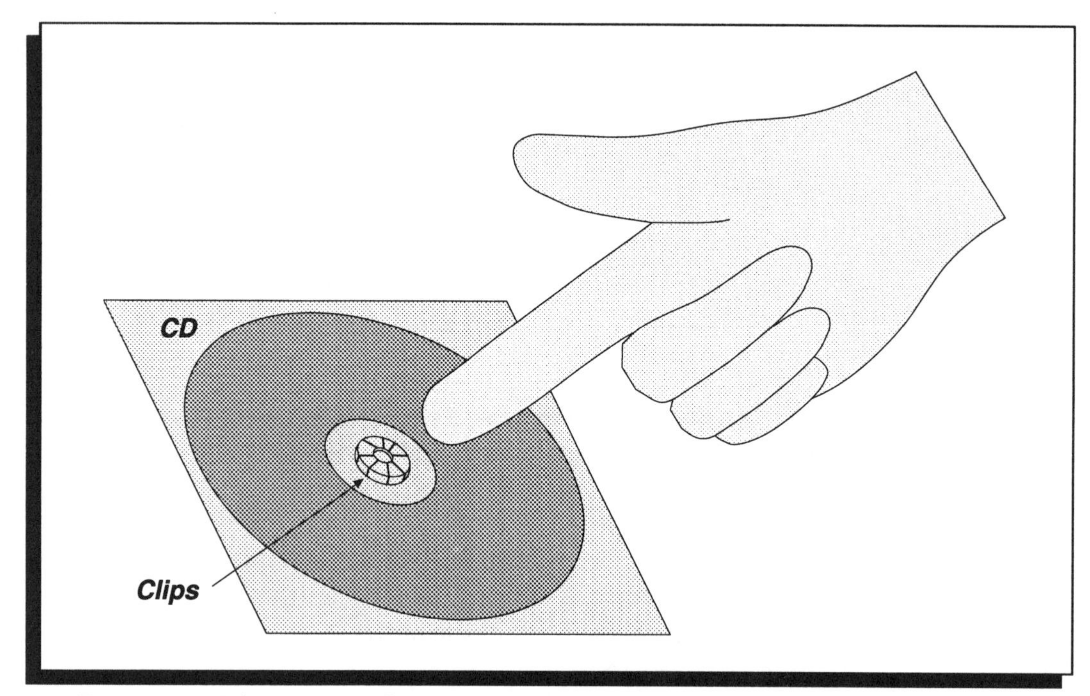

Figure 13.7 Releasing a CD from its case by applying a pressure on the clips with a finger.

pressure on them with a finger, Figure 13.7. The pressure exerted by a finger on the container clips is approximately:

= a force of 1 pound / area of clips
= 1 pound / π (¼)^2 inch^2) $\approx$ 5 lbs/square inch

since the clip unit has a diameter of ½ inch.

The pressure changes in sound waves are much smaller than those produced by the finger on the CD clip. The vibrating diaphragm of a speaker causes pressure changes in the air and the energy is carried away by the sound waves. At the threshold of hearing for 1,000 Hz, the intensity is 10^{-12} watts/m^2. A calculation of this intensity in terms of a pressure change corresponds to a pressure change of 3×10^{-5} newtons/m^2 above and below atmospheric pressure which is about 10^5 newtons/m^2. This means that the ear can detect pressure changes as small as 3 parts in 10^{10}. Our ears are very sensitive! The corresponding displacements are less than the diameter of a molecule by almost a factor of 10. This shows that for sound to be heard it is not necessary to create large pressure changes.

NEWTON'S LAWS OF MOTION

Newton formulated the behavior of objects in motion by three laws which are the basis of mechanics. Here, we will examine these laws and see how they are applied to the performance and development of hi–fi equipment.

(i) First Law of Motion

It states:

Every body continues in its state of rest, or of uniform motion in a straight line, unless it is compelled to change that state by forces impressed upon it.

(ii) Second Law of Motion

It states:

The acceleration of a body is directly proportional to the net force acting on it and inversely proportional to the mass of the body being accelerated.

It can be written as:

Net Applied Force = (acceleration of body) × (mass of body)

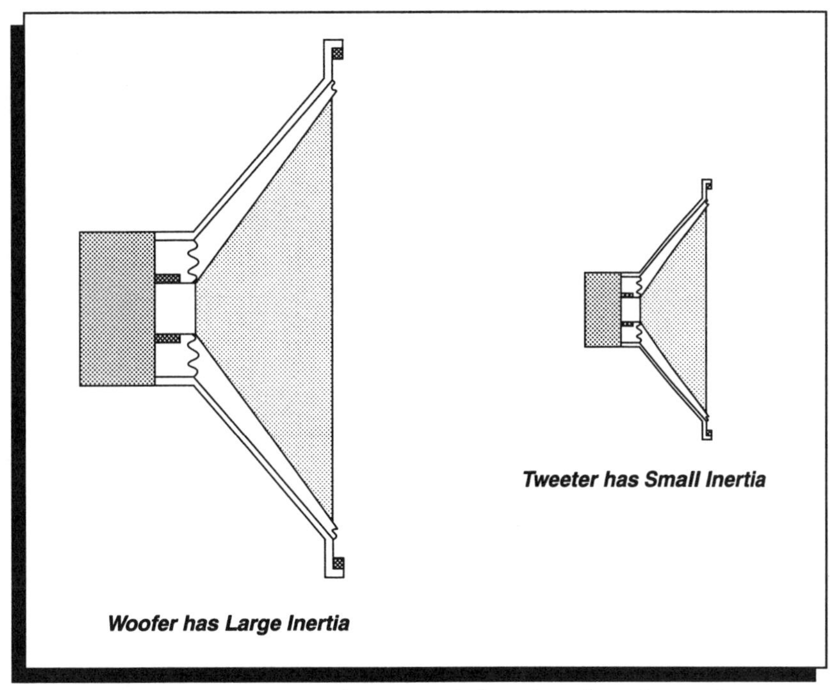

Figure 13.8 Inertia of a tweeter is less than that of a woofer.

(iii) Third Law of Motion

It states:

> Whenever one object exerts a force on a second object, the second object exerts an equal and opposite force on the first.

Now we will see how these laws are used in the mechanics of a hi–fi system.

The tendency of an object to maintain its state of rest or uniform motion in a straight line, as given by the first law, is called its inertia. The harder it is to change the state of an object (at rest or in uniform motion) the more inertia it has. For example Figure 13.8 shows a woofer and a tweeter speaker. It is easier to change the state of uniform motion or rest of a tweeter than that of a woofer. Therefore, the woofer has a larger inertia than the tweeter. For this reason, the woofer cannot be used at high frequencies. Its state of uniform motion or rest cannot be changed fast enough.

We define inertia as:

***inertia* = property of an object that resists changes in its state of rest or of uniform motion in a straight line.**

It is the inertia of our ear drums which limits the response of our ears at high frequencies, especially at 20 kHz and above. Figure 13.9 shows a human outer ear and the ear drum. The measure of inertia is mass. When an object has a large inertia it has a large mass. We define mass as:

mass = quantity of matter in an object

It is measured in grams or kilograms.

The concept of mass should not be confused with weight, which means something very differ-

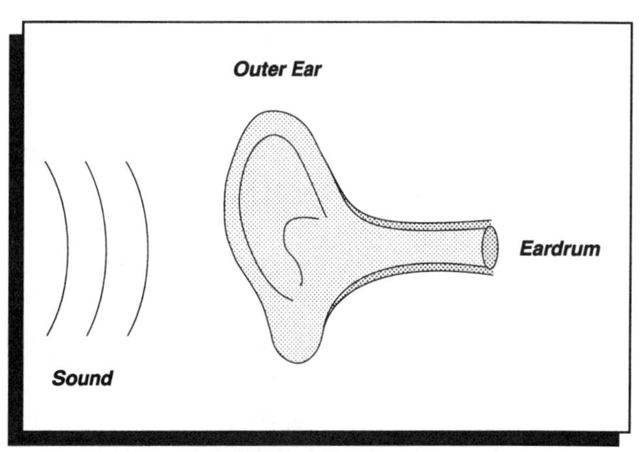

Figure 13.9 Outer ear; ear drum's inertia limits response at frequencies above 20 kHz.

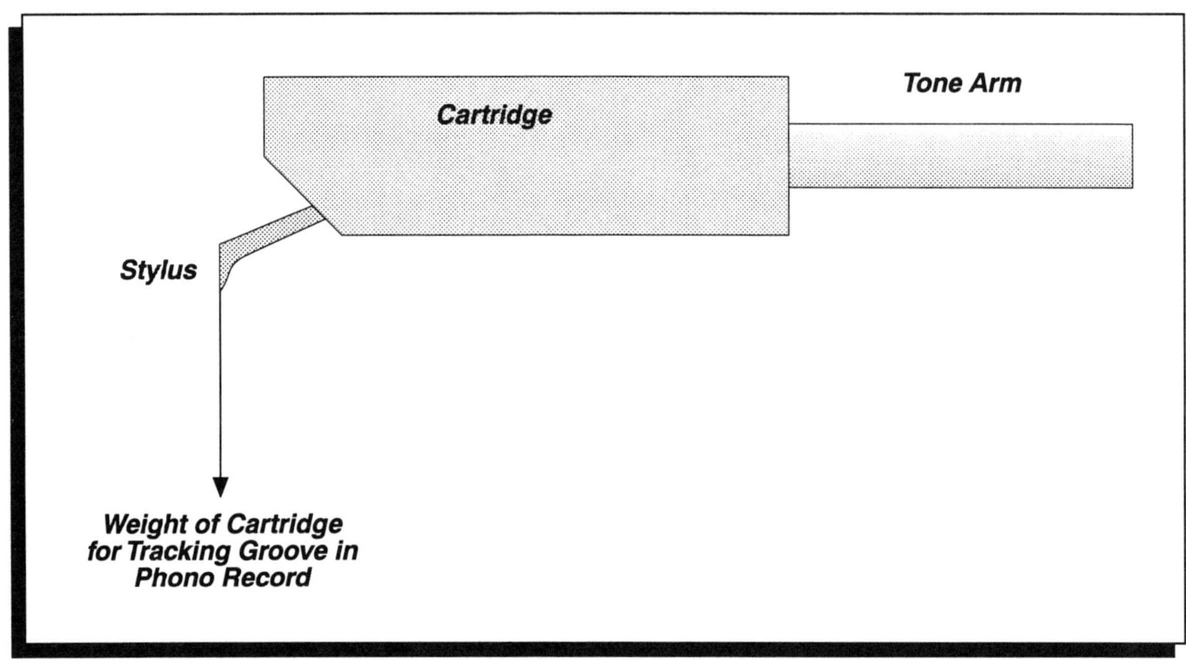

Figure 13.10 Adjusted weight in cartridge for helping the stylus to track the grooves in phono record.

ent. Weight refers to how difficult or easy it is to lift an object against the pull of gravity. It is defined as:

weight = gravitational force acting on an object

A good example is the tracking force of a stylus in a cartridge used with a phono record. Such force is necessary to keep the stylus in the groove of the record. An adjustable weight is applied to the stylus, as shown in Figure 13.10, to help it remain in the groove without excessive force, which would wear out the information in the groove.

The Second Law of Motion tells us what happens to an object when a net force is applied to it; it will accelerate. Two examples of this are the loudspeaker and the laser focus system in a CD player, Figure 13.11. First consider the loudspeaker.

As the current in the voice coil increases, the force of the magnet on the current–carrying coil increases and so the voice coil moves one way or the other, depending on the direction of the current, forcing the diaphragm to push or pull the air. The acceleration of the diaphragm is due to the force between the voice coil carrying a current and the clamped magnet.

As all sorts of frequencies are handled by the speaker, the acceleration will depend on the frequency. Assume that the speaker amplitude is fixed for all frequencies. The speed of the diaphragm will increase with the frequency, since it will take less time to travel a given distance. Moreover, the acceleration will also be affected by the frequency, as the change of speed of the diaphragm will take place in a shorter time at a high frequency. This means that the acceleration will be larger at higher frequencies. From the Second Law of Motion, higher acceleration can be achieved by a larger force on the voice coil (by increasing the voice coil current or even using a more powerful fixed magnet) or by using a diaphragm and voice coil with a smaller mass. For higher frequency range the diaphragm is made smaller.

In the CD laser focus system a similar approach is applied. Because the bit rate is enormous, over 1 MHz, the focus lens is subjected to very large accelerations which continuously change as each pit or flat goes by. According to the Second Law of Motion, this is possible only with a system of lens and focus coil which has a very small mass.

As an example of Newton's third law, consider a string attached to a wall. A pulse is created on the

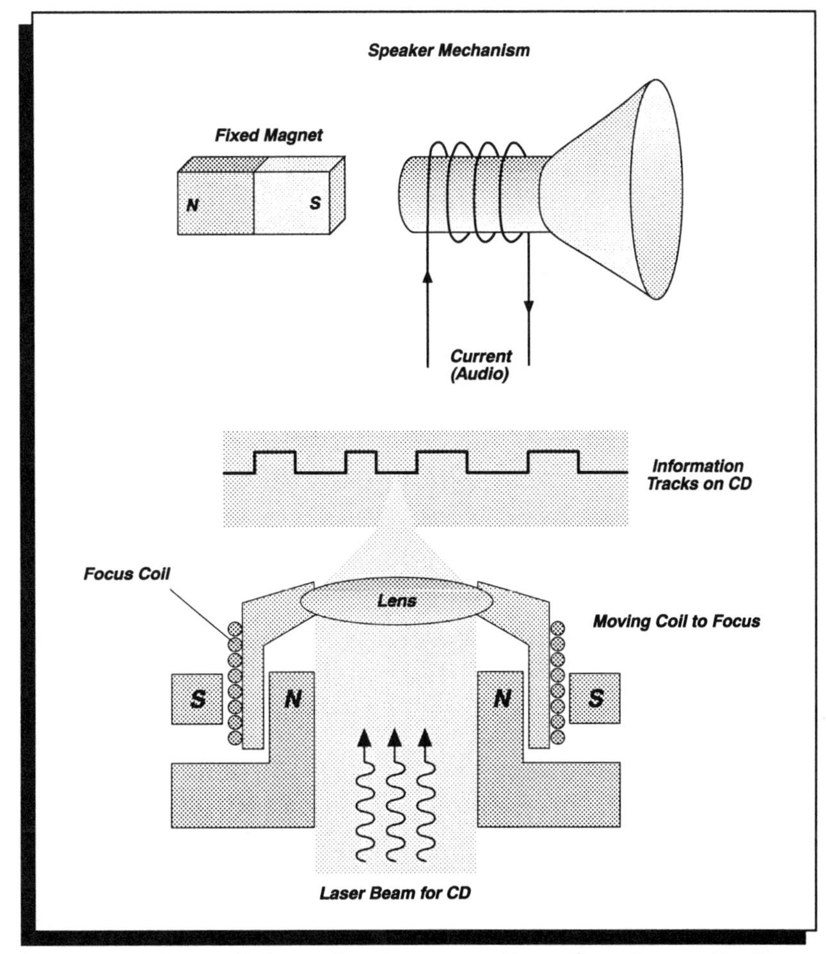

Figure 13.11 Force of clamped magnet on a voice coil accelerates diaphragm in loudspeaker. Force of clamped magnet on focus coil accelerates focus lens in CD player.

string and it travels toward the wall. As the pulse approaches the wall it pulls up and on the wall. But according to Newton's third law, the wall pulls with an equal and opposite force on the string. This causes the string to move downward, initiating an inverted pulse in the opposite direction. This is shown in Figure 13.12. It is important in setting up standing waves. Reflections cause a change of sign of the pulse when it goes back away from the wall.

It is important to note that the two forces in the example above do not cancel each other. Each force acts on a different object. The string pulls on the wall. The wall pulls on the string, initiating the reflected pulse.

Another example of Newton's third law is the speaker voice coil as it feels the force of the magnet next to it. The voice coil in turn exerts a force of

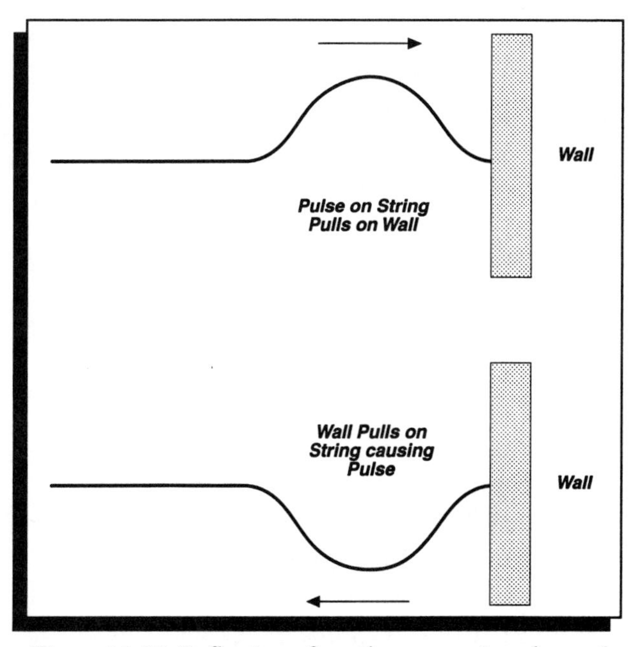

Figure 13.12 Reflection of a pulse on a string clamped at wall and its inversion.

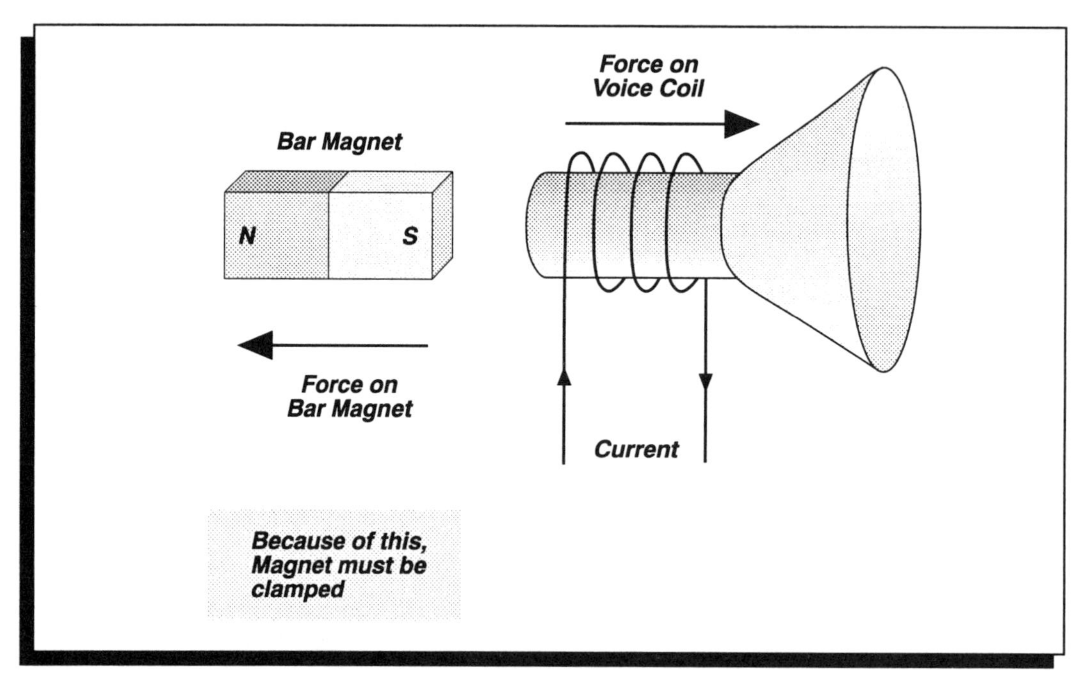

Figure 13.13 Force on voice coil and force on magnet.

equal magnitude but opposite direction on the magnet. Since the magnet is fixed, it does not move. This is illustrated in Figure 13.13.

13.2 ROTATIONAL MOTION

In the previous section Newton's Laws of Motion were applied to motion in one dimension, a straight line. Here, we will extend these laws to rotational motion. All the sources of audio signals in the hi–fi use rotating parts and systems.

■ FREQUENCY OF ROTATION

This is the number of complete revolutions made in each second. However, in the hi–fi, most units deal with low frequencies of revolution, so it is usual to specify the frequency of rotation in revolutions per minute (r.p.m.). Examples are:

— phono record turning at 33⅓ r.p.m.
— CD rotates at 500 to 200 r.p.m.
— mini–Disc rotates at 500 to 350 r.p.m.
— rotating head at 2000 r.p.m. in DAT.

It is instructive to compare the rotational characteristics of a phono record to that of a CD. This is shown in Figure 13.14. The phono record rotates at a constant frequency of rotation. Consequently, the linear speed near the outer edge is larger than near the center. This comes from the fact that the distance around a track near the outer edge is longer than that near the center. The circumference of a circle tells us the distance covered around one track. For a track at a distance r_{inner} from the center, the circumference is $2\pi r_{inner}$, while for an outer track at a distance r_{outer} from the center, that distance around is $2\pi r_{outer}$. Figure 13.15 shows this for two tracks. Because the record rotates at a fixed rate, the speed (or, distance covered per unit time) will be larger for the outer track than the inner one. This means that the recorded wavelength is given as:

$$\textbf{frequency} \times \textbf{wavelength} = \textbf{speed}$$

The speed will be larger on the outer tracks than on the inner ones.

On a CD, an inner track is shorter than an outer track, but the linear speed for all the tracks is kept constant by changing the frequency of rotation. In fact, because of this it is 200 r.p.m. near the outer edge and 500 r.p.m. near the inner edge. This ensures that the length of the pits and flats is the

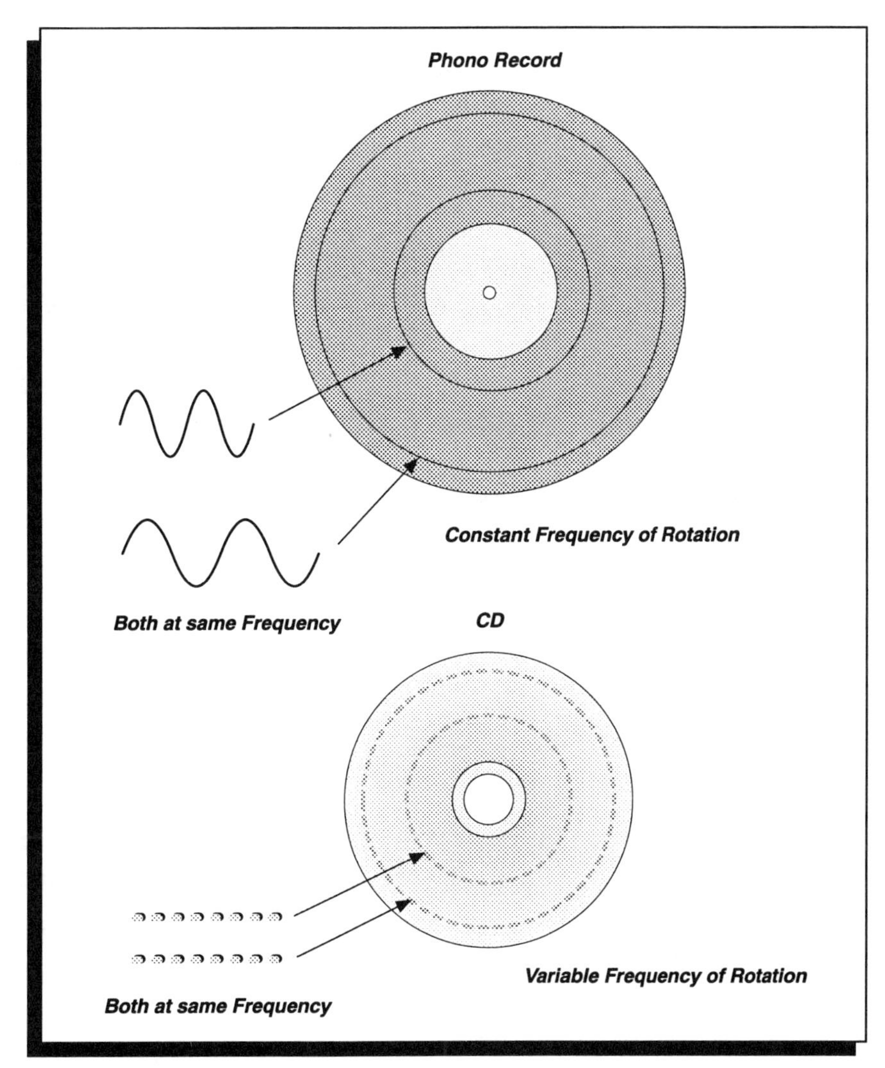

Figure 13.14 Waves recorded on a phono record and a CD.

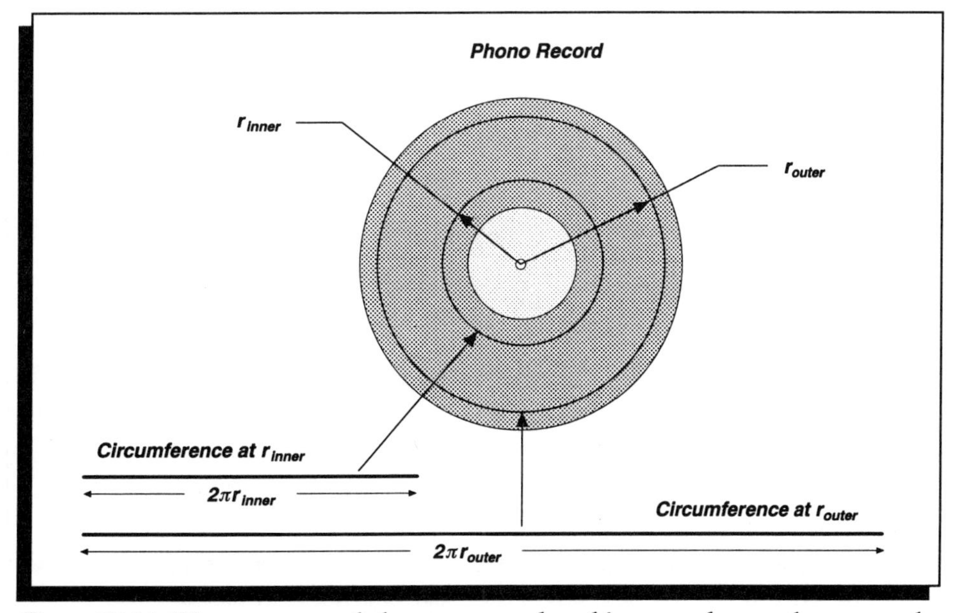

Figure 13.15 Distances covered along outer track and inner track on a phono record.

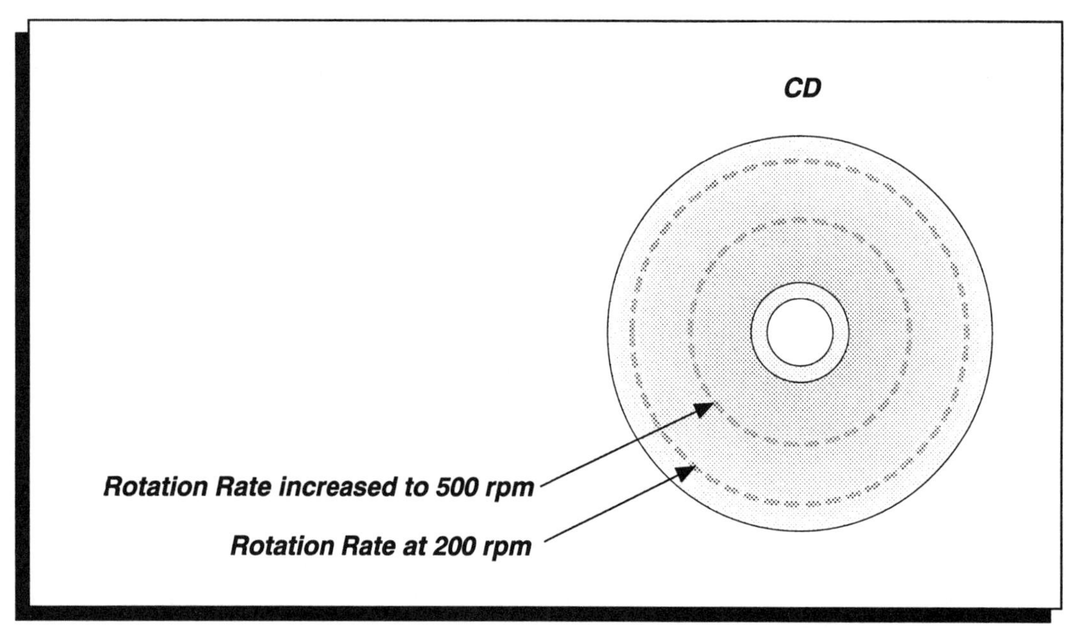

Figure 13.16 Frequency of rotation of a CD is made higher near the inner edge and lower near the outer edge to maintain constant linear speed on all tracks.

same on all tracks. Otherwise the outer tracks would have long waves and the inner ones would have short ones. The advantage of this approach is an increased density of information that can be stored on the CD. Figure 13.16 illustrates this.

Let us now calculate the relative speed between a magnetic tape and the rotating head in a DAT, as shown in Figure 13.17. The record/playback heads are mounted on a rotating drum, 30mm in diameter. In normal operation, this head rotates at a frequency of rotation of 2,000 r.p.m. The speed of this head relative to the tape is calculated from:

$$\textbf{linear speed of head on drum} = \frac{\textbf{distance covered in one rotation}}{\textbf{time for one rotation}}$$

The circumference of the drum is π**d**, where **d** is its diameter. Here, it is $(\pi)(3.0\text{cm}) = 9.43\text{cm}$, since the diameter of the drum is 3.0 cm. The time for one revolution is: $(1/2000\text{rpm})\times(60\ \text{seconds/min}) = 0.03$ second.

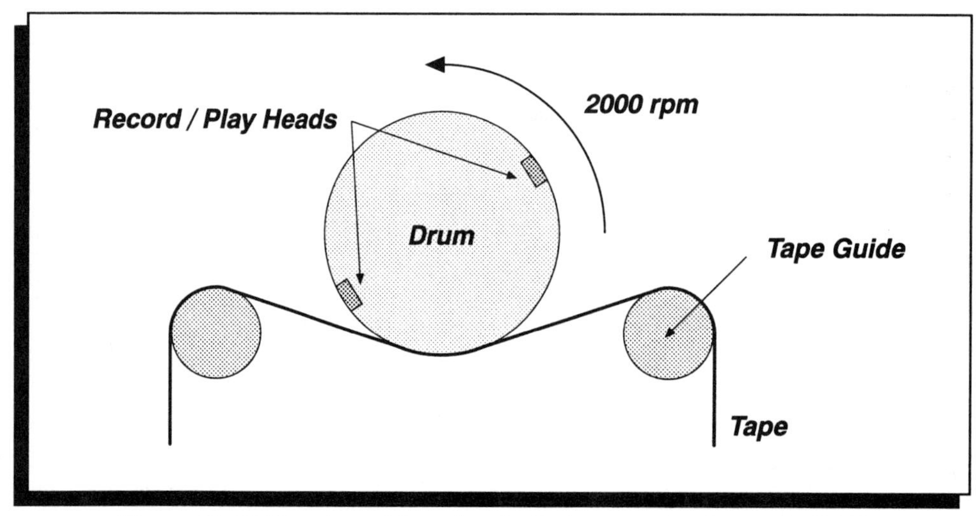

Figure 13.17 Rotation of drum head relative to magnetic tape in DAT.

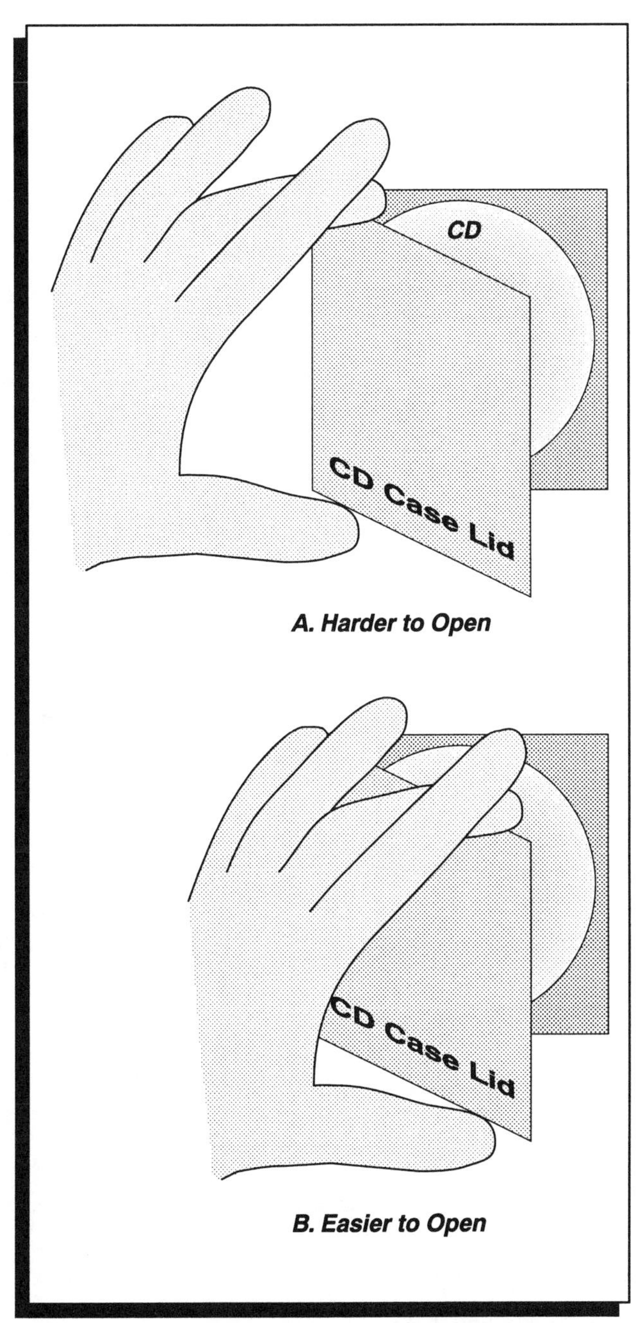

Figure 13.18 When same force is applied to the CD case lid, it is easier to open the lid near the edge because torque is larger there.

The relative speed of the rotating head with respect to the tape (neglecting the tape speed of 0.815 cm/sec at which it is pulled across) is:

speed = 9.43 cm/0.03 sec
= 314.33 cm/sec i.e. 3.14 m/sec

This is a very high speed when compared to the tape speed of an analog tape deck of 4.7625cm/sec ($1\frac{7}{8}$ i.p.s.). Such high speed is necessary to handle the wide range of frequencies for digital recording.

■ ROTATIONAL INERTIA

As mentioned above, there are many applications of rotational motion in the hi–fi. For such cases, Newton's laws of motion can be restated in a form applicable to rotations. For example, the rotational equivalent of Newton's first law states that:

A rotating object continues to rotate uniformly as long as no net torque acts to change this motion.

Here the word torque is the rotational equivalent of force. It will cause the rotational speed to change and in fact it will cause a rotational acceleration.

As an example of torque consider a force applied to the hinged cover of a CD case, shown in Figure 13.18. The same force is applied at the edge of the case and near the hinges.

It is easier to open the case lid near the edge than near the hinges because the torque about the axis of rotation is larger there, even though the applied force is the same. We define torque as:

Torque = Force × perpendicular distance between force and the axis of rotation.

The CD case hinge can be represented schematically as in Figure 13.19. Because the perpendicular distance from the edge is larger, the torque here will be the largest and hence it will be easier to open the lid. This example shows what torque does; it causes a rotation about an axis.

Going back to Newton's First Law of Motion for rotation, the tendency of a rotating object to continue spinning is due to its rotational inertia. It is the equivalent of inertia for linear motion. However, it depends not only on the mass of the rotating object, but also on how that mass is distributed around the point of rotation. The name given to this rotational inertia is moment of inertia.

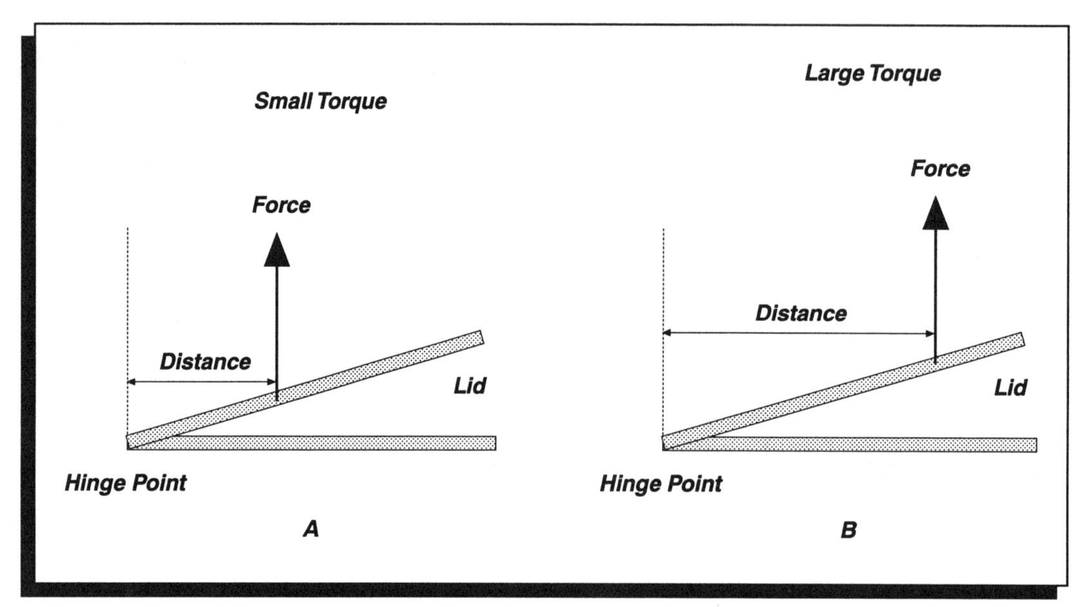

Figure 13.19 For the same force exerted on CD case lid, the torque is larger in B than in A.

We define it as:

moment of inertia = mass of object × (distance to axis of rotation)^2

In fact, the larger the moment of inertia of an object, the more torque is needed to start it rotating, or to change is rotational speed. The moment of rotation does not depend on whether the object is at rest or is rotating; it depends only on the mass and the distance squared to the axis of rotation.

Consider a CD and a mini–Disc as in Figure 13.20. Because the mini–Disc has a smaller mass (it is about the same thickness as the CD), and a smaller diameter than a CD, its moment of inertia is smaller. It will require a smaller torque than a CD to make it rotate at the same frequency of rotation as a CD.

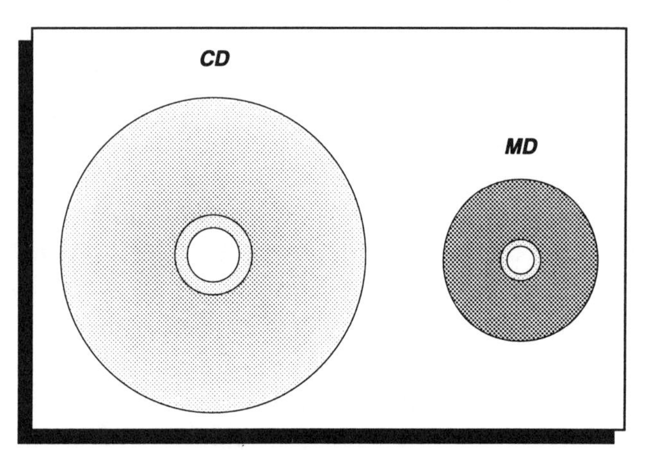

Figure 13.20 Moment of inertia of a CD is larger than that of a mini-Disc.

As a summary, this chapter has presented basic concepts of mechanics which are used in the sources of audio signals and in their reproduction. The impressive achievements in the field of high–fidelity is due, to a large extent, to the precise control of mechanical concepts and to their joint development with electronics and devices.

Summary of Terms

Average Acceleration: Change of an object's velocity per elapsed time.
Average Speed: Distance travelled per elapsed time.
Force: A push or a pull exerted on an object.
Frequency of Rotation: Number of revolutions per second or per minute.
Inertia: Resistance of an object to a change in its motion along a straight line.
Instantaneous speed: Speed at some instant of time.
Instantaneous velocity: Velocity at some instant of time.

Kinetic force of friction: Force due to the sliding of two objects against each other.
Mass: Quantity of matter in an object. It is a measure of an object's inertia.
Moment of inertia: Resistance of an object to a change in its rotational motion.
Newton's First Law of Motion: An object remains at rest or in a state of constant velocity unless a net force acts on it.
Newton's Second Law of Motion: The acceleration of an object depends on the net force acting on it and inversely on its mass.
Newton's Third Law of Motion: When an object exerts a force on another object, this object also exerts a force on the first object, both forces being equal and opposite in direction.
Pressure: Force per unit area.
Rotational Inertia: Resistance of an object to a change in its rotational motion.
Static force of friction: Force between two objects in contact with each other when there is no relative motion between them.
Torque: Force exerted on a body times the distance to the point of rotation. It can cause rotation about a point.
Vector: A quantity which specifies a magnitude (how much) and a direction.
Velocity: A quantity which specifies the speed of an object and its direction.
Weight: Force of gravity on an object.
Wow and Flutter: Fluctuations in tape speed which cause variations in frequency. Wow is a slow speed variation at frequencies up to 10Hz, while flutter is a faster speed variation at frequencies above 10Hz.

NAME ______________________________ DATE ____________

Questions for Review

1 Compare Newton's First Law of Motion as applied to motion in one direction and rotational motion.

2 What is the difference between weight and mass?

3 When will an object acquire a large acceleration?

4 Discuss what would happen to the magnet in a speaker near the voice coil if it were not clamped.

5 Explain the difference between speed and velocity.

6 What will happen to the moment of inertia of a CD when a plastic ring is stretched around its rim?

7 How would you calculate the access time to go from one track to another track in the playback of a CD?

8 Compare a mini–Disc to a CD as to inertia, mass, and weight.

9 Distinguish between a force and a torque.

10 What is the linear speed at the center of a CD? What is its frequency of rotation?

NAME ______________________________ DATE ____________

Please select one answer.

1. A measure of an object's inertia is its:
   ___A. speed
   ___B. velocity
   ___C. weight
   ___D. mass
   ___E. acceleration

2. The torque on the CD case lid can be increased by:
   ___A. increasing the force on the lid, keeping the distance to the point of rotation fixed.
   ___B. keeping the force on the lid constant, but decreasing the distance to the point of rotation.
   ___C. decreasing the force on the lid and decreasing the distance to the point of rotation.
   ___D. doubling the force on the lid and decreasing by a factor of two the distance to the axis of rotation.
   ___E. applying the force to the point of rotation.

3. On a phono record, the linear speed at the outer rim will be ___________ that on a track halfway between the outer rim and the inner rim.
   ___A. the same as
   ___B. two times larger than
   ___C. two times smaller than
   ___D. four times larger than
   ___E. four times smaller than

4. When a car goes around a curve at a constant speed, its ___________ changes.
   ___A. velocity
   ___B. average speed
   ___C. moment of inertia
   ___D. frequency of rotation
   ___E. rotational speed

5. According to Newton's second law of motion, a large acceleration can be achieved by applying a ___________ force to an object which has a ___________ mass.
   ___A. small, small
   ___B. large, large
   ___C. small, large
   ___D. large, small
   ___E. torque and a, large

6. The moment of inertia about the center of a CD is approximately ___________ that of a mini–Disc, also about its center.
   ___A. the same as
   ___B. 16 times larger than
   ___C. 16 times smaller than
   ___D. 4 times larger than
   ___E. 4 times larger than

7. The time that it takes to go around an outer track on a CD is approximately:
   ___A. 500 seconds
   ___B. 200 seconds
   ___C. 0.005 second
   ___D. 0.3 second
   ___E. 0.12 second

8. The spiral pit track on a compact disc makes ___________ laps over a distance of 3.55cm between the outer track and the inner track.
   ___A. 355,000
   ___B. 22,188
   ___C. 44,376
   ___D. 11,094
   ___E. 1,600

9. A CD which can play for 60 minutes is recorded at 1.4 meters/second. If the spiral pit track were unwound, it would be _________ meters long.
   ___A. 84 meters
   ___B. 5040 meters
   ___C. 1.4 meter
   ___D. 60 meters
   ___E. 10,080 meters

10. The frequency of rotation of an outer track on a CD is ___________ that of a phono record.
   ___A. 6 times larger than
   ___B. 6 times smaller than
   ___C. 200 times larger than
   ___D. $33\frac{1}{3}$ times larger than
   ___E. the same as

11. The torque required to spin a compact disc is about ___________ that required for a mini–Disc. Assume both start from rest and reach a frequency of rotation of 200 r.p.m.
   ___A. 4 times larger than
   ___B. 4 times smaller than
   ___C. 16 times larger than
   ___D. 16 times smaller than
   ___E. 8 times larger than

12. What is the difference between speed and velocity?
   ___A. Both have the same meaning.
   ___B. Speed can be used to calculate an object's acceleration, and velocity cannot.
   ___C. Speed tells us how fast an object is moving, while velocity tells us which way it is moving.
   ___D. Speed tells us how fast an object is moving. Velocity tells us its speed and direction of travel.
   ___E. Speed tells us the direction of travel. Velocity tells us how fast an object is going.

13. During the playback of a CD, as the track unwinds toward the outer edge, the speed of the pits and flats relative to the laser beam is kept constant by:
   ___A. increasing the frequency of rotation
   ___B. decreasing the frequency of rotation
   ___C. maintaining the frequency of rotation constant
   ___D. the disc's changing moment of inertia
   ___E. maintaining the disc's velocity constant

14. Since a mini–Disc is recorded at a constant speed of 1.2 to 1.4 m/sec, just like a CD, the frequency of rotation of its outer track will be ___________ that of the outer track of a CD.
   ___A. the same as
   ___B. double
   ___C. half
   ___D. one quarter
   ___E. four times

15. A phono stylus exerts a downward force of 5×10^3 dynes on a record groove. The record groove exerts:
   ___A. an upward force of 5×10^3 dynes on the stylus, in accordance with Newton's first law of motion.
   ___B. an upward force of 10^4 dynes on the stylus, in accordance with Newton's third law of motion.
   ___C. an upward force of 5×10^3 dynes on the stylus, in accordance with Newton's third law of motion.
   ___D. a downward force of 5×10^3 dynes on the stylus, in accordance with Newton's second law of motion.
   ___E. no force on the stylus since it is below the stylus.

NAME ______________________________ DATE ____________

16. In an analog tape deck, the capstan pulls the tape at a constant speed. To help it perform that function a cylinder, called a flywheel, is attached to it. This will:
    ___A. increase the capstan's moment of inertia
    ___B. decrease the capstan's moment of inertia
    ___C. not change the capstan's moment of inertia; it will only increase its mass
    ___D. make the capstan rotate faster
    ___E. give the capstan a larger frequency of rotation.

17. One reason for maintaining a constant linear speed on a CD or a mini–Disc in recording or playback is:
    ___A. to achieve a high density of information
    ___B. to keep the torque on the disc constant
    ___C. to prevent the pit lengths from becoming too short at the outer tracks
    ___D. to prevent the pit lengths from becoming too long at the inner tracks
    ___E. to maintain a constant frequency of rotation

18. Even though the diameter of a mini–Disc is about half that of a CD, it produces music for about 60 to 74 minutes, just like a CD. This must mean that:
    ___A. a mini–Disc has more information than a CD
    ___B. a mini–Disc has less information than a CD
    ___C. a mini–Disc has as much information as a CD
    ___D. information is read by a smaller laser beam
    ___E. spacing between tracks is smaller

19. If the speed of light did not decrease in the plastic of the CD, the pits and flats would then have to be ____________ .
    ___A. closer to the surface where the laser beam enters
    ___B. further from the surface where laser beam enters
    ___C. at the same position since the incoming and reflected beam both travel faster in the plastic
    ___D. smaller
    ___E. spaced closer together

20. The reason for a high speed of the rotating head in a DAT is:
    ___A. recorded waves will be longer
    ___B. recorded waves will be shorter
    ___C. so that frequencies up to 20 kHz can be recorded
    ___D. so that frequencies down to 20 Hz can be recorded
    ___E. so that aliasing will not occur

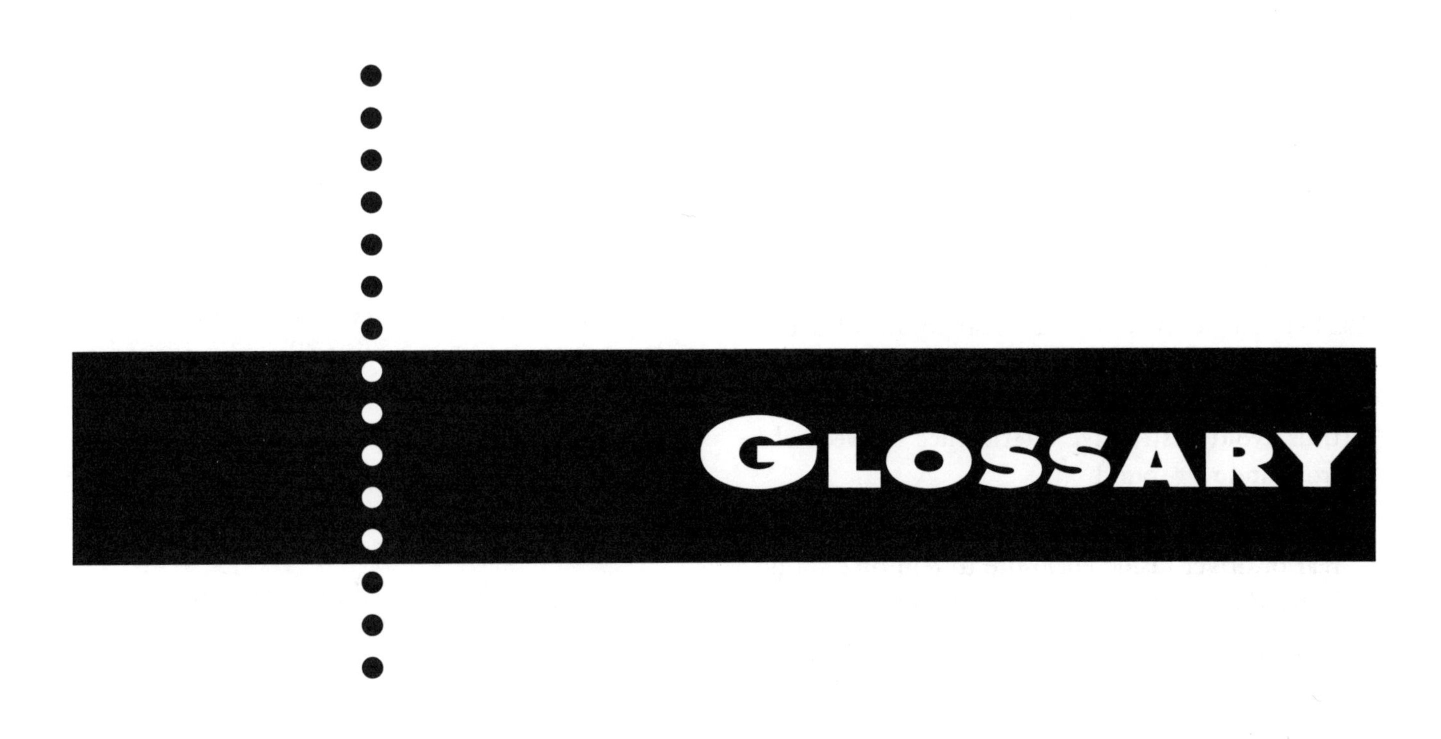

Aliasing: Introduction of erroneous signals in digital processing due to the presence of frequencies higher than ½ the sampling rate.

Ambience: Collection of sonic cues that tell your ear–brain system that you are in a particular place.

Ampere: Unit of electric current specifying how much charge flows per second.

Amplifier: A device where the input signal controls the output signal. It increases the voltage or current or power of an input signal.

Amplitude: Maximum deviation or displacement from equilibrium in a wave.

Amplitude Modulation(AM): A method of making changes on a carrier so that audio information is carried by the carrier; the amplitude of the carrier is changed according to the pattern of audio information.

Analog: Continuous representation of signal so that it looks similar to the wave it represents.

Analog–to–Digital Converter (ADC): An electrical circuit which converts an analog signal into a digital one.

Antenna: A device which converts radio frequency currents along a wire into electromagnetic waves and vice versa.

Antinode: Maximum deviation from equilibrium in a standing wave.

ATRAC: Adaptive Transform Acoustic Coding system, used for encoding audio information on a Mini–Disc. It is based on psychoacoustic principles dealing with the threshold of hearing and masking, so as to reduce the amount of acoustic data recorded.

Audio Signal: Representation of a sound wave as an electrical signal at frequencies ranging from 20 Hz to 20,000 Hz.

Audio Signal Generator: Electronic test unit putting out waveforms in the audio frequency range, the simplest one being a sine wave.

Audio/Video System (A/V): A system that can be used to produce ambience and surround sound with video tapes or sound. Adds an extra dimension to sound from an extra channel of surround sound to existing stereo sound.

Average Acceleration: Change of an object's velocity per elapsed time.

Azimuth Recording: Magnetic recording by heads canted to each other to reduce cross–talk.

Baffle: Partition to separate rear sound of driver from front sound to avoid destructive interference between them or to delay rear sound and thus avoid destructive interference between the front and rear sound.

Bass Reflex: Enclosure which uses sound from rear of driver inside enclosure to reinforce sound in front of speaker by means of a port or duct.

Beats: Intensity variations per unit time when two waves of slightly different frequencies are produced together.

Bias: A high frequency signal added to the audio signal in magnetic recording to reduce distortion due to tape characteristics.

Bimetallic strip: A strip made of two metallic materials which have different thermal expansions. When heated or cooled the strip bends in or out.

Bimorph: Two piezoelectric discs usually connected in parallel in order to produce a push– pull action when a voltage is applied across them.

Binary system: A system of numbers based on two numbers, a "1" and "0". It is much simpler than a decimal system for electronic devices and machines.

Bit: Stands for **Bi**nary digi**ts**. It represents the significant places in a binary number.

Capacitance: Property of a device for storing electrical charge.

Capture Ratio: Ability to reject the weaker of two signals at the same carrier frequency arriving in a tuner. One of the signals is usually caused by reflections off of buildings, bridges and mountains and the other one is the direct signal.

Carrier: Electromagnetic wave which can carry audio information. It is at the frequency of the broadcasting station.

CD player: Compact Disc player. A digital source component that plays compact discs using optical read–out.

Circuit Breaker: A device which protects circuits from excessive electric currents by interrupting the current. Often a bimetallic strip is used to switch the current off; after it is triggered it can be reset.

Coercivity: Property of magnetic material to maintain its magnetic state without being erased by external magnetic fields. It is measured by the magnetic field needed to destroy the magnetic information and reduce the magnetization to zero.

Complex Wave: A wave made up of harmonics.

Compliance: Ability of a mechanical system to move when a force is applied to it. It is a displacement per unit force.

Condensation: Increase in pressure in a sound wave.

Conduction (thermal): Natural flow of heat along an object when a temperature difference exists across it. Flow of energy occurs by vibrations of atoms and electrons that are passed on to neighbors.

Convection: Flow of heat by displacements of material due to temperature differences or other forces acting on it. It usually occurs in fluids and gases when temperature differences cause density changes.

Cross–over Frequency: Common frequency at which signal to one driver is reduced and to the next driver is increased and the two are equal.

Cross–over Network: Electric system which reduces signals going to one driver and sends them to the next one.

Cross–Talk: Amount of signal from one channel which gets mixed up in the other channel.

Curie Temperature: Temperature above which a magnetic material becomes almost non–magnetic and its coercivity goes to zero.

Damping Factor: The ability of an amplifier to stop oscillations of a speaker when output signal has decreased to zero. It is given by the impedance of the speaker divided by the internal resistance of the amplifier.

Decibel (dB): Quantity used to describe the comparison of two powers or two intensities in a logarithmic way. It has no unit since it is a comparison of two quantities. One decibel is approximately the smallest change in loudness level which can be heard.

Density: Property of a material which is its mass per unit volume.

Diaphragm: A vibrating membrane of the driver which pushes air in and out.

Digital: A signal is expressed as a series of numbers each corresponding to an amplitude of the signal.

Digital–to–Analog Converter (DAC): An electrical circuit which converts a digital signal to an analog one.

Diode: A solid state device made of two different semiconductor types of material, n and p.

Dispersion of Speaker: The ability of a speaker to radiate sound in all directions around it at low frequencies and to beam it mainly forward at high frequencies; it is caused by speaker diffraction.

Distortion: The ability of a system to modify an original waveform and hence its spectrum.

Dolby Noise Reduction: Reduces tape hiss from the tape you are copying onto.

Doppler Effect: The change of frequency when there is relative motion between source of sound waves and observer. When the motion between the source and observer is toward each other, there is a frequency increase which depends on relative speed. When the motion between source and observer is away from each other, there is a frequency decrease depending on the relative speed.

Driver: A speaker without an enclosure.

Dynamic Range: Comparison of largest signal to weakest signal in decibels that a system can respond to or store.

Efficiency: The fraction of electrical power sent to speaker which becomes converted to sound; efficiency is expressed in percent.

Electric Field: Energetic region around an electric charge.

Electromagnet: Coil with a magnetic core which produces a magnetic field when a current flows in the coil.

Electromagnetic Wave: A disturbance consisting of an electric and a magnetic field travelling at approximately 3×10^8 m/sec in vacuum.

Electrostatic Speaker: A loudspeaker which operates on the principle that like charges repel and unlike charges attract. It consists of a vibrating membrane.

Enclosure: A structure which separates rear sound from front sound of the driver and adds quality to the radiated sound.

Equalizer: Electronic unit which can modify sound spectrum at selected frequency bands.

Faraday's Law of Induction: Change of magnetic field per unit time creates a voltage across a coil or conductor nearby.

Fifty (50) dB Quieting Sensitivity: Indicates how well a tuner can convert a radio wave signal into sound. It refers to how small a signal from the antenna can produce an audio signal from the tuner at 50dB above noise and distortion.

Filter: A circuit that attenuates, boosts, or removes selected frequencies.

First Left Hand Rule: Indicates direction of magnetic field produced by a coil when current flows. When fingers of left hand point in direction of current flow, the thumb of the left hand indicates the north pole direction.

Fletcher–Munson Curve: The frequency response of the human ear at various levels of loudness.

Force: A push or pull exerted on an object.

Frequency: Number of waves per second; it is measured in Hertz (Hz).

Frequency Modulation (FM): A method of making changes on a carrier so that audio information is carried by the carrier; the change in the carrier frequency corresponds to the audio information.

Frequency of Rotation: Number of revolutions per second or per minute.

Frequency Response: Behavior of a system at different frequencies.

Gain: The factor by which an amplifier increases the level of a signal. It is the amplitude of the output signal divided by the amplitude of the input signal.

Heil Speaker: Corrugated thin plastic sheet with current–carrying wires or strips attached to it. In the presence of a magnetic field, magnetic forces on current conductors cause an accordion–type motion and hence sound. It is based on the second Left Hand Rule.

Helmholtz Resonator: Resonant acoustic cavity whose oscillatory response is well–known and which can be used as a model for more complicated resonant structures, like a speaker enclosure.

Hi–Fi: Faithful reproduction of sound or reproduction of sound with excellent tonal qualities.

Horizontal Polarization: The direction of electric field of electromagnetic wave is horizontal.

Horn: Structure which matches the vibrations of a diaphragm to that of air. When combined with a driver it is capable of achieving very high efficiencies due to the good match it provides.

Hum: Noise picked up in the amplifier and other components from power lines; it is at 60 Hz and its harmonics.

Impedance: Response of a system to a force or a voltage. It determines how much current will flow in an electrical circuit when a voltage is applied to it.

Index of Refraction: Ratio of speed of light in vacuum divided by speed of light in medium.

Inductance: Ability of a circuit to store magnetic energy.

Inertia: resistance of an object to a change in its motion along a straight line.

Infinite baffle: Loudspeaker enclosure which traps rear sound of driver so that it will not interfere destructively with front sound.

Instantaneous Speed: Speed at some instant of time.

Instantaneous Velocity: Velocity at some instant of time.

Insulator (electrical): Material whose electrons are so tightly bound to the atoms that they do not conduct electricity when a voltage is applied across it. It has infinite resistance.

Integrated Amplifier: Electronic unit which consists of a pre–amplifier and a power amplifier.

Intensity: Power from a wave crossing a given surface area. Its unit is watts/m^2.

Kerr Effect: Plane of polarization of light beam is rotated when it is reflected off a magnetized magnetic material.

Kinetic Force of Friction: Force which occurs when two objects slide against each other; this force opposes the relative sliding motion.

Laser: A device which emits a beam of coherent light. The name stands for **L**ight **A**mplification by **S**timulated **E**mission of **R**adiation.

Law of Reflection: When a wave hits an obstacle it goes off in such a direction that the angle of incidence with respect to normal is equal to angle of reflection.

Limiter: Part of a circuit in an FM tuner which suppresses any amplitude variations of the carrier signal.

Longitudinal Wave: Wave where displacements are parallel to direction of travel of wave. Sound in air is an example of such a wave.

Loudness Control: Control in pre–amplifier which makes up for deficiencies of our ears at low levels by increasing the level of the low and high frequencies.

Loudspeaker: A unit that converts electrical signals to sound.

Magnet: Material with magnetic domains which are aligned permanently with resultant north and south poles.

Magnetic Dipole: Elementary unit in magnetism which contains a south pole and a north pole. A current loop has similar magnetic properties.

Magnetic Domain: Aligned group of atoms in a magnetic solid.

Magnetic Field: Energetic region of space around a magnet or a current–carrying conductor where a force will be felt by a test magnet.

Magnetic Hysteresis: Behavior of a magnetic material in various magnetic fields.

Magnetic Planar Speaker: A speaker which has fine wires attached to a plastic sheet in the presence of a magnetic field. A current in wires causes a force to be exerted on the wires, producing motion and hence, sound.

Mass: Quantity of matter in an object. It is a measure of an object's inertia.

Microphone: Electronic device that converts sound into electrical signals.

Mini–Disc: New compact format of digital audio information storage with laser beam readout. It uses ATRAC system for encoding.

Moment of Inertia: Resistance of an object to a change in its rotational motion.

Multipath Interference: Unwanted reflected radio waves which have been reflected by mountains, buildings, and bridges, arriving to receiver slightly later than direct signal.

Multiplexing: Merging of two or more channels together.

Negative Feedback: Feeding back to input a signal 180° out of phase with input signal.

Newton's First Law of Motion: An object remains at rest or in a state of constant speed in a straight line unless a net force acts on it.

Newton's Second Law of Motion: The acceleration of an object depends on the net force acting on it and inversely on its mass.

Newton's Third Law of Motion: When an object exerts a force on another object, this object also exerts a force on the first object, both forces being equal and opposite in direction.

Octave: Interval which consists of ratio of 2:1 in frequency.

Ohm's Law: Determines how much current will flow in a circuit with resistance when a voltage is applied to it.

Operational Amplifier: A high gain amplifier with negative feedback.

Oversampling: Digital filtering technique used in CD player to remove spurious signals from audio range.

Overwriting: Method of erasing digital information by writing directly over previous signal. It works because the signal is digital.

Passive Radiator: A driver which is located in the port of a bass reflex and without any electrical signal going to it. It is driven by the sound from the enclosure.

Piezoelectricity: Production of electricity by certain objects when stressed or twisted. The converse is true.

Playback head: Device on tape deck which reads information on a tape by converting magnetic signal on tape to an electrical signal using Faraday's Law of Induction. It consists usually of a small coil around a core in the shape of a horseshoe with a small gap where the signal is read.

Polarization: Specific direction of electric field in electromagnetic wave.

Poles: Ends of a magnet which are either north or south. A basic unit in magnetism; it always comes in pairs of opposite polarity.

Positive Feedback: Feeding back to input a signal in phase with input signal.

Power: Rate of doing work. In electricity it is the voltage times the current; it is expressed in watts.

Power Amplifier: Electronic unit that boosts the level of signals from a pre–amp so as to provide power for the speakers.

Pre–Amplifier: Electronic unit that raises the level of certain signals, has all the controls, and routes the signal between different components.

Pressure: Force per unit area.

Radiation: Transfer of energy from an object by emission of electromagnetic waves, usually infra–red.

Radio: Compact electronic unit which converts radio frequency signals to audio signals, amplifies it, and drives internal speakers.

Radio Frequency Signal: Electromagnetic wave which is in the tuner frequencies range between 535 kHz and 1605 kHz for AM and 88 MHz to 108 MHz for FM.

Rarefaction: Decrease in pressure in sound wave.

Receiver: Electronic unit containing a tuner, a pre–amplifier, and a power amplifier.

Recording Head: Device on a tape deck which records audio information by aligning magnetic domains on tape. It is a small electromagnet with a gap.

Reflection: Change of direction of a wave by an obstacle or by a medium with different properties than the medium the wave is in, so that the wave does not travel in the new medium.

Refraction: Change of direction of a wave when it enters a medium with different properties because the speed in the new medium is different than in the original medium.

Resistance (electrical): Property of a body that determines how much friction there is to flow of electrons acted upon by an electric field. It causes heating when current flows.

Resonance: Increase in amplitude of vibrations of a body when it is caused to oscillate at its natural frequency or near it.

Retentivity: Ability of magnetic material to retain alignment of domains or magnetization when an external magnetic field is removed. It serves as an element of memory.

Reverberation: Diffuse mixture of reflected sounds in a hall or room.

Reverberation Time: Time for sound power to decay in a hall by a factor of 1,000,000 from its original value when sound power is switched off. It is an important acoustic characteristic of a hall.

Rotary Head: Drum with two (or four) heads for record–playback in magnetic recording, in a mode where drum (and heads) rotates relative to tape.

Rotational Inertia: Resistance of an object to a change in its rotational motion.

Sample and Hold: Circuit which samples a signal and holds amplitude at sampled value for a time long enough for it to be digitized.

Sampling Frequency: The number of times per second the amplitude of signal is measured for digital recording. For a CD it is at 44,100 times per second. It must be at least two times the highest frequency of interest.

Sampling Theorem: To digitize an analog signal it must be sampled at a frequency which is at least twice the highest frequency of interest.

Selector Switch: Determines which specific source of sound is used as a signal input to the hi–fi.

Signal–to–Noise Ratio (S/N): A signal level divided by noise level, expressed in dB.

Sound: Pressure changes at frequencies in the range of 20 Hz to 20,000 Hz.

Specific Heat: Quantity of heat needed to raise the temperature of a unit mass of a substance by one degree.

Speed: Displacement per unit time.

Spring Constant: Force needed to extend or compress a spring by unit distance.

Static Force of Friction: Force between two objects in contact with each other when there is no relative motion between them.

Step–Down Transformer: Transformer which has less turns at the secondary than the primary and which reduces voltage of input signal.

Step–Up Transformer: Transformer which has more turns at secondary than primary, thus raising the voltage level of input signals.

Stereo: Processing of sound in two channels.

Subcarrier: A 38 kHz signal in a studio broadcasting in stereo. It carries the LEFT minus RIGHT channel information before being put on the carrier for broadcasting.

Tape Deck: Electronic unit used for recording on magnetic tapes and playing back the information.

Tape–In Jacks: Provides signals into the pre–amp from recorded sources such as a tape deck or a signal processing unit.

Tape Monitor Switch: A switch in the pre–amp that accepts signals either from a tape deck or sources of sound.

Tape–Out Jacks: Provide signals from sound sources as selected by the selector switch for recording or for further processing.

Tape Saturation: Occurs when most of the magnetic domains are aligned.

Temperature: Measure of how much thermal activity is present in an object. It is measured in Centigrade or Fahrenheit for everyday experiences, and in Kelvin for scientific research.

Thermal Expansion: Change in volume of an object when heated or cooled.

Thermal Resistance: Difficulty offered to the flow of heat along an object. It depends on material and geometric factors.

Threshold of Hearing: Level at which the quietest sound can just be heard. It is called the 0 dB level and corresponds to 10^{-12} watts/m^2 at 1 kHz. It varies with frequency.

Threshold of Pain: Level at which sound causes pain. It is at a level of 130 dB at 1 kHz.

Torque: A force exerted on a body times the distance to the point of rotation. It can cause a rotation about a point.

Total Harmonic Distortion (THD): Amplifier specification which tells how much undesired harmonics have been added to a signal. It is expressed as the ratio of amplitudes of all unwanted harmonics relative to amplitude of fundamental signal, expressed in percent.

Transistor: A solid state device made of three semiconductor elements in a sandwich arrangement. Electrical connections are made to the base, emitter, and collector. It is a building block of amplifiers and electronic circuits.

Transverse Wave: Wave where displacements are perpendicular to direction of travel.

Tuner: Electronic unit that picks up radio stations and extracts the audio information from radio waves.

Vector: A quantity which specifies a magnitude (how much) and a direction.

Velocity: A quantity which specifies the speed of an object and its direction.

Vertical Polarization: The direction of the electric field of an electromagnetic wave is vertical with respect to Earth.

Voice Coil: Part of driver which carries electrical signals that make it move in a magnetic field of permanent magnet. It is attached to the diaphragm and makes that part vibrate.

Voltage: Difference in energy per electrical charge across two points of a circuit.

Wavelength: How long one wave is.

Weight: Force of gravity on an object.

Wow and Flutter: Fluctuations in tape speed (or any other medium) which cause variations in frequency. Wow is a slow speed variation at frequencies up to 10 Hz, while flutter is a faster speed variation at frequencies above 10 Hz.

BIBLIOGRAPHY

Askill, J. "Physics of Musical Sounds," D. Van Nostrand Co, 1979.

Backus, J. "The Acoustical Foundations of Music", W.W. Norton, 1969.

Benade, A.H. "Fundamentals of Musical Acoustics," Oxford University Press, 1976.

Beranek, L.L. "Music, Acoustics, and Architecture," John Wiley and Sons, Inc., 1962.

Berg, R.E.; Stork, D.G. "The Physics of Sound," Prentice–Hall, Inc., 1982.

Borwick, J. "Loudspeaker and Headphone Handbook", Butterworth, 1988.

Cohen, A.B. "Hi–Fi Loudspeakers and Enclosures." Hayden Book Co., Inc., 1968.

Colloms, M. "High Performance Loudspeakers," Halsted Press Book, J. Wiley and Sons 1991.

Goodman, R.L. "Maintaining and Repairing Videocassette Recorders," TAB Books, Inc., 1983.

Hall, D.E. "Musical Acoustics: an Introduction," Wadsworth Publishing Co., 1980.

Horn, T.D. "DAT, the Complete Guide to Digital Audio Tape," TAB Books, Inc., 1991.

Huber, D.M; Runstein, R.E. "Modern Recording Techniques," Sams, 1991.

Institute of High Fidelity "Official Guide to High Fidelity," Howard W. Sams & Co., Inc., 1974.

Johnson, K.W.; Walker, W.C.; Cutnell, J.D. "The Science of Hi–Fidelity," Kendall–Hunt Publishing Co., 1981.

Jorgensen, F. "Magnetic Recording," TAB Books, Inc., 1989.

Klipsch, P.W. "A Low Frequency Horn of Small Dimensions," Journ. Ac. Soc. Am. 13, 137 (1941).

Mallinson, J.C. "The Foundations of Magnetic Recording," Academic Press, 1993.

McComb, G.; Cook, J. "Compact Disc Player, Maintenance and Repair," TAB Books, Inc., 1987.

Pohlmann, K.C. "Principles of Digital Audio," Sams, 1989.

Pohlmann, K.C. "The Compact Disc Handbook," A–R Editions, Inc., 1992.

Rigden, J.S. "Physics and the Sound of Music," John Wiley and Sons, 1977.

Rossing, T.D. "The Science of Sound," Addison–Wesley Publishing Co., 1990.

Sinclair, I.R. "Introducing Digital Audio, CD, DAT, and Sampling," PC Publishing, 1988.

Strong, W.J.; Plitnik, G.R. "Music, Speech, High Fidelity," Soundprint, 1983.

Sweeney, D. "Demystifying Compact Discs, A Guide to Digital Audio," TAB Books, Inc., 1986.

Taylor, C. "Exploring Music," Institute of Physics Publishing, 1992.

Taylor, J.G. "Physics of Stereo/Quad Sound," Iowa State University Press, 1977.

Wagner, R. "Electrostatic Loudspeaker, Design, and Construction," TAB Books, Inc., 1987.

Watkinson, J. "The Art of Digital Audio," Focal Press, 1989.

Weems, D.B. "Great Sound Stereo Speaker Manual," TAB Books, Inc., 1990.

White, H.E.; White, D.H. "Physics and Music, The Science of Musical Sound" Saunders College, 1980.

INDEX

A

W

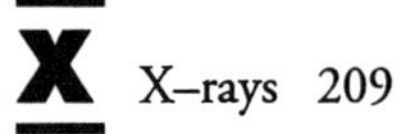